# Theory of Lubrication

# Theory of Lubrication

**Mihir Kumar Ghosh**
*Former Professor, IIT BHU, Varanasi*

**Bankim Chandra Majumdar**
*Former Professor, IIT Kharagpur*

**and**

**Mihir Sarangi**
*Assistant Professor, IIT Kharagpur*

**Tata McGraw Hill Education Private Limited**
NEW DELHI

*McGraw-Hill Offices*

**New Delhi** New York St Louis San Francisco Auckland Bogotá Caracas
Kuala Lumpur Lisbon London Madrid Mexico City Milan Montreal
San Juan Santiago Singapore Sydney Tokyo Toronto

Tata McGraw-Hill

Published by Tata McGraw Hill Education Private Limited,
7 West Patel Nagar, New Delhi 110 008

**Theory of Lubrication**



**ISBN (13): 978-0-07-133292-7**
**ISBN (10): 0-07-133292-8**

Vice President and Managing Director: *Ajay Shukla*
Publishing Manager—Professional: *Mitadru Basu*
Sr. Production Executive: *Rita Sarkar*
DGM—Sales and Business Development—Professional: *S Girish*
Deputy Marketing Manager—Science, Technology and Computing: *Rekha Dhyani*
General Manager—Production: *Rajender P Ghansela*
Manager—Production: *Reji Kumar*



Typeset at White Lotus Infotech Pvt. Ltd., Puducherry and printed at Sanat Printers, 312 EPIP, HSIDC, Kundli, Sonepat, Haryana.

Cover Printer: Sanat Printers

Cover Design: Kapil Gupta

RXBYCDDALLAAC

The McGraw-Hill Companies

# Preface

This book has been written to give a thorough theoretical background to senior undergraduate and graduate students of mechanical engineering and also to practicing engineers involved in the design and analysis of fluid film lubricated bearings for rotating machinery, e.g., power generating turbines, turbo-generators, pumps and compressors, etc. Fluid film lubrication is a very major and vital area of study in the broad subject of tribology which also deals with friction and wear of materials, surface science and technology. However, tribology is a very vast subject which requires knowledge and expertise in materials science, physics and chemistry of interacting surfaces. It is almost impossible to include and do justice to all aspects of tribology in a single text book. With a brief introduction to tribology in general authors have restricted themselves to the broad area of fluid film lubrication keeping in mind its vast application in bearings, seals, gears and variety of other mechanical components which are used in all mechanical devices and machinery. Fluid film lubrication not only reduces friction very significantly but eliminates wear almost completely resulting in long life of components. The topics chosen in the book cover all aspects of fluid film lubrication including rotor-bearing dynamics, thermo-hydrodynamic lubrication, gas lubrication, elastohydrodynamic lubrication, fluid inertia effects and turbulence in fluid film lubrication, etc. which are generally not discussed in the text books available in this area.

The text material has been covered in fifteen chapters.

- **Chapter 1** gives a brief introduction to the subject of tribology.
- **Chapters 2–5** cover the fundamentals of fluid film lubrication, viz., viscosity and rheology of lubricants, mechanics of fluid films, hydrodynamic lubrication dealing with idealized bearings and finite bearings. Both analytical and numerical methods of solutions have been dealt with.
- **Chapters 6–7** address thermo-hydrodynamic lubrication and design aspects of hydrodynamic bearings.
- **Chapter 8** covers all aspects of dynamics of fluid films. Dynamic coefficients and stability of rotor-bearing systems, non-linear analysis of dynamically loaded bearings, squeeze film lubrication, squeeze film dampers, etc. have been discussed adequately.
- **Chapter 9–11** discuss in detail externally pressurized lubrication, fluid inertia effects and turbulence in fluid films, gas lubrication and gas bearings.
- **Chapters 12–15** cover hydrodynamic lubrication of rigid non-conformal rolling contacts and elastically deformable rolling/sliding contacts. Estimation of lubricant film thickness, traction coefficients and thermal effect due to sliding on film thickness, traction coefficients and temperature rise in the contact has been discussed.

Wherever possible, solved examples and problems for practice have been included in the chapters. During the entire course of preparing this manuscript, it was realized that to do full justice to a vast interdisciplinary subject is extremely difficult. The topics covered have been chosen looking into vast teaching, research experience and expertise of the authors. However, it is likely that some topics have not been covered as thoroughly as possible. The total length of the manuscript also prohibits doing this. Attempts have been made to do justice with the topics to the extent possible.

In recent decades very few text books have been written in the area of fluid film lubrication. It is therefore felt that this book will go a long way in filling this void in the availability of an advanced text book in this subject.

**Mihir Kumar Ghosh**
**Bankim Chandra Majumdar**
**Mihir Sarangi**

# Acknowledgements

Authors are indebted to several organizations and individuals who have helped in some way or the other in the course of preparation and development of this manuscript.

Special mention is necessary for Dr. P. Ghosh and Dr. R. K. Pandey associate professors in the Department of Mechanical Engineering at IIT BHU, Varanasi and IIT Delhi respectively, who have contributed to the book through academic discussion with the first author during the preparation of this manuscript. The authors sincerely acknowledge the contribution and help rendered by graduate students of the Department of Mechanical Engineering at IIT BHU and IIT Kharagpur in drafting figures and in preparing the manuscript.

Authors acknowledge with gratitude and extend their sincere thanks to technical societies for granting permission to print some figures from technical journals in the book. Specifically thanks are due to the American Society of Mechanical Engineers (ASME), the Institution of Mechanical Engineers (I Mech. E, London, U.K.) for granting permission to print some figures from the Transactions of ASME Journal of Lubrication Technology, Journal of Tribology and Proc. I Mech. E, Engineering Tribology Journal and Proceedings. S. Chand & Co. New Delhi, publisher of technical books for granting permission to extract some material from the "Introduction to Tribology of Bearings" authored by Dr. B. C. Majumdar.

**Mihir Kumar Ghosh**
**Bankim Chandra Majumdar**
**Mihir Sarangi**

# Contents

# List of Symbols

| | |
|---|---|
| $\alpha$ | Pressure-viscosity coefficient of lubricant |
| $B_x, B_y, B_z$ | Body forces in $x, y, z$ directions respectively |
| B | Thrust pad length |
| $B$ | Damping of the oil film, $b$ (dimensionless damping) |
| $\beta$ | Dynamic load ratio (Chapter 12) |
| β | Temperature-viscosity coefficient of lubricant |
| β | Bulk modulus of fluid |
| $c_p$ | Specific heat of lubricant at constant pressure |
| $c_v$ | Specific heat of lubricant at constant volume |
| $C$ | Radial clearance |
| $\delta_{ij}$ | Kronecker delta |
| e | Eccentricity of the journal centre |
| **e** | Strain rate |
| ε | Eccentricity ratio of the journal |
| $\varepsilon = \frac{e}{C}$ | Eccentricity ratio |
| $E$ | Young's modulus |
| f | Coefficient of friction |
| $\phi, \varphi$ | Attitude angle |
| F | Friction force |
| F | Fluid film force acting on the journal (Chapter 8) |
| $F$ | Shear force |
| Φ | Dissipation function |
| $G$ | Shear modulus of fluid |
| $\gamma, \dot{\gamma}$ | Shear strain and shear strain rate respectively |
| $\gamma = \frac{v}{\omega}$ | Whirl frequency ratio |
| $h$ | Film thickness |
| $\bar{h} = \frac{h}{C}$ | Dimensionless film thickness |
| $\eta$ | Absolute viscosity of lubricant |
| $\eta$ | Viscosity of fluid |

| | |
|---|---|
| $I$ | Moment of inertia |
| $k, k_f$ | Thermal conductivity of fluid/lubricant |
| $k_c, \delta_c$ | Capillary design parameter (dimensionless) |
| $k_o, \delta_o$ | Orifice design parameter (dimensionless) |
| $k_s$ | Thermal conductivity of solid |
| $K$ | Stiffness of bearing |
| $K$ | Stiffness of the oil film, $k$ (dimensionless stiffness) |
| L | Bearing length |
| $L/D$ | Length to diameter ratio |
| $\lambda$ | Second coefficient of viscosity |
| $\Lambda$ | Bearing number |
| $m$ | Mass flow rate |
| $m_x, m_y$ | Mass flow rate per unit length along $x$ and $y$ coordinates |
| $\mu$ | Coefficient of friction |
| M | Rotor mass per bearing, $\overline{M} = \dfrac{MC\omega^2}{W}$ (dimensionless) |
| $M$ | Frictional torque |
| **M, D, K, F** | Global assembled mass, damping, stiffness and force matrices of rotor-bearing system |
| $n$ | Ploytropic gas-expansion coefficient |
| $\nu$ | Kinematic viscosity |
| $N_U$ | Nusselt number |
| $N$ | Speed of the journal |
| $p$ | Hydrodynamic pressure of lubricant |
| $P_e$ | Peclet number |
| $q, Q$ | Flow rate |
| $Q$ | Volume flow rate of fluid |
| $Q$ | Heat transfer rate (Chapter 6) |
| $\overline{Q}$ | Dimensionless flow rate |
| $r, \theta, z$ | Cylindrical coordinate axes |
| $\rho$ | Density of fluid |
| R | Journal radius |
| $R$ | Radius of journal |
| $Re$ | Reynolds number |
| $\sigma$ | Normal stress |
| $\sigma$ | Squeeze number (dimensionless) |
| $S$ | Sommerfeld number (dimensionless) |
| $S'$ | Modified Sommerfeld number |
| $t$ | Time |
| $T$ | Temperature of the lubricant |
| $\tau$ | Shear stress |
| $u, v, w$ | Flow velocitiy components in $x$, $y$, and $z$ directions respectively |

| | |
|---|---|
| $u_1, u_2$ | Velocities of moving surfaces 1 and 2 respectively |
| $\upsilon$ | Frequency of vibration of the journal centre |
| $U$ | Surface velocity |
| $v_r, v_\theta, v_z$ | Flow velocity components in $r$, $\theta$, $z$ coordinates respectively |
| V | Volume of fluid |
| $\vec{V}$ | Flow velocity vector |
| $\omega$ | Angular velocity of the journal |
| $W$ | Total load carrying capacity of bearing |
| W | External load on the journal |
| $\Omega$ | Whirl ratio |
| $x, y, z$ | Cartesian coordinate axes |
| $\xi$ | Peak pressure ratio (Chapter 12) |
| $\xi$ | Fluidity |

Chapter 1

# Introduction

## 1.1 | History of Lubrication

Tribology generally deals with the studies of the phenomena related to surfaces rubbing together. In other words, it is a systematic and scientific study of interacting surfaces. It encompasses the study of physics, chemistry, and mechanics of the interacting surfaces which include friction, lubrication, and wear of materials. From the dawn of civilization, attempts were made to transport men and materials from one place to the other. Greatest challenges faced were related to minimizing friction. In this process of development, rolling wheel was invented, which reduced friction significantly and formed the core of all surface transportation systems of today, for example, railroad vehicles and automobiles. The survival of these transportation systems depended on minimizing friction and wear. The Industrial Revolution of the nineteenth and twentieth centuries also posed similar challenges. Lubrication of interacting surfaces became an essential requirement of all sliding systems. The lubricant may be considered a third body which is put between the interacting surfaces to avoid direct contact between the surfaces with the objective of reducing friction. It forms an interface between the interacting surfaces. Therefore, if the rubbing surfaces can be properly lubricated by a suitable lubricant, both friction and wear can be reduced significantly.

The genesis of fluid film lubrication and its development can be traced back to the experiments of Tower (1883). Tower, a railroad engineer, conducted a series of experiments on lubrication of railroad bearings to minimize friction. In order to lubricate the bearing, a hole was drilled at the center of the bearing to feed the oil. However, when the shaft was rotated Tower observed that oil was oozing out of the hole. He tried to prevent this by putting a plug in the hole. However, he noticed that the plug was being thrown out of the hole. Tower believed that an oil film was formed between the journal or the shaft and the bearing which generated sufficiently high pressure to throw the plug out. It was also noticed that the hole was drilled in the loaded region where the gap between the shaft and the bearing was minimum.

Petroff (1883) at the same time was interested in estimating friction in journal bearings. He conducted experiments to measure frictional torque needed to run a shaft supported by a journal bearing with the clearance space filled with viscous oil under concentric position. Petroff developed a relationship between the frictional force and operating parameters of the bearing as:

$$\text{Friction force} = \text{viscosity of oil} \times \text{surface speed of the journal} \\ \times \text{bearing surface area/film thickness of oil}$$

However, he did not notice that fluid film could also generate pressure under eccentric operation. Tower can thus be given the credit for developing the concept of hydrodynamic/fluid film lubrication. These developments motivated Reynolds (1886) to look into the mechanics of fluid film lubrication. He explained the mechanics of formation of hydrodynamic film in journal bearings and developed the well-known Reynolds equation for hydrodynamic lubrication. It was also explained that presence of converging wedge-shaped film, surface motion, and viscosity of the oil were responsible for the development of hydrodynamic pressure. Reynolds gave solutions for the infinitely long journal bearing and for squeeze film lubrication of two approaching elliptic plates. He also suggested the boundary condition for film rupture in the divergent portion of the bearing as that both pressure and pressure gradient must be zero at the film rupture boundary. However, the solutions given by him were approximate and he did not go for integration of the Reynolds equation.

Sommerfeld (1904) integrated the Reynolds equation and obtained analytical expressions for pressure distribution, load capacity, locus of journal center for various loads, and friction factor for full $2\pi$ film and half $\pi$ film using a substitution known as Sommerfeld substitution. These developments formed the basis for engineers to design and develop bearings for different types of machinery. Kingsbury (1897) and Mitchell (1905) both developed models of tilting pad bearings. Patents were eventually granted to both. Kingsbury also developed gas bearing in 1913 and showed that air film can also be used to carry load.

Stodola (1927) and Newkirk and Taylor (1925) dealt with the dynamics of lubricant films in bearings and rotors supported on such a film. The concept of half frequency whirl instability was derived by Newkirk. Half frequency whirl instability occurs in bearings when the rotor is run at a speed twice the critical speed of the system.

Foundation of modern lubrication theory was developed during 1945–1965 after the Second World War. Noteworthy among these are the short bearing solutions of Reynolds equation for journal bearings developed by DuBois and Ocvirk (1953). Experiments of Cole and Hughes (1956) revealed incomplete oil film at the inlet of the journal bearing and the presence of finger-like striations in the divergent region of the oil film. Experiments of Jacobson and Floberg (1957) on cavitations and film rupture and the work of Floberg (1961) delineated the correct boundary conditions for film rupture and also developed film reformation boundary condition in the divergent portion of the film at the inlet region.

Raimondi and Boyd (1958) developed a numerical procedure to solve Reynolds equation for pressure distribution using finite difference method. Design charts were also developed for the design of partial arc and full journal bearings. Rippel (1963) also developed design procedure and design charts for capillary and orifice compensated hydrostatic bearings.

Parallel to these, Hertz (1881) developed the theory of contact mechanics and gave the solution for contact stresses and deformation of cylindrical and spherical bodies in contact under a load. This formed the basis for the development of elastohydrodynamic lubrication theory of rolling/sliding solid contacts. Martin's (1916) solution to determine hydrodynamic film thickness in gears assuming the surfaces to be rigid and lubricated by an isoviscous lubricant, Grubin's (1949) analysis for film thickness in elastic contacts lubricated by a piezoviscous oil and Petrusevich's (1951) simultaneous solution of elasticity equations with Reynolds equation giving pressure distribution and deformed shape of the surfaces proved to be milestones toward the development of present day elastohydrodynamic lubrication theory of rolling/sliding contacts.

Dowson and Higginson (1959) produced a series of publications in which numerical solution procedures were developed to solve elastohydrodynamic lubrication problems. Simultaneous solutions of Reynolds and elasticity equations, often coupled with energy equation were also obtained. Along with these, new machine elements, viz., pumping rings, rubber bearings, and foil bearings, appeared which required the application of elastohydrodynamic lubrication theory to accurately estimate minimum film thickness in these machine elements as discussed by Pinkus (1987). An account of Hertz theory of solid–solid contact and its application to mechanical engineering problems over 100 years since its development has been given by Johnson (1982).

Hydrodynamic lubrication theory has also been extended to metal forming processes where bulk plastic deformations occur, for example, in metal forming processes such as cold rolling, extrusion, forging, etc.

## 1.2 | History of Friction and Wear

Friction and wear of materials of solid surfaces in relative sliding motion under dry or unlubricated condition of operation is also of tremendous technological importance. In many applications, where the Stribeck parameter is small, fluid film lubrication is not possible. Rough surfaces often rub against each other under unlubricated condition resulting in high friction and wear. Fundamental laws of friction, viz., (i) friction force is proportional to normal load, (ii) friction force is independent of apparent area of contact, (iii) static friction is greater than kinetic friction, and (iv) friction is not depended on sliding velocity were enunciated by Leonardo Da Vinci (1452–1519), Amonton (1699), and Coulomb (1785). Interlocking of asperities and its resistance to relative sliding was considered to be the major factor causing friction. These basic laws are mostly accepted even today except the independence of kinetic friction on sliding velocity. Hardy (1936) proposed molecular attraction as the reason for friction between two interacting surfaces. Bowden and Tabor (1950) proposed the theory of cold welding, shearing, and ploughing of the soft metal by hard asperities as responsible factor for friction. Molecular attraction theory of Tomlinson (1929) attributed frictional resistance to rupture of molecular bond between two interacting surfaces. In fact, Bowden and Tabor's welding and shearing is similar to this. Molecular bond is similar to cold welding or adhesion between two interacting surfaces. Bowen and Tabor's theory is widely accepted now for metal to metal contacts, whereas molecular bond rupture theory is accepted for polymers. Wear is removal or loss of material during the process of sliding between the two rubbing surfaces. Major effort made in the study of wear had been to identify the modes of wear and to quantitatively assess the magnitude of wear. Burwell and Strang (1952) and Archard and Hirst (1956) gave empirical relations of sliding wear. Rabinowicz (1965) identified four main types of wear, viz., sliding or adhesive wear, abrasive wear, corrosion, and surface fatigue. Noteworthy among recent development in understanding the mechanism of formation of wear particles are delamination theory developed by Suh (1973) and shake down theory of Kapoor *et al.* (1994).

## 1.3 | Regimes of Lubrication

Broadly speaking there are four main regimes of lubrication:

1. Boundary lubrication
2. Partial or mixed lubrication
3. Elastohydrodynamic lubrication
4. Hydrodynamic lubrication

However, lubrication of solid surfaces in relative motion can be classified into the following regimes depending on the commonly known modes of lubrication in mechanical components:

- Hydrodynamic lubrication
- Elastohydrodynamic lubrication
- Hydrostatic/externally pressurized lubrication
- Squeeze film lubrication
- Partial or mixed lubrication
- Boundary lubrication and lubrication by solid lubricants

However, it is appropriate to mention that squeeze film lubrication, elastohydrodynamic lubrication, externally pressurized lubrication can also be visualized as a form of hydrodynamic lubrication. Broadly all these modes of lubrication may be called as fluid film lubrication.

It is well known that surface roughness plays an important role in deciding the regime of lubrication in a mechanical component. The relationship between coefficient of friction and film thickness in various regimes of lubrication can be qualitatively and quantitatively visualized through Fig. 1.1 and Table 1.1, respectively. Coefficient of friction and film thickness are plotted against a parameter known as 'Stribeck parameter' which is equal to $\eta U/P$, where $\eta$ – viscosity of the oil, $U$ – relative sliding speed and $P$ – load per unit projected surface area of the contact. Various regimes of lubrication and film thickness as related to Stribeck parameter can be seen in Fig. 1.1. Relationship between regimes of lubrication and surface roughness can be seen in Fig. 1.2. In this figure, the coefficient of friction has been plotted against a film parameter which is the ratio of minimum film thickness in the contact to the composite root mean square value of roughness of the two surfaces.

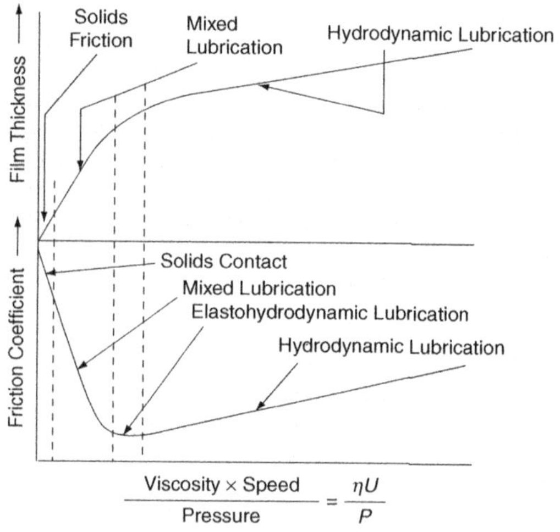


**Figure 1.1** | Stribeck Curves Showing Modes of Lubrication

**Table 1.1** | Operating Film Thickness, Coefficient of Friction, and Wear for Various Lubrication Modes

| **Lubrication mode** | **Film thickness ($\mu$m)** | **Friction coefficient** |
|---|---|---|
| • Hydrodynamic lubrication | 1–10 | $10^{-2}$–$10^{-3}$ |
| • Elastohydrodynamic lubrication | 0.1–1 | $10^{-2}$–$10^{-3}$ |
| • Externally pressurized lubrication | 5–50 | $10^{-3}$–$10^{-6}$ |
| **Lubrication type** | **Friction coefficient** | **Degree of wear** |
| • Thick film | 0.001 | None |
| • Boundary and mixed lubrication | 0.05–0.15 | Mild |
| • Unlubricated or dry | 0.5–2.0 | Severe |

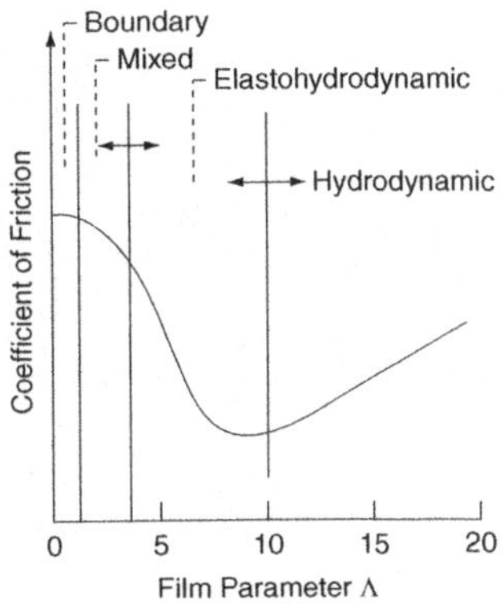


**Figure 1.2** | Friction Coefficient versus Film Parameter

In dry/solid lubrication or boundary lubrication in the contact, the coefficient of friction is generally high. As the film parameter increases, the coefficient of friction decreases proportional to inverse of the film thickness, whereas for higher values of film parameter which occur in the hydrodynamic lubrication regime, the coefficient of friction increases in proportion to increase in film thickness.

In hydrodynamic lubrication, $W \propto \frac{1}{h^2}$, whereas in elastohydrodynamic lubrication load $W$ has little effect on film thickness $h$. In both hydrodynamic and elastohydrodynamic lubrication friction, force $F$ can be expressed as $F \propto \frac{1}{h}$. Hence, $\mu_{HL} \propto \frac{1/h}{1/h^2} \propto h$ and $\mu_{EHL} \propto \frac{1/h}{\text{constant}} \propto \frac{1}{h}$.

$$\Lambda = \frac{\text{minimum film thickness}}{\text{rms value of composite surface roughness}}$$

## 1.4 | Hydrodynamic Lubrication

Hydrodynamic lubrication regime is generally observed in the lubrication of conformal surfaces where one surface envelope the other and the clearance space between the two surfaces is filled with a viscous fluid. This is typically observed in hydrodynamic journal and thrust bearings as shown in Figs 1.3 and 1.4, respectively. Flow of a viscous fluid through the convergent passage due to relative motion between the surfaces is responsible for the generation of positive pressure in hydrodynamic journal and thrust bearings. Since the contact area is usually large, very high pressure is not generated. Usually the maximum pressure generated is less than 20 MPa. Minimum film thickness in these bearings is given as:

$$h_{\min} \, \alpha \left( \frac{U}{W} \right)^{0.5} > 1 \ \mu\text{m}$$

where $U$ is relative surface speed or surface speed of the rotor, $W$ is load.

Coefficient of friction usually varies between $10^{-2}$ and $10^{-3}$. The flow is generally laminar. However, in case of high speed bearings the flow may become turbulent.

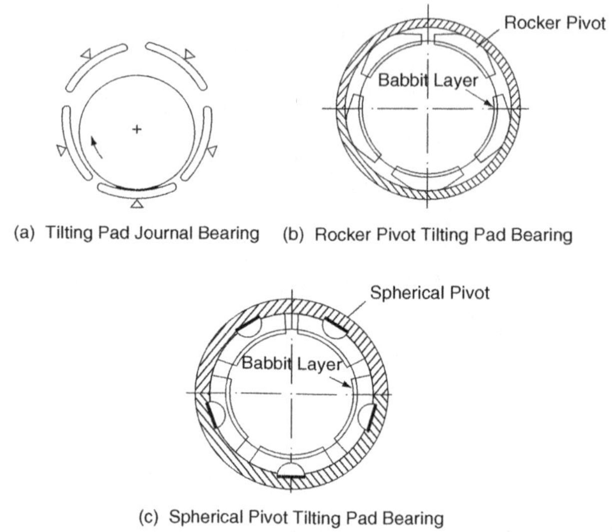


**Figure 1.3** | Hydrodynamic Journal Bearing

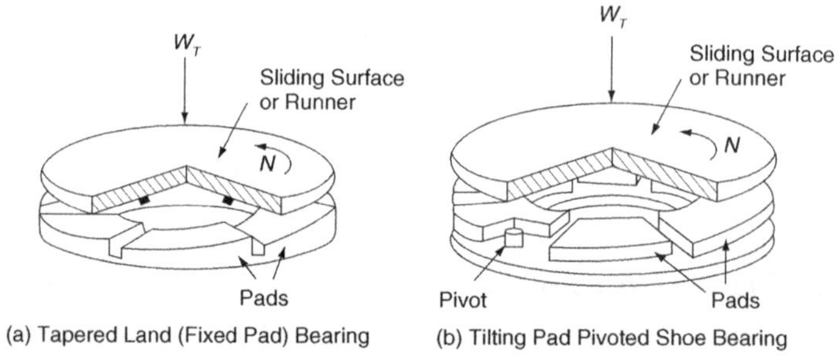


**Figure 1.4** | Hydrodynamic Thrust Bearing

## 1.5 | Hydrostatic or Externally Pressurized Lubrication

Externally pressurized lubrication is characterized by the supply of lubricant at high pressure into a deep groove or recess of the bearing through a flow restrictor, viz., capillary, orifice, or a constant flow valve. Recess area is usually large and pressure is constant in the recess area. Generation of load carrying capacity is due to high pressure in the recess area. Film thickness and recess pressure adjust itself depending on the load. A hydrostatic bearing is shown schematically in Fig. 1.5. A hydraulic pump and a pressure control system are necessary for the bearing. Film thickness varies between 5 and 50 μm and the coefficient of friction is usually between $10^{-3}$ and $10^{-5}$.

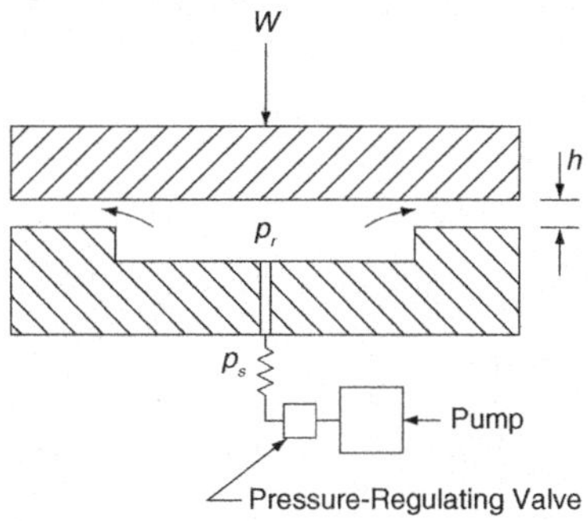


**Figure 1.5(a)** | Hydraulic Thrust Bearing

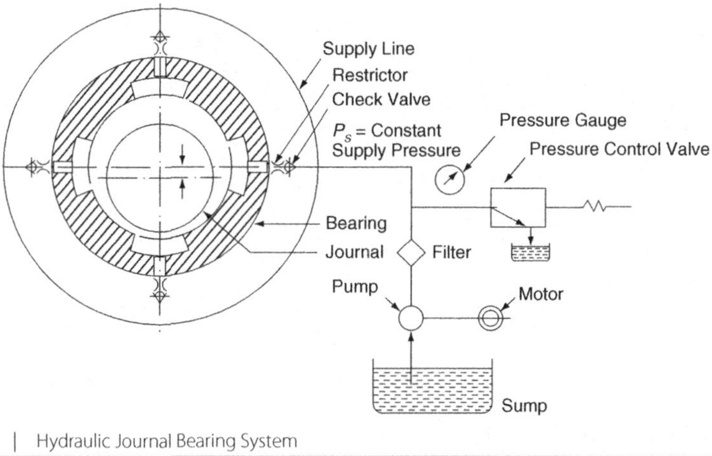


**Figure 1.5(b)** | Hydraulic Journal Bearing System

## 1.6 | Elastohydrodynamic Lubrication

Elastohydrodynamic lubrication regime exists in nonconformal contacts, for example, in rolling element bearings, gears, cams, etc. It is usually a line or a point contact, which under loaded condition becomes a rectangular or an elliptic contact, respectively. A typical rolling element bearing is shown in Fig. 1.6. Maximum pressure in the contact can vary between 0.2 and 2.0 GPa for contacts made of hard materials of high modulus of elasticity. Elastic deformation of the surfaces is significant and comparable with the film thickness of the lubricant. Minimum film thickness usually lies between 0.1 and 1.0 μm and the coefficient of friction between $10^{-3}$ and $10^{-2}$. Elastohydrodynamic regime of lubrication also occurs in the contacts

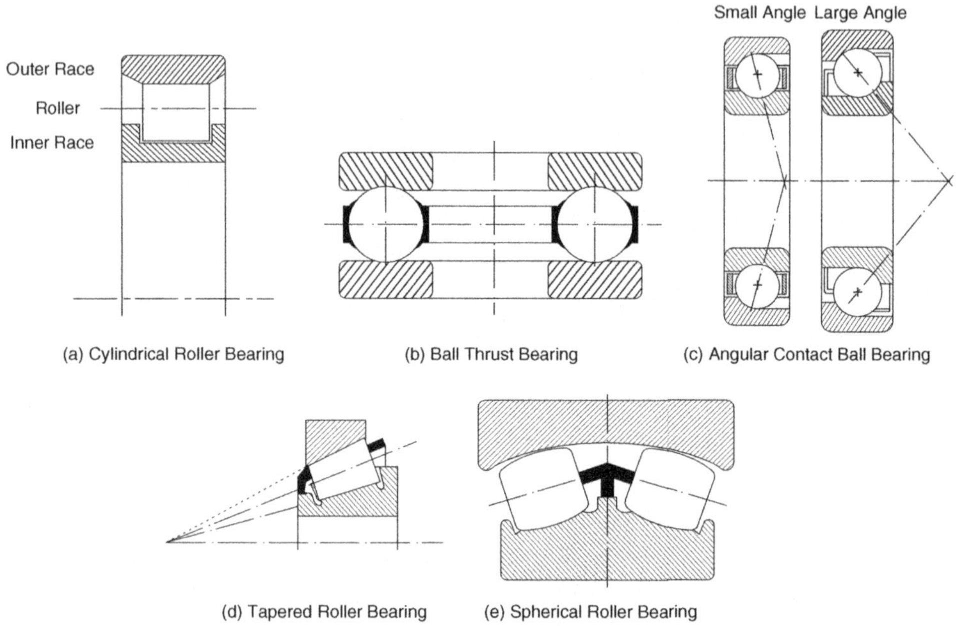


**Figure 1.6** | Typical Rolling Element Bearing

made of soft materials with low modulus of elasticity. Significant elastic deformation of the surfaces can take place at relatively low pressure in the range of 2–10 MPa. Contrary to solid surfaces made of high modulus materials in which the viscosity of lubricant increase manifold due to high pressures, the contacts made of soft materials does not witness change in the viscosity. Human joints which are made of soft materials and lubricated by synovial fluid as shown in Fig. 1.7 typically operate in soft elastohydrodynamic lubrication regime.

## 1.7 | Squeeze Film Lubrication

Squeeze film lubrication shown in Fig. 1.8 is characterized by squeezing out of a viscous lubricant trapped between two approaching surfaces. Pressure is generated due to the normal relative approach velocity of the two surfaces when the film thickness decreases due to squeeze flow of the viscous fluid from the sides. This mode of lubrication is very vital in many practical situations. In I.C. engines, fluid film lubrication of piston pin and small end of the connection rod is possible because of squeeze action. Squeeze film also plays important role in dynamically loaded bearings. Damping coefficient is obtained in rotor-bearing systems due to squeeze film action.

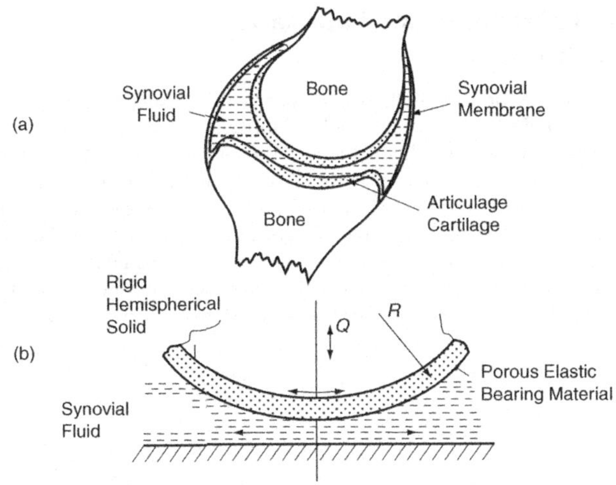


**Figure 1.7** | Human Joint Lubrication Model

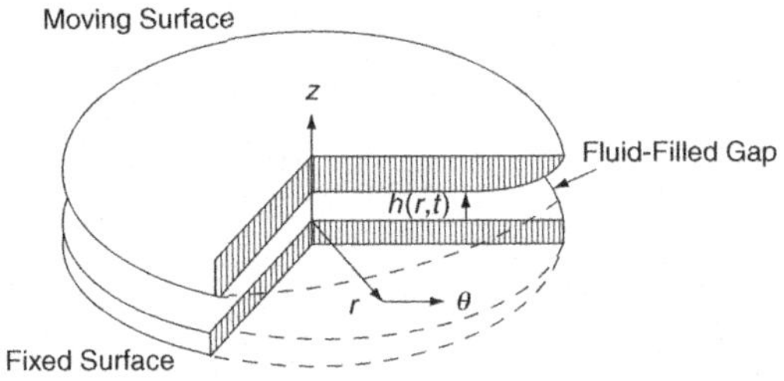


**Figure 1.8** | Schematic Diagram of Squeeze Bearing

## 1.8 | Partial and Mixed Lubrication

Partial and or mixed lubrication regime is encountered when the Stribeck parameter is less than a critical value as can be seen in Fig. 1.1. Mixed or partial lubrication regime is characterized by some asperity to asperity interaction depending on roughness of the surfaces in the presence of a lubricant. Thus, load is carried by both hydrodynamic effect due to lubricant film and asperity contact. Friction coefficient increases due to this. It is also some times referred to as partial lubrication regime. Mixed lubrication can thus be visualized as a combination of hydrodynamic/squeeze film and asperity to asperity contact at macro level.

## 1.9 | Boundary Lubrication and Lubrication by Solid Lubricants

Real surface of a mechanical component is shown in Fig. 1.9. Solid films are usually oxides of the metal, and superimposed over the solid films are layers of boundary lubricant. Boundary lubrication is a mode of lubrication provided by the boundary films which are usually layers of chemisorbed films of fatty acids or films of chemical reactant products such as iron sulphides or chlorides. These films provide protection to the metal surfaces, exhibit low friction coefficient, and adhere to the surfaces at increased temperatures too.

Solid lubrication is obtained when certain solid materials known as solid lubricants are coated or smeared over the metal surfaces to reduce friction and wear. For example, materials like graphite, molybdenum disulphide, tungsten disulphide, Teflon, etc. when coated over metals exhibit low friction coefficient. Solid lubricants possess a typical layer lattice or lamellar structure as shown in Fig. 1.10. Due to this, these materials

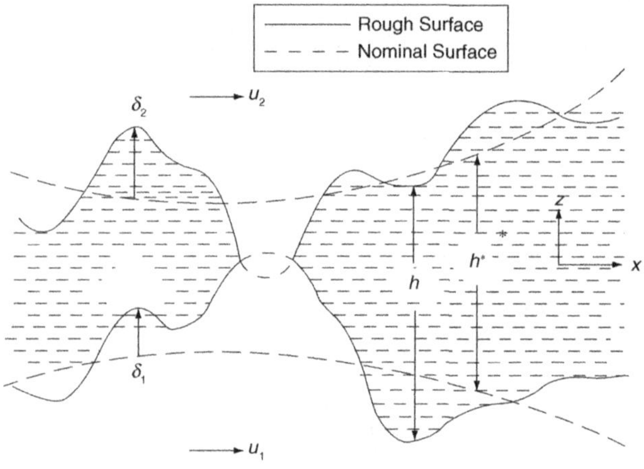


**Figure 1.9(a)** | Lubrication of Rough Surfaces

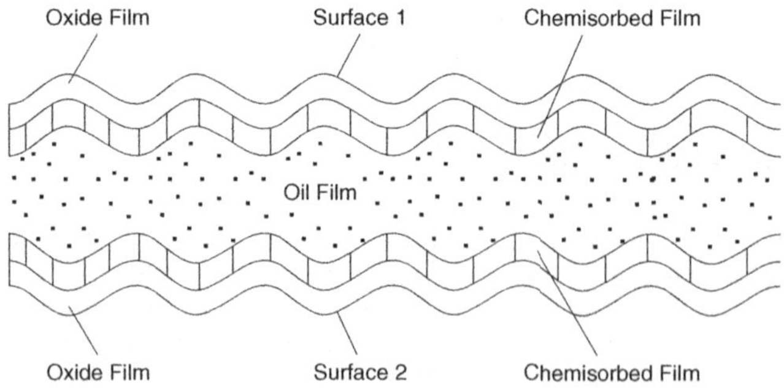


**Figure 1.9(b)** | Asperity with Surface Films

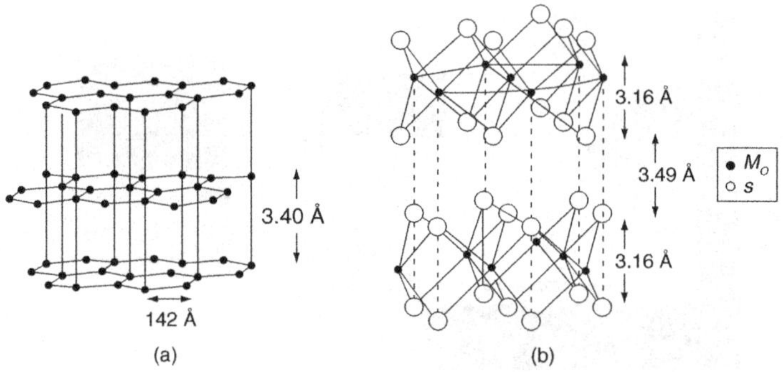


**Figure 1.10** | Structure of **(a)** Graphite and **(b)** Molybdenum Disulphide

are strong in compression and weak in shear which accounts for their low friction coefficient. In applications where other forms of lubrication are not possible, for example, in vacuum and at high temperatures solid lubricants are the only alternative that can provide lubrication. Usually solid lubricants are coated over the metal substrate by either physical vapor deposition or chemical vapor deposition techniques to obtain better adhesion of the layer to the substrate. This increases life as well as load carrying capacity. Some ceramic materials like zirconia, alumina, titanium carbide, silicon nitride, boron nitride, etc. are coated over metals to reduce wear and provide lubrication at high temperatures.

## 1.10 | Micro and Nano Tribology

With advances in technology, the size of mechanical, electrical, and optical components is reducing at a very fast pace. Development of micromachines and mechanical systems has reduced the size of components to the nano scale. Rapid actuation requires fast moving interacting surfaces. The chemical and mechanical stability of moving surfaces require study of friction, adhesion, and lubrication and wear at atomic dimensions and time scales. This has led to development of 'nano tribology'.

Nano tribology is a new discipline in which friction, adhesion, wear, and lubrication studies are put in a unified framework at the micro and nano level. Nano tribology uses new tools and instruments designed for such studies, viz., atomic force microscope (AFM), surface force apparatus (SFA), and scanning tunneling microscope (STM). Atomic force microscopy is a very widely used tool for studies related to micro and nano tribology. AFM is used for measurement of friction force, surface texture, lubricant film thickness, etc.

The AFM basically relies on a scanning technique to produce very high resolution, three-dimensional images of sample surfaces. It also measures ultrasmall forces of the order of less than 1 nN present between the AFM tip surface mounted on a flexible cantilever beam, and a sample surface as shown in Fig. 1.11. Such small forces are measured by measuring the deflection of a nanosized flexible cantilever beam having an ultrasmall mass by optical interference, capacitance, or tunneling current measurement methods. The deflection of the order of 0.02 nm can be measured. For a typical cantilever, stiffness of 10 N/m a force of the order of 0.02 nN can be measured. Some important findings from studies done using AFM are as follows:

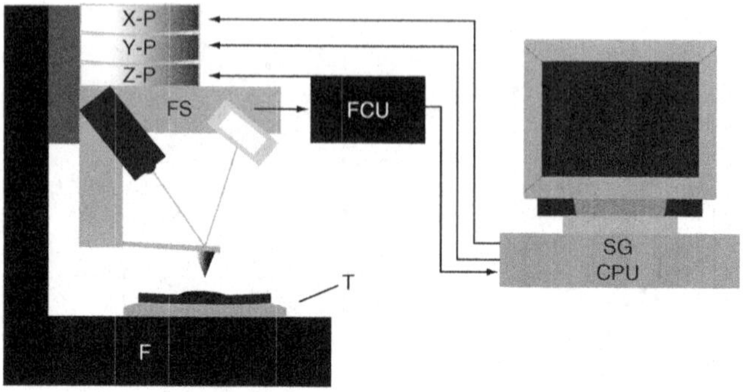


**Figure 1.11** | Atomic Force Microscope

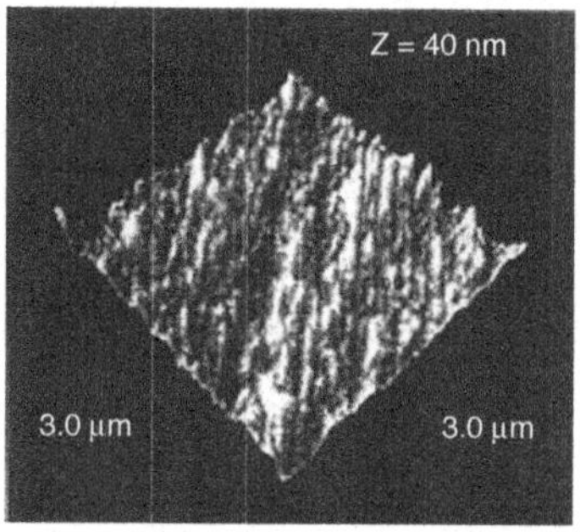


**Figure 1.12** | AFM Image of a Surface Roughness

- Coefficient of friction in regions which are not smooth is higher. Therefore, friction force values vary from point to point in the scan.
- For small contact areas and very low loads used in microscale studies, indentation hardness and modulus of elasticity are higher than macro scale measurements.
- The coefficient of friction values on the microscale are much lower than that on the macro scale.

Surface topography image is derived from monitoring the vertical forces on the cantilever and friction image is acquired simultaneously by monitoring the lateral motion of the cantilever.

Lubricant film thickness of the order of less than 100 nm can be measured using AFM. It is done by determining force/distance curve for a surface with no lubricant and one with a lubricant. The film thickness is determined from changes in these curves. The changes depend on cantilever stiffness, lubricant density, probe geometry, and lubricant thickness. A typical surface topography obtained using AFM is shown in Fig. 1.12.

## 1.11 | Biotribology

Biotribology is a multidisciplinary field consisting of mechanical engineering, bio-engineering, materials engineering, and bio-chemistry. It includes all effects created in the interface areas between two biological materials or between biological material and nonbiological materials. The term 'biotribology' was introduced by Dowson and Wright (1973) to cover all aspects of tribology related to biological systems. The best-known example of the subject is the numerous studies of natural synovial joint lubrication and the design, manufacture, and performance of various forms of total joint replacements. Wear of bearing surfaces in humans and animals can result in pain and restricted movement. The consequences of excessive wear of the bearing material, viz., articular cartilage in synovial joints are well known.

Typical examples of tribology applied to biological systems include:

- Wear of dentures
- Friction of skin and garments affecting the comfort of clothes, socks, shoes, and slipperiness
- Tribology of contact lenses and ocular tribology
- Tribology at microlevels inside cells, vessels, and capillaries such as lubrication by plasma of red blood cells in narrow capillaries
- The wear of replacement heart valves
- The lubrication of the pump in total artificial hearts
- The wear of screws and plates in bone fracture repair
- Lubrication in pericardium and pleural surfaces
- Tribology of natural synovial joints and artificial replacements

Therefore, it is important to eliminate or minimize wear and friction in biological environment to improve quality of life of human beings. For details, the reader is suggested to look into the references listed at the end of this chapter (refer to Jin *et al.,* 2006).

## 1.12 | Tribology and Economic Gain

Rapid growth of industries after the Second World War particularly in automobile and rail road sector, air transportation sector, energy sector, and computers/electronics and automation sectors called for improvements in 'tribology' related issues so that loss of energy due to friction and materials loss due to wear could be reduced to a minimum level. Higher speeds and loads at which new generation machines started operating demanded utmost reliability of tribo-components. To assess the impact of tribology-related loss on the economy, the Institution of Mechanical Engineers, London, UK, formed a committee widely known as Jost Committee. The committee in its report identified wear as the single most important factor responsible for failure and breakdown of a machinery thereby causing severe financial loss. The total loss due to friction and wear was estimated as 750 million pounds for UK in 1981. Similar studies were also conducted in the United States, and the loss to the US economy was estimated to be approximately $21 billion in 1980. It can be imagined that for every nation similar loss to economy occur in proportion to the size of their economy. Countries like the United Kingdom, the United States, Japan, Germany, France, and others realized that through better practice of 'tribology', potential savings can be achieved in the spending. Research and development activities were increased and 'tribology' emerged as an academic discipline in academic institutions.

Newer disciplines have emerged in 'tribology' in recent years, viz., 'magnetic storage tribology' which addresses the tribology problem of computer and data storage systems 'biotribology' which deal with issues related to biological and medical systems, 'micro and nano tribology' with particular reference to MEMS and NEMS, mechatronic and robotic systems. Extreme environment tribology addresses the issues related to tribology of systems that operate at high vacuum, cryogenic, and high temperature conditions. All such systems require special materials and lubricants to overcome tribological problems.

## 1.13 | Summary

Tribology plays a very significant role in giving reliability and long life to mechanical component. The issues involved in this context can be depicted through the chart as shown below:

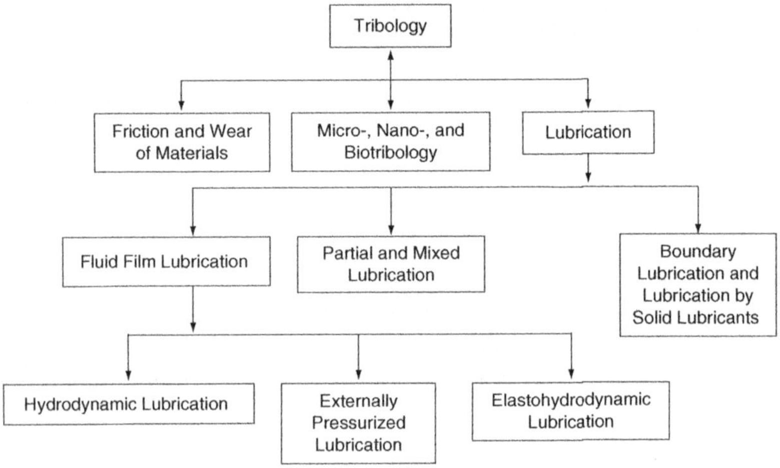


## REFERENCES

*Amonton, G.* (1699), De la Resistance Causee dams les Machiner, Royal Society (Paris), 206.

*Archard, J.F. and Hirst, W.* (1956), The wear of metals under unlubricated conditions, *Proc. Royal Soc. London*, Ser. **236**, 397–410.

*Bowden, F.P. and Tabor, D.* (1950), *The Friction and Lubrication of Solids*, Oxford Press, Cambridge.

*Burwell, J.T. and Strang, C.D.* (1952), On empirical law of adhesive wear, *J. Applied Physics*, **23**, 18–28.

*Cole, J.A. and Hughes, C.J.* (1956), Oil flow and film extent in complete journal bearings, *Proc. Inst. Mech. Engrs.*, London, U.K., **170**, 17, 499–510.

*Coulomb, C.A.* (1785), Theorie des Machines Simples, *Mem. Math. Phys. Acad. Sci.*, **10**, 161.

*DuBois, G.B. and Ocvirk, F.W.* (1953), Analytical derivation and experimental evaluation of short-bearing approximation for full journal bearing, NACA Rep. 1157.

*Dowson, D. and Higginson, G.R.* (1959), A numerical solution of the elasto-hydrodynamic problem, *J. Mech. Engg. Sci.*, **1**(1), 6.

*Dowson, D. and Wright, V.* (1973), Biotribology, *Proceedings of the Conference on the Rheology of Lubrication*, The Institute of Petroleum, The Institution of Mechanical Engineers and the British Society of Rheology, 81–88.

*Floberg, L.* (1961), Boundary conditions of cavitation regions in journal bearings, ASLE Transactions, **4**, 262–266.

*Grubin, A.N.* (1949), *Fundamentals of Hydrodynamic Theory of Lubrication of Heavily Loaded Cylindrical Surfaces*, Translation of Russian Book No.30, Central Scientific Institute for Technology and Mechanical Engineering, Moscow, Chap. 2.

*Hertz, H.* (1881), The contact of elastic solids, *J. Reine Angew. Math.*, **92**, 156–171.

*Hardy, W.B.* (1950), *Collected Scientific Papers*, Cambridge University Press, 1936.

*Jin, Z.M., Stone, M., Ingham, E., and Fisher, J.* (2006), Biotribology, Current Orthopedics, **20**, 32–40.

*Jacobson, B. and Floberg, L.* (1957), The finite journal bearing considering vaporization, Trans. Chalmers Univ. Tech., Gothenburg, 189 and 190.

*Johnson, K.L.* (1982), One hundred years of Hertz contact, *Proc. Inst. Mech. Engrs.*, London, U.K., **196** (39), 363–378.

*Kapoor, A., Williams, J.A. and Johnson, K.L.* (1994), *Int. J. of Wear*, **81**.

*Kingsbury. A.* (1897), Experiments with an air lubricated bearing, *Journal of American Society of Naval Engineers*, 9.

*Leonardo da Vinci* (1938), *The Notebooks of Leonardo da Vinci*, E. Macardy (ed.), Reynal & Hitchcock, New York.

*Martin, H.M.* (1916), Lubrication of gear teeth, *Engineering* (London), **102**, 119–121.

*Mitchell, A.G.M.* (1905), The lubrication of plane surfaces, *Z. Math. U Physik*, **132**, 123.

*Newkirk, B.L. and Taylor, H.D.* (1925), Shaft whipping due to oil action in journal bearing, *Gen. El. Review*, **28**, 559.

*Petroff, N.* (1883), Friction in machines and the effect of the lubricant (in Russian), *Inzh. Zh. St. Petersburg*, **1–4** (71), 227, 377, 535.

*Petrusevich, A.I.* (1951), Fundamental conclusions from the contact—hydrodynamic theory of lubrication, *IZV. Nauk, SSSR. Otd. Tekh. Nauk.* **2**, 209–233.

*Pinkus, O.* (1987), The Reynolds centennial: a brief history of the theory of hydrodynamic lubrication, *Trans. of ASME, Journal of Tribology*, **109** (Jan.), 2–20.

*Rabinowicz, E.* (1965), *Friction and Wear of Materials*, Wiley Publication, New York.

*Raimondi, A.A. and Boyd, J.* (1958), A solution for the finite journal bearing and its applications to analysis and design—Part I, II and III, ASLE Transactions, **1**(1), 159–209.

*Reynolds, O.* (1886), On the theory of lubrication and its application to Mr. Beauchamp Tower's experiments, including an experimental determination of the viscosity of olive oil, *Phil. Trans. Royal Society*, **177**, 157–234.

*Rippel, H.C.* (1963), Design of hydrostatic bearings, in 10 Parts, *Machine Design*, **35**, 108–158.

*Sommerfeld, A.* (1904), Zur Hydrodynamischen Theorie der Schmiermittelreibung, *Z. Angew. Math. Phys.*, **50**, 97–155.

*Stodola, A.* (1927), *Steam and Gas Turbines*, McGraw Hill, New York, USA.

*Suh, N.P.* (1973), The delamination theory of wear, *Int. J. of Wear*, **25**, 111–124.

*Tower, B.* (1983), First and second report on friction experiments (friction of lubricated bearings), *Proc. Inst. Mech. Engrs.* London, UK. p. 659 and pp. 58–70.

Chapter 2

# Viscosity and Rheology of Lubricants

## 2.1 | Introduction

All fluids offer resistance to flow. The physical property of the fluid that characterizes this resistance to flow is called the viscosity. Viscosity and rheology of fluids are concerned with issues related to the resistance to flow of the fluids under various operating conditions such as high pressures and temperatures and also the laws that govern the flow of viscous fluids. Based on this, fluids are characterized as Newtonian or non-Newtonian fluids.

## 2.2 | Newtonian Behavior of Fluids

As shown in Fig. 2.1, let us consider a fluid either liquid or gas contained between two parallel plates of area $A$ separated by a small gap $h$. One of the plates is held stationary while the other is set in motion in the $x$ direction at a constant velocity $U$.

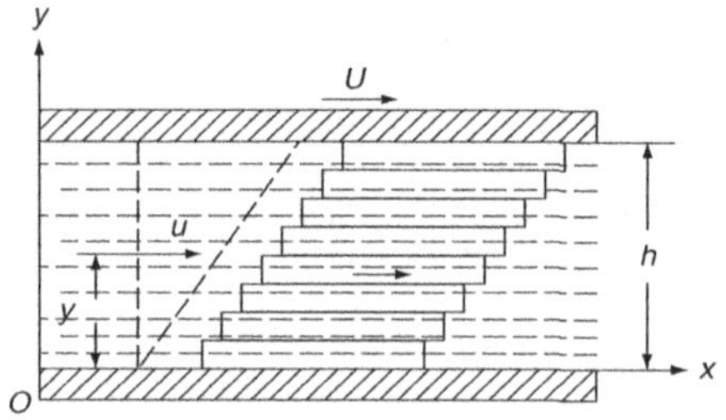


**Figure 2.1** | Flow of a Viscous Fluid Between Parallel Plates

The fluid gains momentum and attains a steady state velocity profile in due course. Fluid layer adhering to the moving plate has a velocity *U*, whereas the layer adhering to the stationary plate has zero velocity. A shear force *F* is necessary to move the nonstationary plate to overcome the viscous resistance of the fluid. When steady state condition is reached, the velocity profile is linear as shown in Fig. 2.1.

Newtonian fluids obey Newton's law of flow which states that the shear force per unit area or shear stress is proportional to the local velocity gradient or shear strain rate as expressed below:

$$\tau = \eta \frac{du}{dz} = \eta \dot{\gamma}; \ \tau = \frac{F}{A} \text{ and } \dot{\gamma} = \frac{du}{dz} \tag{2.1}$$

where the constant of proportionality $\eta$ is called the coefficient of absolute viscosity of the fluid. Alternately coefficient of dynamic viscosity $\nu$ is defined as the ratio of absolute viscosity to density of the fluid.

$$\nu = \eta / \rho \tag{2.2}$$

## 2.3 | Non-Newtonian Fluids

Fluids that are not described by Equation (2.1) and do not obey Newton's law of flow are called non-Newtonian fluids. Non-Newtonian behavior is generally characterized by a nonlinear relationship between shear stress $\tau$ and shear strain rate $\dot{\gamma}$ and fluids are classified according to this relationship as shown in Figs 2.2 and 2.3.

### 2.3.1 | Power Law or Oswald-de Walle Model

Power law fluids are generally characterized by a power law relationship between shear stress and shear strain rate as described below:

$$\tau = m\dot{\gamma}^n \tag{2.3}$$

where $m$ and $n$ are constants for a particular fluid, $m$ is a measure of consistency of fluid and $n$ is a measure of the deviation of the fluid from Newtonian behavior.

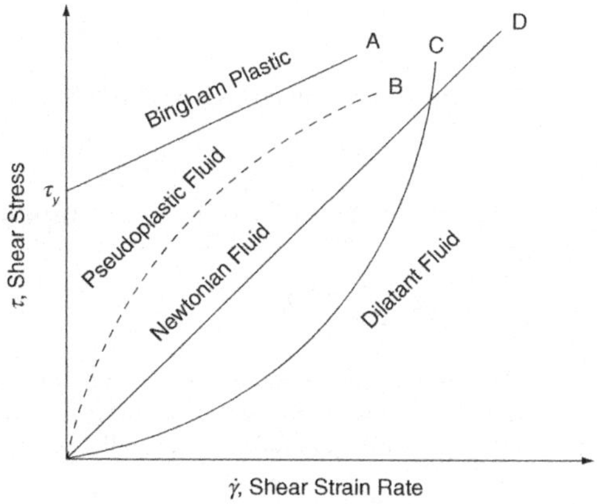


**Figure 2.2** | Shear Stress-strain Rate Curve for Non-Newtonian Fluids

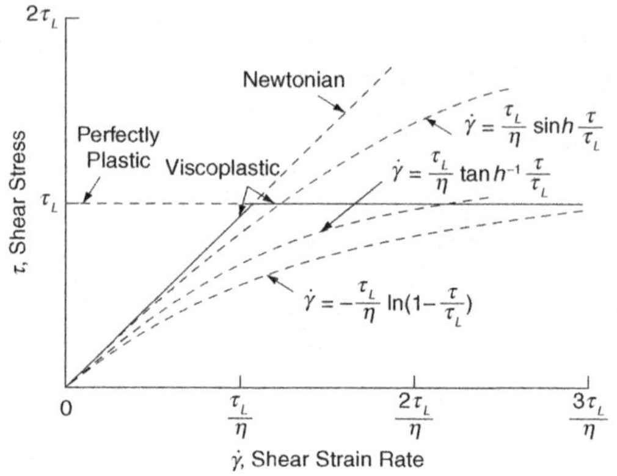


**Figure 2.3** | Rheological Models for Non-Newtonian Lubricants

Fluids for which shear stress versus shear rate curve is characterized by a progressively decreasing slope, i.e., when $n < 1$, are called pseudo plastic fluids. On the contrary, fluids which are characterized by a progressively increasing slope of the curve are called dilatants fluids, i.e., for $n > 1$.

### 2.3.2 | Bingham Plastic

Fluids which exhibit a yield stress at zero shear rate require a shear stress equal to yield stress to be applied prior to the flow. The relationship between shear stress and shear rate thereafter is similar to Newtonian fluids, i.e., linear. The relationship can be expressed as:

$$\tau = \tau_y + \eta_p \dot{\gamma} \tag{2.4}$$

$\tau_y$ is yield stress and $\eta_p$ is the plastic viscosity

Greases, plastics, emulsions (e.g., paints), etc. exhibit this type of behavior.

### 2.3.3 | Viscoelastic Fluids

Viscoelastic fluids exhibit both elastic and viscous characteristic. Linear elastic or Hookean and Newtonian viscous behavior can be described as:

$$\dot{\gamma} = \frac{\tau}{\eta} + \frac{\dot{\tau}}{G} \tag{2.5}$$

where '$G$' is shear modulus of the fluid. In steady state condition $\dot{\tau} = 0$ and the fluid behaves as a Newtonian fluid. Maxwell first proposed the above relationship. It can be visualized as a combination of a spring and dashpot in series. On the other hand, Kelvin or Voigt model is the parallel arrangement of a spring and dashpot and the stress-strain relationship for Kelvin model is expressed as:

$$\tau = G\gamma + \eta\dot{\gamma} \tag{2.6}$$

Above models are shown schematically in Fig. 2.3. Maxwell described the constant $\left(\frac{\eta}{G}\right)^{-1}$ as relaxation time or the time constant for exponential decay of stress at a constant strain when the motion is stopped. Newtonian lubricant also behaves as viscoelastic fluids often at high pressures. Some fluids also exhibit time dependent behavior and are known as either thixotropic fluids or rheopectic fluids. Shear stress decreases with time for a rheopectic fluids. Nonlinear viscoelastic behavior has also been observed of Newtonian lubricants at high pressure and shear rates.

### 2.3.4 | Nonlinear Constitutive Relationships

Nonlinear viscoelastic constitutive relationship was proposed by Johnson and Tevaarwerk (1977) based on two disc machine experiments. At high pressure and shear rates it is given as:

$$\dot{\gamma} = \dot{\gamma}_e + \dot{\gamma}_\vartheta = \frac{\dot{\tau}}{G} + \frac{\tau_e}{\eta}\sin h\frac{\tau}{\tau_e} \tag{2.7}$$

where $\tau_e$ is shear stress at which nonlinear behavior is observed. G is the limiting elastic shear modulus. Elastic behavior is important only at low shear–strain ratio.

Bair and Winer (1979a) observed in their experiments that lubricants exhibit limiting shear stress characteristics which means that shear stress cannot exceed the limit shear stress. Shear stress approaches limiting shear stress value at infinitely high shear strain rates. Following rheological equations were proposed by Bair and Winer (1979b):

$$\dot{\gamma} = \dot{\gamma}_e + \dot{\gamma}_v = \frac{\dot{\tau}}{G} + \frac{\tau_L}{\eta}\tan h^{-1}\left(\frac{\tau}{\tau_L}\right) \tag{2.8a}$$

$$\dot{\gamma} = \left(\tau_L / \eta\right)\ln\left(\frac{1}{1-\tau/\tau_L}\right) \tag{2.8b}$$

$\tau_L$ is the limiting shear stress and is dependent on pressure and temperature, i.e., it increases linearly with pressure and decreases linearly with temperature.

### 2.3.5 | Emulsions as Lubricants

Liquid–liquid and liquid–gas mixtures have also been used as lubricants in emulsion form. Water in oil emulsion has been used as lubricant in rolling contact bearings and rolling processes. Air bubbles finely dispersed in oil called bubbly oil has also been used in bearings. Mass ratio and density of a gas–liquid mixture can be expressed as a function of the individual densities of the gas and liquid from the basic definition of density:

$$\frac{1}{\rho_m} = \frac{V_g + V_l}{m_g + m_l} = \frac{V_g m_g}{m_g\left(m_g + m_l\right)} + \frac{V_l m_l}{m_l\left(m_g + m_l\right)} = \frac{X}{\rho_g} + \frac{1-X}{\rho_l} \tag{2.9}$$

Subscripts *m*, *l*, and *g* stand for mixture, liquid, and gas, respectively.

Therefore, if the mass ratio $X$ and the densities of the gas and liquid are known, the density of mixture may be determined from Equation 2.9. The density of the liquid can be considered constant at a certain temperature and independent of dissolved gas molecules. However, density of gas $\rho_g$ and mass ratio $X$ need to be specified as a function of pressure.

Similarly, relationships have given by several researchers for viscosity of mixtures in terms of viscosities of liquid and gas. However, these relationships are empirical in nature and thus uncertainties are involved. Following correlations are often used in a two phase system as given in the book by Hewitt and Hall–Taylor (1970):

$$\eta_m = \eta_l \quad \text{(Owens)} \tag{2.10a}$$

$$\eta_m = \frac{1}{\dfrac{X}{\eta_g} + \dfrac{(1-X)}{\eta_l}} \quad \text{(Isbin)} \tag{2.10b}$$

$$\eta_m = X\eta_g + (1-X)\eta_l \quad \text{(Cicchitti)} \tag{2.10c}$$

$$\eta_m = X\eta_g \frac{\rho_m}{\rho_g} + (1-X)\eta_l \frac{\rho_m}{\rho_l} \quad \text{(Dukler)} \tag{2.10d}$$

## 2.4 | Units of Viscosity

Units of absolute viscosity in various systems are given below:

SI system–Pa-s or N s/$m^2$, kgf s/$m^2$
cgs system–dyn.s/$cm^2$ or Poise (P), 1 cP = $10^{-2}$ P
British system–1bf.s/$in^2$ or Reyn

Viscosity is usually expressed in cP and factors to convert cP into other units are as follows:

1cP = $10^{-3}$ N.s/$m^2$ = $1.45 \times 10^{-7}$ Reyn
1cP = $1.02 \times 10^{-4}$ kgf . s/$m^2$

Units of kinematic viscosity are:

SI system–$m^2$/s
cgs system–$cm^2$/s or a Stoke(S), 1cS = $10^{-2}$S
British system–$in^2$/s

## 2.5 | Pressure–Temperature Effects on Viscosity

Viscosity of liquids significantly increases with pressure and decreases with temperature. Extensive experimental data are available in the open literature. Bridgeman (1926, 1949) studied the effects of pressure on viscosity of liquids.

In many practical applications where fluid is in shear at very high pressure, it has been observed that viscosity increase by many orders of magnitude. Very high pressures of the order of 1GPa or more are often encountered in concentrated contacts, e.g., in gears and rolling element bearings. Lubricant viscosity increases manifold when it passes through these contacts. It has been of interest to express this variation in the form of an empirical relationship which can be used in calculations of pressures and lubricant film thickness in these contacts.

Barus (1893) proposed an exponential relationship for viscosity pressure variation as follows:

$$\eta = \eta_0 e^{\alpha p} \tag{2.11}$$

where $\eta$ is viscosity at pressure $p$ and $\eta_0$ is the viscosity at ambient pressure and $\alpha$ is the pressure–viscosity coefficient of the fluid which has a unit m²/N in SI system. It is known that Barus' relationship overestimates viscosity at high pressures and does not agree well with the experimental results at high pressures. However, it gives a reasonably good estimation of viscosity at moderate pressures and is widely used because of its simple relationship with pressure.

On the other hand, Roelands (1966) developed an empirical expression to determine viscosity of fluids at elevated pressures given as follows:

$$\log_{10}\eta + 1.2000 = \left(\log_{10}\eta_0 + 1.2000\right)\left(1 + \frac{p}{2000}\right)^{z_1} \tag{2.12}$$

where $\eta$ is absolute viscosity of lubricant in cP at pressure $p$ in kgf/cm² and $\eta_0$ is viscosity at atmospheric pressure in cP, $z_1$ is the viscosity–pressure index and is dimensionless. The above equation can also be expressed as:

$$\bar{\eta} = \frac{\eta}{\eta_0} = 10^{-\left(1.2 + \log_{10}\eta_0\right)\left[1 - \left(1 + \frac{p}{2000}\right)^{z_1}\right]} \tag{2.13}$$

and can be rewritten as

$$\bar{\eta} = \frac{\eta}{\eta_0} = \left(\frac{\eta_\infty}{\eta_0}\right)^{1 - \left(1 + \frac{p}{c_p}\right)^{z_1}} \tag{2.14}$$

where $\eta_\infty = 6.31 \times 10^{-5}$ N.s/m²

$c_p = 1.96 \times 10^{8}$ N/m²

$\eta_\infty$, $c_p$ are constants for which same dimensions are to be used.

It can be shown that both Barus' and Roelands' formulae give the same result as the pressure tends to zero or atmospheric pressure, i.e.,

$$\frac{\partial}{\partial p}\left(\ln \eta_{Roelands}\right)_{p \to 0} = \frac{\partial}{\partial p}\left(\ln \eta_{Barus}\right)_{p \to 0} = \alpha \tag{2.15}$$

Also there is a correlation between $\alpha$ and $z_1$. $z_1$ can be expressed in terms of $\alpha$ as:

$$z_1 = \frac{\alpha}{\left(1/c_p\right)\left(\ln \eta_0 - \ln \eta_\infty\right)} \tag{2.16}$$

where $1/c_p = 5.1 \times 10^{-9}$ m²/N and $\ln \eta_\infty = 9.67$

This gives

$$z_1 = \frac{\alpha}{5.1 \times 10^{-9}\left(\ln \eta_0 + 9.67\right)} \tag{2.17}$$

Viscosity–pressure effect is shown in Fig. 2.4(a).

Viscosity of liquids decreases with increase in temperature significantly as can be seen in Fig. 2.4(b). Similar to viscosity–pressure equations both Barus and Roelands gave viscosity–temperature relationships to calculate viscosity of fluids at elevated temperatures as follows:

Barus' equation:

$$\eta(T)=\eta_0 e^{-\beta(T-T_0)} \tag{2.18}$$

where $T$ is the fluid temperature, $T_0$ is the temperature for base viscosity and $\beta$ is the temperature–viscosity coefficient which has a dimension as reciprocal of temperature.

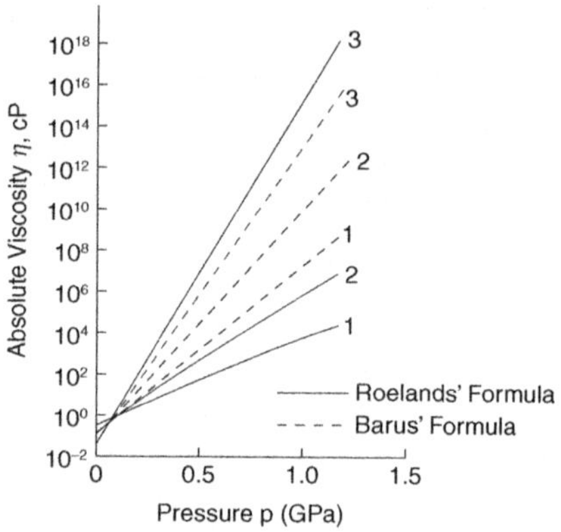


**Figure 2.4(a)** | Viscosity–Pressure Graphs of Three Fluids

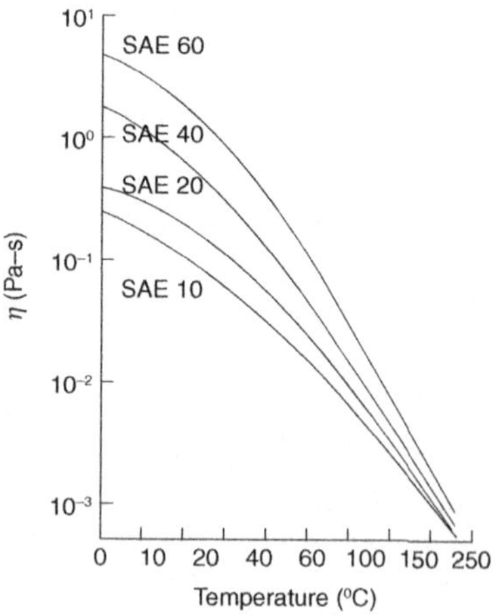


**Figure 2.4(b)** | Viscosity as a Function of Temperature at Atmospheric Pressure of Several SAE Petroleum-based Oils

Roelands' equation:

$$\log_{10}\left(\log_{10}\eta(T)+1.2000\right)=-S_0\log_{10}\left(1+\frac{T}{135}\right)+\log_{10}G_o \tag{2.19}$$

where $T$ is in °C, $G_0$ and $S_0$ are dimensionless constants indicative of viscosity grade of the lubricant and slope of viscosity–temperature relationship, respectively.

Using Equation (2.12), the above equation can be written as:

$$\bar{\eta}=\frac{\eta(T)}{\eta_0}=\left(\frac{\eta_\infty}{\eta_0}\right)10^{G_o\left(1+\frac{T}{135}\right)^{-S_0}} \tag{2.20}$$

Table 2.1 give the viscosities of some fluids, values of pressure–viscosity coefficients $\alpha$, $z_1$, and viscosity–temperature relationship may be found from the available open literature.

Viscosity–temperature variation is also represented quite accurately by Vogel's equation as written below:

$$\eta = ke^{b/(T+\theta)} \tag{2.21}$$

where $k$, $b$, and $\theta$ are constants. $k$ has the units of viscosity and gives the inherent viscosity of oil, $b$ is viscosity variation term which has the unit of temperature. $\theta$ shows the infinite viscosity temperature and for mineral oils $\theta = 115$. Mobil Limited, an oil company, provided data for viscosity and density of engine and gear oils at 25°C, 40°C, and 100°C, respectively, which are shown in Table 2.2. Vogel constants $b$ and $\theta$ in degree centigrade calculated using the given data are also shown in the table. $\theta$ values range between 100 and 123 and usually an average value of 115 is taken in Equation (2.20).

### 2.5.1 | ASTM Chart

Viscosity–temperature relationship most commonly used in practical and industrial applications is based on Walther's law and is given as follows:

$$\log_{10}\left(\nu+\gamma\right)=\log_{10}n+\frac{1}{T^c}$$

where $T$ is absolute temperature, $\gamma$, $n$, $c$ are constants. $\nu$ is kinematic viscosity in centistokes. $\gamma$ is usually taken as 0.6. Viscosity–temperature chart known as ASTM chart employs the above relationship and works very well for all mineral oils.

**Table 2.1** | Viscosity, Pressure–Viscosity Coefficient ($\alpha$, m$^2$/N), and Viscosity Pressure Index ($z_1$) of Three Different Oils [Jones *et al.* 1975]

| | **Abs. viscosity, $\eta$(cP) at zero pressure** | | | **$\alpha$(m$^2$/N) × 10$^{-8}$** | | | **$z_1$(Eq. 2.16)** | | |
|---|---|---|---|---|---|---|---|---|---|
| | **Temperature, °C** | | | **Temperature, °C** | | | **Temperature, °C** | | |
| **Oil** | **38** | **99** | **149** | **38** | **99** | **149** | **38** | **99** | **149** |
| Synthetic paraffinic oil (1) | 414 | 34.3 | 10.9 | 1.77 | 1.51 | 1.09 | 0.40 | 0.47 | 0.42 |
| Super-refined napthenic oil (2) | 68.1 | 6.86 | 2.74 | 2.51 | 1.54 | 1.27 | 0.71 | 0.64 | 0.66 |
| Synthetic hydrocarbon, i.e., traction fluid (3) | 34.3 | 3.53 | 1.62 | 3.12 | 1.71 | 0.939 | 0.97 | 0.83 | 0.57 |

**Table 2.2** | Kinematic Viscosity and Density of SAE Grade Oils

| | | Kinematic viscosity cSt | | | Density | | |
|---|---|---|---|---|---|---|---|
| **Oil** | **SAE** | **25°C** | **40°C** | **100°C** | **25°C** | **40°C** | **100°C** |
| 5W | | 48.2 | 25.45 | 4.994 | 0.886 | 0.851 | 0.812 |
| 10W | | 86.9 | 42.75 | 6.786 | 0.877 | 0.867 | 0.829 |
| 15W | | 102 | 47.82 | 7.302 | 0.879 | 0.869 | 0.831 |
| 20W | | 259 | 112.7 | 12.52 | 0.886 | 0.876 | 0.839 |
| 20 | | 129 | 59.57 | 8.387 | 0.880 | 0.870 | 0.832 |
| 30 | | 259 | 112.7 | 12.52 | 0.886 | 0.876 | 0.839 |
| 40 | | 361 | 151.1 | 15.27 | 0.891 | 0.881 | 0.884 |
| 50 | | 639 | 250.3 | 21.11 | 0.899 | 0.889 | 0.852 |

## 2.6 | Viscosity Grades of Oils

Society of Automotive Engineers of USA graded oils on the basis of the kinematic viscosity measured at 100°F (38°C). Dynamic viscosity is expressed in terms of SUV unit or Saybolt seconds which is the time taken in seconds for 60 cm³ of oil to flow through the orifice of Saybolt universal viscometer at 38° C (100°F). Dynamic viscosity is expressed in centistokes as

$$\eta_k = 0.22t - \frac{180}{t} \tag{2.22}$$

where $t$ is in Saybolt seconds. Multigrade oils which are usually thin base oil thickened by addition of a polymer less than 5%. Multigrade oil viscosity changes much less with temperature in comparison to base oil. However, polymers suffer from shear degradation with prolonged use and permanent loss of viscosity may take place. Viscosities of some SAE graded oils are shown in Fig. 2.4(b) varying with temperature.

## 2.7 | Viscosity Index

Oils produced by various refineries vary in respect of viscosity temperature behavior. Pennsylvania oils (1920) was supposed to be good in the sense that its viscosity varied less with temperature, whereas viscosity of Gulf coast oils varied much with temperature.

The viscosities of Pennsylvania oils and Gulf coast oils were measured at 210°F and 100°F. Pennsylvania oils were given a viscosity index (VI) of 100 and California oils a VI of 0 as shown in Fig. 2.5. For unknown oil viscosity is measured at 100°F. Viscosity index of the oil is given by

$$VI = \frac{\eta_l - \eta_u}{\eta_l - \eta_h} \times 100 \tag{2.23}$$

where $\eta_l$ = Viscosity of low VI oil at 38°C
$\eta_h$ = Viscosity of high VI oil at 38°C
$\eta_u$ = Viscosity of unknown oil at 38°C

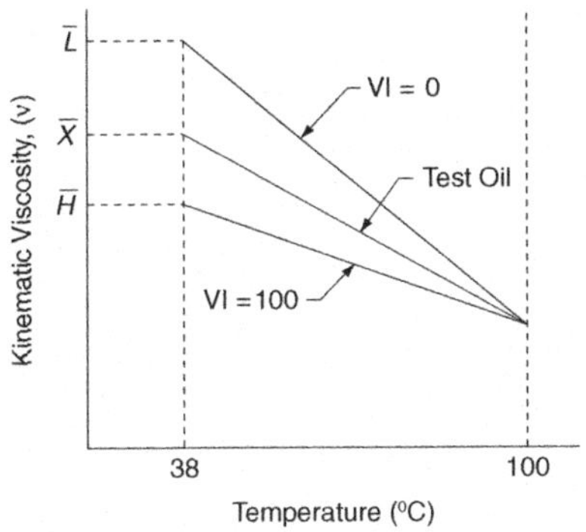


**Figure 2.5** | Viscosity Index (VI)

Viscosities $\eta_l$, $\eta_h$ and $\eta_u$ are measured using Saybolt universal viscometer and expressed in SUV units. Values of viscosities of oils are assumed to be equal at 100°C.

## 2.8 | Viscosity Measurement

Viscosity of a fluid is measured with the help of an instrument called viscometer. There are various types of viscometers commonly employed to measure viscosity. These are devices which use viscous flow theory to relate the flow rate with pressure drop across the standardized flow path and viscosity of the fluid.

Following types of viscometer are commonly used to measure viscosity:

- Capillary tube viscometer
- Concentric cylinder viscometer
- Cone and plate viscometer
- Falling sphere viscometer

### 2.8.1 | Capillary Tube Viscometer

In this type of viscometer as shown in Fig. 2.6, the fluid flows through a long capillary tube of small diameter; the flow rate is measured and related to the pressure drop across the two points on the tube. Usually length to diameter ratio of the capillary tube is greater than 20, i.e., $l_c > 20d_c$.

The volume flow rate, $q$, is the product of area and average flow velocity, thus

$$q = \frac{\pi d_c^4}{128\eta}\Delta p = \frac{\pi d_c^4}{128\eta l_c}\rho g h \; ; \; \Delta p = \frac{\rho g h}{l_c} \tag{2.24}$$

where $\Delta p$ is the pressure drop across the length of the capillary tube and $p = \rho g h$, $\rho$ is the density of the fluid. Thus, viscosity of the fluid is given by:

$$\eta = \frac{\pi d_c^4 \rho g h}{128 l_c q} = \frac{\pi d_c^4 \rho g h}{128 l_c (V/t)} \tag{2.25}$$

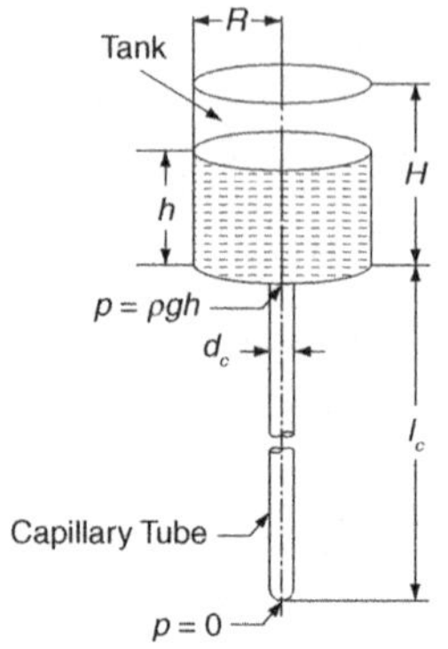


**Figure 2.6** | Capillary Tube Viscometer

since $q = \frac{V}{t}$, where $V$ = known volume of liquid and $t$ is the time taken for this volume to flow through the capillary. Thus, $\nu = \frac{\eta}{\rho} = ct$, where $c = \frac{\pi d_c^4 gh}{128 l_c V}$, is a constant for a given viscometer. This viscometer measures dynamic viscosity of the liquid.

The above relationship is true for incompressible laminar flow only and for Reynolds number less than 2100.

## 2.8.2 | Concentric Cylinder Viscometer

In this type of viscometer as shown in Fig. 2.7, two cylinders one enveloping the other is taken. Usually the inner cylinder is held stationary and outer cylinder is rotated at a constant angular speed with clearance space filled with the viscous fluid for which the viscosity is to be determined. Clearance space or the film of fluid is thin relative to the radius of the cylinders. In steady state condition when the flow is laminar and the Reynolds number is low, the radial and axial flow velocities are zero, and there is no pressure gradient in the circumferential direction.

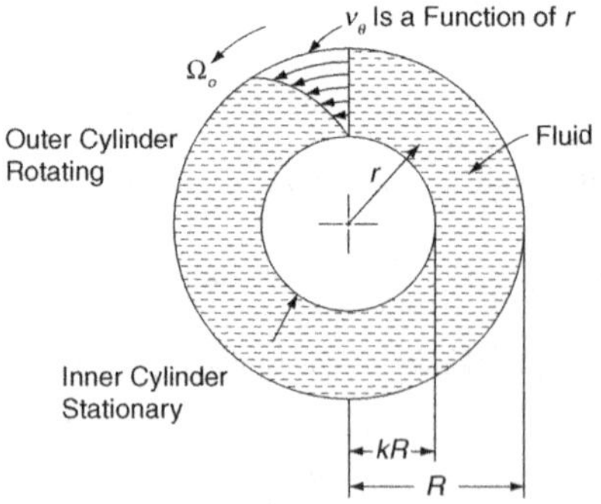


**Figure 2.7** | Flow of an Incompressible Fluid Between Two Concentric Cylinders

The momentum equations in $r$, $\theta$, $z$, i.e., radial, circumferential, and axial directions, respectively, can be written as:

$r-$ component:

$$-\rho\frac{v_\theta^2}{r} = -\frac{\partial p}{\partial r} \tag{2.26}$$

$\theta-$ component:

$$0 = \frac{\partial}{\partial r}\left(\frac{1}{r}\frac{\partial}{\partial r}(rv_\theta)\right) \tag{2.27}$$

$z-$component:

$$0 = \frac{\partial p}{\partial z} + \rho g \tag{2.28}$$

$v_r$, $v_\theta$, $v_z$ are fluid velocity components in $r$, $\theta$, $z$, directions, respectively.

Integrating Equation (2.26) with respect to $r$ with following boundary conditions at the inner and outer peripheries of the cylinder as:

$v_\theta = 0, \quad r = r_i; \quad v_\theta = \Omega r_o, \quad r = r_o$ would yield

$$v_\theta = \Omega r_o \frac{\left(\frac{r_i}{r} - \frac{r}{r_i}\right)}{\left(k - \frac{1}{k}\right)} \tag{2.29}$$

where $k = \frac{r_i}{r_o}$ and the expression for the shear stress can be determined as:

$\tau_{r\theta} = -\eta r\frac{\partial}{\partial r}\left(\frac{v}{r}\right)$; Since the flow is in the circumferential direction only $v_r = v_z = 0$.

Thus, shear stress is obtained using Equation (2.28) as:

$$\tau_{r\theta} = -2\eta\Omega r_o^2\left(\frac{1}{r^2}\right)\left(\frac{k^2}{1-k^2}\right) \tag{2.30}$$

The torque required to run the outer cylinder at an angular speed $\Omega$ can be determined as the moment of the shear force at the cylinder surface and is written as:

$$M = 2\pi r_o^2 L\left(-\tau_{r\theta}\right)_{r=r_o} = 4\pi\eta L r_o^2\Omega\left(\frac{k^2}{1-k^2}\right) \tag{2.31}$$

Viscosity of the fluid can be determined from the measured torque to rotate the outer cylinder at a required speed. This viscometer measures absolute viscosity of the liquid.

### 2.8.3 | Cone and Plate Viscometer

Cone and plate viscometer is shown in Fig. 2.8. In this, an inverted cone is rotated on a plate with space between the cone and plate filled with a viscous fluid of which viscosity is to be determined. The angle which

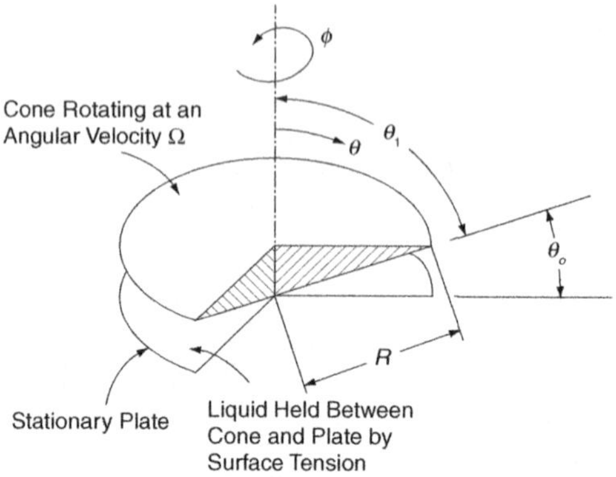


**Figure 2.8** | Cone and Plate Viscometer Configuration

the cone makes with the plate is usually very small. The torque required to rotate the cone at an angular speed Ω is a measure of the viscosity of the fluid. The angle between cone and flat is usually kept very small, about a degree or two to ensure that the flow is such that only shear stress component $\tau_{\theta\phi}$ is significant, whereas other shear stress components practically do not exist. The magnitude of this shear stress component is almost constant in the entire fluid film between the surfaces.

Thus, the flow is assumed to be entirely tangential and therefore flow velocity component $v_\phi$ is a function of $r$, $\theta$ only, whereas flow velocities in radial and conical directions, i.e., $r$, $\theta$ directions, i.e., $v_r$, $v_\theta$, are zero. Thus, momentum equations can be written as:

$r-$ component:

$$-\rho\frac{v_\phi^{\,2}}{r} = -\frac{\partial p}{\partial r} \tag{2.32}$$

$\theta-$ component:

$$-\rho\cot\theta\frac{v_\phi^{\,2}}{r} = -\frac{1}{r}\frac{\partial p}{\partial \theta} \tag{2.33}$$

$\phi-$ component:

$$0 = \frac{1}{r^2}\frac{\partial}{\partial r}\left(r^2\tau_{r\phi}\right) + \frac{1}{r}\frac{\partial \tau_{\theta\phi}}{\partial \theta} + \frac{\tau_{r\phi}}{r} + 2\cot\theta\frac{\tau_{\theta\phi}}{r} \tag{2.34}$$

However, in case of slow viscous flow centrifugal inertia terms can be set equal to zero, i.e., terms containing $\frac{v_\phi^{\,2}}{r}$ can be set to zero. Other shear stress components are zero.

Assuming $v_\phi\,(r,\theta) = rf(\theta)$ which satisfy the boundary conditions at $\theta = \theta_1$ and $\theta = \frac{\pi}{2}$. Since the angular velocity $\frac{v_\phi}{r}$ is independent of $r$ means that $\tau_{r\phi} = 0$. Thus, Equation (2.33) reduces to an ordinary differential equation as given below:

$$\frac{d\tau_{\theta\phi}}{d\theta} = -2\tau_{\theta\phi}\cot\theta \tag{2.35}$$

Equation (2.34) can be integrated to give

$$\tau_{\theta\phi} = \frac{C_1}{\sin^2\theta} \tag{2.36}$$

where $C_1$ is a constant of integration which can be evaluated from the torque transmitted to the stationary plate by the fluid by using the boundary condition for torque at $\theta = \frac{\pi}{2}$. Thus, the torque at the stationary plate is given by:

$$M = \int_0^{2\pi}\int_0^{R} \tau_{\theta\phi}\big|_{\theta=\frac{\pi}{2}} r^2 dr d\phi = \frac{2\pi R^3}{3}\left(\frac{C_1}{\sin^2\frac{\pi}{2}}\right) \tag{2.37}$$

Equations (2.35) and (2.36) can be combined to give:

$$\tau_{\theta\phi} = \frac{3M}{2\pi R^3 \sin^2\theta}.$$

However, for small values of cone angle $\alpha$, $\sin^2\theta$ would be equal to 1 and the torque would be independent of cone angle. Since only velocity component $v_\phi$ exist while other velocity components are zero shear stress $\tau_{\theta\phi}$ would be given by:

$$\tau_{\theta\phi} = -\eta\sin\theta\frac{d}{d\theta}\left(\frac{v_\phi/r}{\sin\theta}\right) = \frac{3M}{2\pi R^3\sin^2\theta} \tag{2.38}$$

Integrating with respect to $\theta$ one would obtain the velocity component as:

$$\frac{v_\phi}{r} = \frac{3M}{4\pi R^3\eta}\left[\cot\theta + \frac{1}{2}\left(\ln\frac{1+\cos\theta}{1-\cos\theta}\right)\sin\theta\right] + C_2 \tag{2.39}$$

where $C2$ is a constant of integration. For very small cone angle $\alpha$, $\theta = \frac{\pi}{2}$. Since $v_\varphi = 0$ for $\theta = \frac{\pi}{2}$ when substituted in the above equation would give $C_2 = 0$. Therefore, $\theta = \frac{\pi}{2} - \alpha$, $v_\phi = \Omega r \sin\left(\frac{\pi}{2} - \alpha\right)$.

Equation (2.38) would reduce to:

$$\Omega r \sin\alpha_1 = \frac{3M}{4\pi R^3\eta}\left[\cot\alpha_1 + \left(\frac{1+\cos\alpha_1}{1-\cos\alpha_1}\right)\sin\alpha_1\right] \tag{2.40}$$

where $\alpha_1 = \frac{\pi}{2} - \alpha$.

Equation (2.39) can be used to determine the viscosity of the fluid if the torque $M$ is required to rotate the viscometer at an angular speed $\Omega$ is measured.

Alternatively, a simplified expression can be determined for small cone angle $\alpha$ with shear stress $\tau_{\theta\phi} = \frac{\eta v_\phi}{h}$ where $h = r\tan\alpha \cong r\alpha$, is film thickness at a radius $r$ and the surface velocity $v_\phi = \Omega r$. The torque $M$ required to rotate the viscometer at an angular speed $\Omega$ can be determined as:

$$M = \int_0^R \tau_{\theta\phi}.2\pi r^2 dr = \int_0^R \frac{\eta\Omega}{\alpha}.2\pi r^2 dr = \frac{2\pi\eta\Omega R^3}{3\alpha} \tag{2.41}$$

Measuring torque viscosity of the fluid can be determined from the above equation also fairly accurately.

### 2.8.4 | Falling Sphere Viscometer (Fig. 2.9)

A sphere is allowed to fall from rest in a viscous fluid; it will reach a terminal constant velocity after it accelerates in the initial stage of falling. When this steady state condition is reached, the sum of force of gravity on the solid which acts in the direction of fall, the sum of buoyancy force, and the force due to fluid motion which act in opposite direction balance each other.

Component of normal force $F_n$ in the vertical direction is expressed in the integral form as:

$$F_n = \int_0^{2\pi}\int_0^{\pi} (-p\cos\theta) R^2 \sin\theta d\theta d\phi$$

Pressure on the surface of the sphere is given as:

$$p = -\rho g R\cos\theta - \frac{3\eta v_t}{2R}\cos\theta$$

Substituting for pressure and integrating the normal force against the falling sphere is obtained as:

$$F_n = \frac{4}{3}\pi R^2 \rho g + 2\pi\eta R v_t \tag{2.42}$$

where the first term is due to buoyancy effect and second term is due to form drag effect.

Tangential force due to shear stress on the surface of the sphere is expressed as:

$$F_s = \int_0^{2\pi}\int_0^{\pi} \tau_{r\theta} \sin\theta\, R^2 \sin\theta\, d\theta\, d\phi$$

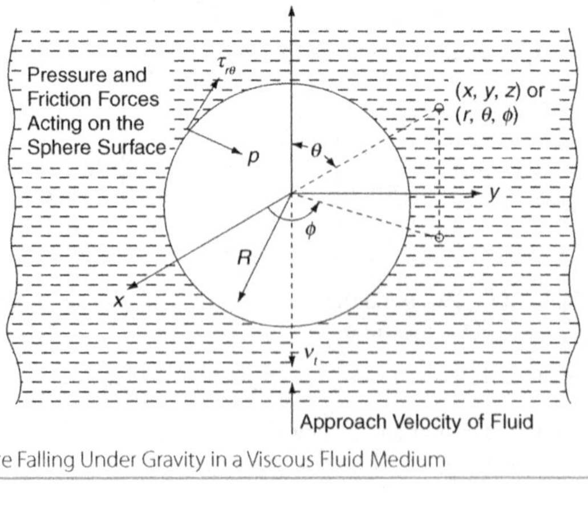


**Figure 2.9** | Rigid Sphere Falling Under Gravity in a Viscous Fluid Medium

Shear stress on the surface of the sphere is given by

$$\tau_{r\theta} = \frac{3}{2}\eta v_t R \sin\theta$$

Substituting for shear stress and integrating, the tangential force acting on the sphere is obtained as:

$$F_t = 4\pi\eta R v_t \tag{2.43}$$

Therefore, total force acting on the sphere can obtained by adding Equations (2.42) and (2.43).

Force–balance relationship for the sphere is obtained by equating the gravity force due to falling sphere to the total force acting on the sphere that resists the motion and can be written as:

$$\frac{4}{3}\pi R^3 \rho_s g = \frac{4}{3}\pi R^3 \rho_f + 6\pi\eta v_t R \tag{2.44}$$

where $R$ = radius of sphere
$v_t$ = terminal velocity of sphere in the fluid
$\rho_s$ = density of the sphere
$\rho_f$ = density of the fluid
$\eta$ = viscosity of the fluid

or $$\eta = 2R^2(\rho_s - \rho_f)g/9v_t \tag{2.45}$$

However, this result is valid when $2Rv_t\rho/\eta$ is less than 0.1.

Thus, measuring the terminal velocity of the sphere in the fluid, viscosity of the fluid can be determined using the above relationship. For further details related to viscometers, readers are referred to Bird, Stewart and Lightfoot (1960).

### 2.8.5 | High-Pressure Rheometers

Various types of rheometers have been designed and developed by many researchers to study the behavior of viscous fluids at high pressure of the order of 1 GPa or more and at high shear rates as well.

High-pressure shear stress apparatus where pressures up to 1.2 GPa were generated was developed by Bair and Winer (1979) and is shown in Fig. 2.10. In this apparatus, an intensifier piston is driven into

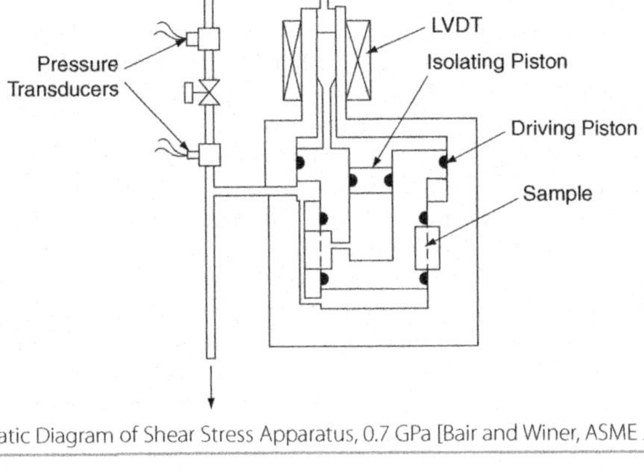


**Figure 2.10** | Schematic Diagram of Shear Stress Apparatus, 0.7 GPa [Bair and Winer, ASME JOLT, 1979]

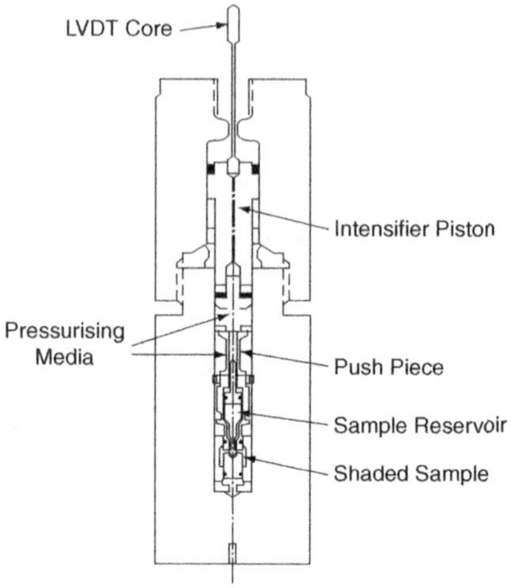


**Figure 2.11** | Schematic Diagram of High-Pressure Shear Stress Apparatus, 1.2 GPa [Bair and Winer ASME, JOLT, 1979]

a high-pressure chamber which increases the pressure of the driving fluid and test sample. The driving piston displacement and velocity were measured by LVDT. The device is assembled and filled so that the intensifier piston reaches the push piece when the desired pressure level is reached.

High shear stress viscometer also developed by Bair and Winer (1979) is shown in Fig. 2.11. In this case, the fluid sample is sheared between a central rod and a cylindrical hole in which the rod is pushed or moved axially through the hole by a pressure difference imposed across a driving piston attached to the rod. The pressure difference is measured and is proportional to the shear stress. The rod displacement and velocity are measured by a LVDT permitting determination of the strain and strain rate. Based on these measurements, rheological models given in Equation (2.8) were proposed by Bair and Winer (1979).

Rheological behavior of lubricants at high pressures and shear rates have also been investigated by Johnson and Cameron (1967), Johnson and Roberts (1974), and Johnson and Tevaarverk (1977) using two disc rolling contact apparatus in which rolling with spin and rolling with side slip could be obtained, the apparatus was designed by Poon and Hains (1966). Nonlinear viscoelastic rheological model given by Equation (2.7) was developed using the data obtained through measurements in the disc apparatus. Details can be obtained in the reference cited above.

Earlier, Hirst and Moore (1974) also presented an investigation based on two disc experiments for pressures up to 1 GPa and proposed an Erying type (sine hyperbolic) relationship.

Pressure shear plate impact experiments were also conducted by Ramesh and Clifton (1987, 1992) under high pressures and at high shear rates of the order of $10^6$/s to study shear rheology of lubricants. The details of the experimental technique are available in the above references.

Jacobson (1984) used a new impact type of rheometer where a spherical ball is made to impact on two parallel flat surfaces to study the rheology of lubricants under transient pressure and to measure the shear strength increase with pressure for pressurization times of 138 μs. Maximum pressure of the order of 5.5 GPa were generated during the impact time. Hoglund and Jacobson (1985) also developed an experimental apparatus to measure the shear strength of lubricants subjected to high pressure and temperature. Maximum pressure of 2.5 GPa was generated in the experimental stage. Details of experimentation are available in the above references. Knowledge about high-pressure and temperature rheological behavior of lubricants is extremely important for successful lubrication of concentrated rolling/sliding solid contacts. Rheological models of lubricants are also required to estimate lubricant film thickness, traction coefficient, and temperature rise due to sliding in these contacts.

## 2.9 | Chemistry of Lubricants

The emulsifiable oils are either based on paraffin or naphthenic structures with sulfonated derivatives as emulsifiers, with or without extra pressure additives. These include fats, fatty acids other esters, sulfurized fats, mineral oils, and chlorinated fats and mineral oils. The so called 'heavy duty products' have increasing amounts of extra pressure additives and the expense of the mineral oil content (Rossmoore).

### 2.9.1 | Water-based and Oil-based Systems

Water- and oil-based systems are used in cutting tool operations to reduce the cutting force and to increase the life of the tool. Water is used as a coolant because of its high thermal capacity, but it cannot act as a good lubricant which increases the cutting force. Water will cause problems of corrosion to cutting tool and materials. To overcome the lubrication effect and corrosion problems of water, oil is used in cutting operations. Oil gives lower cooling effect and its cost is higher than water. Oil added to water with good emulsifier (soluble oils) gives good lubrication and cooling effect (Machodo and Wallbank, 1997). Fluid supply system in cutting tool operation is developed in which flow parameters (pressure, temperature, flow velocity, and humidity) and cooling distance (the distance between nozzle and cutting zone) are controllable (Junyan Liu *et al.*, 2005).

### 2.9.2 | Synthetic Lubricants

Synthetic lubricants are derived from animal fats and plant oils as raw materials and are thus renewable raw materials par excellence. Compared to mineral-based products oleo-chemical products are 2–5 times more expensive. The higher price of oleo-chemical esters is due to the costly multistep synthesis of these chemicals. Oleo-chemicals will not affect the carbon dioxide balance in the atmosphere. Due to the high cost of these lubricants, they are used in closed lubrication systems (Willing, 2001).

### 2.9.3 | Suspensions and Pastes

Suspensions are considered to be dispersed when the repulsive potential between the particles is of sufficient magnitude that the attractive Van der Waals potentials are counterbalanced or exceeded. Coagulation can be achieved by adding salt to dispersed suspension (Davies and Binner, 1999). Paste systems behave like particle suspensions because they share many of the same characteristics. Suspensions consist of a homogeneously distributed particulate solid phase carried by a liquid solution. Electrostatic, stearic, Van der Waals, or a combination of these forces can result in a stable dispersion. Pastes consist of a mixture of solid phase mechanically separated by a fluid solution (Hurysz and Cochran, 2003).

### 2.9.4 | Solid Lubricants

Soft noble metals, inorganic fluorides, and some metal oxides have been successfully employed as solid lubricants. These materials generally possess stable thermo-chemistry at elevated temperatures as well as low shear strength properties which make them good solid lubricants. Use of solid lubricants depends on selection of appropriate counterface. Solid lubricants will form a transfer film on the sliding counterface in order to reduce friction and wear (DellaCorte, 1996).

### 2.9.5 | Coatings and Additives

Coatings on the surfaces reduce the wear of the material. Composite coating (PS300) is a plasma sprayed nichrome bonded $Cr_2O_3$ coating with silver and $BaF_2/CaF_2$ eutectic solid lubrication additions. Coatings provide good tribological properties at temperatures as high as 900°C. Chrome oxide is used in the composite coating instead of chrome carbide, because of its low machinable cost. Chrome oxide is a high temperature lubricant which cannot oxidize (DellaCorte, 1996).

Lubricant additives improve tribological properties and reduce oxidation of petroleum lubricants. Considering environmental and biological effects, lubricant additives must be zincless and do not have phosphorous. The derivatives of heterocyclic compounds are antiwear and extreme pressure additives. Heterocyclic compounds possess excellent anticorrosive, antirust, copper deactivating, and antiwear properties (Huang *et al.*, 2000).

### 2.9.6 | Toxicological and Biological Aspects of Metal Working Lubrication

The fluids, after their usage have to be disposed. The disposal of fluids will affect the environment. When fluid is used as cutting fluid in tool operations, it will cause several reactions on the skin and other parts of the body of the operator. There are four types of oil induced skin diseases: dermatitis, oil folliculitis, oil acne, and finally keratoses and warts (Baradie, 1996). By considering environmental and toxicological problems, lubricating oil additives are added, which are zincless and do not contain phosphorous (Huang *et al.*, 2000).

### 2.9.7 | Lubrication Clarification, Recycling, and Disposal

The recycling of the lubricant is necessary to remove the contaminants in it. The contaminants are removed by selecting a clarification process. The clarification process is selected based on the type of the cutting fluid, fineness of filtration required, proportion of contaminants in the coolant and volume flow of the fluid. The straight oil fluids can be removed by stirring and water contaminants were removed by heat treatment. Recycling of cutting fluids will reduce the wastage of fluid and disposal of fluid to environment. Recycling system is a part of central fluid system which can store and pump the fluid to the machine parts. After the disposal of lubricant, the chemicals polluting the environment are oils, phenols, phosphates, and heavy metals. To reduce the disposal of fluid, settling and addition of base concentrates usually restore the adequate quality for continuous use. When necessary, oil-based fluids should be disposed by burning (Baradie, 1996).

## REFERENCES

*Andreas Willing* (2001), Lubricants based on renewable resources—environmentally compatible alternative to mineral oil products, Chemosphere, Elsevier Science Ltd, **43**, 89–98.

*Bair, S. and Winer, W.O.* (1979a), A rheological model for elastohydrodynamic contacts based on laboratory data, *Trans. ASME, J. Lub. Tech.*, **101**, 258–265.

*Bair, S. and Winer, W.O.* (1979b), Shear strength measurements of lubricants at high pressure, *Trans. ASME, J. Lub. Tech.*, **101**, 251–257.

*Baradie M. A. EI* (1996), Cutting fluids: Part II: Recycling and clean machining, *Journal of Materials Processing Technology*, 798–806.

*Barus, C.* (1893), Isotherms, isopiestic and isometrics relative to viscosity, *Am. J. Sci.*, **45**, 87–96.

*Bird, R.B., Stewart, W.E. and Lightfoot, E.N.* (1960), *Transport Phenomena*, Wiley International Edition, John Wiley & Sons Inc.

*Bridgeman, P.W.* (1926), The effect of pressure on viscosity of forty three pure liquids, *Proc. Am. Acad.*, **61**, 631–643.

*Bridgeman, P. W.* (1949), Viscosities to 30,000 Kg/cm$^2$, *Proc. Am. Acad. Arts and Sci.*, **17**, 117–127.

*Christopher DellaCorte (1996)*, The effect of counterface on the tribological performance of high temperature solid lubricant composite from 25 to 650°C, *Surface and Coatings Technology*, Elsevier Science Ltd, **86–87**, 486–492.

*Davies, J. and Binner, J.G.P.* (1999), Coagulation of electrosterically dispersed concentrated alumina suspensions for paste production, *J. Euro. Cer. Soc.*, 1555–1567.

*Hewitt, G.F., and Hall-Taylor, N.S.* (1970), *Annular Two Phase Flow*, I Ed., p.30,

*Hirst, W. and Moore, A.J.* (1974), Non-Newtonian behaviour in elastohydrodynamic lubrication, *Proc. Roy. Soc.*, London, *Ser. A*, **337**, 101–121.

*Hoglund, E. and Jacobson, B.* (1985), Experimental investigation of the shear strength of lubricants subjected to high pressure and temperature, *ASME*, Paper No. 85-Trib-34.

*Huang W., Dong J., Li F. and Chen B.* (2000), The performance and antiwear mechanism of (2-sulfurone-benz othiazole)-3-methyl esters as additives in synthetic lubricant, *Tribology International*, 553–557.

*Hurysz, K. M. and Cochran, J. K.* (2003), The application of models for high solids content suspensions to pastes, *J. Euro. Cer. Soc.*, 2047–2052.

*Jacobson, B.* (1984), A high pressure-short time shear strength analyzer for lubricants, *ASME*, Paper No. 84, Trib-3.

*Johnson, K.L. and Cameron, R.* (1967). Shear behaviour of elastohydrodynamic oil films at high rolling contact pressures, *Proc. Inst. Mech. Engrs.*, London, UK, **182**(14), 307–327.

*Johnson, K.L. and Roberts, A.D.* (1974), Observations of viscoelastic behavior of an EH film, *Proc. Roy. Soc.*, London, *Ser A*. **337**, 217–242.

*Johnson, K.L. and Tevaarwerk, J.L.* (1977), Shear behaviour of EHD oil film, *Proc. Roy. Soc, London, Ser. A*, **356**, 215–236.

*Jones, W.R., et al.* (1975), Pressure viscosity measurement of several lubricants to $5.5 \times 10^8$ N/m$^2$, *ASLE Trans.*, **18**, 249–262.

*Junyan Liu, Rongdi Han, and Yongfeng Sun* (2005), Research on experiments action mechanism with water vapour as coolant and lubricant in green cutting, *Int. J. Machine Tools & Manufacture*, 687–694.

*Machodo, A. R. and Wallbank, J.* (1997), 'The effect of extremely low lubricant volumes in machining', *Wear*, 76–82.

*Ramesh, K. T. and Clifton, R, J.* (1987), A pressure-shear plate impact experiment for studying the rheology of lubricants at high pressures and high shearing rates, *Trans. ASME, J. Tribology*, **109**, 215–222 .

*Ramesh, K.T. and Clifton, R.J.* (1992), Finite deformation analysis of pressure-shear plate impact experiments on elastohydrodynamic lubrication, *Trans. ASME, J. Appl. Mech.*, **59**, 754–761.

*Roelands, C.J.A.* (1966), *Correlational Aspects of Viscosity-Temperature-Pressure Relationship of Lubricating Oils*: Druk., V.R.B., Graomingen, Netherlands.

*Rossmoore, H. W.* (1989), Microbial Degradation of Waterbased Metalworking Fluids, in *Comprehensive Biotechnology: The Principles, Applications and Regulations of Biotechnology in Industry, Agriculture and Medicine*, Moo-Young, M. (ed.), Pergamon Press, New York, 249–269.

Chapter 3

# Mechanics of Lubricant Films and Basic Equations

## 3.1 | Introduction

Flow of lubricants through narrow gaps or clearance space of bearings, seals, gears, etc. is governed by the laws of flow of fluids. Following four basic laws are concerned with flow of fluids.

1. Conservation of mass which is expressed in the form of continuity equation.
2. Newton's second law of motion or conservation of momentum usually expressed as momentum equation.
3. Conservation of energy expressed in the form of energy equation derived using first law of thermodynamics.
4. Equation of state for a perfect gas. It is usually required for compressible flow or in gas/air as a lubricant.

The above laws apply to a control volume or a fixed quantity of matter, i.e., a definite system.

## 3.2 | Momentum Equations

To derive momentum equations, we consider an elemental cube or volume of fluid as shown in Fig. 3.1. Stresses acting on the cube faces are shown with directions. The stresses must be symmetric, that is:

$$\tau_{xy} = \tau_{yx},\ \tau_{xz} = \tau_{zx},\ \tau_{yz} = \tau_{zy}$$

The total force $F$ is made up of the total surface force $F_s$ due to stresses and body force $B$ which is a force per unit volume. The momentum equation for a control volume then becomes:

$$F_s + \oint_{volume} B dv = \frac{\partial}{\partial t}\oint \vec{V}\, p dv + \oint_{surface} \vec{V}\,\rho\vec{V}.dA \tag{3.1}$$

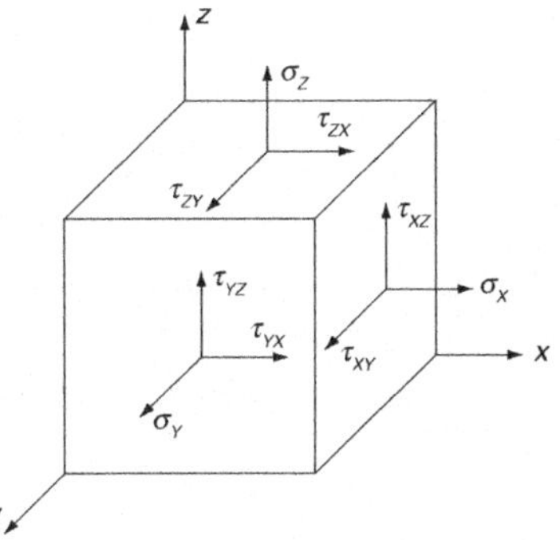


**Figure 3.1** | Stresses on the Surface of a Fluid Volume Element

where $\vec{V}$ is the velocity vector and $dv$ the elemental volume. This equation is valid only for noninertial reference axes.

Momentum balance in the $x$ direction can be written as:

$$\rho\frac{Du}{Dt} = \rho\left(\frac{\partial u}{\partial t} + u\frac{\partial u}{\partial x} + v\frac{\partial u}{\partial y} + w\frac{\partial u}{\partial z}\right) = \frac{\partial \sigma_x}{\partial x} + \frac{\partial \tau_{xy}}{\partial y} + \frac{\partial \tau_{xz}}{\partial z} + B_x \tag{3.2}$$

Similarly, momentum equations in $y$ and $z$ directions, respectively, are given as:

$$\rho\frac{Dv}{Dt} = \rho\left(\frac{\partial v}{\partial t} + u\frac{\partial v}{\partial x} + v\frac{\partial v}{\partial y} + w\frac{\partial v}{\partial z}\right) = \frac{\partial \tau_{yx}}{\partial x} + \frac{\partial \sigma_y}{\partial y} + \frac{\partial \tau_{yz}}{\partial z} + B_y \tag{3.3}$$

$$\rho\frac{Dw}{Dt} = \rho\left(\frac{\partial w}{\partial t} + u\frac{\partial w}{\partial x} + v\frac{\partial w}{\partial y} + w\frac{\partial w}{\partial z}\right) = \frac{\partial \tau_{zx}}{\partial x} + \frac{\partial \tau_{zy}}{\partial z} + \frac{\partial \sigma_z}{\partial z} + B_z \tag{3.4}$$

$\frac{Du}{Dt}, \frac{Dv}{Dt}$ and $\frac{Dw}{\partial t}$ are components of inertia forces in $x$, $y$, and $z$ directions, respectively, and $B_x$, $B_y$, and $B_z$ are components of body forces in $x$, $y$, and $z$ directions, respectively.

## 3.3 | Stress–Strain Relationship for Fluids

For a Newtonian fluid, it is assumed that stress and strain rate are linearly related and the general relationship can be expressed in Cartesian coordinates as:

$$\text{Strain rate; } e_{ij} = \partial u_i/\partial x_j \text{ for nonrotational flow} \tag{3.5}$$

$$\text{Stress: } \sigma_{ij} = -p\delta_{ij} + 2\eta e_{ij} + \delta_{ij}\lambda\phi \tag{3.6}$$

where $\phi$ is the fluid dilatation, i.e., the rate at which fluid flows out from each or it measures the expansion of the fluid and $\lambda$ is a second coefficient of viscosity, $\delta_{ij}$ is the Kronecker delta.

Hydrostatic pressure p in the fluid is the mean of the three normal stresses and therefore $\sigma_{xx} = \sigma_{yy} = \sigma_{zz} = -p$ since it is compressive in nature.

Also

$$\sigma_{xx} + \sigma_{yy} + \sigma_{zz} = -3p \tag{3.7}$$

Further

$$\sigma_{xx} = -p + \lambda\phi + 2\eta\frac{\partial u}{\partial x} \tag{3.8.a}$$

$$\sigma_{yy} = -p + \lambda\phi + 2\eta\frac{\partial v}{\partial y} \tag{3.8.b}$$

$$\sigma_{zz} = -p + \lambda\phi + 2\eta\frac{\partial w}{\partial z} \tag{3.8.c}$$

where

$$\phi = \left(\frac{\partial u}{\partial x} + \frac{\partial v}{\partial y} + \frac{\partial w}{\partial z}\right) \tag{3.9}$$

Thus,

$$\sigma_{xx} + \sigma_{yy} + \sigma_{zz} = -3p + (3\lambda + 2\eta)\phi \tag{3.10}$$

therefore,

$$3\lambda + 2\eta = 0 \tag{3.11}$$

or $\lambda = -\frac{2\eta}{3}$ and this is known as Stoke's condition and ensures that pressure $p$ is defined as the average of normal stresses for a compressible fluid at rest. The term $\left(\lambda + \frac{2}{3}\eta\right)$ is called the coefficient of bulk viscosity and is equal to zero.

It is known for fluids that normal component of strain rate $e_{ii}$ is directly identified with true normal strain rate. Therefore,

$$e_{xx} = \frac{\partial u}{\partial x}, e_{yy} = \frac{\partial v}{\partial y} \text{ and } e_{zz} = \frac{\partial w}{\partial z} \tag{3.12}$$

Whereas, shear rate components are equal to one half the true rate of shear strain components which is denoted as $\gamma_{ij}$, i.e., $e_{ij} = \frac{1}{2}\gamma_{ij}(i = j)$

Thus,

$$e_{xy} = e_{yx} = \frac{1}{2}\left(\frac{\partial u}{\partial y} + \frac{\partial v}{\partial x}\right) \tag{3.13a}$$

$$e_{yz} = e_{zy} = \frac{1}{2}\left(\frac{\partial v}{\partial z} + \frac{\partial w}{\partial y}\right) \tag{3.13b}$$

$$e_{xz} = e_{zx} = \frac{1}{2}\left(\frac{\partial w}{\partial x} + \frac{\partial u}{\partial z}\right) \tag{3.13c}$$

Therefore, shear stress $\tau_{ij}$ can be related to shear strain rate for a Newtonian fluid as:

$$\tau_{ij} = \eta\left(\frac{\partial u_i}{\partial x_j} + \frac{\partial u_j}{\partial x_i}\right) \tag{3.14}$$

Thus,

$$\tau_{xy} = \eta\left(\frac{\partial u}{\partial y} + \frac{\partial v}{\partial x}\right) \tag{3.15a}$$

$$\tau_{yz} = \eta\left(\frac{\partial v}{\partial z} + \frac{\partial w}{\partial y}\right) \tag{3.15b}$$

$$\tau_{xz} = \eta\left(\frac{\partial w}{\partial x} + \frac{\partial u}{\partial z}\right) \tag{3.15c}$$

We arrive at Navier–Stokes equations for flow of a Newtonian fluid by substituting the above relationships for normal and shear stresses in terms of strain rates given by Equations (3.8) and (3.15), respectively, into momentum balance Equations (3.2), (3.3), and (3.4) as:

## 3.4 | Navier–Stokes Equations

$$\rho\frac{Du}{Dt} = B_x - \frac{\partial p}{\partial x} - \frac{2}{3}\frac{\partial}{\partial x}(\eta\phi) + 2\frac{\partial}{\partial x}\left(\eta\frac{\partial u}{\partial x}\right) + \frac{\partial}{\partial y}\left[\eta\left(\frac{\partial u}{\partial y} + \frac{\partial v}{\partial x}\right)\right] + \frac{\partial}{\partial z}\left[\eta\left(\frac{\partial u}{\partial z} + \frac{\partial w}{\partial x}\right)\right] \tag{3.16}$$

$$\rho\frac{Dv}{Dt} = B_y - \frac{\partial p}{\partial y} - \frac{2}{3}\frac{\partial}{\partial y}(\eta\phi) + 2\frac{\partial}{\partial y}\left(\eta\frac{\partial v}{\partial y}\right) + \frac{\partial}{\partial x}\left[\eta\left(\frac{\partial u}{\partial y} + \frac{\partial v}{\partial x}\right)\right] + \frac{\partial}{\partial z}\left[\eta\left(\frac{\partial v}{\partial z} + \frac{\partial w}{\partial y}\right)\right] \tag{3.17}$$

$$\rho\frac{Dw}{Dt} = B_z - \frac{\partial p}{\partial z} - \frac{2}{3}\frac{\partial}{\partial z}(\eta\phi) + 2\frac{\partial}{\partial z}\left(\eta\frac{\partial w}{\partial z}\right) + \frac{\partial}{\partial x}\left[\eta\left(\frac{\partial u}{\partial z} + \frac{\partial w}{\partial x}\right)\right] + \frac{\partial}{\partial y}\left[\eta\left(\frac{\partial v}{\partial z} + \frac{\partial w}{\partial y}\right)\right] \tag{3.18}$$

The above equations are valid for viscous flow of a compressible, Newtonian fluid with variable viscosity, and form the basis for all analytical work in fluid mechanics. Therefore, these equations also form the base for the development of mechanics of lubrication in bearings, seals, and almost all mechanical components.

## 3.5 | Continuity Equation

Integral form of mass continuity relationship can be developed beginning write

$$\oint_{c.s} \rho\vec{V}.dA = -\oint_{c.v} \rho dv \tag{3.19}$$

The above relationship can be rewritten applying Gauss' theorem as:

$$\oint_{c.v} \nabla.\left(\rho\vec{V}\right)dv + \frac{\partial}{\partial t}\oint_{c.v} \rho dv = \oint_{c.v}\left[\nabla.\rho\vec{V} + \frac{\partial \rho}{\partial t}\right]dv \tag{3.20}$$

Since the volume is arbitrary, the integrand must be zero so the differential form of continuity equation is written as

$$\frac{\partial \rho}{\partial t} + \nabla.\left(\rho\vec{V}\right) = 0 \tag{3.21}$$

For steady state flow $\frac{\partial \rho}{\partial t} = 0$ and we get

$$\nabla.\left(\rho\vec{V}\right) = 0 \tag{3.22}$$

In Cartesian coordinates, it can be written as:

$$\frac{\partial \rho}{\partial t} + \frac{\partial(\rho u)}{\partial x} + \frac{\partial(\rho v)}{\partial y} + \frac{\partial(\rho w)}{\partial z} = 0 \tag{3.23}$$

It can also be written in Cartesian tensor notation as:

$$\frac{\partial \rho}{\partial t} + \frac{\partial(\rho u_i)}{\partial x_i} = 0 \tag{3.24}$$

Continuity equation can also be developed from the mass flux balance in an elemental control volume shown in Fig. 3.2.

Therefore, net mass flow rate out is:

$$\left[\rho u + \frac{\partial(\rho u)}{\partial x}dx\right]dydz + \left[\rho v + \frac{\partial(\rho v)}{\partial y}dy\right]dxdz$$
$$+\left[\rho w + \frac{\partial(\rho w)}{\partial z}dz\right]dxdy - \rho udydz - \rho vdxdz - \rho wdxdy$$

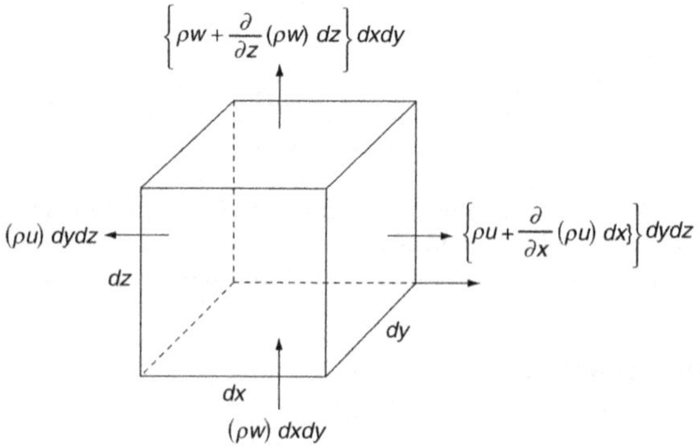


**Figure 3.2** | Mass Flow Through a Fluid Volume Element Shown in Two Directions Only

or

$$\left[\frac{\partial(\rho u)}{\partial x}+\frac{\partial(\rho v)}{\partial y}+\frac{\partial(\rho w)}{\partial z}\right].dxdydz$$

and this must be equal to rate of mass decrease in the elemental volume, i.e., $\frac{\partial \rho}{\partial t}dxdydz$

Therefore, mass continuity equation is obtained as given by Equation (3.23), i.e.,

$$\frac{\partial \rho}{\partial t}+\frac{\partial(\rho u)}{\partial x}+\frac{\partial(\rho v)}{\partial y}+\frac{\partial(\rho w)}{\partial z}=0$$

Fluid flow problems require solution of Navier–Stokes equations satisfying appropriate boundary conditions and also require simultaneously satisfying mass flow continuity equation. In the case of thermal fluid problems, thermal energy balance equation also needs to be solved satisfying appropriate thermal boundary conditions.

## 3.6 | Energy Equation

Energy equation is basically a thermal or heat balance relationship in a control volume, i.e., increase of total energy including kinetic, internal, and potential energy in the control volume is equal to the heat flow into the control volume, plus the rate of internal heat generation minus the work done by the fluid on its surroundings.

It is expressed in the integral form as:

$$\frac{\partial}{\partial t}\oint_{c.v.}\rho E dv+\oint_{c.s}\rho E u_i dA_i=-\oint_{c.s}q_i dA_i+\int u_i\sigma_{ji}dA_j+\oint_{c.v.}q^* dv \tag{3.25}$$

where $E$ is the total energy per unit mass, $q^*$ is the internal heat generation rate per unit volume, $q_i$ is rate of heat flux by conduction and radiation flowing out and $u_i\sigma_{ji}$ is the rate of total work done by the fluid in the control volume per unit surface area, respectively. Because $\sigma_{ij}=-p\delta_{ij}+\tau_{ij}$ using Gauss theorem for an arbitrary volume the integrand must be zero. Therefore, the energy equation becomes:

$$\oint_{c.v.}\left[\frac{\partial}{\partial t}(\rho E)+\frac{\partial}{\partial x_i}(\rho E u_i)+\frac{\partial}{\partial x_i}q_i+q^*+\frac{\partial}{\partial x_i}(p u_i)-\frac{\partial}{\partial x_i}(u_j\tau_{ij})\right]dv=0 \tag{3.26}$$

Thus, the differential form of energy equation combined with the continuity equation becomes:

$$\rho\frac{DE}{Dt}=-\frac{\partial q_i}{\partial x_i}-p\frac{\partial u_i}{\partial x_i}-u_i\frac{\partial p}{\partial x_i}+u_i\frac{\partial \tau_{ji}}{\partial x_j}+\Phi+q^* \tag{3.27}$$

where $\Phi=\tau_{ij}\frac{\partial u_j}{\partial x_j}$ is the dissipation function and is the rate of irreversible work done by shear stress. Since,

$$E(\text{Total Energy})=E_k\ (\text{Kinetic Energy})+E_p\ (\text{Potential Energy})+E_i\ (\text{Internal Energy})$$

and

$$\rho\frac{D}{Dt}\left(E_k+E_p\right)=-u_i\frac{\partial p}{\partial x_i}+u_i\frac{\partial \tau_{ji}}{\partial x_j}$$

The energy equation reduces to

$$\rho\frac{DE_i}{Dt} = -p\frac{\partial u_i}{\partial x_i} - \frac{\partial q_i}{\partial x_i} + \Phi + q^* \tag{3.28}$$

It can be shown that the energy equation becomes:

$$\rho c_v \frac{DT}{Dt} = -p\left(\nabla.\ \vec{V}\right) + \nabla.\left(k\nabla T\right) - \nabla.q^r + \Phi + q^* \tag{3.29}$$

where $k$ is the thermal conductivity of the fluid, $q^r$ is the radiation heat flux vector and $c_v$ is the specific heat of the fluid at constant volume.

Since the dilatational work $-p\nabla.\vec{V} = 0$ for an incompressible fluid, the energy equation is written neglecting radiation heat flux as:

$$\rho c_v \left(\frac{\partial T}{\partial t} + u\frac{\partial T}{\partial x} + v\frac{\partial T}{\partial y} + w\frac{\partial T}{\partial z}\right) = \frac{\partial}{\partial x}\left(k\frac{\partial T}{\partial x}\right) + \frac{\partial}{\partial y}\left(k\frac{\partial T}{\partial y}\right) + \frac{\partial}{\partial z}\left(k\frac{\partial T}{\partial z}\right) + \eta\Phi \tag{3.30}$$

where dissipation function is expressed as:

$$\begin{aligned}\Phi = {} & 2\left[\left(\frac{\partial u}{\partial x}\right)^2 + \left(\frac{\partial v}{\partial y}\right)^2 + \left(\frac{\partial w}{\partial z}\right)^2\right] + \left(\frac{\partial u}{\partial y} + \frac{\partial v}{\partial x}\right)^2 + \left(\frac{\partial v}{\partial z} + \frac{\partial w}{\partial y}\right)^2 \\ & + \left(\frac{\partial w}{\partial x} + \frac{\partial u}{\partial z}\right)^2 - \frac{2}{3}\left(\frac{\partial u}{\partial x} + \frac{\partial v}{\partial y} + \frac{\partial w}{\partial z}\right)\end{aligned} \tag{3.31}$$

Energy equation for compressible flow is written as:

$$\rho c_p \frac{DT}{Dt} = \frac{Dp}{Dt} + \nabla.\left(k\nabla T\right) - p\nabla.\vec{V} - \nabla.q^r + \Phi + q^* \tag{3.32}$$

for $q^* = 0$ and $q^r = 0$, i.e., no radiation it is expressed in Cartesian coordinates as:

$$\rho c_p \frac{DT}{Dt} = -p.\nabla\left(\nabla.\vec{V}\right) + \nabla.\left(k\nabla T\right) + \Phi + \gamma T\frac{Dp}{Dt} \tag{3.33}$$

or

$$\begin{aligned}\rho c_p \left(\frac{\partial T}{\partial t} + u\frac{\partial T}{\partial x} + v\frac{\partial T}{\partial y} + w\frac{\partial T}{\partial z}\right) = {} & \frac{\partial}{\partial x}\left(k\frac{\partial T}{\partial x}\right) + \frac{\partial}{\partial y}\left(k\frac{\partial T}{\partial y}\right) + \frac{\partial}{\partial z}\left(k\frac{\partial T}{\partial z}\right) \\ & + \gamma T\left(\frac{\partial p}{\partial t} + u\frac{\partial p}{\partial x} + v\frac{\partial p}{\partial y} + w\frac{\partial p}{\partial z}\right) + \eta\Phi\end{aligned} \tag{3.34}$$

where $\gamma = \frac{1}{v}\left(\frac{\partial v}{\partial T}\right)_p = -\frac{1}{\rho}\left(\frac{\partial \rho}{\partial T}\right)_p$ is the coefficient of thermal expansion, the term $\gamma T\frac{Dp}{Dt}$ is the heating or cooling of the fluid by compression or expansion. $c_p$ is the specific heat of the fluid at constant pressure.

Navier–Stokes equation, continuity equation and energy equation in cylindrical $(r, \theta, z)$ coordinates are given in Appendix I. For further reading refer to Hughes and Brighton (1967).

## 3.7 | Reynolds Equation

Osborne Reynolds (1886) derived the Reynolds equation for incompressible flow of lubricants in the clearance space of bearings. It owes its genesis to the experiments conducted by Tower (1883) which demonstrated the generation of pressure in the fluid film formed in the clearance space of the journal bearing due to wedge action, i.e., due to relative sliding of the surfaces with a convergent or wedge-shape clearance space filled with a viscous fluid. Reynolds equation is a differential equation that governs the pressure distribution in the fluid film formed due to self-action.

From the point of view of fluid mechanics, the flow of a viscous fluid in the clearance space of a bearing or seal is considered as a slow viscous flow in which viscous forces dominate over the inertia forces. Reynolds equation combines Navier–Stokes and continuity equations into a single equation which can be derived from the Navier–Stokes equation by making basic assumptions given below.

Reynolds equation can now be derived under the following assumptions known as basic assumptions in the theory of lubrication that:

- Flow of lubricant is laminar
- Inertia effects are negligible, i.e., both temporal and convective inertia are negligible
- Viscosity and density of the lubricant does not vary across the film thickness, i.e., viscosity across the film thickness is constant and not dependent on temperature or pressure. However, it can vary along the direction of the film, i.e., $\eta = f_1(x, y)$ and $\rho = f_2(x, y)$
- Pressure gradient across the film thickness is zero or alternately pressure remains constant across the film thickness, i.e., $\frac{\partial p}{\partial z} = 0$
- All the velocity gradients along the fluid flow direction or along the film are negligible in comparison to the velocity gradients across the film thickness which can be expressed referring to Fig. 3.3 as:

$$\frac{\partial u}{\partial x}, \frac{\partial u}{\partial y}, \frac{\partial v}{\partial x}, \frac{\partial v}{\partial y}, \frac{\partial w}{\partial x}, \frac{\partial w}{\partial x} << \frac{\partial u}{\partial z}, \frac{\partial v}{\partial z}, \frac{\partial w}{\partial z},$$

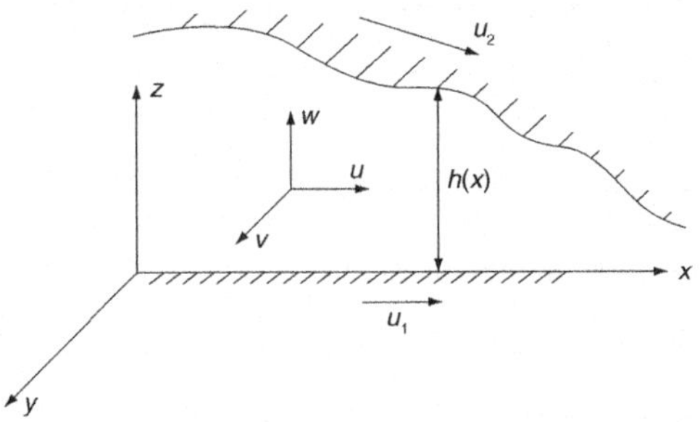


**Figure 3.3** | Coordinate Axes and Bearing Surfaces

Since the film thickness is usually $10^{-2}$–$10^{-3}$ times lower than the length dimensions the above condition is generally satisfied

- Body forces are negligible
- There is no-slip at the fluid-solid boundaries, i.e., fluid layers adhere to the surfaces and will have the velocities of the surfaces

When the above assumptions are applied to the Navier–Stokes, Equations (3.16), (3.17), and (3.18) reduce to following equations:

$$0 = -\frac{\partial p}{\partial x} + \frac{\partial}{\partial z}\left(\eta \frac{\partial u}{\partial z}\right) \tag{3.35}$$

$$0 = -\frac{\partial p}{\partial y} + \frac{\partial}{\partial z}\left(\eta \frac{\partial v}{\partial z}\right) \tag{3.36}$$

$$0 = \frac{\partial p}{\partial z} \tag{3.37}$$

Integrating Equations (3.35) and (3.36) twice with respect to $z$ we get:

$$u = \frac{z^2}{2\eta}\frac{\partial p}{\partial x} + C_1 \frac{z}{\eta} + C_2 \tag{3.38}$$

$$v = \frac{z^2}{2\eta}\frac{\partial p}{\partial y} + C_3 \frac{z}{\eta} + C_4 \tag{3.39}$$

where $C_1$, $C_2$, $C_3$, $C_4$ are constants of integrations.

The boundary conditions for fluid velocities $u$ and $v$ are

$$\text{at } z = 0,\ u = u_1,\ v = 0$$

$$\text{at } z = h,\ u = u_2,\ v = 0$$

Substituting the above boundary condition in Equations (3.37) and (3.38), we get the lubricant velocity distributions in the film as:

$$u = \frac{1}{2\eta}\frac{\partial p}{\partial x}\left(z^2 - zh\right) + \left(1 - \frac{z}{h}\right)u_1 + \frac{z}{h}u_2 \ ; \ v = \frac{1}{2\eta}\frac{\partial p}{\partial y}\left(z^2 - zh\right) \tag{3.40}$$

Substituting Equations (3.38) and (3.39) into continuity Equation (3.23), we get

$$\begin{aligned} &\frac{\partial \rho}{\partial t} + \frac{\partial}{\partial x}\left[\frac{\rho}{2\eta}\frac{\partial p}{\partial x}\left(z^2 - zh\right) + \rho\left(1 - \frac{z}{h}\right)u_1 + \rho\frac{z}{h}u_2\right] \\ &+ \frac{\partial}{\partial y}\left[\frac{\rho}{2\eta}\frac{\partial p}{\partial y}\left(z^2 - zh\right)\right] + \frac{\partial(\rho w)}{\partial z} = 0 \end{aligned} \tag{3.41}$$

Integrating Equation (3.40) across the film thickness gives:

$$\begin{aligned} &\frac{\partial}{\partial x}\left[\frac{\rho}{2\eta}\frac{\partial p}{\partial x}\int_0^h \left(z^2 - zh\right)dz\right] + \frac{\partial}{\partial y}\left[\frac{\partial p}{\partial y}\int_0^h \left(z^2 - zh\right)dh\right] \\ &\quad + \frac{\partial}{\partial x}\left[\rho\int_0^h \left(1 - \frac{z}{h}\right)u_1\, dh + \int_0^h \frac{\partial \rho}{\partial t}dz = 0\right] \end{aligned} \tag{3.42}$$

using the following general rule for integration

$$\int_0^h \frac{\partial}{\partial x}\left[f(x,y,z)\right]dz = -f(x,y,h)\frac{dh}{dx} + \frac{\partial}{\partial x}\int_0^h f(x,y,z)\,dz \tag{3.43}$$

we get Reynolds equation for compressible flow as:

$$\frac{\partial}{\partial x}\left(\frac{\rho h^3}{12\eta}\frac{\partial p}{\partial x}\right) + \frac{\partial}{\partial y}\left(\frac{\rho h^3}{12\eta}\frac{\partial p}{\partial y}\right) = \frac{\partial}{\partial y}\left(\rho h\frac{u_1+u_2}{2}\right) + \rho(w_2 - w_1) - \rho u_2\frac{\partial \rho}{\partial x} + h\frac{\partial \rho}{\partial t} \tag{3.44}$$

where $(w_2 - w_1) = \frac{\partial h}{\partial t}$

Equation (3.44) is rewritten as:

$$\frac{\partial}{\partial x}\left(\frac{\rho h^3}{12\eta}\frac{\partial p}{\partial x}\right) + \frac{\partial}{\partial y}\left(\frac{\rho h^3}{12\eta}\frac{\partial p}{\partial y}\right) = \frac{\partial}{\partial x}\left[\rho h\left(\frac{u_1+u_2}{2}\right)\right] + \left(\rho\frac{\partial h}{\partial t} - \rho u_2\frac{\partial h}{\partial x}\right) + h\frac{\partial \rho}{\partial t} = 0 \tag{3.45}$$

For incompressible flow the Reynolds equation becomes:

$$\frac{\partial}{\partial x}\left(\frac{h^3}{12\eta}\frac{\partial p}{\partial x}\right) + \frac{\partial}{\partial y}\left(\frac{h^3}{12\eta}\frac{\partial p}{\partial y}\right) = \frac{\partial}{\partial x}\left\{(u_1+u_2)h\right\} + \frac{\partial h}{\partial t} + u_2\frac{\partial h}{\partial x} \tag{3.46}$$

- The terms on the left hand side of Equation (3.46), i.e.,

  $$\frac{\partial}{\partial x}\left(\frac{\rho h^3}{12\eta}\frac{\partial p}{\partial x}\right), \frac{\partial}{\partial y}\left(\frac{\rho h^3}{12\eta}\frac{\partial p}{\partial y}\right)$$

  are known as Poiseuille flow or pressure induced flow term.
- The term $\frac{\partial}{\partial x}\left[\rho h\left(\frac{u_1+u_2}{2}\right)\right]$ on the right is known as Couette flow term. This term is split into three different terms, viz.,
  - $h\left(\frac{u_1+u_2}{2}\right)\frac{\partial h}{\partial x}$ is known as the physical wedge term or the flow term due to wedge action. It is due to surface velocity and film thickness variation in the direction of surface motion. Positive pressures are generated due to decrease in film thickness in the direction of surface motion or due to convergent film shape.
  - $h\left(\frac{u_1+u_2}{2}\right)\frac{\partial \rho}{\partial x}$ is known as density wedge term. Positive pressures would be generated if density varies along the flow direction. It is necessary that density decreases in the flow direction which is possible due to increase in temperature. This may result into pressure generation even in case of parallel film shape, i.e., for constant film thickness.

- $\rho \frac{h}{2}\frac{\partial}{\partial x}(u_1 + u_2)$ is called stretch term. Generation of positive pressure is possible if the bearing surfaces are elastic or stretchable. Velocities must decrease in the flow direction. Since bearing surfaces are usually rigid this term is zero.

- The term $\left(\rho \frac{\partial h}{\partial t} - \rho u_2 \frac{\partial h}{\partial x}\right)$ is known as squeeze flow term and can lead to positive pressure generation if the surfaces approach each other.
- The term $h\frac{\partial \rho}{\partial t}$ is known as local expansion term and is governed by local time rate of density. This term is related to thermal expansion of the lubricant between the bearing surfaces. Since it is a transient process, it is generally of no consequence in lubrication analysis.

## 3.8 | Lubricant Flow

Mass flow rate per unit length of the lubricant through a section of the fluid film is given as:

$$m_x = \int_0^h (\rho u)\, dz \tag{3.47}$$

using Equation (3.40) it can be written as

$$m_x = \int_0^h \left[\frac{\rho}{2\eta}\frac{\partial p}{\partial x} z(z-h) + \rho\left(1 - \frac{z}{h}\right)u_1 + \rho u_2\right] dz \tag{3.48}$$

$$m_x = -\frac{\rho h^3}{12\eta}\frac{\partial p}{\partial x} + \frac{\rho h}{2}(u_1 + u_2) \tag{3.49}$$

Similarly, flow rate per unit length in $y$ direction is expressed as:

$$m_y = \int_0^h (\rho v) dz \tag{3.50}$$

which reduces to using the expression for $v$ from Equation (3.40) as

$$m_y = \int_0^h \frac{\rho}{2\eta}\left\{\frac{\partial p}{\partial y} z(z-h)\right\} dz \tag{3.51}$$

Integrating we get

$$m_y = -\frac{\rho h^3}{12\eta}\frac{\partial p}{\partial y} \tag{3.52}$$

In case of incompressible lubricants corresponding volume flow rates per unit length are given as:

$$q_x = -\frac{h^3}{12\eta}\frac{\partial p}{\partial x} + \frac{h}{2}(u_1 + u_2) \tag{3.53}$$

$$q_y = -\frac{h^3}{12\eta}\frac{\partial p}{\partial x} \tag{3.54}$$

## 3.9 | Shear Forces

The shear stresses on the surface of solids are given for a Newtonian fluid as:

$$\tau_{xz} = \eta \frac{\partial u}{\partial z}$$
$$\tau_{yz} = \eta \frac{\partial v}{\partial z} \tag{3.55}$$

Velocity gradients are obtained from Equation (3.40) by differentiating with respect to z as:

$$\frac{\partial u}{\partial z} = \frac{1}{2\eta}\frac{\partial p}{\partial x}(2z - h) + \frac{(u_2 - u_2)}{h}$$
$$\frac{\partial v}{\partial z} = \frac{1}{2\eta}\frac{\partial p}{\partial y}(2z - h) \tag{3.56}$$

Therefore,

$$\tau_{yz} = \frac{1}{2}\frac{\partial p}{\partial x}(2z - h) + \frac{\eta}{h}(u_2 - u_1)$$
$$\tau_{yz} = \frac{1}{2}\frac{\partial p}{\partial y}(2z - h) \tag{3.57}$$

Shear stresses at $z = 0$ are given as:

$$\tau_{xz} = -\frac{h}{2}\frac{\partial p}{\partial x} + \frac{\eta}{h}(u_2 - u_1)$$
$$\tau_{yz} = -\frac{h}{2}\frac{\partial p}{\partial y} \tag{3.58}$$

and the shear stresses at $z = h$ are expressed as:

$$\tau_{xz} = \frac{h}{2}\frac{\partial p}{\partial x} + \frac{\eta}{h}(u_2 - u_1)$$
$$\tau_{yz} = \frac{h}{2}\frac{\partial p}{\partial y} \tag{3.59}$$

The shear force can be evaluated on the solid surfaces as:

$$F = \iint_{S.A} \tau dA \tag{3.60}$$

Thus, shear force at $z = 0$ and at $z = h$ can be determined as:

$$F_{0,h} = \iint_{S.A} \left[ \pm \frac{1}{2}\frac{\partial p}{\partial x} h + \frac{\eta}{h}(u_2 - u_1) \right] dxdy \tag{3.61}$$

Reynolds equations for compressible and incompressible fluids are given by Equations (3.45) and (3.46), respectively. Expressions for flow rates and shear force are also given in Equations (3.53), (3.54), (3.59), and (3.60), respectively.

These equations will be directly used henceforth wherever applicable in this form. The above equations are valid for smooth surfaces only. For rough surfaces, Reynolds equation will have to be modified to account for surface roughness effect. In a similar manner for non-Newtonian fluids, Reynolds equation has to be modified taking into consideration the non-Newtonian relationship between shear stress and shear strain rate. Reynolds equation given by (3.46) and (3.47) can incorporate piezoviscous and thermal effects of the lubricant in the flow directions ($x$, $y$), and thus thermohydrodynamic lubrication problems can be solved in conjunction with energy equation. However, variation of viscosity and density across the fluid film cannot be accounted. A fully thermal Reynolds equation needs to be developed and used for addressing temperature variations across the film. Similarly, for turbulent lubrication, turbulent Reynolds equation has to be derived and used. Fluid inertia effects in laminar and turbulent flow can be incorporated for high speed bearings by suitably including fluid inertia effects in the Reynolds equation.

## 3.10 | Reynolds Equation Assumptions Justified

An order of magnitude analysis is necessary to ensure that the terms neglected in the Navier–Stokes equations are couple of orders of magnitude lower that the terms retained to derive the Reynolds equation. It is commonly known that film thickness in the fluid films in lubricated contacts is usually 1000 times smaller than the radius of the journal and axial length of bearings or seals. Therefore, terms neglected must be at least $10^{-3}$ times lower than the terms retained to derive the Reynolds equation. To establish this, order of magnitude of various terms have to be determined of Equations (3.16) to (3.18). Equation (3.16) can be rewritten as:

$$\rho\left\{\frac{\partial u}{\partial t}+u\frac{\partial u}{\partial x}+v\frac{\partial u}{\partial y}+w\frac{\partial u}{\partial z}\right\}=B_x-\frac{\partial p}{\partial x}-\frac{1}{2}\frac{\partial}{\partial y}\left\{\eta\left(\frac{\partial u}{\partial x}+\frac{\partial v}{\partial y}\frac{\partial w}{\partial z}\right)\right\}$$
$$+2\frac{\partial}{\partial x}\left(\eta\frac{\partial u}{\partial x}\right)+\frac{\partial}{\partial y}\left[\eta\left(\frac{\partial u}{\partial y}+\frac{\partial v}{\partial x}\right)\right]+\frac{\partial}{\partial z}\left[\eta\left(\frac{\partial u}{\partial z}+\frac{\partial w}{\partial x}\right)\right]$$

Following dimensionless parameters are defined to rewrite the above equation in the nondimensional form as:

$$\bar{x}=\frac{x}{l},\ \bar{y}=\frac{y}{l},\ \bar{z}=\frac{z}{h},\ \tau=vt,\ \bar{u}=\frac{u}{u_0},\ \bar{v}=\frac{v}{u_0},\ \bar{w}=\frac{w}{u_0},\ \bar{\rho}=\rho/\rho_0,$$
$$\bar{\eta}=\eta/\eta_0 \text{ and } \bar{p}=\mathrm{h}^2 p/\eta_0 u_0 l,\ B_x=\rho g \text{ (gravity force)}$$

It can be shown that the Equation (3.16) becomes after manipulations as written below:

$$\sigma\frac{\partial\bar{u}}{\partial\tau}+R^*\bar{u}\frac{\partial\bar{u}}{\partial\bar{x}}+R^*\bar{v}\frac{\partial\bar{u}}{\partial\bar{y}}+R^*\left(\frac{h}{l}\right)\bar{w}\frac{\partial\bar{w}}{\partial\bar{z}}=-\frac{1}{\bar{\rho}}\frac{\partial\bar{p}}{\partial\bar{x}}+\frac{1}{\bar{\rho}}\frac{\partial}{\partial\bar{z}}\left(\bar{\eta}\frac{\partial\bar{u}}{\partial\bar{z}}\right)$$
$$+2\left(\frac{h}{l}\right)^2\frac{1}{\bar{\rho}}\frac{\partial}{\partial\bar{x}}\left(\bar{\eta}\frac{\partial\bar{u}}{\partial\bar{x}}\right)+\frac{1}{\bar{\rho}}\left(\frac{h}{l}\right)\frac{\partial}{\partial\bar{x}}\left(\bar{\eta}\frac{\partial\bar{w}}{\partial\bar{x}}\right)$$

$$-\frac{2}{3}\left(\frac{h}{l}\right)^2 \frac{1}{\bar{\rho}}\frac{\partial}{\partial \bar{x}}\left[\bar{\eta}\left(\frac{\partial \bar{u}}{\partial \bar{x}}+\frac{\partial \bar{v}}{\partial \bar{y}}+\frac{l}{h}\frac{\partial \bar{w}}{\partial \bar{z}}\right)\right]$$

$$+\left(\frac{h}{l}\right)^2 \frac{1}{\bar{\rho}}\frac{\partial}{\partial \bar{y}}\left[\bar{\eta}\left(\frac{\partial \bar{u}}{\partial \bar{y}}+\frac{\partial \bar{v}}{\partial \bar{x}}\right)\right]+g\frac{l}{u_0^{\,2}}R^* \tag{3.62}$$

where Reynolds number $R$ is given as:

$$R=\frac{\text{inertia}}{\text{viscous}}=\left(\frac{\rho_0 u_0 l}{\eta_0}\right)$$

and film Reynolds numbers are expressed as:

$$R^*=R\left(\frac{h}{l}\right)^2$$

and the squeeze number $\sigma$ is given by:

$$\sigma=\left(\frac{\rho_0 \upsilon h^2}{\eta_0}\right)$$

It can be easily examined that all the terms in Equation (3.55) except the pressure term

$$\frac{1}{\bar{\rho}}\frac{\partial \bar{p}}{\partial \bar{x}} \text{ and } \frac{1}{\bar{\rho}}\frac{\partial}{\partial \bar{z}}\left(\bar{\eta}\frac{\partial \bar{u}}{\partial \bar{z}}\right) \text{ are of the order of } \frac{h}{l} \text{ or } \left(\frac{h}{l}\right)^2$$

since $\frac{h}{l}$ is usually of the order of $10^{-3}$, all terms other than $\frac{1}{\bar{\rho}}\frac{\partial p}{\partial \bar{x}}$ and $\frac{1}{\bar{\rho}}\frac{\partial}{\partial \bar{z}}\left(\bar{\eta}\frac{\partial \bar{u}}{\partial z}\right)$ are at least $10^{-3}$ times these terms and therefore can be neglected. In a similar manner, it can be shown that in Equation (3.17) also only the pressure terms $\frac{1}{\bar{\rho}}\frac{\partial \bar{p}}{\partial \bar{y}}$ and $\frac{1}{\bar{\rho}}\frac{\partial}{\partial \bar{z}}\left(\bar{\eta}\frac{\partial \bar{v}}{\partial \bar{z}}\right)$ would be of the order of 1 and rest of the terms would be of the order of $10^{-3}$ to $10^{-6}$ and can be neglected.

Thus, we see that Navier–Stokes equations reduce to Equations (3.35), (3.36), and (3.37) with the assumptions made to derive the Reynolds equation and the assumption are valid as shown above. For more details refer to Hamrock (1994).

## 3.11 | Derivation of Thermal Reynolds Equation

In the derivation of generalized Reynolds equation, several assumptions were made. One of the assumptions made was that fluid properties, i.e., viscosity and density do not vary across the film thickness. This assumption also implies that temperature and pressure do not vary across the film. But there are situations when viscous heat generation is very significant and the heat carried away by the lubricant is significantly lower than the heat generation then the balance heat is transferred to the surrounding solids by conduction. This would result in significant temperature variation across the fluid film resulting in variation of both viscosity and density across the film. Pressure variation across the film is insignificant though and therefore pressures across the lubricant film remain constant.

Thermal Reynolds equation is derived retaining all the assumptions made to derive the Reynolds Equations (3.45) and (3.46) except that viscosity and density are now allowed to vary across the film thickness.

Thermal Reynolds equation was developed by Dowson (1962) and was later given in simpler form by Fowles (1970).

To derive generalized thermal Reynolds equation, we rewrite Equations (3.35) and (3.36) as below:

$$\frac{\partial p}{\partial x} = \frac{\partial}{\partial z}\left(\eta \frac{\partial u}{\partial z}\right)$$

$$\frac{\partial p}{\partial y} = \frac{\partial}{\partial z}\left(\eta \frac{\partial v}{\partial z}\right)$$

The gradients of flow velocities $u$ and $v$ across the film can now be found by integrating the above equations with respect to $z$ as:

$$\frac{\partial u}{\partial z} = \frac{z}{\eta}\frac{\partial p}{\partial x} + \frac{B(x,y)}{\eta} \tag{3.63}$$

$$\frac{\partial v}{\partial z} = \frac{z}{\eta}\frac{\partial p}{\partial y} + \frac{C(x,y)}{\eta} \tag{3.64}$$

where $B$ and $C$ are constants of integration. Integrating once again with respect to $z$ and introducing following boundary conditions.

$$\begin{aligned} z = 0,\ u = u_1, \text{ and } v = 0 \\ z = h,\ u = u_2, \text{ and } v = 0 \end{aligned} \tag{3.65}$$

Expressions for velocity components are obtained as:

$$u = u_1 + \frac{\partial p}{\partial x}\int_0^z \frac{z}{\eta}dz + \left(\frac{u_2 - u_1}{f_0} - \bar{z}\frac{\partial p}{\partial x}\right)\int_0^z \frac{dz}{\eta} \tag{3.66}$$

$$v = \frac{\partial p}{\partial y}\int_0^z \frac{z}{\eta}dz - \bar{z}\frac{\partial p}{\partial y}\int_0^z \frac{dz}{\eta} \tag{3.67}$$

where

$$f_0 = \int_0^h \frac{dz}{\eta};\ f_1 = \int_0^h \frac{zdz}{\eta} = \bar{z}f_0$$

$$\bar{z} = \int_0^h \frac{z}{\eta}dz \Big/ \int_0^h \frac{dz}{\eta}$$

Integrating the continuity equation with respect to $z$ between limits 0 and $h$ gives the following expression:

$$\int_0^h \frac{\partial p}{\partial t}dz + \int_0^h \frac{\partial(\rho u)}{\partial x}dz + \int_0^h \frac{\partial(\rho v)}{\partial y}dz + [\rho w]_0^h = 0 \tag{3.68}$$

The above equation can be expanded according to the general integration rule, i.e.,

$$\int_{h1}^{h2} \frac{\partial}{\partial x} f(x,y,z)\,dz = \frac{\partial}{\partial x}\int_{h1}^{h2} f(x,y,z)\,dz - f(x,y,h_2)\frac{\partial h_2}{\partial x} + f(x,y,h_1)\frac{\partial h_1}{\partial x} \tag{3.69}$$

which gives

$$\int_0^h \frac{\partial \rho}{\partial t} dz + \frac{\partial}{\partial x}\int_0^h (\rho u)\, dz + \frac{\partial}{\partial y}\int_0^h (\rho v)\, dz - (\rho u)_2 \frac{\partial h}{\partial x} + [\rho w]_0^h \tag{3.70}$$

The integrals of $(\rho u)$ and $(\rho v)$ can be evaluated by parts to give

$$\int_0^h \frac{\partial \rho}{\partial t} dz + h\left[\frac{\partial (\rho u)_2}{\partial x}\right] - \frac{\partial}{\partial x}\int_0^h \left[\rho z \frac{\partial u}{\partial z} + zu\frac{\partial \rho}{\partial z}\right] dz - \frac{\partial}{\partial y}\int_0^h \left[\rho z \frac{\partial v}{\partial z} + zv\frac{\partial \rho}{\partial z}\right] dz + [\rho w]_0^h = 0 \tag{3.71}$$

The expressions for $u$ and $v$ can now be introduced from Equations (3.66), (3.67), and Equation (3.71) then becomes:

$$\frac{\partial}{\partial x}\left[(f_2 + g_1)\frac{\partial p}{\partial x}\right] + \frac{\partial}{\partial y}\left[(f_2 + g_1)\frac{\partial p}{\partial y}\right] = h\frac{\partial}{\partial x}\left[\frac{\partial}{\partial x}(\rho u)_2\right] - \frac{\partial}{\partial x}\left[\frac{(u_2 - u_1)(f_3 + g_2)}{f_0} + u_1 g_3\right]$$
$$+ \int_0^h \frac{\partial \rho}{\partial t} + (\rho w)_2 - (\rho w_1) \tag{3.72}$$

where

$$f_0 = \int_0^h \frac{dz}{\eta};\ f_1 = \int_0^h \frac{z\partial z}{\eta} = \bar{z} f_0$$

$$f_2 = \int_0^h \frac{\rho z(z - \bar{z})}{\eta} dz;\ f_3 = \int_0^h \frac{\rho z}{\eta} dz$$

$$g_1 = \int_0^h \left[z\frac{\partial \rho}{\partial z}\left(\int_0^z \frac{z}{\eta} dz - \bar{z}\int_0^z \frac{dz}{\eta}\right)\right] dz$$

$$g_2 = \int_0^h \left[z\frac{\partial \rho}{\partial z}\int_0^z \frac{dz}{\eta}\right] dz$$

$$g_3 = \int_0^h z\frac{\partial \rho}{\partial z} dz$$

Equation (3.73) represents the thermal Reynolds equation applicable to fluid film lubrication problems which include variation of lubricant properties along and across the film. It may be noted that all the '$g$' functions contain $\frac{\partial \rho}{\partial z}$ and since the density is more often constant across the fluid film in majority of lubrication conditions, these can be neglected.

Fowles (1970) gave the simpler form of the thermal Reynolds equation as:

$$\frac{\partial}{\partial x}\left(I_2 \frac{\partial p}{\partial x}\right) + \frac{\partial}{\partial y}\left(I_2 \frac{\partial p}{\partial x}\right) = u_1 \frac{\partial (I_3)}{\partial x} + \frac{\partial}{\partial x}\left[\frac{I_1}{f_0}(u_2 - u_1)\right]$$
$$+ \int_0^h \frac{\partial \rho}{\partial t} dz + (\rho w)_2 - (\rho w)_1 \tag{3.73}$$

where

$$I_2 = \frac{f_1}{f_0} I_1 - \int_0^h \rho \left( \int_0^z \frac{z}{\eta} dz \right) dz$$

$$I_1 = -\int_0^h \rho \left( \int_0^z \frac{dz}{\eta} \right) dz$$

$$I_3 = \int_0^h \rho dz; f_1 = \int_0^h \frac{z}{\eta} dz; f_0 = \int_0^h \frac{dz}{\eta}$$

## 3.12 | Reynolds Equation for Lubrication with Non-Newtonian Fluids

Several non-Newtonian rheological relationships have been developed for thin film lubricant flows. Most of these use explicit rheological relationships which are given below:

- Newtonian model:

$$\dot{\gamma}_{ij} = \frac{1}{\eta} \tau_{ij} \tag{3.74}$$

- Linear viscoelastic model: Maxwell's model

$$\dot{\gamma}_{ij} = \frac{1}{G} \frac{d\tau_{ij}}{dt} + \frac{\tau_{ij}}{\eta} \tag{3.75}$$

- Nonlinear viscoelastic model I: Johnson and Tevaarwerk (1975)

$$\dot{\gamma}_{ij} = \frac{1}{G} \frac{d\tau_{ij}}{dt} + \left( \frac{\tau_0}{\tau_e} \right) \frac{\tau_{ij}}{\eta} \sinh \left( \frac{\tau_e}{\tau_e} \right) \tag{3.76}$$

- Nonlinear viscoelastic model II: Bair and Winer (1979)

$$\dot{\gamma}_{ij} = \frac{1}{G} \frac{d\tau_{ij}}{dt} + \left( \frac{\tau_0}{\tau_e} \right) \frac{\tau_{ij}}{\eta} \tanh \left( \frac{\tau_e}{\tau_0} \right) \tag{3.77}$$

- Viscoelastic plastic model: Bair and Winer (1979)

$$\dot{\gamma}_{ij} = \frac{1}{G} \frac{d\tau_{ij}}{dt} - \frac{\tau_{ij}}{\tau_e} \frac{\tau_L}{\eta} \ln \left| 1 - \frac{\tau_e}{\tau_L} \right| \tag{3.78}$$

- Oswald-de Walle power law model:

$$\tau_{ij} = m \dot{\gamma}_{ij}^n = m \left| \left( \dot{\gamma}_{ij} \right) \right|^{n-1} \dot{\gamma}_{ij} \tag{3.79}$$

and

$$\eta = m \left\{ \left( \dot{\gamma}_{ij} \right)^2 \right\}^{\frac{n-1}{2}} = I^{\frac{n-1}{2}} \tag{3.80}$$

where viscosity $\eta$ is dependent on the second invariant of strain rate tensor which reduces to

$$I = \left(\frac{\partial u}{\partial z}\right)^2 + \left(\frac{\partial v}{\partial z}\right)^2 \tag{3.81}$$

The constitutive relation then becomes

$$\eta = \eta(I) \tag{3.82}$$

and is suitable for incompressible inelastic fluid.

In case of Newtonian fluids, the velocities and their derivatives are linearly dependent on the pressure gradient. However, in the case non-Newtonian fluids, it is assumed that the strain rates within the fluid are principally generated by relative surface velocities. Therefore, it is applicable to Coquette-dominated highly non-Newtonian fluids.

In the above relationships described in Equations (3.74) to (3.80)

$\dot{\gamma}_{ij}$ = Shear strain rate
$\tau_{ij}$ = Shear stress
$G$ = Shear modulus of the fluid
$\tau_e$ = Equivalent shear stress up to which linear relationship holds
$\tau_L$ = Limiting shear stress
$\tau_0$ = Reference shear stress at ambient pressure and temperature
$\tau_{xz}$ and $\tau_{yz}$ = Shear stresses in $x$ and $y$ directions, respectively,

$u, v, w$ are velocity components in $x, y,$ and $z$ directions, respectively.

It is necessary to know the pressure, velocity, shear field, and temperature field in order to determine shear forces and friction in lubrication of surfaces with non-Newtonian fluids. With the usual assumptions made to derive Reynolds equation for Newtonian fluids, i.e., fluid inertia and body forces are negligible; length dimension across the film is small in comparison to other length dimensions. Thus, the equilibrium equations reduce to following forms:

$$\left.\begin{aligned} \frac{\partial p}{\partial x} &= \frac{\partial \tau_{xz}}{\partial z} \\ \frac{\partial p}{\partial y} &= \frac{\partial \tau_{yz}}{\partial z} \\ \frac{\partial p}{\partial z} &= 0 \end{aligned}\right\} \tag{3.83}$$

In addition to this, following equations have also to be satisfied to describe the flow and temperature fields, viz.,

- Equation of continuity of flow
- Equation of rheological behavior of fluid
- The energy equation for thermal energy balance

The velocity gradients for non-Newtonian fluids are written in terms of several types of rheological relationship, i.e., in terms of $f(\tau_e)$ as:

$$\frac{\partial u}{\partial z} = a.\frac{\partial \tau_{xz}}{dt} + \tau_{xz}\frac{f(\tau_e)}{\tau_e} \tag{3.84}$$

$$\frac{\partial v}{\partial z} = a\frac{\partial \tau_{yz}}{dt} + \tau_{yz}\frac{f(\tau_e)}{\tau_e}$$

where $a = 0$ if the fluid elasticity is neglected and $a = \frac{1}{G}$ if fluid elasticity is considered and $\tau_e = \left(\tau_{xz}^2 + \tau_{yz}^2\right)^{1/2}$; $\tau_e$ is equivalent shear stress when flow is two dimensional.

Lubricant characteristics in terms of pressure ($p$) and temperature ($T$) are described by the following relationship:

$$\eta(p,T) = \eta_0 \exp\left(\alpha p + \beta\left(\frac{1}{T} - \frac{1}{T_0}\right)\right) \tag{3.85}$$

where $\eta_0$ is the viscosity of fluid at reference temperature $T_0$ and at ambient pressure, $\alpha$, $\beta$ are viscosity pressure and viscosity temperature coefficients, respectively.

Similar relationships can be written for limiting shear stress of the fluid, e.g.,

$$\tau_L(p,T) = \tau_{L0} + ap + b(T - T_0) \tag{3.86}$$

where $a$ and $b$ are constants that describe pressure and temperature dependence of the limiting shear stress.

Generalized Reynolds equation for non-Newtonian fluids was developed by Najji *et al.* (1989) for incompressible lubrication, and by Wolff and Kubo (1996) for compressible lubrication.

The energy equation for non-Newtonian fluids is similar to the energy equation for Newtonian fluids as given in Equations (3.30) and (3.34), except that the dissipation function, $\phi$, is to be incorporated for non-Newtonian fluids as:

$$\phi = \tau_e f(\tau_e) \tag{3.87}$$

Generalized Reynolds equation for non-Newtonian lubricating films can be developed following Najji *et al.* (1989) and Wolff and Kubo (1996).

The Rheological behavior of the fluid can be expressed as:

$$\begin{aligned}\tau_{xz} &= \frac{\tau_e}{f(\tau_e)}\left(\frac{\partial u}{\partial z} - a\frac{d\tau_{xz}}{dt}\right)\\ \tau_{yz} &= \frac{\tau_e}{f(\tau_e)}\left(\frac{\partial v}{\partial z} - a\frac{d\tau_{yz}}{dt}\right)\end{aligned} \tag{3.88}$$

Substituting $\tau_{xz}$ and $\tau_{yz}$ as given above in equations of equilibrium (3.83), we get

$$\begin{aligned}\frac{\partial p}{\partial x} &= \frac{\partial}{\partial z}\left[\frac{\tau_e}{f(\tau_e)}\left(\frac{\partial u}{\partial z} - a\frac{d\tau_{xz}}{dt}\right)\right]\\ \frac{\partial p}{\partial y} &= \frac{\partial}{\partial z}\left[\frac{\tau_e}{f(\tau_e)}\left(\frac{\partial v}{\partial z} - a\frac{d\tau_{yz}}{dt}\right)\right]\\ \frac{\partial p}{\partial z} &= 0\end{aligned} \tag{3.89}$$

Integrating Equation (3.89) twice with respect to z and substituting following boundary conditions for no slip at the boundaries:

$$u = u_1,\ v = v_1,\ w = w_1 \text{ for } z = 0$$
$$u = u_2,\ v = v_2,\ w = w_2 \text{ for } z = h$$

Velocity distributions are obtained as:

$$u = u_1 + \frac{\partial p}{\partial x}\left[\int_0^z \frac{f(\tau_e)}{\tau_e} z dz - \frac{f_1}{f_0}\int_0^z \frac{f(\tau_e)}{\tau_e} dz\right] + \frac{(u_2 - u_1 - k_x)}{f_0}\int_0^z \frac{f(\tau_e)}{\tau_e} dz + \int_0^z a\frac{d\tau_{xz}}{dt} dz \tag{3.90}$$

$$v = v_1 + \frac{\partial p}{\partial y}\left[\int_0^z \frac{f(\tau_e)}{\tau_e} z dz - \frac{f_1}{f_0}\int_0^z \frac{f(\tau_e)}{\tau_e} dz\right] + \frac{(v_2 - v_1 - k_y)}{f_0}\int_0^z \frac{f(\tau_e)}{\tau_e} dz + \int_0^z a\frac{d\tau_{yz}}{dt} dz \tag{3.91}$$

Velocity gradients are obtained by the differentiation of Equations (3.90) and (3.91) with respect to z

$$\frac{\partial u}{\partial z} = \frac{f(\tau_e)}{\tau_e}\frac{\partial p}{\partial x}\left(z - \frac{f_1}{f_0}\right) + \frac{f(\tau_e)}{\tau_e}\frac{(u_2 - u_1 - k_x)}{f_0} + a\frac{d\tau_{xz}}{dt} \tag{3.92}$$

$$\frac{\partial v}{\partial z} = \frac{f(\tau_e)}{\tau_e}\frac{\partial p}{\partial y}\left(z - \frac{f_1}{f_0}\right) + \frac{f(\tau_e)}{\tau_e}\frac{(v_2 - v_1 - k_y)}{f_0} + a\frac{d\tau_{yz}}{dt} \tag{3.93}$$

where

$$k_x = \int_0^h a\frac{d\tau_{xz}}{dt} dz;\ f_0 = \int_0^h \frac{f(\tau_e)}{\tau_e} dz$$

$$k_y = \int_0^h a\frac{d\tau_{yz}}{dt} dz;\ f_1 = \int_0^h \frac{f(\tau_e)}{\tau_e} z dz$$

Flow continuity equation can be written in the integral form as:

$$\frac{\partial}{\partial x}\int_0^h u dz + \frac{\partial}{\partial y}\int_0^h v dz + (w_2 - w_1) = 0 \tag{3.94}$$

Using Leibnitz's rule for integration as given in Equation (3.69) one can write

$$\frac{\partial}{\partial x}\int_0^h u dz + \frac{\partial}{\partial y}\int_0^h v dz = u_2\frac{\partial h}{\partial x} + v_2\frac{\partial h}{\partial y} + w_2 - w_1 \tag{3.95}$$

Integrating by parts the left hand side of above equation gives:

$$\int_0^h u dz = u_2 h - \int_0^h\left(z\frac{\partial u}{\partial z}\right) dz$$

and

$$\int_0^h v dz = v_2 h - \int_0^h\left(z\frac{\partial v}{\partial z}\right) dz \tag{3.96}$$

Substituting Equations (3.92) and (3.93) into the above equations and integrating across the film thickness, generalized Reynolds equation for non-Newtonian fluids is obtained as:

$$\frac{\partial}{\partial x}\left(I_2\frac{\partial p}{\partial x}\right)+\frac{\partial}{\partial y}\left(I_2\frac{\partial p}{\partial y}\right)=h\left[\frac{\partial(u_2h)}{\partial x}+\frac{\partial(v_2h)}{\partial y}\right]-u_2\frac{\partial h}{\partial x}-v_2\frac{\partial h}{\partial y}-$$
$$\frac{\partial}{\partial x}\left[(u_1-u_2-k_x)\frac{I_1}{f_0}+(u_2-u_1)h\right]$$
$$\frac{\partial}{\partial y}\left[(v_1-v_2-k_y)\frac{I_1}{f_0}+(v_2-v_1)h\right]-\left(\frac{\partial I_x}{\partial x}+\frac{\partial I_y}{\partial y}\right)+(w_2-w_1) \quad (3.97)$$

where

$$I_1=-\int_0^h\left(\int_0^z\frac{f(\tau_e)}{\tau_e}dz\right)dz$$

$$I_2=-\frac{f_1I_1}{f_0}-\int_0^h\left(\int_0^z\frac{f(\tau_e)}{\tau_e}zdz\right)dz$$

$$I_x=-\int_0^h\left(\int_0^z a\frac{d\tau_{xz}}{dt}dz\right)dz$$

$$I_y=-\int_0^h\left(\int_0^z a\frac{d\tau_{yz}}{dt}dz\right)dz$$

Usually surface velocities in $y$ direction, i.e., $v_1$ and $v_2$ are zero or in other words surfaces are not moving in $y$ direction and $v_1=v_2=0$. Only one of the surfaces moves in $x$ direction, i.e., either $u_1=0$ or $u_2=0$.

Equation (3.94) is the generalized Reynolds equation for non-Newtonian fluids developed by Wolff and Kubo (1996) for elastic viscoplastic lubrication. When $a=0$, the fluid is viscoplastic and in that case $I_x$ and $I_y$ are zero. In case of $\frac{f(\tau_e)}{\tau_e}=\frac{1}{\eta}$ and $a=0$, it becomes the Equation (3.73) developed by Fowles (1970) as generalized thermal Reynolds equation. For $\eta=$ constant, it reduces to the Reynolds equation for isoviscous and incompressible lubricants developed by Najji *et al.* (1989). Regarding derivation of Reynolds' equation for bubbly fluids and emulsions readers can refer to Szeri (1998).

## 3.13 | Reynolds Equation for Power Law Fluids

Dien and Elrod (1983) adopted perturbation procedure to develop a generalized Reynolds equation for power law fluids. With the usual assumptions made for the derivation of Reynolds equation, the simplified momentum equations are written as:

$$\frac{\partial p}{\partial x}=\frac{\partial}{\partial z}\left(\eta\frac{\partial p}{\partial z}\right)$$
$$\frac{\partial p}{\partial y}=\frac{\partial}{\partial z}\left(\eta\frac{\partial v}{\partial z}\right) \quad (3.98)$$

The viscosity $\eta$ is dependent on the second invariant of the strain rate tensor which reduces to:

$$I = \left(\frac{\partial u}{\partial z}\right)^2 + \left(\frac{\partial v}{\partial z}\right)^2 \tag{3.99}$$

The relationship for viscosity then becomes:

$$\eta = \eta(I) \tag{3.100}$$

To proceed further, the pressure gradient is expressed in the form:

$$\nabla = \lambda \nabla p_1 \tag{3.101}$$

where $\lambda$ is the amplitude parameter for expansion and $p_1$ is the reference pressure.

Following first order perturbations we can write in terms of '$\lambda$' as:

$$\begin{aligned} u &= u_0(x, y) + \lambda u_1(x, y) + \ldots\ldots\ldots\ldots \\ v &= v_0(x, y) + \lambda v_1(x, y) + \ldots\ldots\ldots\ldots \end{aligned} \tag{3.102}$$

The second invariant of strain rate tensor then expands to

$$I = \left(\frac{\partial u}{\partial z}\right)^2 + \left(\frac{\partial v}{\partial z}\right)^2 + 2\lambda\left(\frac{\partial u_0}{\partial z}\frac{\partial u_1}{\partial z} + \frac{\partial v_0}{\partial z}\frac{\partial v_1}{\partial z}\right) \tag{3.103}$$

$$I = I_0 + \lambda I_1 \tag{3.104}$$

Since the viscosity depends only on the second invariant, it is written:

$$\eta = \eta(I_0 + \lambda I, + \ldots\ldots\ldots\ldots) = \eta(I_0) + \lambda\left(\frac{\partial \eta}{\partial I}\right) I_1 + \ldots\ldots I = I_0$$

$$\eta_0 = \eta(I_0); \eta_1 = \left(\frac{\partial \eta}{\partial I}\right)_{I=I_0} = I_1 \tag{3.105}$$

Substituting Equation (3.105) into Equation (3.98) the zeroth order equations are obtained as:

$$\begin{aligned} \frac{\partial}{\partial z}\left(\eta_0 \frac{\partial u_0}{\partial z}\right) = 0 \\ \frac{\partial}{\partial z}\left(\eta_0 \frac{\partial v_0}{\partial z}\right) = 0 \end{aligned} \tag{3.106}$$

Integration of Equation (3.106) and substitution of following boundary conditions

$$u_0 = U \text{ and } v_0 = V \text{ at } z = h$$

$$u_0 = v_0 = 0 \text{ at } z = 0$$

we get

$$u_0 = U\frac{z}{h}, v_0 = V\frac{z}{h} \tag{3.107}$$

The first order equations, being the coefficients of $\lambda$ are:

$$\frac{\partial}{\partial z}\left(\eta_0\frac{\partial u_1}{\partial z}\right)+\frac{\partial}{\partial z}\left(\eta_1\frac{\partial u_0}{\partial z}\right)=\frac{\partial p_1}{\partial x}$$

and

$$\frac{\partial}{\partial z}\left(\eta_0\frac{\partial v_1}{\partial z}\right)+\frac{\partial}{\partial z}\left(\eta_1\frac{\partial v_0}{\partial z}\right)=\frac{\partial p_1}{\partial y} \tag{3.108}$$

$\eta_1$ in Equations (3.108) is given by Equations (3.105) and using Equation (3.105) the above equations reduce to:

$$\eta_0\frac{\partial^2 u_1}{\partial z^2}+\frac{U}{h}.2\frac{\partial \eta}{\partial I}\frac{\partial}{\partial z}\left(\frac{U}{h}\frac{\partial u_1}{\partial z}+\frac{V}{h}\frac{\partial v_1}{\partial z}\right)=\frac{\partial p_1}{\partial x}$$

$$\eta_0\frac{\partial^2 u_1}{\partial z^2}+\frac{V}{h}.2\frac{\partial \eta}{\partial I}\frac{\partial}{\partial z}\left(\frac{U}{h}\frac{\partial u_1}{\partial z}+\frac{V}{h}\frac{\partial v_1}{\partial z}\right)=\frac{\partial p_1}{\partial y} \tag{3.109}$$

The above equations are linear with constant second derivatives of first order velocities $u_1$ and $v_1$. Solutions of above equations give:

$$\frac{\partial^2 u_1}{\partial z^2}=\frac{1}{\eta_0}\frac{\partial p_1}{\partial x}-\frac{U}{h}\left(\frac{U}{h}\frac{\partial p_1}{\partial x}+\frac{V}{h}\frac{\partial p_1}{\partial y}\right).2\frac{\dfrac{\partial \eta}{\partial I}}{\eta_0^2\left(1+\dfrac{\partial \ln \eta}{\partial \ln\left(I^{1/2}\right)}\right)} \tag{3.110}$$

$$\frac{\partial^2 v_1}{\partial z^2}=\frac{1}{\eta_0}\frac{\partial p_1}{\partial y}-\frac{V}{h}\left(\frac{U}{h}\frac{\partial p_1}{\partial x}+\frac{V}{h}\frac{\partial p_1}{\partial y}\right).2\frac{\dfrac{\partial \eta}{\partial I}}{\eta_0^2\left(1+\dfrac{\partial \ln \eta}{\partial \ln\left(I^{1/2}\right)}\right)} \tag{3.111}$$

Usually the surface velocity $V=0$ since the surface moves in only one direction. Therefore, the above equations reduce to

$$\frac{\partial^2 u_1}{\partial z^2}=\frac{1}{\eta_0}\frac{\partial p_1}{\partial y}-\frac{U^2}{h^2}\frac{\partial p_1}{\partial x}2\frac{\dfrac{\partial \eta}{\partial I}}{\eta_0^2\left(1+\dfrac{\partial \ln \eta}{\partial \ln\left(I^{1/2}\right)}\right)} \tag{3.112}$$

$$\frac{\partial^2 v_1}{\partial z^2}=\frac{1}{\eta_0}\frac{\partial \pi}{\partial y} \tag{3.113}$$

Having both zeroth and first order velocities, the total velocities are given as:

$$u = u_0 + \lambda u_1$$
$$v = v_0 + \lambda v_1 \quad (3.114)$$

The mass flux is determined by integrating $u$ and $v$ across the film thickness and is written in general as:

$$M = \rho\frac{Sh}{2} - \rho\frac{h^3}{12\eta_o}\left\{\nabla p - \frac{\hat{s}(\hat{s}.\nabla p)}{\left(1+\dfrac{\partial \ln I^{1/2}}{\partial \ln \eta}\right)}\right\} \quad (3.115)$$

where $\hat{s}$ is the direction unit vector of local surface velocity $S$.

In the case of Newtonian fluids, $\eta$ is independent of $I$ and the last term in the above equation varnishes. In general case of non-Newtonian fluids, $\eta_0$ and $\dfrac{\partial \ln I^{1/2}}{\partial \ln \eta}$ should be evaluated from the local Couette strain rate.

Since, in case of a power law fluid

$$\tau_{xz} = m\left(\frac{\partial u}{\partial z}\right)^n = m\left(\frac{\partial u}{\partial z}\right)^{n-1}\frac{\partial u}{\partial z} = \eta\frac{\partial u}{\partial z}$$

and similarly

$$\tau_{yz} = m\left(\frac{\partial v}{\partial z}\right)^n = m\left(\frac{\partial v}{\partial z}\right)^{n-1}\frac{\partial v}{\partial z} = \eta\frac{\partial v}{\partial z} \quad (3.116)$$

and

$$\eta = m\left\{\left(\frac{\partial u}{\partial z}\right)^2\right\}^{\frac{n-1}{2}} = I^{\frac{n-1}{2}} \quad (3.117)$$

Therefore,

$$\frac{\partial \ln I^{1/2}}{\partial \ln \eta} = \frac{1}{n-1}$$

Alternately mass flux $M$ is written as:

$$M = \rho\frac{Sh}{2} - \frac{\rho h^3}{12\eta_o}\left\{\nabla p - \left(\frac{n-1}{n}\right)\hat{s}(\hat{s}.\nabla p)\right\} \quad (3.118)$$

where $n$ is the local power law exponent

$$\eta_0 = mI_0^{\frac{n-1}{2}}$$

since

$$I_0 = \left\{\left(\frac{\partial u_0}{\partial z}\right)^2 + \left(\frac{\partial v_0}{\partial z}\right)^2\right\} \quad (3.119)$$

Using Equation (3.97) and for $V = 0$,

$$I_0 = \left(\frac{U}{h}\right)^2 \text{ and } \eta_0 = m\left(\frac{U}{h}\right)^{n-1} \tag{3.120}$$

Mass flux in $x$-direction is given as:

$$M_x = -\frac{\rho h^3}{12\eta_0}\frac{1}{\eta}\frac{\partial p}{\partial x} \tag{3.121}$$

Mass flux in $y$-direction is given as:

$$M_y = -\frac{\rho h^3}{12\eta_0}\frac{1}{\eta}\frac{\partial p}{\partial y} \tag{3.122}$$

The generalized Reynolds equation for power law fluids can be determined using mass flow continuity relationship as given below:

$$\frac{\partial}{\partial x}(M_x) + \frac{\partial}{\partial y}(M_y) + h\frac{\partial p}{\partial t}\rho\frac{\partial h}{\partial t} = 0 \tag{3.123}$$

Thus, the generalized Reynolds equation is determined as:

$$\frac{\partial}{\partial x}\left(\frac{\rho h^{2+n}}{nm}\frac{\partial}{\partial y}\right) + \frac{\partial}{\partial y}\left(\frac{\rho h^{2+n}}{m}\frac{\partial p}{\partial t}\right) = 6U^n\frac{\partial}{\partial x}(\rho h) + 12\frac{\partial}{\partial t}(\rho h) \tag{3.124}$$

For incompressible lubrication it reduces to

$$\frac{\partial}{\partial x}\left(\frac{h^{2+n}}{n}\frac{\partial}{\partial x}\right) + \frac{\partial}{\partial y}\left(h^{2+n}\frac{\partial p}{\partial y}\right) = 6m\ U^n\frac{\partial h}{\partial x} + 12\frac{\partial h}{\partial t} \tag{3.125}$$

## 3.14 | Examples of Slow Viscous Flow

### 3.14.1 | Flow of Two Immiscible Liquids Down an Inclined Plane

Figure 3.4 shows schematically the flow of two immiscible liquids designated as '$a$' and '$b$' down an incline plane. The flow is due to gravity force acting on the fluids.

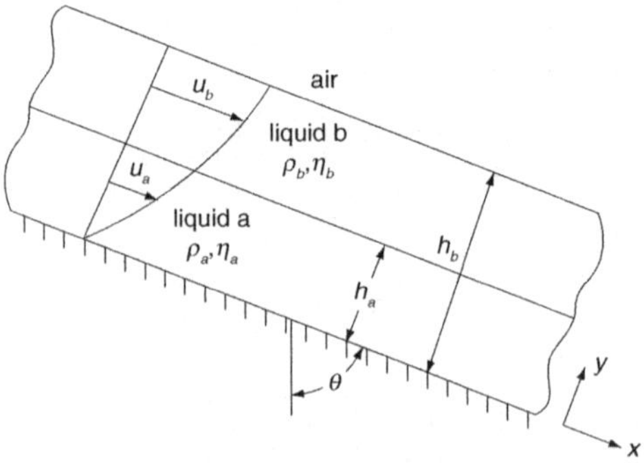


**Figure 3.4** | Falling Film on Inclined Plane Pressure $p_2$

There is an interface between the layers of fluids and an interface with air at the top surface of the fluid $b$. The viscosities and densities of the fluids are defined as $\eta_a, \eta_b$ and $\rho_a, \rho_b$, respectively. The possible nature of velocity distributions are also shown in the figure. The liquid film thicknesses are $h_a$, $h_b$, respectively. Inclination of the plane with vertical is $\theta$.

The generic momentum equation in direction normal to the plane, i.e., in $z$ direction can be written as:

$$0 = -\frac{\partial p}{\partial y} - \rho g \sin\theta \tag{3.126}$$

$\rho$ can be replaced by $\rho_a$ or $\rho_b$ for fluids $a$ and $b$.

A partial integration with respect to $y$ gives:

$$p = -y\rho g \sin\theta + f(x) \tag{3.127}$$

The arbitrary function $f(x)$ is evaluated using the fact that atmospheric pressure $p_a$ exists on top of liquid $b$. Thus, using this condition we obtain the expression for pressure in the liquid layer $b$ as:

$$p = \rho_b g(h_b - y)\sin\theta + p_a; h_a \le y \le h_b \tag{3.128}$$

Interface pressure $p_i = \rho_b g(h_b - h_a)\sin\theta + p_a$

The pressure in the liquid '$a$' is written as:

$$p = \rho_a g(h_a - y) g \sin\theta + p_i; 0 \le y \le h_a \tag{3.129}$$

$x$-directional momentum equation for the flow of a viscous fluid is expressed as:

$$0 = \eta \frac{\partial^2 u}{\partial y^2} + \rho g \cos\theta \tag{3.130}$$

No slip condition is applicable at the wall and also at the interface. Thus,

$$u_a(0) = 0 \text{ and } u_a(h_a) = u_b(h_b - h_a) \tag{3.131}$$

A second condition is that the shear stress is continuous across the interface. Thus,

$$\eta_a \frac{du_a}{dy} = \eta_b \frac{du_b}{dy} \text{ at } y = h_a \tag{3.132a}$$

Since air exerts negligible shear stress on the liquid.

$$\eta_b \frac{du_b}{dy} \cong 0 \text{ at } y = h_b \tag{3.132b}$$

Integrating Equation (3.130) twice with respect to $y$ one would obtain the velocity distributions of the flow of fluids '$a$' and '$b$' as given below:

$$u_a = -\frac{g}{\nu_a} y^2 \cos\theta + C_1 y + C_2 \tag{3.133}$$

$$u_b = -\frac{g}{\nu_b} y^2 \cos\theta + C_3 y + C_4 \tag{3.134}$$

The constants of integration $C_1$, $C_2$, $C_3$, and $C_4$ can be determined using Equations (3.131) and (3.132). Thus,

$$C_1 = \frac{gh_a}{\nu_a}\cos\theta\left[\frac{\rho_b}{\rho_a}h_a + 1\right];\ C_2 = 0\ ;\ C_3 = \frac{h_b}{\nu_b}g\cos\theta;$$

$$C_4 = \frac{gh_b\cos\theta}{2\nu_b}\left[h_b - 2h_a\right] + \frac{gh_a^2\cos\theta}{\nu_a}\left[1 + \frac{\rho_b}{\rho_a}h_a\right] - \frac{gh_b\cos\theta}{\nu_b}\left(h_b - h_a\right)$$

The fluid velocities are thus obtained as:

$$u_a = -\frac{g}{\nu_a}y^2\cos\theta + y\frac{gh_a}{\nu_a}\cos\theta\left[\frac{\rho_b}{\rho_a}h_a + 1\right] \tag{3.135}$$

$$u_b = -\frac{g}{\nu_b}y^2\cos\theta + y\frac{h_b}{\nu_b}g\cos\theta + \frac{gh_b\cos\theta}{2\nu_b}\left[h_b - 2h_a\right] + \frac{gh_a^2\cos\theta}{\nu_a}\left[1 + \frac{\rho_b}{\rho_a}h_a\right] - \frac{gh_b\cos\theta}{\nu_b}\left(h_b - h_a\right) \tag{3.136}$$

### 3.14.2 | Flow of a Viscous Fluid Through a Vertical Annulus

Figure 3.5 shows schematically a vertical annulus through which a viscous fluid flows due a pressure gradient along the flow direction. It is desired to find out an expression of velocity distribution within the clearance space of the annulus.

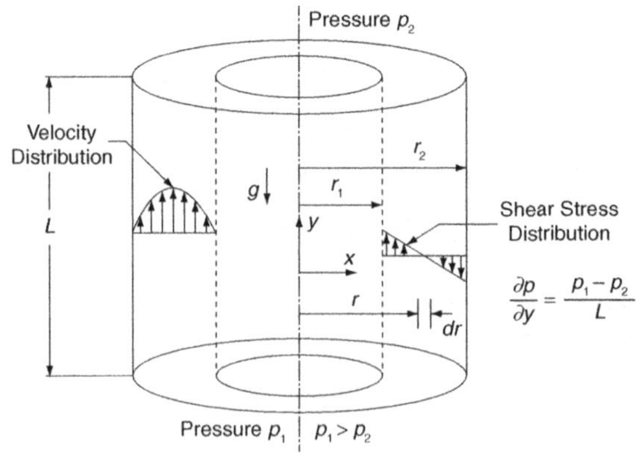


**Figure 3.5** | Flow Through an Annulus

Force acting on the elemental area due to pressure gradient $\frac{\partial p}{\partial y}$ in the flow direction, i.e., along the length of the annulus is written as:

$$p(2\pi rdr)-\left[p+\left(\frac{\partial p}{\partial y}\right)dy\right].2\pi rdr+2\rho g\pi rdrdy=\left(-\frac{\partial p}{\partial y}+\rho g\right)2\pi rdrdy \tag{3.137}$$

Net viscous force acting on the element from right to left is obtained as:

$$\left[\tau+\frac{\partial \tau}{\partial r}dr\right]2\pi(r+dr)dy-\tau 2\pi rdy=2\pi\left[\tau+r\frac{\partial \tau}{\partial r}\right]drdy \tag{3.138}$$

For fully developed laminar flow through the annulus, Equations (3.137) and (3.138) can be combined to give:

$$\left(-\frac{\partial p}{\partial y}+\rho g\right)2\pi rdrdy+2\pi\left(\tau+r\frac{\partial \tau}{\partial r}\right)drdy=0$$

This reduces to:

$$\tau+r\frac{\partial \tau}{\partial r}=r\frac{\partial p}{\partial y}-\rho gr \tag{3.139}$$

Since shear stress $\tau=\eta\frac{\partial u}{\partial r}$, above equation reduces to,

$$\frac{d}{dr}\left(r\eta\frac{\partial u}{\partial r}\right)=r\frac{\partial p}{\partial y}-\rho gr \tag{3.140}$$

Integrating Equation (3.130) with respect $r$ successively it can be shown that flow velocity can be expressed as:

$$u=\frac{r^2}{2\eta}\frac{dp}{dy}-\frac{r^2}{4\eta}\rho g+\frac{A}{\eta}\ln r+B \tag{3.141}$$

where $A$ and $B$ are constants of integration which can be determined by substituting the boundary conditions for fluid velocity as mentioned below:

$$u=0 \text{ at } r=r_1 \text{ and } r_2$$

The constants of integration are obtained using above boundary conditions as:

$$A=\frac{\left(\rho g-\frac{dp}{dy}\right)\left\{\frac{r_2^2-r_1^2}{4}\right\}}{\ln(r_2/r_1)};\quad B=\frac{r_1^2}{4\eta}\left(\rho g-\frac{dp}{dy}\right)-A\ln r_1$$

Thus, velocity distribution can be written as:

$$u = \frac{1}{4\eta}\left(\rho g - \frac{dp}{dy}\right)\left[\frac{r_2^{\,2} - r_1^{\,2}}{\ln(r_2/r_1)}\ln r - r^2\right] + \frac{r_1^{\,2}}{4\eta}\left(\rho g - \frac{dp}{dy}\right) - A\ln r_1 \tag{3.142}$$

Maximum velocity occurs when $\frac{du}{dr} = 0$ and it can be shown that it occurs at a radius given by:

$$r = \sqrt{\frac{2A}{\rho g - \frac{dp}{dy}}} \tag{3.143}$$

For further examples of flow of viscous fluids through narrow gaps readers can refer to Bird, Stewart and Lightfoot (1960) and Hamrock (1994).

## Examples

***E.3.1*** Consider the system shown in Fig. 3.6 in which the cylindrical rod is being moved with an axial velocity $V$. The rod and the cylinder are coaxial. Determine the expressions of the steady state velocity distribution and volume rate of flow. Determine the force per unit length to move the rod. Pressure gradient is zero. Proceed from the fundamentals.

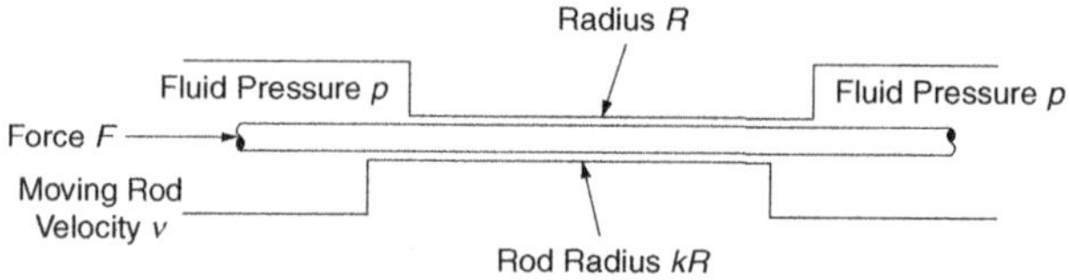


**Figure 3.6 |** A Cylindrical Rod Driven through a Viscous Fluid in a Pipe

***Solution:***

Navier–Stokes equation in the direction of flow is written in cylindrical coordinates as:

$$\rho\left(\frac{\partial v_z}{\partial t} + v_r\frac{\partial v_r}{\partial r} + \frac{v_\theta}{r}\frac{\partial v_z}{\partial \theta} + v_z\frac{\partial v_z}{\partial z}\right) = B_z - \frac{\partial p}{\partial z} + \eta\nabla^2 v_z \tag{3.1a}$$

where

$$\nabla^2 = \frac{\partial^2}{\partial r^2} + \frac{1}{r}\frac{\partial}{\partial r} + \frac{1}{r^2}\frac{\partial^2}{\partial \theta^2} + \frac{\partial^2}{\partial z^2}$$

Following assumptions are made to get the solution:

- Inertia forces and body forces are negligible
- Velocity gradients in the direction of flow are also negligible and therefore

$$\frac{\partial v_z}{\partial r} >> \frac{\partial v_z}{\partial z}, \frac{\partial v_\theta}{\partial \theta}$$

- Density and viscosity are constant

Given that pressure gradient $\frac{dp}{dz} = 0$, the equation of motion reduces to

$$\eta\left(\frac{\partial^2 v_z}{\partial r^2} + \frac{1}{r}\frac{\partial v_z}{\partial r}\right) = 0 \quad \text{or} \quad \frac{\partial}{\partial r}\left(r\frac{\partial v_z}{\partial r}\right) = 0 \tag{3.2a}$$

Integrating above equation twice with respect to $r$, we can get

$$v_z = c_1 \ln r + c_2 \tag{3.3a}$$

Boundary conditions for the axial velocity are

$$v_z = V \text{ at } r = kR \text{ and } v_z = 0 \text{ at } r = R$$

Substituting the boundary conditions, constants of integration $c_1$ and $c_2$ are determined as

$$c_1 = \frac{V}{\ln k},\ c_2 = -\frac{V}{\ln k}\ln R$$

and the expression of axial velocity is obtained by substituting the constants in Equation (3.3a) as

$$v_z = \frac{V}{\ln k}\ln\left(\frac{r}{R}\right) \tag{3.4a}$$

Volume flow rate of the fluid $Q$ is determined as:

$$Q = \int_{kR}^{R} v_z 2\pi r dr$$

Using the expression of axial velocity given in Equation (3.4a), the volume flow rate is obtained after integration as:

$$Q = \frac{\pi R^2 V}{2\ln(1/k)}\left[\frac{1-k^2}{\ln(1/k)} - 2k^2\right] \tag{3.5a}$$

Force per unit length required to move the rod is determined as:

$$F = \int_{kR}^{R} \tau 2\pi r dr = \int_{kR}^{R} \eta\frac{\partial v_z}{\partial r}2\pi r dr \tag{3.6a}$$

Substituting for $\frac{\partial v_z}{\partial r}$ using the expression for $v_z$ and integrating the force required to move the rod is obtained as:

$$F = \frac{2\pi\eta V R^2}{\ln k}(1-k) \tag{3.7a}$$

***E.3.2*** A thin film of viscous fluid is held between two parallel circular plates that are separated by a distance $h$ apart as shown in Fig. 3.7. One plate is held stationary while the other rotates at an angular speed $\omega$ about its axis. Fluid flows out in the radial direction due to a pressure gradient in the radial direction. Determine the expression of the radial velocity distribution. Analyze for following flow conditions:

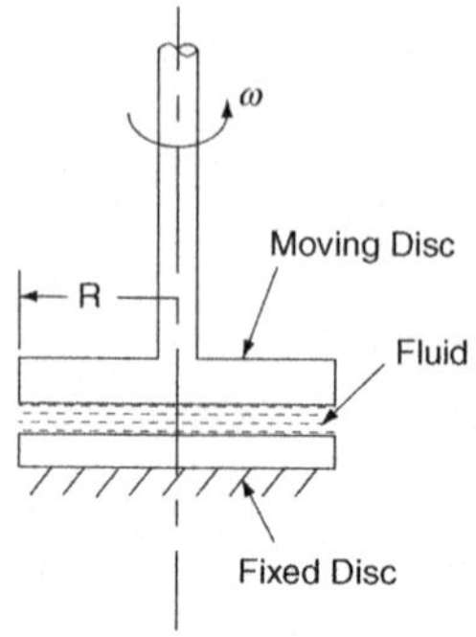


**Figure 3.7** | A Circular Disc Rotating on a Film of Viscous Fluid

1. Couette flow
2. Poiseuille flow
3. A flow for which volume flow is zero
4. Zero shear stress on the moving plate
5. Zero shear stress on the fixed plate

***Solution:***

Assume that the flow is laminar and the fluid is Newtonian. All inertia and body forces are neglected and pressure gradient in the circumferential direction is zero for a parallel film. Fluid film is thin so that only velocity gradient across the film is considered and all other velocity gradients are negligible.

Navier–Stokes equation reduces to:

$$\frac{\partial p}{\partial r} = \eta \frac{\partial^2 v_r}{\partial z^2} \tag{3.8a}$$

Integrating the above equation and satisfying following velocity boundary conditions $v_r = 0$ at $z = 0$ and $v_r = R\omega$ at $z = h$, the velocity distribution is determined as:

$$v_r = \frac{1}{2\eta}\frac{\partial p}{\partial r} z(z-h) + \frac{R\omega z}{h} \tag{3.9a}$$

Thus, mean velocity can be determined as

$$v_m = \frac{1}{h}\int_0^h v_r \, dz \tag{3.10a}$$

Using Equation (3.9a) mean velocity is obtained as

$$v_m = -\frac{h^2}{2\eta}\frac{\partial p}{\partial r}$$

$$v_r = -\frac{h^2}{2\eta}\frac{\partial p}{\partial r}\left(\frac{z}{h} - \left(\frac{z}{h}\right)^2\right) + \frac{R\omega z}{h} \tag{3.11a}$$

or

$$\frac{v_r}{R\omega} = \frac{6v_m}{R\omega}(\xi - \xi^2) + \xi = \lambda(\xi - \xi^2) + \xi \tag{3.12a}$$

where

$$\lambda = \frac{6v_m}{R\omega} \text{ and } \xi = \frac{z}{h}$$

1. In case of Couette flow, pressure gradient is taken as zero, i.e., $\lambda = 0$, therefore

$$\frac{v_r}{R\omega} = \xi$$

2. In Poiseuille flow,

$$\frac{v_r}{R\omega} = \lambda(\xi - \xi^2), \quad \lambda = \frac{v_r}{R\omega} \cdot \frac{1}{(\xi - \xi^2)}$$

3. Volume flow rate is determined as:

$$Q = \int_0^h v_r \, dz$$

Substituting for $v_r$ using Equation (3.11a), flow rate is obtained after integrating as

$$Q = -\frac{h^3}{12\eta} \frac{\partial p}{\partial r} + \frac{R\omega h}{2}$$

$$Q = v_m h + \frac{R\omega h}{2}$$

For zero volume flow rate $Q = 0$ and thus $v_m = -R\omega/2$
Therefore,

$$\lambda = \frac{6v_m}{R\omega} = -3$$

4. Flow with zero shear stress at the fixed surface

Shear stress on the fixed surface $\tau_{z=0} = \left.\frac{\partial v_r}{\partial z}\right|_{z=0} = 0$

Therefore, using Equation (3.12a) and substituting $\xi = 0$, gives,

$$\tau_{\xi=0} = \eta(\lambda + 1) = 0$$

Thus,

$$\lambda = -1$$

5. Flow with zero shear stress at the moving surface

Shear stress at the moving surface $\tau_{z=h} = \left.\frac{\partial v_r}{\partial z}\right|_{z=h} = 0$

Using Equation (3.12a) and substituting $\xi = 1$, gives,

$$\tau_{\xi=1} = \eta(-\lambda + 1) = 0$$

Thus,

$$\lambda = 1$$

## Problems

***P.3.1*** Two immiscible incompressible fluids are flowing through a horizontal thin slit of length $L$ and width $W$ under the influence of a pressure gradient. The flow rates are so adjusted that the slit is half filled with fluid 1(denser) and half filled with fluid 2(less dense). It is desired to analyze the system in terms of velocity distributions and momentum flux. Refer to Fig. P.3.1. Determine the flow rates, velocity at the interface, and drag on the walls when density of fluid A is twice the density of fluid B. Both fluids have same viscosity.

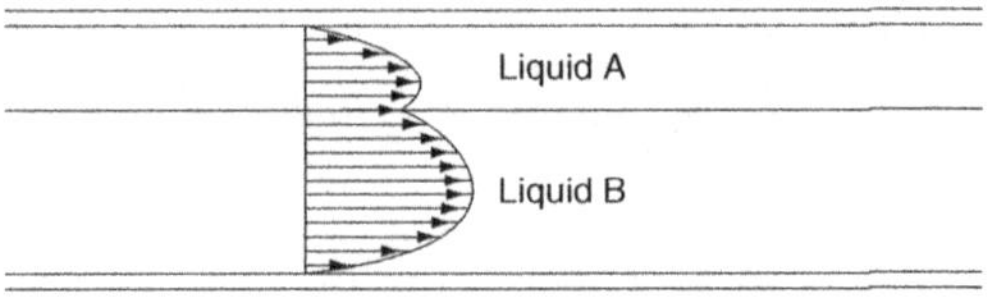


**Figure P.3.1** | Flow of Two Immiscible Fluids Through a Slit

***P.3.2*** A shaft of radius $r_1$ concentric with a sleeve of radius $r_2$ is held stationary when the sleeve moves with an axial velocity $V$. The clearance space between the shaft and the sleeve is filled with a fluid of viscosity $\eta$. Determine the expression of velocity profile of the fluid and volume flow rate when there is pressure gradient

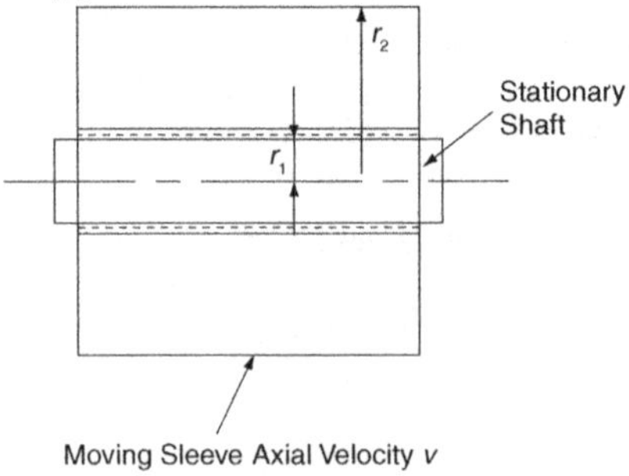


**Figure P.3.2** | Flow of a Viscous Fluid Between a Shaft and a Sleeve

in the axial direction. Density and the viscosity of the fluid are assumed constant. Fluid inertia effects and body forces are negligible. Assume $\left(\frac{r_2 - r_1}{r_1} = 10^{-3}\right)$.

**P.3.3** Develop the expression of fluid velocity for the flow of a Bingham plastic fluid through an annulus and determine the expression of volume flow rate. Extend the analysis for the flow of a power law fluid through an annulus.

**P.3.4** Determine the expression of velocity profile of axial flow of a fluid between two coaxial stationary cylinders under an axial pressure gradient. Determine the expressions of flow rate and shear stresses at both surfaces. Take $\kappa = r_2/r_1$. Determine the maximum velocity of fluid between two coaxial cylinders given $r_2 = 7.5cm$ and $r_1 = 5cm$ and flow rate $Q = 5.4 \times 10^{-3}$ m³/sec. Determine (i) pressure drop, (ii) maximum value of flow velocity, and (iii) shear stress at the wall of both cylinders.

**P.3.5** Analyze the flow of a Newtonian fluid between two concentric rotating cylinders. Assume the flow to be peripheral so that only tangential velocity component $v_\theta$ exists. Assume the inner and outer cylinders to be rotating at angular velocities $\omega_1$ and $\omega_2$, respectively.

## REFERENCES


*Bair, S. and Winer, W.O.* (1979), A rheological model for elastohydrodynamic contacts based on primary laboratory data, *ASME, Journal of Lubrication Technology*, **101**, 258–265.

*Bird, R.B., Stewart, W.E. and Lightfoot, E.N.* (1960), *Transport Phenomena*, Wiley International Edition, John Wiley & Sons Inc.

*Dien, I.K., and Elrod, H.G.* (1983), A generalized steady state Reynolds equation for non-Newtonian fluids, with application to journal bearings, *ASME, Journal of Lubrication Technology*, **105**, 385–390.

*Dowson, D.* (1962), A generalized Reynolds equation for fluid film lubrication, *Int. J. of Mechanical Sciences*, Pergamon Press Ltd, **4**, 159–170.

*Fowles, P.E.* (1970), A simpler form of the general Reynolds equation, *Trans. A.S. M. E. Journal of Lubrication Tech.*, **92**, 662,

*Hamrock, B.J.* (1994), *Fundamentals of Fluid Film Lubrication*, McGraw Hill International Edition, Singapore.

*Hughes, W.F. and J.A. Brighton* (1967), *Theory and Problems of Fluid Dynamics*, Schaum's Outline Series, McGraw Hill Book Company, New York.

*Najji, B., Bou-Said, B., and Berthe, D.* (1989), New formulation for lubrication with non-Newtonian fluids, *ASME, Journal of Tribology*, **111**, 29–34.

*Pinkus, O. and Sterhlicht, B.* (1961), *Theory of Hydrodynamic Lubrication*, McGraw Hill Book Co. Inc., New York.

*Reynolds, O.* (1886), On the theory of lubrication and its application to beauchamp tower's experiments, *Phil. Trans. Roy Soc.*, London, **177**, 157–234.

*Szeri, A.Z.* (1998), *Fluid film Lubrication Theory and Design*, Cambridge University Press, U.K.

*Tevaarwerk, J. and Johnson, K.L.* (1975), A simple non-linear constitutive equation for elastohydrodynamic oil film, Wear, **35**, 345–346.

*Wolff R. and Kubo, A.* (1996), Generalized non-Newtonian fluid model incorporated into elastohydrodynamic lubrication, *ASME, Journal of Tribology*, **118**, 74–82.

Chapter 4

# Hydrodynamic Lubrication

## 4.1 | Introduction

Hydrodynamic films in mechanical components are formed due to self-action or wedge action which results in the generation of pressure to carry the load avoiding metal to metal contact. Film thickness in nonconformal contacts, e.g., line and point contacts, which are encountered in rolling element bearings, gears, etc., is usually 1 μm or less, and the lubrication is referred to as thin film lubrication. On the contrary, film thickness in conformal contacts, viz., in journal bearings, thrust bearings and seals, etc. are thick usually between 2–10 μm and the lubrication is generally referred to as thick film lubrication.

This chapter deals with the analytical solutions of hydrodynamic lubrication in journal and thrust bearings. Classical solutions usually fall in the category of isoviscous and isothermal regime of lubrication. This implies that viscosity–pressure–temperature effects are not considered. It also implies that pressures and temperatures are not high enough to have significant influence on the viscosity of the lubricant which can be considered as constant. However, isothermal conditions are idealizations which in practice seldom exist.

## 4.2 | Hydrodynamic Journal Bearings

Journal bearings are mechanical components used to support shafts of a machine. Journal bearings are therefore designed to carry radial loads. The load carrying capacity is developed due to the generation of pressure by the fluid film formed in the clearance space between the bearing and journal. The lubricant is dragged or entrained into the clearance space because of rotation of the journal or the shaft. The clearance space is convergent divergent when the journal center is eccentric with respect to the bearing center subject to a radial load that is acting either due to the weight of the rotor or due to an external force. A journal bearing-rotor system is shown schematically in Fig. 4.1.

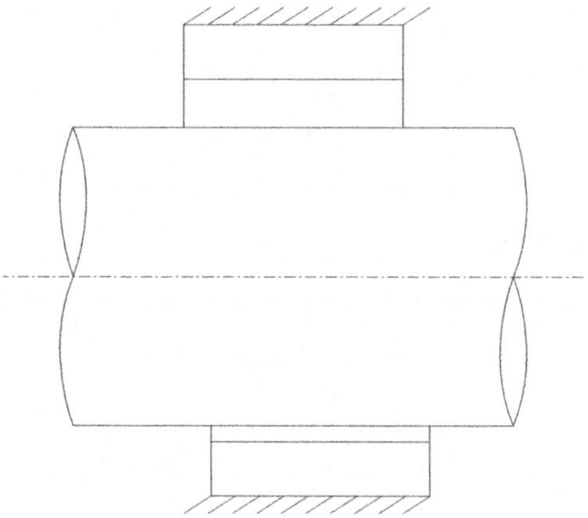

Figure 4.1 | Journal Bearing and Shaft

The journal is eccentric with respect to the bearing under loaded condition. The shape of the clearance space is shown in Fig. 4.2.

The expression for film thickness can be easily determined from the geometry of the film shape. Figure 4. 2 is referred to determine the expression of film thickness.

Film thickness $h = AB$ which is the film thickness at $B$ at angle $\theta$ from the line of center, i.e., $h = BO_b - AO_b$;

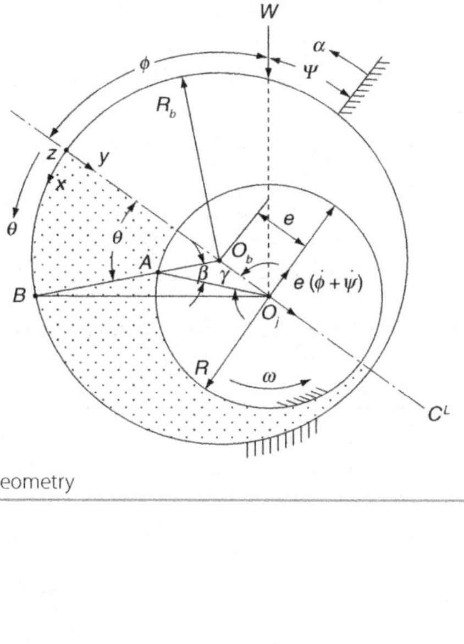


Figure 4.2 | Journal Bearing Geometry

From the triangle $O_jAO_b$, one can write

$$\frac{AO_b}{\sin\gamma} = \frac{R}{\sin\theta} = \frac{e}{\sin\beta} \tag{4.1}$$

where

$R = O_jA$, radius of the journal

$e = O_jO_b$, eccentricity of the journal center with respect to bearing center since,

$$\gamma = \theta - \beta$$

$$AO_b = \frac{R\sin\gamma}{\sin\theta} = \frac{R}{\sin\theta}\sin[\theta - \beta]$$

since,

$$\beta = \sin^{-1}\left(\frac{e}{R}\sin\theta\right)$$

and

$$AO_b \frac{R}{\sin\theta}\sin\left[\theta - \sin^{-1}\left(\frac{e}{R}\sin\theta\right)\right]$$

Film thickness is expressed as:

$$h = (R + C) - \frac{R}{\sin\theta}\sin\left[\theta - \sin^{-1}\left(\frac{e}{R}\sin\theta\right)\right] \tag{4.2}$$

Since $R_b = BO_b$ radius of the bearing, $R_b = R + C$, where $C$ is radial clearance between the bearing and journal centers in concentric position.

The above expression on simplification with the assumption that $\frac{e}{R} << 1$ yields the expression of the film thickness

$$h = C + e\cos\theta$$

$$h = C(1 + \varepsilon\cos\theta) \tag{4.3}$$

where $\varepsilon = \frac{e}{C}$ which lies between 0 and 1 and is called the eccentricity ratio. $\frac{C}{R}$ is usually of the order of $10^{-3}$ in liquid lubricated bearings.

When angle $\theta$ is measured from a fixed reference, the film thickness is written as:

$$h = C + e\cos[\alpha - (\phi + \psi)] \tag{4.4}$$

Time rate of change in film thickness or squeeze velocity $\frac{\partial h}{\partial t}$ can be determined by differentiating $h$ with respect to time, $t$ as:

$$\frac{\partial h}{\partial t} = \frac{de}{dt}\cos\left[\alpha - (\phi + \psi)\right] + e\left\{\frac{d}{dt}(\phi + \psi)\right\}.\sin\left[\alpha - (\phi + \psi)\right];$$

$$\frac{\partial h}{\partial t} = \dot{e}\cos\theta + e\left(\dot{\phi} + \dot{\psi}\right)\sin\theta \tag{4.5}$$

where $\dot{\psi} = \frac{\partial \psi}{\partial t}$ is angular velocity of the rotation of the load and $\dot{\phi} = \frac{\partial \phi}{\partial t}$ is the rate of change of attitude angle with respect to the load.

The Reynolds equation for an incompressible lubricant can now be written for a journal bearing as:

$$\frac{\partial}{\partial x}\left(\frac{h^3}{\eta}\frac{\partial p}{\partial x}\right) + \frac{\partial}{\partial y}\left(\frac{h^3}{\eta}\frac{\partial p}{\partial x}\right) = 6R\omega\frac{\partial h}{\partial x} + 12\left[\dot{e}\cos\theta + e\left(\dot{\phi} + \dot{\psi}\right)\sin\theta\right] \tag{4.6}$$

where $h = C(1 + \varepsilon\cos\theta)$

Under steady state condition, the terms $\dot{e} = \dot{\phi} = \dot{\psi} = 0$ and Equation (4.6) reduces to

$$\frac{\partial}{\partial x}\left(\frac{h^3}{\eta}\frac{\partial p}{\partial x}\right) + \frac{\partial}{\partial y}\left(\frac{h^3}{\eta}\frac{\partial p}{\partial x}\right) = 6R\omega\frac{\partial h}{\partial x} \tag{4.7}$$

Hydrodynamic lubrication of journal bearings require solution of the above equation satisfying appropriate boundary conditions, to determine load carrying capacity for a known eccentricity ratio, oil flow rate, and frictional power loss in terms of known viscosity of lubricant, geometry, e.g., diameter of the journal, bearing length, and operating speed of the journal.

Analytical solution of hydrodynamic journal bearings will be discussed in the following sections.

## 4.3 | Long Bearing Solution

Long bearing solution assumes that the axial length of bearing is very large in comparison to its diameter, i.e., *L/D >>1*. In this case, it is assumed that film pressure in the axial direction does not vary and therefore the term $\frac{\partial p}{\partial y}$ in the Reynolds equation can dropped, i.e., $\frac{\partial p}{\partial y} = 0$. It also implies that the flow of the lubricant from the bearing ends is also negligible or zero.

The Reynolds Equation (4.7) can be written neglecting $\frac{\partial p}{\partial y}$ term for a constant viscosity lubricant as:

$$\frac{\partial}{dx}\left(h^3\frac{dp}{dx}\right) = 6\eta U\frac{dh}{dx} \quad \text{where } U = \omega R \tag{4.8}$$

Substituting $x = R\theta$, the above equation becomes

$$\frac{\partial}{d\theta}\left(h^3\frac{dp}{d\theta}\right) = 6\eta UR\frac{dh}{d\theta} \tag{4.9}$$

where $h = C(1 + \varepsilon\cos\theta)$

The relationship between the load carrying capacity of the oil film and the eccentricity ratio can be determined by integrating the pressure distributed over the film domain area. Pressure distribution in the oil film can be determined by integrating the above equation satisfying appropriate boundary conditions.

Oil or the lubricant is usually fed to the bearing through an oil hole or a groove made at a location coincident with the maximum film position, i.e., at $\theta = 0$. It is also assumed that pressure is equal to ambient pressure at $\theta = 0$ or otherwise equal to oil feed pressure if the oil is fed at a pressure higher than the atmospheric pressure. Integration of the Reynolds Equation (4.9) is necessary to determine the pressure distribution and load capacity for a known eccentricity ratio. Sommerfeld (1904) gave what is known as Sommerfeld substitution which made it possible to integrate the Reynolds equation.

Integrating Equation (4.9) with respect to $\theta$ once yields:

$$\frac{dp}{d\theta} = 6\eta UR \frac{h - \bar{h}}{h^3} \tag{4.10}$$

where $\bar{h}$ is the film thickness at which $\dfrac{\partial p}{\partial \theta} = 0$

Equation (4.10) is sometimes known as integrated Reynolds equation.

Integrating Equation (4.10) once again with respect to $\theta$ gives

$$p = \frac{6\eta UR}{C^2}\left[\int \frac{d\theta}{(1+\varepsilon\cos\theta)^2} - \frac{\bar{h}}{C}\int \frac{d\theta}{(1+\varepsilon\cos\theta)^3}\right] + A \tag{4.11}$$

$A$ is the constant of integration which can be determined satisfying the following boundary conditions:

$$p = 0 \text{ at } \theta = 0$$

Integrals $\displaystyle\int \frac{d\theta}{(1+\varepsilon\cos\theta)^2}$ and $\displaystyle\int \frac{d\theta}{(1+\varepsilon\cos\theta)^3}$ can be determined by invoking Sommerfeld substitution as stated below:

$$(1+\varepsilon\cos\theta) = \frac{1-\varepsilon^2}{1-\varepsilon\cos\gamma} \tag{4.12}$$

$\gamma$ is the Sommerfeld variable

Alternately,

$$\cos\gamma = \frac{\varepsilon + \cos\theta}{(1+\varepsilon\cos\theta)}\ ; \ \sin\gamma = \frac{(1-\varepsilon^2)^{1/2}\sin\theta}{(1+\varepsilon\cos\theta)}$$

$$\sin\theta = \frac{(1-\varepsilon^2)^{1/2}\sin\gamma}{1+\varepsilon\cos\gamma} \tag{4.13}$$

$$\cos\theta = \frac{\cos\gamma - \varepsilon}{1+\varepsilon\cos\gamma} \text{ and}$$

$$d\theta = \frac{(1-\varepsilon^2)^{1/2}\, d\gamma}{1-\varepsilon\cos\gamma} \tag{4.14}$$

using the substitutions given by Equations (4.12) and (4.14) in the Equation (4.11), the pressure distribution is obtained as:

$$p = \frac{6\eta U R\varepsilon}{C^2}\frac{(2+\varepsilon\cos\theta)\sin\theta}{(2+\varepsilon^2)(1+\varepsilon\cos\theta)^2} \tag{4.15}$$

The first boundary condition $p = 0$ at $\theta = 0$ yields the constant of integration $A = 0$ in Equation (4.11).

The second boundary conditions $\frac{dp}{d\theta} = 0$ at $h = \bar{h}$ gives the value of film thickness $\bar{h}$ as:

$$\bar{h} = \frac{2C(1-\varepsilon^2)}{(1+\varepsilon^2)} \tag{4.16}$$

Thus, the expression of maximum pressure $p_m$ is obtained as:

$$p_m = (\eta U R / C^2)\frac{6\varepsilon\sin\theta_m(2+\varepsilon\cos\theta_m)}{(2+\varepsilon^2)(1+\varepsilon\cos\theta_m)^2} \tag{4.17}$$

which occurs at a location, $\theta_m$ and is determined as:

$$\sin\theta_m = \frac{(4-5\varepsilon^2+\varepsilon^4)^{1/2}}{(2+\varepsilon^2)} \tag{4.18}$$

Load carrying capacity, attitude angle and flow rate of the lubricant in the circumferential direction, coefficient of friction or frictional loss of energy etc can be determined using the pressure distribution given by Equation (4.15). Sommerfeld assumed that the entire bearing clearance is full of lubricant and thus while determining load capacity the integration process was carried between $\theta = 0$ and $2\pi$. This is known as Sommerfeld full film boundary condition. However, Sommerfeld boundary condition leads to generation of negative pressure in the divergent portion of the lubricant film, i.e., between π to 2π which are equal in magnitude to the positive pressures generated in the convergent portion of the film between *0* and π. Pressure distribution thus obtained is shown in Fig. 4.3 schematically.

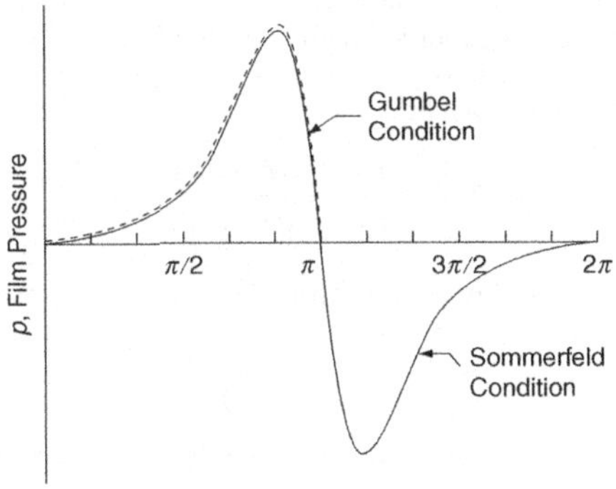


**Figure 4.3** | Circumferential Pressure Distribution in Journal Bearing Film

### 4.3.1 | Load Capacity

Components of load along the line of centers and perpendicular to it, respectively, are given as:

Load component along the line of centers of the bearing is determined as:

$$W_r = -L\int_0^{2\pi} p.R\cos\theta d\theta = W\cos\phi \tag{4.19}$$

Load component of the bearing in the direction normal to the line of centers is given by:

$$W_t = L\int_0^{2\pi} pR\sin\theta d\theta = W\sin\phi \tag{4.20}$$

Substituting for $p$ from Equation (4.15) gives

$$W_r = \frac{6\eta UR^2L}{C^2}\int_0^{2\pi}\frac{\varepsilon(2+\varepsilon\cos\theta)\sin\theta\cos\theta}{(2+\varepsilon^2)(1+\varepsilon\cos\theta)^2}d\theta \tag{4.21}$$

and

$$W_t = \frac{6\eta UR^2L}{C^2}\int_0^{2\pi}\frac{\varepsilon(2+\varepsilon\cos\theta)\sin^2\theta}{(2+\varepsilon^2)(1+\varepsilon\cos\theta)^2}d\theta \tag{4.22}$$

Integrations yield

$$W_r = 0;\ W_t = \frac{12\pi\eta U\varepsilon(R/C)^2 L}{(2+\varepsilon^2)(1-\varepsilon^2)^{1/2}} \tag{4.23}$$

Attitude angle $\phi$ is determined as

$$\tan\phi = \frac{W_t}{W_r};\ \phi = \tan^{-1}\frac{W_t}{W_r} = \frac{\pi}{2}$$

This means that displacement of the journal takes place in the direction normal to the load. This is because of inclusion of negative pressures of the divergent portion of the film due to use of full film Sommerfeld boundary condition. It is well known that lubricating oils/fluids can only with stand very small negative pressure or in other words low vacuum condition and therefore the oil film ruptures in the divergent region of the clearance space. Sommerfeld boundary condition or full film condition is therefore unrealistic and should not be used to estimate bearing performance parameters.

The total load capacity $W$ is thus given by:

$$W = W_t = \frac{12\pi\eta UL\varepsilon(R/C)^2}{(2+\varepsilon^2)(1-\varepsilon^2)^{1/2}} \tag{4.24}$$

This can be expressed in terms of a dimensionless parameter $S$ known as Sommerfeld number, $S$ as given below:

$$S = \frac{\eta N}{P}\left(\frac{R}{C}\right)^2 = \frac{(2+\varepsilon^2)(1-\varepsilon^2)^{1/2}}{12\pi^2\varepsilon} \tag{4.25}$$

where $P = \dfrac{W}{LD}$, $N$ is rps of journal

### 4.3.2 | Friction Force

Friction force at the journal surface can be determined by integrating shear stress on the journal surface over the journal surface area. Therefore, the friction force is written as:

$$F_j = \int_0^{2\pi} \tau_j LRd\theta \tag{4.26}$$

$$\tau_j = \frac{\eta U}{h} + \frac{h}{2R}\frac{dp}{d\theta}$$

Substituting $\tau_j$ in Equation (4.26) and integrating we get

$$F_j = \eta UL\left(\frac{R}{C}\right)\frac{4\pi\left(1+2\varepsilon^2\right)}{\left(2+\varepsilon^2\right)\left(1-\varepsilon^2\right)^{1/2}} \tag{4.27}$$

and coefficient of friction on the journal surface is given as:

$$\mu_j = \frac{F_j}{W} = \left(\frac{R}{C}\right)\frac{1+2\varepsilon^2}{3\varepsilon} \tag{4.28}$$

Similarly, friction force at the bearing surface can be determined as:

$$F_b = \int_0^{2\pi} \tau_b LRd\theta \tag{4.29}$$

where $\tau_b = -\frac{h}{2}\frac{dp}{d\theta}$ substituting this in Equation (4.29) and integrating the friction force at the bearing surface is obtained as:

$$F_b = \frac{\eta ULR}{C}\frac{4\pi\left(1-\varepsilon^2\right)^{1/2}}{\left(2+\varepsilon^2\right)} \tag{4.30}$$

since $F_j > F_b$ the relationship between load and friction forces is given by:

$$RF_j = RF_b + We \tag{4.31}$$

Therefore, it is appropriate to evaluate friction force at the journal surface only.

It has been seen that Sommerfeld's full film condition does not exist in practice and also leads to incorrect result, a more realistic boundary condition must be used to determine bearing performance parameters such as load capacity, oil flow rate, friction coefficient, etc.

Experimental observations of Cole and Hughes (1956) confirmed that oil film breaks down or ruptures in the divergent portion of the clearance space. Continuous oil film does not exist beyond the rupture boundary and oil film in this region breaks into finger like striations separated by air film until it builds up again at the reformation boundary near the position of oil entry at $\theta = 0$ where convergent shape of the clearance space starts. This is shown in Fig. 4.4. This phenomenon is known as cavitation. Therefore, proper film rupture boundary condition must be used to identify the film rupture boundary that satisfies the continuity of flow of lubricant in the circumferential direction.

Alternatively, the bearing performance parameters were determined assuming that film ruptures at $\theta = \pi$ where pressure becomes equal to zero or ambient. This boundary condition was suggested by Gümbel and is

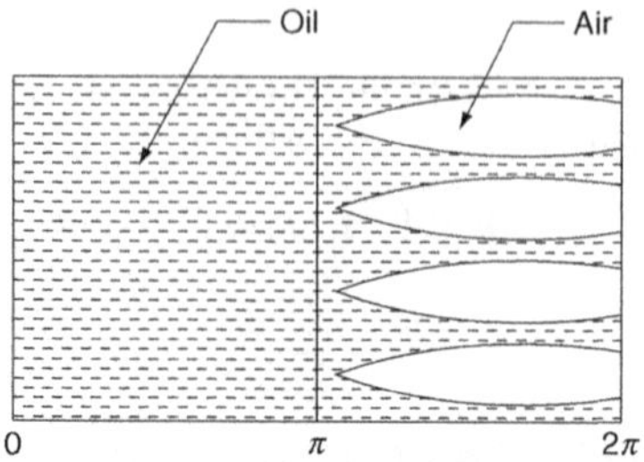


**Figure 4.4** | Open View Journal Bearing Surface Showing Cavitation Striations

more commonly known as half Sommerfeld boundary condition or π –film boundary condition. Therefore, load components in radial and normal directions are expressed, respectively, as:

$$W_r = -L\int_0^{\pi} pR\cos\theta d\theta \text{ and } W_t = L\int_0^{\pi} pR\sin\theta d\theta$$

Substituting for $p$ from Equation (4.15) and carrying out above integrations, we get

$$W_r = 12\eta U\left(\frac{R}{C}\right)^2 L\frac{\varepsilon^2}{(2+\varepsilon^2)(1-\varepsilon^2)} \tag{4.32}$$

$$W_t = 6\eta U\left(\frac{R}{C}\right)^2 L\frac{\pi\varepsilon}{(2+\varepsilon^2)(1-\varepsilon^2)^{1/2}} \tag{4.33}$$

and the total load capacity $W$ is given as:

$$W = \sqrt{W_r^2 + W_t^2};$$

Using Equations (4.32) and (4.33) load capacity of the bearing is obtained as:

$$W = 6\eta UL\left(\frac{R}{C}\right)^2 \frac{\varepsilon\left[\pi^2 - \varepsilon^2(\pi^2-4)\right]^{1/2}}{(2+\varepsilon^2)(1-\varepsilon^2)} \tag{4.34}$$

and attitude angle, $\phi$, is determined as:

$$\phi = -\tan^{-1}\left(\frac{W_t}{W_r}\right)$$

$$\phi = -\tan^{-1}\left(\frac{\pi}{2\varepsilon}\sqrt{1-\varepsilon^2}\right) \tag{4.35}$$

Although half Sommerfeld boundary condition yields satisfactory and realistic results, it does not satisfy the requirement of flow continuity in the cavitation region. Flow continuity condition is fulfilled at the film rupture or cavitation boundary satisfying following boundary conditions, i.e.,

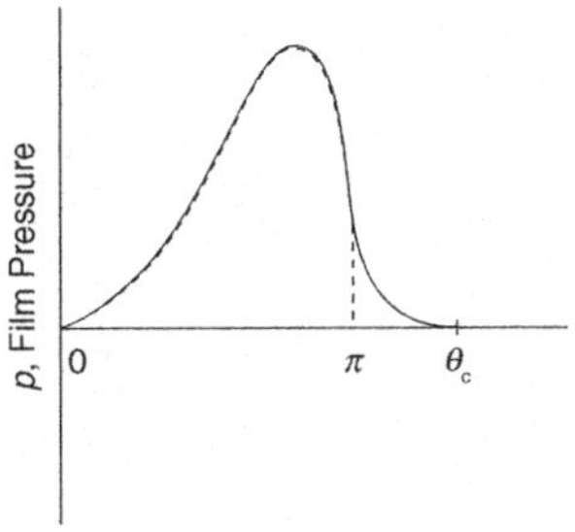


**Figure 4.5** | Circumferential Pressure Distribution for Reynolds Boundary Condition

$$p = 0 \text{ at } \theta = 0^0 \text{ and } p = \frac{dp}{d\theta} = 0 \text{ at } \theta = \theta_c$$

where $\theta_c$ is location of film rupture boundary, and usually occurs beyond $\pi$.

The above boundary condition is known as Swift–Steiber or Reynolds boundary condition and the nature of pressure distribution obtained using above boundary conditions can be seen in Fig. 4.5.

The film rupture location, $\theta_c$, can be determined following Reynolds boundary conditions at $\theta_c$, i.e.,

$$p = \frac{dp}{d\theta} = 0 \text{ at } \theta = \theta_c \tag{4.36}$$

Equation (4.15) gives the pressure variation with $\theta$ and satisfies the inlet boundary condition of $p = 0$ at $\theta = 0$. Equations (4.9) and (4.10) give

$$\frac{\partial p}{\partial \theta} = \frac{6UR\eta}{C^2}\left[\frac{1}{(1+\varepsilon\cos\theta)^2} - \frac{h^*}{C}\frac{1}{(1+\varepsilon\cos\theta)^3}\right]$$

Substituting $\frac{dp}{d\theta} = 0$ at $\theta = \theta_c$ gives the film thickness at $\theta = \theta_c$ as:

$$\frac{h^*}{C} = (1+\varepsilon\cos\theta_c) = \frac{1-\varepsilon^2}{1-\varepsilon\cos\gamma_c} \tag{4.37}$$

where $\gamma_c$ is the Sommerfeld variable at the cavitation or film rupture boundary. Pressure distribution can be expressed in terms of Sommerfeld variable $\gamma$ as:

$$p(\gamma) = \frac{6\eta UR}{C^2(1-\varepsilon^2)^{3/2}}\left\{\gamma - \varepsilon\sin\gamma - \frac{(2+\varepsilon^2)\gamma - 4\varepsilon\sin\gamma + \varepsilon^2\sin\gamma\cos\gamma}{2[1+\varepsilon\cos(\gamma_c - \pi)]}\right\} \tag{4.38}$$

putting $p(\gamma_c) = 0$ we get,

$$\varepsilon[\sin\gamma'\cos\gamma' - \gamma_c] + 2[\gamma_c\cos\gamma' - \sin\gamma'] = 0 \tag{4.39}$$

where $\gamma' = \gamma_c - \pi$

Therefore, for a given $\varepsilon$, $\gamma_c$ can be determined from Equation (4.39) and thereby $\theta_c$ is also determined. The two load components are obtained as:

$$W_r = -3\eta UL\left(\frac{R}{C}\right)^2 \frac{\varepsilon\left(1+\cos\gamma_c'\right)^2}{\left(1-\varepsilon^2\right)\left(1+\varepsilon\cos\gamma_c'\right)} \quad (4.40)$$

$$W_t = 6\eta UL\left(\frac{R}{C}\right)^2 \frac{\left(\gamma_c\cos\gamma_c' - \sin\gamma_c'\right)}{\left(1-\varepsilon^2\right)^{1/2}\left(1+\varepsilon\cos\gamma_c'\right)} \quad (4.41)$$

The load $W$, attitude angle $\phi$ and coefficient of friction $\mu_j$ can be determined as:

$$W = \frac{3\eta UL(R/C)^2}{\left(1-\varepsilon^2\right)^{1/2}\left(1+\varepsilon\cos\gamma_c'\right)}\left[\frac{\varepsilon^2\left(1+\cos\gamma_c'\right)^4}{1-\varepsilon^2} + 4\left(\gamma_c\cos\gamma_c' - \sin\gamma_c'\right)^2\right]^{1/2} \quad (4.42)$$

Attitude angle is given as:

$$\tan\phi = -\frac{2\left(1-\varepsilon^2\right)^{1/2}\left(\sin\gamma_c' - \gamma_c\cos\gamma_c'\right)}{\varepsilon\left(1+\cos\gamma_c'\right)^2} \quad (4.43)$$

Friction factor is obtained as:

$$\left(\frac{R}{C}\right)\mu = \frac{\varepsilon\sin\phi}{2} + \frac{2\pi^2 S}{\left(1-\varepsilon^2\right)^{1/2}} \quad (4.44)$$

where the Sommerfeld number $S$ is defined as:

$$S = \frac{\eta N(R/C)^2}{P} \text{ and } P = \frac{W}{2LR} \text{ and it is obtained as:}$$

$$S = \frac{\left(1-\varepsilon^2\right)^{1/2}\left(1+\varepsilon\cos\gamma_c'\right)}{3\pi\left[\dfrac{\varepsilon^2\left(1+\cos\gamma_c'\right)^4}{\left(1-\varepsilon^2\right)+4\left(\gamma_c\cos\gamma_c' - \sin\gamma_c'\right)^2}\right]^{1/2}} \quad (4.45)$$

## 4.4 | Boundary Conditions

We have seen that usually Reynolds or Swift–Steiber (1931, 1933) boundary condition is used for film rupture near the divergent portion of the film. It has been observed that divergent portion near the film formation region close to the lubricant entry causes upstream cavitations too.

Therefore, correct boundary condition at the start of the hydrodynamic film in divergent space formulated by Floberg (1961), i.e., film reformation boundary condition as given below should be used to satisfy flow continuity requirement at the inlet of the oil.

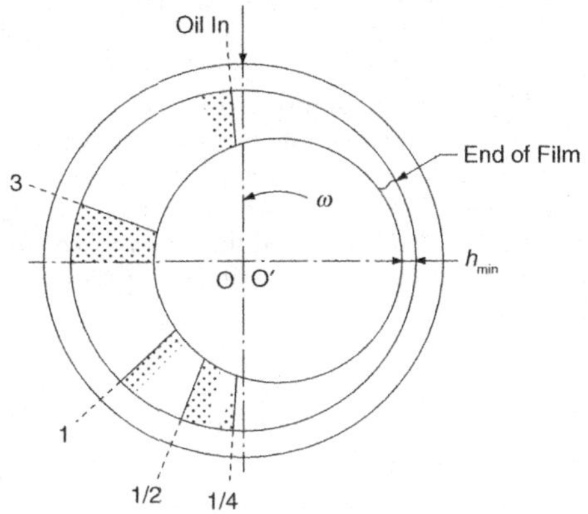


**Figure 4.6** | Beginning of Film for Various L/D Ratios $\theta_1 = f\{(L/D), \varepsilon, \theta_s, \theta_E\}$ [Pinkus, ASME JOT, 1987]

$$\frac{Uh_1}{2} - \frac{h_1^3}{12\eta}\left\{\frac{\partial p}{\partial x} - \frac{\partial p}{\partial y}\left(\frac{dx}{dy}\right)\right\}\Bigg|_1 = \frac{Uh_s}{2} \tag{4.46}$$

The location of starting line will thus depend on *L/D* ratio, $\theta_s$, $\varepsilon$ and $\theta_E$ as seen in Fig. 4.6, which shows how widely such starting lines may differ from each other. It is known that Reynolds or Swift–Steiber boundary conditions at the film rupture do not predict the subambient pressures that occur just prior to cavitation boundary. Since, most liquids can withstand only a small magnitude of tensile stresses or subambient pressures, and therefore a very small region of subambient pressures will exist in reality. The boundary conditions proposed by Floberg and Coyne–Elrod (1971) can predict subambient pressure regions too. Pressure profiles based on these boundary conditions are of the type shown in Fig. 4.7.

The boundary condition given by Coyne–Elrod (1970) at the film-cavity interface is as given below:

$$p = -\frac{\sigma}{\gamma} + \Delta p \tag{4.47}$$

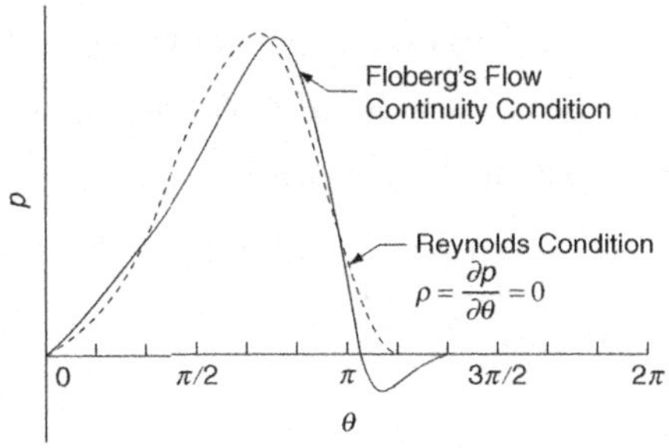


**Figure 4.7** | Circumferential Pressure Distributions for Various Cavitation Boundary Conditions

where $\gamma$ is the radius of curvature of the interface and $\sigma$ is the surface tension of the lubricant. The pressure correction $\Delta p$ is dependent on the surface tension parameter $\eta U/\sigma$ and is negligible for small values of surface tension parameter.

It was observed by Cole and Hughes (1956) in their experiment that in the cavitation region, the oil flow breaks into oil streams in the form of narrow strips adhering to the runner. Cavities between strips are occupied by air or gases at constant pressure equal to saturation pressure or vapor pressure of dissolved gases. It is usually referred to as gaseous cavitations. Subcavity pressures are usually negligible and the pressure is equal to vapor pressure in the cavity. Floberg (1965) suggested the following condition at the end of the pressure build up and in the cavitations region, i.e., the oil flow leaving the upstream film region:

$$\theta_{cav}^{down} = a\frac{Uh}{2} \tag{4.48}$$

where $a$ is the fractional width of the oil in the cavitations zone.

Therefore, for flow balance

$$(1-a)\frac{Uh}{2} - \frac{h^3}{12}\frac{\partial p}{\partial x} = 0 \tag{4.49}$$

The above equation is satisfied only for

$$\frac{\partial p}{\partial x} = 0 \text{ and } a = 1, \text{ where as for } p \geq p_{cav} \text{ and } 0 < a < 1$$

which means that oil fills the whole width at the film cavity interface. However, there are usually a finite number of oil strips in the cavitation zone separated by gaseous regions. According to Floberg, the following relationship is applicable at both film region and cavitation region:

$$\left(\frac{\partial p}{\partial x} - \frac{\partial p}{\partial y}\cdot\frac{\partial x}{\partial y}\right) = \frac{6\eta U}{h^2} \tag{4.50}$$

## 4.5 | Short Bearing Solution

If the $L/D \ll 1$, i.e., the length of the bearing is very small in comparison to its diameter, the bearing is referred to as a short bearing. This results in increased axial flow and decreased circumferential flow. Usually, when $\frac{L}{D} \cong 0.25$, the axial pressure gradient is far greater than circumferential pressure gradient which is approximately zero. Therefore, axial flow dominates. Dubois and Ocvirk (1953) assumed that circumferential pressure gradient $\frac{\partial p}{\partial \theta} = 0$ and proposed a theoretical solution for full journal bearing.

With the above assumption, the Reynolds equation reduces to:

$$\frac{\partial}{\partial y}\left(h^3\frac{\partial p}{\partial y}\right) = 6\eta U\frac{dh}{dx} \tag{4.51}$$

Substituting $x = R\theta$, the above equation becomes

$$\frac{\partial}{\partial y}\left(h^3\frac{\partial p}{\partial y}\right) = 6\eta U\frac{dh}{Rd\theta} \tag{4.52}$$

Integrating Equation (4.52) twice with respect to y pressure distribution in the lubricant film is obtained as:

$$p = \frac{6\eta U}{Rh^3}\frac{dh}{d\theta}\frac{y^2}{2} + A_1 y + A_2 \tag{4.53}$$

where constants $A_1$ and $A_2$ has to be determined by applying the following boundary conditions:

$$p = 0 \text{ at } y = \pm L/2$$

Pressure distribution is thus given by:

$$p = \frac{3\eta U}{RC^2}\left(\frac{L^2}{4} - y^2\right)\frac{\varepsilon \sin\theta}{(1+\varepsilon\cos\theta)^3} \tag{4.54}$$

The load capacity can be determined by integrating the pressure over the fluid film area for the film extent of $\pi$. Radial component of the load capacity is determined as:

$$W_r = -2\int_0^{\pi}\int_0^{L/2} p\cos\theta R d\theta dy \tag{4.55}$$

$$W_r = -\frac{\eta U L^3}{2C^2}\int_0^{\pi}\frac{\varepsilon\sin\theta\cos\theta}{(1+\varepsilon\cos\theta)^3}d\theta \tag{4.56}$$

using Sommerfeld substitution and integrating the radial component of the load capacity is obtained as:

$$W_r = -\frac{\eta U L^3}{C^2}\frac{\varepsilon^2}{(1-\varepsilon^2)^2} \tag{4.57}$$

Load component perpendicular to the line of centers is given by:

$$W_t = 2\int_0^{\pi}\int_0^{L/2} p\sin\theta \, R \; d\theta \, dy \tag{4.58}$$

and is obtained substituting $p$ given by Equation (4.15) and integrating using Sommerfeld substitution as:

$$W_t = \frac{\eta U L^3}{4C^2}\frac{\pi\varepsilon}{(1-\varepsilon^2)^{3/2}} \tag{4.59}$$

The load capacity $W$ is

$$W = \sqrt{W_r^2 + W_t^2} = \frac{\eta U L^3}{4C^2}\frac{\varepsilon}{(1-\varepsilon^2)^2}\left[\pi^2(1-\varepsilon^2) + 16\varepsilon^2\right]^{1/2} \tag{4.60}$$

and the attitude angle $\phi$ is

$$\phi = \tan^{-1}\frac{W_t}{W_r} = \tan^{-1}\left\{\frac{\pi}{4}\frac{(1-\varepsilon^2)^{1/2}}{\varepsilon}\right\} \tag{4.61}$$

The friction on the journal is

$$F_j = \int_0^{2\pi} \eta \frac{U}{h} LR d\theta = \frac{\eta ULR}{C} \frac{2\pi}{\left(1-\varepsilon^2\right)^{1/2}} \tag{4.62}$$

and normalized coefficient of friction is

$$\left(\frac{R}{C}\right)\mu = \frac{2\pi^2 S}{\left(1-\varepsilon^2\right)^{1/2}} \tag{4.63}$$

where $S = \frac{\eta N}{P}\left(\frac{R}{C}\right)^2$

The volume flow rate of the lubricant from bearing sides is

$$Q = -2\int_0^{\pi} \frac{Rh^3}{12\eta} \frac{\partial p}{\partial y}\bigg|_{y=\frac{L}{2}} d\theta = \varepsilon ULC \tag{4.64}$$

Short bearing theory is applicable for low $L/D$ ratios only usually $L/D \leq 0.25$. However, the solution can be extended up to $L/D = 0.5$ when very accurate results are not required.

Load capacity of the long bearing solution can be compared with the load capacity obtained from short bearing theory and is given below for $\pi$ film:

$$\frac{W(\text{Short bearing})}{W(\text{Long bearing})} = \left(\frac{L}{D}\right)^2 \frac{\left(2+\varepsilon^2\right)\left[16\varepsilon^2 + \pi^2\left(1-\varepsilon^2\right)\right]^{1/2}}{6\left(1-\varepsilon^2\right)\left[\pi^2 - \varepsilon^2\left(\pi^2-4\right)\right]^{1/2}} \tag{4.65}$$

It can be seen that the ratio is dependent on $\frac{L}{D}$ and $\varepsilon$. It is known that short bearing theory gives better estimate for bearings with $\frac{L}{D} < 0.5$, whereas long bearing theory generally overestimates load capacity and should be used for $\frac{L}{D} > 2.0$.

## 4.6 | Oil Flow

Oil flow generally consists of two components:

1. Oil flow from the sides of the bearing due to hydrodynamic pressure of the oil film and can be calculated using hydrodynamic pressure gradients due to relative rotation of the journal. This is referred to as hydrodynamic flow or circumferential flow $Q_c$.
2. Usually the lubricant is fed to the bearing through a hole or a groove at a pressure higher than the atmosphere pressure. Pressure feeding of oil results in flow of oil through the bearing sides also. It is more through the region where film thickness is high, i.e., around the region of maximum film thickness and gradually reduces as the feed pressure effect reduces in the region of minimum film thickness. This is referred so as hydrostatic flow or flow due to feed pressure ($Q_p$).

Therefore, total oil flow rate is the sum of circumferential flow and feed pressure flow.

$$Q = Q_c + Q_p$$

### 4.6.1 | Circumferential Oil Flow

The oil flow through the bearing sides or side leakage is determined by the axial pressure gradient and can be determined using short bearing theory as:

$$Q_c = Q_s = 2\int_{\theta_i}^{\theta_c} \frac{h^3}{12\eta}\frac{\partial p}{\partial y}\bigg|_{\pm L/2} Rd\theta = ULC\varepsilon \tag{4.73}$$

where $\theta_i$ angular coordinate at which the pressure build up begins

$\theta_c$ angular coordinate at which the film ruptures or cavitations begin

Alternately, it can also be determined as the difference of the oil flow rate entering into the clearance space at the beginning of the film and that flowing out at the film rupture boundary in the circumferential direction. Therefore, it can be expressed as:

$$Q_c = \int_{-L/2}^{L/2}\left(Q_x d_y\right)_{\theta=\theta_i} - \int_{-L/2}^{L/2}\left(Q_x d_y\right)_{\theta=\theta_c} \tag{4.74}$$

Since the pressure gradient at the start of the film pressure, i.e., $\frac{\partial p}{\partial \theta}\Big|_{\theta=\theta_i} \cong 0$ and pressure gradient at the film rupture boundary, i.e., $\frac{\partial p}{\partial \theta}\Big|_{\theta=\theta_c}$ is also zero, the equation can be written as:

$$Q_c = \left(\frac{UhL}{2}\right)_{\theta=\theta_i} - \left(\frac{UhL}{2}\right)_{\theta=\theta_c} \tag{4.75}$$

Since $\qquad h(\theta_i) = C(1+\varepsilon\cos\theta_i)$ and $h(\theta_c) = C(1+\varepsilon\cos\theta_c)$

$$Q_c = \frac{ULC\varepsilon}{2}\left(\cos\theta_i - \cos\theta_c\right) \tag{4.76}$$

The film rupture takes place at a location down stream of minimum film thickness, therefore $\theta_c = \pi + \gamma$ where $\gamma$ is measured from the location of minimum film thickness.

For $\theta_i = 0$ $\theta_c = \pi$, i.e., for $\pi$ film, the flow rate is obtained as:

$$Q_c = ULC\varepsilon \tag{4.78}$$

### 4.6.2 | Feed Pressure Flow in Journal Bearings

Bearing and the groove geometry is shown in Fig. 4.8. Feed pressure flow in plain journal bearings for various groove geometries have been determined by Martin and Lee (1982) as feed pressure flow equations by curve fitting of the computed results from an accurate finite difference model. Very useful data on flow from oil holes, few selected axial groove shapes have also been presented prior to Martin and Lee's work by Wilock and Booser [1975] and Hirano and Shodai [1958]. Various equations for flow due to feed pressure were also given by Shaw and Macks [1949] and Cameron [1981].

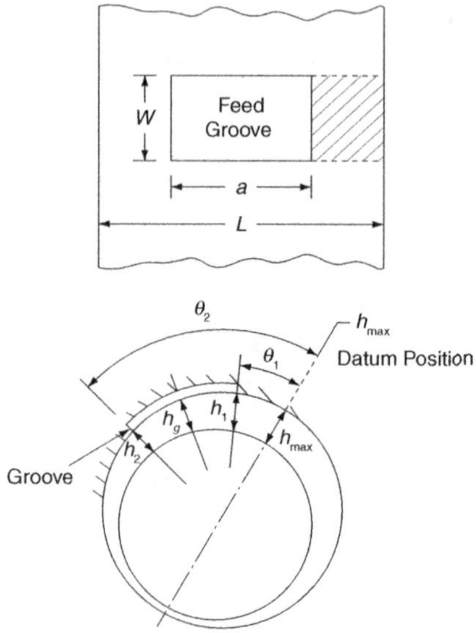


**Figure 4.8** | Groove Geometry (Martin and Lee, ASLE, 1982)

Martin and Lee (1982) developed equation for feed pressure flow through a rectangular groove at any position in the bearing and for any journal eccentricity ratio and angular position of the journal center given as below:

- Rectangular groove (concentric journal)

$$\frac{Q_p \eta}{C^3 p_f} = \frac{1}{3}\frac{\left(1.25 - 0.25\dfrac{a}{L}\right)}{\left(\dfrac{L}{a}-1\right)^{0..333}} + \frac{1}{3}\frac{b/L}{\left(1-\dfrac{a}{L}\right)} \tag{4.79}$$

where $p_f$ = feed pressure, a-axial length of the groove, b = circumferential length of the groove, $C$ = radial clearance, $L$ = bearing length.

- Eccentric position of journal for groove extent up to 270°

$$\frac{Q_p \eta}{C^3 p_f} = \left[\frac{1.25 - 0.25a/L}{6\left(\dfrac{L}{a}-1\right)^{0..333}} \cdot f_1\right] + \left[\frac{D/L}{6\left(1-\dfrac{a}{L}\right)} \cdot f_2\right] \tag{4.80}$$

where $f_1 = (1+\varepsilon\cos\theta_1)^3 + (1+\varepsilon\cos\theta_2)^3$

and $$f_2 = \left[\theta + 3\varepsilon \sin\theta + \varepsilon^2\left(1.5\theta + \frac{3}{4}\sin 2\theta\right) + \varepsilon^3\left(\sin\theta - \frac{\sin^3\theta}{3}\right)\right]_{\theta_1}^{\theta_2}.$$

However, there is a small loss of accuracy at large angular-groove extents. For bearings of $\frac{L}{D}$ ratio up to 0.5 above equations can be used with confidence for any groove extent up to 270°. For long bearings of $\frac{L}{D}$ ratio equal to 1.0, the equation can be applied upto a groove extent of 180°. For steadily loaded bearings, the angular extent of the groove is more commonly less than 30°. For circumferentially grooved eccentric journal bearings, $\theta_1 = 0^0$ and $\theta_2 = 360^0$, the groove flow is given as:

$$\frac{Q_p\eta}{C^3 p_f} = \frac{\pi D}{3(L-a)}(1+1.5\varepsilon^2) \tag{4.81}$$

Dimensionless feed pressure flow through a single circular oil hole is given as:

$$\frac{Q_p\eta}{C^3 p_f} = 0.675\left(\frac{h_g}{C}\right)^3\left(\frac{d_h}{L}+0.4\right)^{1.75} \tag{4.82}$$

where $d_h$ = hole diameter, $h_g$ is film thickness at the groove mid point, $L$ is bearing length

Cameron had also developed similar equations for feed pressure flow through feed grooves (18) which are given below:

- Single circular oil hole:

$$\frac{Q_p\eta}{C^3 p_f} = \frac{1}{12}\left(\frac{h_g}{C}\right)^3\left(1.2+\frac{11d_h}{l}\right) \tag{4.83}$$

- Rectangular feed groove of small angular extent

$$\frac{Q_p\eta}{C^3 p_f} = \left[\frac{1}{3}\frac{M}{6\left(\frac{l}{a}-1\right)^{0.333}}\right] + \left[\frac{1}{3}\left(\frac{b}{L}\right)\frac{M}{6(1-a/l)}\cdot f_2\right] \tag{4.84}$$

- Expressions for rectangular feed groove of large angular extent for zero eccentricity are not available
- Circumferential oil groove (360°)

$$\frac{Q_p\eta}{C^3 p_f} = \pi D(1+1.5\varepsilon^2)/(3l) \tag{4.85}$$

where $M = (\frac{h_g}{C})^3$, $l$ is sum of the axial land lengths

For two grooves at 90° to load line

$$\mathrm{M} = \left(\frac{h_{g1}}{C}\right)^3 + \left(\frac{h_{g2}}{C}\right)^3 \tag{4.86}$$

## 4.7 | Hydrodynamic Thrust Bearings

Hydrodynamic thrust bearings meant to carry the axial load of a rotor in its simplest design comprises of two inclined plane surfaces that slide relative to each other. The bearing surface is usually fixed while the rotor surface moves at a speed relative to the bearing surface. The bearing surface may be a fixed inclined surface or otherwise it can be a plane surface pivoted at a point through which the center of pressure or the load passes. These are also called fixed inclined pad or pivoted pad thrust bearings. These pads can also be of the shape of a sector in which case the bearing is called sector pad thrust bearing as shown in Figs 4.9 and 4.10. Various profiles may also be provided to the pad surface of thrust bearings, viz., exponential, cycloid, catenary, polynomial, and parallel steps which may improve the performance of the bearings.

### 4.7.1 | Inclined Pad Thrust Bearing

The geometry of an inclined pad thrust bearing is shown in Fig 4.11. To generate pressure, the runner has to feed oil into a converging wedge. The expression of film thickness can be written as:

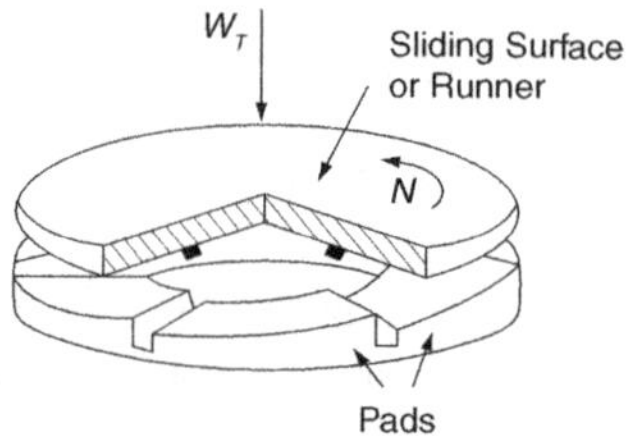


**Figure 4.9** | Tapered Land (Fixed Pad)

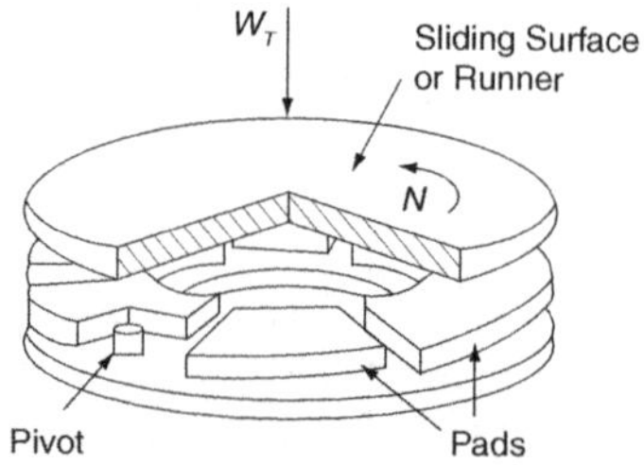


**Figure 4.10** | Tilting Pad on Levelling Linkage

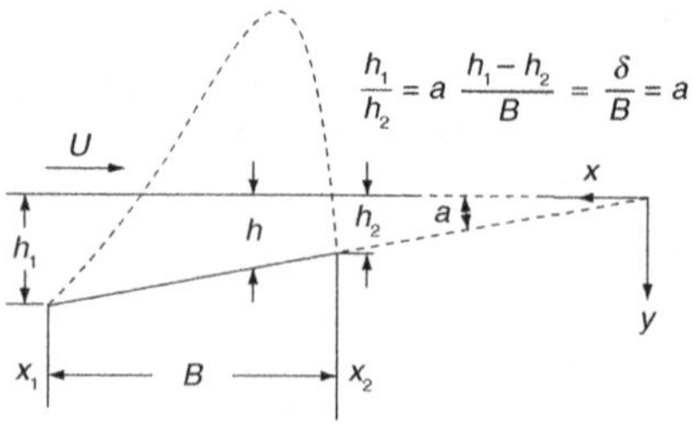


**Figure 4.11** | Inclined Pad Thrust Bearing

$$h(x) = (h_1 - \alpha x) \text{ where } \alpha = \frac{h_1 - h_2}{B}$$

The Reynolds equation for the plane slider can be written in one dimensional form for infinite width of the plane as:

$$\frac{d}{dx}\left(h^3 \frac{dp}{dx}\right) = 6\eta U \frac{dh}{dx} \tag{4.87}$$

which can be expressed after integrating with respect to $x$ as:

$$\frac{dp}{dx} = 6\eta U\left(\frac{h - \bar{h}}{h^3}\right) \tag{4.88}$$

where $\bar{h}$ is the film thickness at which maximum pressure is obtained, i.e.,

$$\frac{dp}{dx} = 0 \text{ at } h = \bar{h}$$

The boundary conditions at the inlet and exit of the pad are as:

$$p = 0 \text{ at } x = 0 \text{ and } p = 0 \text{ at } x = B$$

Integrating Equation (4.88) with respect to $x$, the pressure distribution is obtained as:

$$p = 6\eta U\left[\int \frac{1}{(h_1 - \alpha x)^2} dx - \int \frac{\bar{h}}{(h_1 - \alpha x)^3} dx\right] + A \tag{4.89}$$

$A$ is the constant of integration which is determined by substituting boundary conditions. After integrating and applying the boundary conditions, one can get the pressure distribution as:

$$p = \frac{6\eta UB}{h_2^2} \frac{1}{(a^2 - 1)}\left[\left(\frac{h_1}{h_1 - \alpha x} - 1\right)\left(1 - \frac{h_2}{h_1 - \alpha x}\right)\right] \tag{4.90}$$

where

$$\bar{h} = \frac{2ah_2}{(a+1)}; \; a = \frac{h_1}{h_2} \tag{4.91}$$

The maximum pressure in the pad is determined from Equation (4.90) as:

$$p_m = \frac{3\eta U(a-1)^2}{2\alpha a(a+1)} = \frac{3\eta UB(h_1 - h_2)}{2h_1 h_2 (h_1 t h_2)} \tag{4.92}$$

Load capacity of the pad obtained using the pressure distribution over the pad length is shown in Fig. 4.12 (b) for various values of film thickness ratio '$a$'.

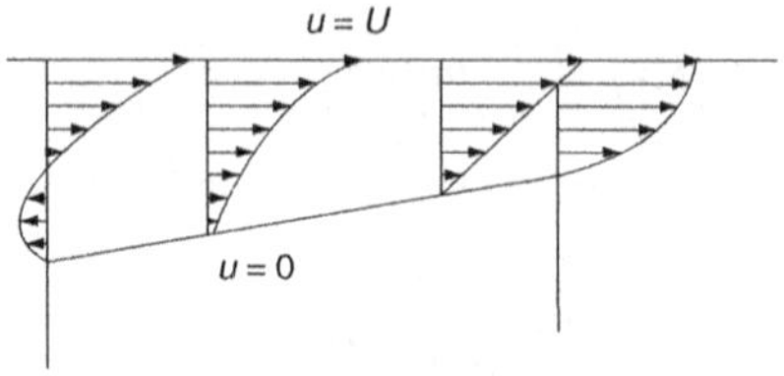


**Figure 4.12(a)** | Velocity Distribution Across the Film of Inclined Pad Thrust Bearing

## 4.7.2 | Bearing Performance Parameters

- Load carrying capacity is calculated using the pressure distribution given by Equation (4.90) as:

$$W = L\int_0^B p dx \tag{4.93}$$

where $L$ is the width of the pad. Substituting $p$ given by Equation (4.90) and carrying out the integration load capacity is obtained as:

$$W = \frac{6\eta UB^2L}{h_2^2}\frac{1}{(a-1)^2}\left[\ln\ a - \frac{2(a-1)}{a+1}\right] \tag{4.94}$$

- Dimensionless load capacity,

$$\bar{W} = \frac{Wh_2^2}{6\eta UB^2L}$$

- Friction force on the moving plane is determined as:

$$F = L\int_0^B \tau\ dx \tag{4.95}$$

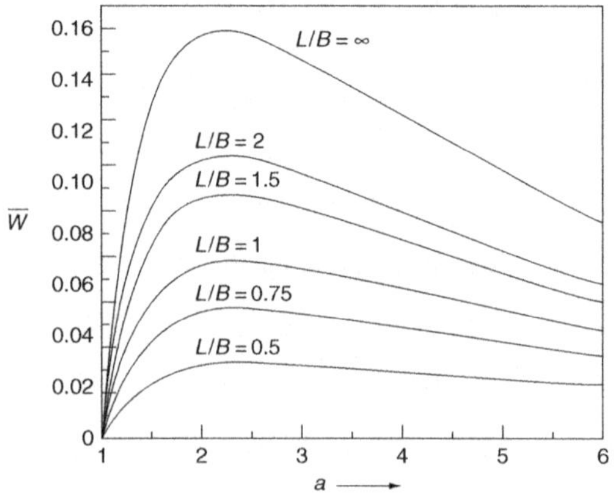


**Figure 4.12 (b)** | Dimensionless Load Capacity Versus Film Thickness Ratio of Inclined Pad Thrust Bearing

where $\tau$ is the shear stress on the moving surface which is determined as:

$$\tau = \frac{h}{2}\frac{dp}{dx} + \frac{\eta U}{h} \tag{4.96}$$

Substituting $\tau$ in Equation (4.95) and integrating one gets friction force as:

$$F = \frac{\eta UBL}{h_2}\frac{1}{(a-1)}\left[4\ln a - \frac{6(a-1)}{a+1}\right] \tag{4.97}$$

The coefficient of friction can be determined as:

$$\mu = \frac{F}{W} = \frac{h_2}{B}\left[\frac{2(a^2-1)\ln a - 3(a-1)^2}{3(a+1)\ln a - 6(a-1)}\right] \tag{4.98}$$

- Center of pressure is the location along the pad at which the resultant force acts. If $x_{cp}$ is the distance of the center of pressure from the inlet then it can be expressed as:

$$x_{cp} = \frac{1}{W}\int_0^B pxdx \tag{4.99}$$

Substituting $p$ from Equation (4.90) and integrating $x_{cp}$ is obtained in the dimensionless form as:

$$\frac{x_{cp}}{B} = \frac{a(2+a)\ln a - (a-1)[2.5(a-1)+3]}{(a^2-1)\ln a - 2(a-1)^2} \tag{4.100}$$

Film thickness ratio '*a*' for which the load capacity is maximum can be determined by differentiating expression of load capacity given by Equation (4.94) with respect to $a$ and equating it to zero, i.e., $\frac{dW}{da} = 0$. This gives a = 2.18. Using Equation (4.94) and substituting for a = 2.18 in Equation (4.94) one obtains $W_{max}$ as:

$$W_{max} = 0.1602\frac{\eta ULB^2}{h_2^2} \tag{4.101}$$

- Velocity profile and oil flow rate can be determined as:
  Flow velocity is given below as:

$$u = \frac{z(h-z)}{2\eta}\frac{dp}{dx} + \frac{U(h-z)}{h} \tag{4.102}$$

Velocity profiles obtained using above equation is shown in Fig. 4.12 (a).

- Oil flow rate per unit width of the bearing can be expressed as:

$$Q_x = \frac{h^3}{12\eta}\frac{\partial p}{\partial x} + \frac{Uh}{2} \tag{4.103}$$

Oil flow rate can be evaluated at the location, where $\frac{dp}{dx} = 0$ i.e. at $h = \bar{h}$ as:

$$Q_x = \frac{U\bar{h}}{2} = \frac{2ah_2U}{(a+1)} \tag{4.104}$$

### 4.7.3 | Pad Shape Variations

Lord Rayleigh in 1918 proposed step bearing consisting of two parallel regions which he showed to give biggest load carrying capacity (Fig. 4.13). It is commonly known as Rayleigh step bearing. The pressure gradients in two parallel regions are constant and maximum pressure occurs at the step (Fig. 4.13).

To maintain continuity of flow throughout the bearing

$$Q_x = \frac{Uh_1}{2} - \frac{h_1^3}{12\eta}\left(\frac{\partial p}{\partial x}\right)_1 = \frac{Uh_2}{2} - \frac{h_1^3}{12\eta}\left(\frac{\partial p}{\partial x}\right)_2 \tag{4.105}$$

where $\left(\frac{\partial p}{\partial x}\right)_1 = p_m / B_1$ and $\left(\frac{\partial p}{\partial x}\right)_2 = p_m / B_2$

Substituting the values of $\left(\frac{\partial p}{\partial x}\right)_1$ and $\left(\frac{\partial p}{\partial x}\right)_2$ as in Equation (4.105), $p_m$ is evaluated as:

$$p_m = \frac{6U\eta}{h_2^2}\left(\frac{h_1}{h_2} - 1\right) \Bigg/ \left[\frac{\left(h_1/h_2\right)^3}{B_1} + \frac{1}{B_2}\right]$$

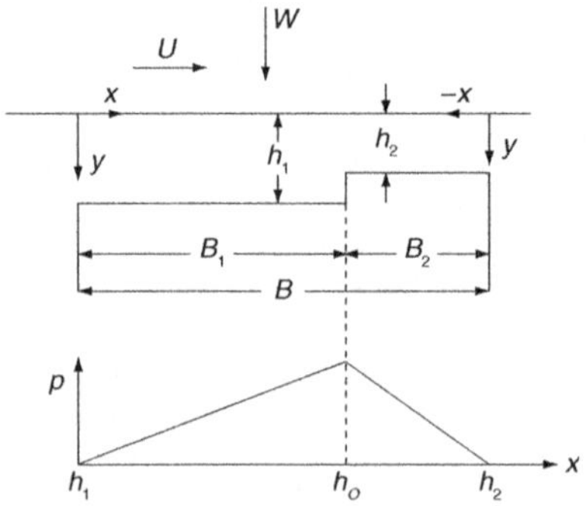


**Figure 4.13** | Rayleigh Step Thrust Bearing

Load capacity per unit width is determined as the area under the triangle as:

$$\frac{W}{L} = \frac{p_m B}{2} \tag{4.106}$$

Both $\frac{h_1}{h_2}$ and $\frac{B_1}{B_2}$ ratios can be varied and optimum configuration works out as $\frac{h_1}{h_2} = 1.87$ and $\frac{B_1}{B_2} = 2.549$.

The optimum load capacity of bearing per unit width of the pad is determined as:

$$\frac{W}{L} = 0.206\eta UB^2 / h_2 \tag{4.107}$$

Thus, it is seen that for the same length of the pad, a Rayleigh step bearing gives higher load capacity than fixed inclined pad bearing. However, this advantage is lost in the case of square pad bearing when $L = B$.

It can be shown that for square inclined pad bearing the optimum configuration gives load coefficient of 0.072, whereas square Rayleigh step results in load coefficient of 0.0725. There is practically no difference between the two.

An alternative configuration known as tapered land bearing as shown in Fig. 4.14 is also often used. The optimum configuration for infinite width bearing results in the load coefficient of 0.192 for optimum configuration of $\frac{h_1}{h_2} = 2.25$ and $\frac{B_1}{B_2} = 0.8$.

Any design problem in bearings usually begins with certain given data, e.g., diameter of the journal, load acting, and operating speed. Design solution boils down to choosing a length to diameter ratio, selecting an oil of appropriate viscosity, proper clearance ratio to ensure a minimum film thickness during operation, and achieving a proper heat balance to obtain an effective oil temperature during operation to ensure an effective viscosity of the oil to develop necessary load carrying capacity satisfying minimum film thickness requirement. To ensure a proper heat balance, the following relationship must be satisfied:

Heat generated due to viscous friction = Heat carried by the outgoing lubricant + Heat dissipation to surrounding media including heat conduction to the solids.

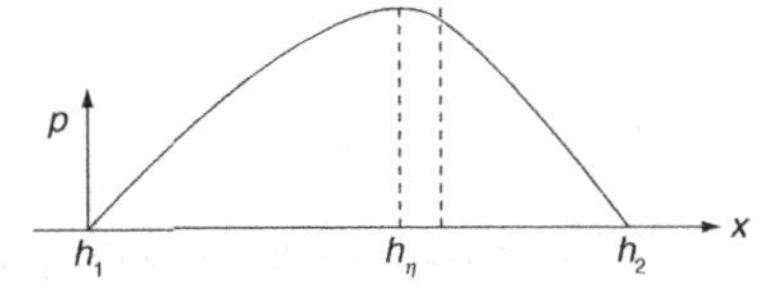


**Figure 4.14** | Tapered Land Thrust Bearing

## Examples

**E.4.1** A journal bearing is operating under following operating conditions: Journal diameter = 20 cm, bearing length = 10 cm and journal speed = 600r.p.m. Clearance ratio may be chosen between 0.5, 1, 1.5, and 2.0 mm/m. Select a clearance ratio and determine load carrying capacity, oil flow rate, power loss, and effective operating temperature when the oil entry temperature is 55°C while the viscosity of the oil at 38°C is 100cS and at 100°C is 12cS. Specific gravity of the oil is 0.9. The bearing is designed to run at an eccentricity ratio 0.6.

***Solution:***

**Given data:** Journal diameter, D = 20 cm, bearing length, L = 10 cm. Since L/D ratio = 0.5, $\nu = 100cS$ at $38^0C$ and $12cS$ at $100^0C$. Short bearing theory may be used to seek solution to this problem.

Using ASTM relationship given in Chapter 2, kinematic viscosity of the oil is determined at inlet temperature of 55°C. Thus,

$$\log_{10}\log_{10}(\nu + 0.6) = n\log_{10} T + c$$

$n$ and $c$ are constants to be determined using given data. T is absolute temperature in degree Kelvin and $\nu$ is kinematic viscosity in centistokes. Therefore,

$$\log_{10}\log_{10}(100 + 0.6) = n\log_{10} 311 + c$$

$$\log_{10}\log_{10}(12 + 0.6) = n\log_{10} 373 + c$$

Solving the above equations, $n$ and $c$ are determined as $n = -3.2$ and $c = 8.28$

Kinematic viscosity of oil is determined at T = 328°K, i.e., 55°C from the following equation $\log_{10}\log_{10}(\nu + 0.6) = -3.2\log_{10} 328 + .28$, kinematic viscosity is obtained as:

$$\nu = 48.76\ cS = 48.76 \times 10^{-6}\frac{m^2}{s}$$

Absolute viscosity of the oil at 55°C can be determined from the above value of $\nu$ as

$$\eta = \nu\rho = 48.76 \times 10^{-6} \times 900 = 0.044 \text{ Pas}$$

For short bearing approximation load carrying capacity $W$ is determined as:

$\varepsilon = 0.6$, Selecting $\frac{C}{R} = 2 \times 10^{-3}$, $C = 2 \times 10^{-4}$, shaft speed, $N = 600$ rpm

$$U = \frac{2\pi NR}{60} = \frac{2\pi \times 600 \times 0.1}{60} = 6.28 \text{ m/s}$$

- $$W = \frac{\eta U L^3 \varepsilon\sqrt{\pi^2 - \varepsilon^2(\pi^2 - 16)}}{4C^2(1-\varepsilon^2)^2}$$

$$W = \frac{0.044 \times 6.28 \times 0.1 \times 0.1 \times 0.1 \times 0.6 \times \sqrt{9.87 - 0.36(9.87 - 16)}}{4 \times 0.2 \times 0.2 \times 10^{-6} \times (1 - 0.36)^2} = 8.79 \times 10^3 \text{N}$$

- Oil flow rate,
  $Q = \varepsilon ULC = 0.6 \times 6.28 \times 0.1 \times 0.2 \times 10^{-3} = 0.07536 \times 10^{3}$ Nm$^3$/sec

  Frictional power loss, $P = FU$ where $F$ is friction force at the journal surface and using short bearing theory it is determined as

- $F = \dfrac{\eta ULR}{C} \times \dfrac{2\pi}{\sqrt{1-\varepsilon^2}} = \dfrac{0.044 \times 6.28 \times 0.1 \times 0.1}{0.2 \times 10^{-3}} \times \dfrac{2\pi}{\sqrt{1-0.36}} = 0.1355 \times 10^{3}$ N

- $P = 0.1355 \times 10^{3} \times 6.28$ W $=$ 0.85 kW

- Effective operating temperature ($T_e$) is determined as:
  Rise in temperature of the oil $(\Delta T)$ is determined using heat balance relationship. Therefore,
  Heat generated $= \dfrac{FU}{J}$, Heat carried by the flowing lubricant $= \rho Q c_p \Delta T$, Thus

$$\Delta T = \frac{FU}{J \rho Q c_p}$$

where $J$ is mechanical equivalent of heat and $c_p$ is specific heat of lubricant at constant pressure. In SI unit $J = 1$ *Nm* and $c_p = 2000$ *Joule/kg/°K*
Temperature rise of the lubricant is determined as

$$\Delta T = \frac{FU}{J \rho Q c_p} = \frac{0.1355 \times 10^{3} \times 6.28}{0.9 \times 10^{3} \times 0.07536 \times 10^{-3} \times 2000} = 6.25\,^{0}\text{C}$$

using 0.8 rule effective operating temperature is determined as

$$T_e = T_i + 0.8\Delta T = 55 + 0.8 \times 6.25 = 60^{0}\text{C}$$

**E.4.2** A full journal bearing of width 20 cm with a journal of diameter 10 cm has diametric clearance of 100 micro meters. The journal rotates at 1200 rpm. The absolute viscosity of lubricant at 20° C is 0.04 Pas. For an eccentricity ratio of 0.6, determine the minimum film thickness, load carrying capacity, attitude angle, Sommerfeld number, friction factor. Mass density, and specific heat of the oil at constant pressure may be taken as 900 kg/m$^3$ and 2.0 J/g/$^0$K, respectively.

***Solution:***

**Given data:** Journal diameter, D = 10 cm, Bearing length, L = 20 cm, diametric clearance = 100 $\mu m$, viscosity of oil at 20° C = 0.04 Pas, oil density, $\rho = 900$ kg/m$^3$, specific heat of oil, $c_p = 2.0$ J/g/$^0$K. Since L/D ratio = 2.0 long bearing approximation may be used.

Radial clearance, C = 50 $\mu m$, Eccentricity ratio, $\varepsilon = 0.6$, R = 5.0cm, N = 1200 rpm, $\dfrac{R}{C} = 10^3$

Minimum film thickness $h_{\min} = C(1-\varepsilon) = 50 * 10^{-6}(1-0.6) = 20 * 10^{-6}$ or 20 $\mu m$

Load carrying capacity $W = \dfrac{6\eta UL}{(2+\varepsilon^2)(1-\varepsilon^2)} \left(\dfrac{R}{C}\right)^2 \varepsilon \sqrt{\pi^2 - \varepsilon^2(\pi^2 - 4)}$ where $U = \dfrac{2\pi RN}{60}$

$$U = \frac{2\pi \times .05 \times 1200}{60} = 6.28 \text{ m/s}$$

$$W = \frac{6 \times 0.04 \times 6.28 \times 0.2 \times 10^6}{(2+0.36)(1-0.36)} \times 0.6\sqrt{\pi^2 - 0.36(\pi^2 - 4)} = 0.3335 \times 10^6 \text{ N}$$

$$\text{Sommerfeld number } S = \frac{\eta NLD}{W}\left(\frac{R}{C}\right)^2 = \frac{0.04 \times 1200 \times 0.2 \times 0.1 \times 10^6}{60 \times 0.3335 \times 10^6} = 0.04797$$

$$\text{Attitude angle, } \phi = \tan^{-1}\left(\frac{\pi\sqrt{1-\varepsilon^2}}{2\varepsilon}\right) = \tan^{-1}\frac{\pi \times 0.64}{1.2} = 59.157^0$$

$$\text{Friction factor } \mu\left(\frac{R}{C}\right) = \frac{\varepsilon \sin\phi}{2} + \frac{2\pi^2 S}{(1-\varepsilon^2)^{1/2}} = \frac{0.6 \times \sin 59.157^0}{2} + \frac{2\pi^2 \times 0.04797}{(1-0.36)^{1/2}} = 1.4412$$

**E.4.3** An inclined pad thrust bearing is operating with following data:
Pad width = 0.1 m, pad length = 0.15 m, sliding speed = 3m/s, absolute viscosity of the oil = 0.02 Pas, minimum oil film thickness = 25 × $10^{-6}$m, maximum oil film thickness = 50 × $10^{-6}$m. Determine the load carrying capacity, oil flow rate, friction factor, and center of pressure. Determine also the value of maximum load carrying capacity.

***Solution:***

**Given data:** Pad width, $L = 0.1\ m$, Pad length, $B = 0.15\ m$, minimum film thickness, $h_2 = 25\ \mu m$, maximum film thickness, $h_1 = 50\ \mu m$, $a = h_1/h_2 = 2$, sliding speed, $U = 3 m/s$.

Load carrying capacity is determined as:

$$W = \frac{6\eta UB^2 L}{h_2^2}\frac{1}{(a-1)^2}\left[\ln a - \frac{2(a-1)}{a+1}\right] = \frac{6 \times 0.02 \times 3 \times 0.15^2 \times 0.1}{(25 \times 10^{-6})^2 \times 1}\left[\ln 2 - \frac{2}{3}\right]$$

$$= 35 \times 10^3 \text{ N}$$

$$\text{Oil flow rate } Q_x = \frac{2ah_2 U}{(a+1)} = \frac{2 \times 2 \times 25 \times 10^{-6} \times 3}{3} = 10^{-4} \text{ m}^3/\text{s}$$

Friction factor $\frac{\mu B}{h_2} = \left[\frac{2(a^2-1)\ln a - 3(a-1)^2}{3(a+1)\ln a - 6(a-1)}\right]$ where $\mu = \frac{F}{W}$ is coefficient of friction and F is friction force. Thus,

$$\frac{\mu B}{h_2} = \left[\frac{2(2^2-1)\ln 2 - 3(2-1)^2}{3(2+1)\ln 2 - 6(2-1)}\right] = \frac{6\ln 2 - 3}{9\ln 2 - 6} = \frac{1.15888}{0.2383} = 4.863$$

$$\text{Coefficient of friction } \mu = \frac{4.863 \times 25 \times 10^{-6}}{0.15} = 0.81 \times 10^{-3}$$

Center of pressure $x_{cp} = B\left[\dfrac{a(a+2)\ln a-(a-1)[2.5(a-1)+3]}{(a^2-1)\ln a-2(a-1)^2}\right]$

$$x_{cp} = 0.15\times\left[\frac{2(2+2)\ln 2-(2-1)[2.5(2-1)+3]}{(2^2-1)\ln 2-2(2-1)^2}\right]$$

$$= 0.15\times\frac{8\ln 2-5.5}{3\ln 2-2} = 0.0853m$$

Center of pressure is located at a distance of 0.0853m from the inlet end, i.e., maximum film thickness end

Maximum load capacity is obtained at film thickness ratio a = 2.18 and can be determined from the following expression

$$W_{max} = 0.1602\frac{\eta ULB^2}{h_2^2} = 0.1602\frac{0.02\times 3\times 0.15^2\times 0.1}{(25\times 10^{-6})^2} = 216\times 10^3 \text{ N}$$

## Problems

**P.4.1** A turbine rotor operates at a speed of 9000 rev/min and weighs 35kN. The shaft diameter is 150mm. What *L/D* ratio and radial clearance is to be chosen for a pair of journal bearings to support the load. The operating eccentricity ratio should not go below 0.6. The oil recommended is SAE 30 with an absolute viscosity of 112.7cS at 40°C and 12.5cS at 100°C. Oil inlet temperature is 50°C. The Sommerfeld variable $S = \dfrac{\eta ULR^2}{WC^2}$ for eccentricity ratio of 0.6 is given below for three different *L/D* ratios.

| *L/D* | 1 | 0.5 | 0.25 |
|---|---|---|---|
| *S* | 0.4 | 1.0 | 3.4 |

**P.4.2** A 15 cm diameter bearing, L/D = 0.5 of clearance ratios 1 and 2 mm/m supports a rotor at a speed of 10000 rev/min. Oil used is SAE 30 with an absolute viscosity of 112.7*cS* at 40°*C* and 12.5*cS* at 100°*C*. The oil inlet temperature is 50°*C*. Determine the load carrying capacity, oil flow rate, temperature rise using 0.8 rule, and power loss.

**P.4.3** A thrust pad of infinite width and length *B* has an exponential film shape given by following expression: $h = h_0 e^{ax/B}$ where $h_0$ is the minimum film thickness at the exit point of the lubricant. Show that value of film thickness where the pressure gradient $\dfrac{dp}{dx} = 0$ is: $\bar{h} = \dfrac{3}{2}\left(\dfrac{1-e^{-2a}}{1-e^{-3a}}\right)$ where $\bar{h} = h/h_0$. Find the value of dimensionless load $\bar{W} = \dfrac{h_0^2(W)}{U\eta B^2}$ and *W* is load per unit width of the thrust pad, *U* is surface speed and $\eta$ is the viscosity of the oil. Also prove that the coefficient of friction at the moving surface is obtained as: $\mu = k\sqrt{\dfrac{U\eta}{W}}$, k is a constant. Determine the value of *k* for $a = 1$.

**P.4.4** A thrust pad of short width has wave shape oil film where film thickness is given by:

$$h = h_m\left(1 + a\sin\frac{\pi}{2}\left(1 - \frac{2x}{B}\right)\right),$$

where $a$, is the wave amplitude and $h_m$ is the mean film thickness. Determine the expression of pressure distribution, load capacity, oil flow rate, and friction coefficient using short bearing theory. Bearing width $L << B$, when $B$ is length of the pad in the flow direction. Assume moving surface speed as $U$, $a = 0.2$. Determine the performance variables when $U = 5$ m/s, $h_m = 20\,\mu m$, $B = 10$ cm, and viscosity of the oil is 0.03 *N.s/m*².

**P.4.5** An infinitely wide hydrodynamic thrust bearing can be designed with Rayleigh step or tapered land pad shapes as shown in Figs. 4.13 and 4.14, respectively. Analyze the bearings and determine optimum configurations to carry maximum load. Determine the expressions for load capacity, oil flow rate, and friction coefficient. Determine the value of parameters when the moving surface speed is 8 m/s, pad length is 30 cm, minimum film thickness is $15\,\mu m$, and viscosity of the oil is 0.03 *N.s/m*².

## REFERENCES


*Ausman, J.S.* (1963), An improved analytical solution for self acting gas-lubricated journal bearings, *Trans. ASME, Ser. D Journal of Basic Engineering*, **83**, 188–194.

*Cameron, A.* (1981), *Basic Lubrication Theory*, Wiley Eastern Ltd., New Delhi, India.

*Castelli, V. and Pirvics, J.* (1968), Review of numerical methods in gas bearing film analysis, *Trans ASME, JOLT*, **90**, 777–792.

*Castelli, V. Stevenson, C.H. and Gunter, E.J. Jr.* (1964), Steady state characteristics of gas lubricated self acting partial arc journal bearings of finite width, *ASLE Transactions*, **7**, 153–167.

*Christophernon, D.G.* (1942), A new mathematical model for solution of film lubrication problems, *Proc., Inst. of Mech. Engineers London*, U.K., **146**, 126–135.

*Cole, J.A. and Hughes, C.J.* (1956), Oil flow and film extent in complete journal bearings, *Proc. Inst. Mech. Engineers*. London, U.K., **170**, 499–510.

*Coyne, J.C. and Elrod, H.* (1970), Conditions for rupture of a lubricating film, Part I: theoretical model, *ASME Trans, Ser F*, **92**, 451–457.

*Coyne, J.C. and Elrod, H.* (1971), Conditions for rupture of a lubricating film, Part II: new boundary conditions for Reynolds equation, *ASME, Trans. Ser F*, **93**, 156–167.

*Dubois, G.B. and Ocvirk, F.W.* (1953), Analytical derivation and experimental evaluation of short-bearing approximation for full journal bearings, NASA Report No. **1157**.

*Floberg, L.* (1961), On hydrodynamic lubrication with special reference to cavitations in bearings, *Chalmers Teckniska Hogskola Doctora Vahandlinger*, Gothenberg, Sweden, No. **30**.

*Floberg, L.* (1965), On hydrodynamic lubrication with special reference to sub-cavity pressures and number of streams in cavitations regions, *Acta Polytech. Scandinavia, Mech. Engineering Ser*, **19**, 3–35.

*Hamrock, B.J.* (1994), *Fundamentals of Fluid Film Lubrication*, McGraw Hill International Edition, McGraw Hill Inc., New York.

*Majumdar, B.C.* (1986), *Introduction to Tribology of Bearings*, A. H. Wheeler & Co. Pvt. Ltd., Allahabad, India.

*Martin, F.A. and Lee, C.S.* (1982), Feed-pressure flow in plain journal bearings, *ASLE Trans.*, **26**, 381–392.

*Pinkus, O.* (1987), The Reynolds centennial: a brief history of the theory of hydrodynamic lubrication, *Trans. ASME, Journal of Tribology*, **109**, 2–20.

*Pinkus, O. and Sternlicht, B.* (1961), *Theory of Hydrodynamic Lubrication*, Mc Graw Hill Book Co. Inc., New York.

*Raimondi, A.A. and Szeri, A.Z.* (1984), Journal and Trust Bearings, Handbook of Lubrication, E.R. Booser (ED), Vol. II, CRC Press.

*Raimondi, A.A. and Boyd, J.* (1958), A solution for the finite journal bearing and its applications to analysis and design—Part I, II, III, *ASLE Transactions*, **1**, 159–209.

*Sommerfeld, A.* (1904), Zur Hydrodynamic chon Theorie der Schmiermittel- reibung, *Z. Angew. Math. Phys*, **50**, 97–155.

*Steiber, W.* (1933), Das Schwimn Lager, Krayam, VDI, Berlin.

*Swift, H.W.* (1931), The stability of lubricating films in journal bearings, *Proc. Inst. Civil Engineers London*, U.K., **233**, 267–288.

*Szeri, A.Z.* (1998), *Fluid Film Lubrication Theory and Design*, Cambridge University Press, U.K.

Chapter 5

# Finite Bearings

## 5.1 | Introduction

In Chapter 4, the Reynolds equation for idealized (infinitely long and short) journal bearings was solved. These gave closed form solutions for load carrying capacity and coefficient of friction. Practical journal bearings have length to diameter ratio (*L/D*) between 0.5 and 1.5. The infinitely short or narrow bearing theory is valid for *L/D* of about 0.25 and that of infinitely long bearing can be applied for bearing having *L/D* equal to or greater than 4. Hence the theories (developed in Chapter 4) which yielded analytical solutions cannot be applied in practice, although these indicate the trends of the results. However, for accurate results, the Reynolds equation in two dimensions is to be solved. Due to difficulty in obtaining analytical solution, a method using electric analogy was devised by Kingsbury (1930). But this method is very approximate. Approximate analytical solutions due to Boegli (1947) and numerical solutions using finite difference and finite element methods using digital computer are given. At the end of this chapter, cavitation and cavitation boundary conditions are discussed.

## 5.2 | Analytical Solution

### 5.2.1 | Approximate Solution

The solution given by Boegli (1947) makes two approximations:

1. The pressure functions along the length and width of the bearing are independent.
2. The pressure function along the bearing in the direction of motion is the same as that of an infinitely long bearing. Although the method of solution given here is for a finite slider bearing, it can be easily applied to a finite journal bearing also.

The governing differential equation for a finite oil bearing having constant viscosity can be written as:

$$\frac{\partial}{\partial x}\left(h^3\frac{\partial p}{\partial x}\right)+\frac{\partial}{\partial y}\left(h^3\frac{\partial p}{\partial y}\right)=6\eta U\frac{dh}{dx} \tag{5.1}$$

Equation (5.1) can be nondimensionalized as

$$\frac{\partial}{\partial \bar{x}}\left(\bar{h}^3\frac{\partial \bar{p}}{\partial \bar{x}}\right)+\left(\frac{B}{L}\right)^2\frac{\partial}{\partial \bar{y}}\left(\bar{h}^3\frac{\partial \bar{p}}{\partial \bar{y}}\right)=\frac{d\bar{h}}{d\bar{x}} \tag{5.2}$$

where $\bar{h}=\frac{h}{h_2}, \bar{x}=\frac{x}{B}, \bar{y}=\frac{y}{L}$ and $\bar{p}=\frac{ph_2^2}{6\eta UB}$

Here $h_2$ is the minimum film thickness, and $L$ and $B$ are length and width of the rectangular slider. Let $\bar{p}=f(\bar{x})f(\bar{y})$ and substituting this into Equation (5.2) and using assumption (1), we get

$$\frac{\partial}{\partial \bar{x}}\left(\bar{h}^3 f(\bar{y})\frac{df(\bar{x})}{d\bar{x}}\right)+\left(\frac{B}{L}\right)^2\bar{h}^3 f(\bar{x})\frac{d^2 f(\bar{y})}{d\bar{y}^2}=\frac{d\bar{h}}{d\bar{x}} \tag{5.3}$$

In Equation (5.3) $\bar{h}$ is assumed to be a function of $\bar{x}$ only implying there is no misalignment. Equation (5.3) is solved at the point where the pressure is maximum, *i.e.*,

where $$\frac{df(\bar{x})}{d\bar{x}}=0$$

The following equation is the result

$$\frac{f(\bar{y})}{\left(\frac{B}{L}\right)^2 f(\bar{x})}\frac{d^2 f(\bar{x})}{d\bar{x}^2}+\frac{d^2 f(\bar{y})}{d\bar{y}^2}=\frac{\frac{d\bar{h}}{d\bar{x}}}{\left(\frac{B}{L}\right)^2\bar{h}^3 f(\bar{x})} \tag{5.4}$$

with the use of assumption (2), one can write

$$3\bar{h}^2\frac{df(\bar{x})}{d\bar{x}}\frac{d\bar{h}}{d\bar{x}}+\bar{h}^3\frac{d^2 f(\bar{x})}{d\bar{x}^2}=\frac{d\bar{h}}{d\bar{x}} \tag{5.5}$$

At the point of maximum pressure $\frac{df(\bar{x})}{d\bar{x}}=0$

hence

$$\frac{d^2 f(\bar{x})}{d\bar{x}^2}=\frac{\frac{d\bar{h}}{d\bar{x}}}{\bar{h}^3} \tag{5.6}$$

Equation (5.6) is now substituted into Equation (5.4) and one obtains

$$\frac{d^2 f(\bar{y})}{d\bar{y}^2} + \frac{f(\bar{y})}{\left(\frac{B}{L}\right)^2} \frac{\frac{d\bar{h}}{d\bar{x}}}{\bar{h}^3 f(\bar{x})} = \frac{\frac{d\bar{h}}{d\bar{x}}}{\left(\frac{B}{L}\right)^2 \bar{h}^3 f(\bar{x})} \tag{5.7}$$

Putting $M = \frac{\frac{d\bar{h}}{d\bar{x}}}{\bar{h}^3 f(\bar{x})}$, Equation (5.7) can be written as

$$\frac{d^2 f(\bar{y})}{d\bar{y}^2} + \frac{M}{\left(\frac{B}{L}\right)^2} f(\bar{y}) = \frac{M}{\left(\frac{B}{L}\right)^2} \tag{5.8}$$

The solution of Equation (5.8) is given by

$$f(\bar{y}) = -\left(\frac{e^k - 1}{e^k - e^{-k}}\right)\left(e^{k\bar{y}} + e^{-k(1-\bar{y})}\right) \tag{5.9}$$

where
$$k = -\frac{M}{\left(\frac{B}{L}\right)^2}$$

The value of $M$ can be evaluated at $\bar{h} = \bar{h}_m$ and $\bar{p} = \bar{p}_{max}$

When this method is to be applied in a journal bearing, $M$ is to be calculated at the position of $\bar{h} = \bar{h}_m$ where $\bar{p} = \bar{p}_{max}$.

The ratio of $\frac{W}{W_\infty}$ can be calculated from the foregoing analysis. As $f(\bar{x})$ is $\bar{p}_\infty$, the pressure function of the infinite bearing, the ratio $\frac{W}{W_\infty}$ is simply the integral of $f(\bar{y})$ of Equation (5.9).

Hence,
$$\frac{W}{W_\infty} = 1 - \frac{2\left(1 - e^{-k}\right)^2}{k\left(1 - e^{-2k}\right)} \tag{5.10}$$

Knowing the load capacity of an infinitely long bearing, the actual load capacity can be computed from Equation (5.10).

## 5.3 | Numerical Solution

### 5.3.1 | Finite Difference Method

In this section, a numerical solution of two-dimensional Reynolds equation for a finite journal bearing using finite difference method is given.

Equation (5.1) when nondimensionalizing with the following substitutions:

$$\theta = \frac{x}{R}, \bar{y} = \frac{y}{\left(\frac{L}{2}\right)}, \bar{h} = \frac{h}{C}, \bar{p} = \frac{pC^2}{6\eta UR}$$

results in

$$\frac{\partial}{\partial\theta}\left(\bar{h}^3 \frac{\partial p}{\partial\theta}\right) + \left(\frac{D}{L}\right)^2 \bar{h}^3 \frac{\partial^2 \bar{p}}{\partial \bar{y}^2} = \frac{d\bar{h}}{d\theta} \tag{5.11}$$

Equation (5.11) assumes that $\bar{h}$ is only a function of $\theta$, so can be expressed as

$$\frac{\partial^2 \bar{p}}{\partial\theta^2} + \left(\frac{D}{L}\right)^2 \frac{\partial^2 \bar{p}}{\partial \bar{y}^2} + \frac{3}{\bar{h}} \frac{\partial \bar{p}}{\partial\theta} \frac{\partial \bar{h}}{\partial\theta} = \frac{\left(\frac{d\bar{h}}{d\theta}\right)}{\bar{h}^3} \tag{5.12}$$

and $\bar{h} = 1 + \varepsilon\cos\theta$, $\varepsilon$ being eccentricity ratio. (5.13)

A developed view for the half of the bearing is drawn (see Fig. 5.1). The area is divided into a number of mesh sizes $(\Delta\theta \times \Delta\bar{y})$ and using the central difference quotients, Equation (5.12) can be written in the form as

$$\left[\frac{p_{i+1,j} - 2p_{i,j} + p_{i-1,j}}{(\Delta\theta)^2}\right] + \left(\frac{D}{L}\right)^2 \left[\frac{p_{i,j+1} - 2p_{i,j} + p_{i,j-1}}{(\Delta\bar{y})^2}\right]$$
$$-\frac{3\varepsilon}{2}\left[\frac{p_{i+1,j} - p_{i-1,j}}{h_i(\Delta\theta)}\sin\theta_i\right] = -\varepsilon\frac{\sin\theta_i}{h_i^3} \tag{5.14}$$

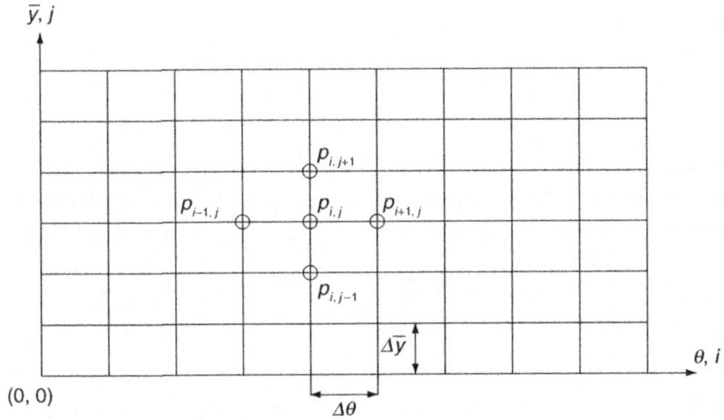


**Figure 5.1** | A Developed View of a Bearing Showing the Mesh Size $(\Delta\theta \times \Delta\bar{y})$

where $p_{i,j}$ and $h_i$ are the pressure and film thickness at any mesh point $(i,j)$, $p_{i+1,j}$, $p_{i-1,j}$, $p_{i,j+1}$, and $p_{i,j-1}$ are pressures at the four adjacent points, and $\theta_i = \frac{2(\Delta\theta)_i}{D}$, $(i,j)$ is the numerical coordinate system.

On simplification Equation (5.14), it boils down to

$$p_{i,j} = \left[ \begin{array}{l} p_{i+1,j} + p_{i-1,j} + \left(\frac{D}{L}\right)^2 \left(\frac{\Delta\theta}{\Delta\bar{y}}\right)^2 \left(p_{i,j+1} + p_{i,j-1}\right) \\ -\frac{3}{2}\varepsilon \frac{\left(p_{i+1,j} - p_{i-1,j}\right)}{h_i}(\Delta\theta)\sin\theta_i + \frac{\varepsilon \sin\theta_i}{h_i^3}(\Delta\theta)^2 \end{array} \right] \Bigg/ \left[ 2\left\{1 + \left(D/L\right)^2 \left(\frac{\Delta\theta}{\Delta\bar{y}}\right)^2\right\}\right] \tag{5.15}$$

For a square mesh, i.e., $\Delta\theta = \Delta\bar{y} = \Delta$ (say), Equation (5.15) reduces to

$$p_{i,j} = \left[ \begin{array}{l} p_{i+1,j} + p_{i-1,j} + \left(\frac{D}{L}\right)^2 \left(p_{i,j+1} + p_{i,j-1}\right) \\ -\frac{3}{2}\varepsilon \frac{\left(p_{i+1,j} - p_{i-1,j}\right)}{h_i}\Delta\sin\theta_i + \frac{\varepsilon \sin\theta_i}{h_i^3}\Delta^2 \end{array} \right] \Bigg/ \left[ 2\left\{1 + \left(D/L\right)^2\right\}\right] \tag{5.16}$$

It is seen that the pressure at any mesh point $(i, j)$ is expressed in terms of pressures at the four adjacent points. To start with iteration method, the pressures at all the mesh points are assumed and those at the boundaries are set. Equation (5.15) is then solved for all the mesh points. As the pressures were assumed in the beginning, Equation (5.15) will not be satisfied. The error at the point $(i,j)$ is

$$(Error)_{i,j} = RHS \text{ of Equation (5.15)} - p_{i,j} \tag{5.17}$$

The new pressure can be computed using a successive over-relaxation scheme (SOR) as

$$\left(p_{i,j}\right)_{new} = \left(p_{i,j}\right)_{old} + (error)_{i,j}.\text{orf} \tag{5.18}$$

where 'orf' is the over-relaxation factor.

The use of over-relaxation factor in Equation (5.18) accelerates the convergence of the numerical process. It is, however, very difficult to estimate the optimum value of this factor. From experience it has been found *orf* generally varies from 1.2 to 1.5.

The process will be repeated till the specified accuracy is attained by a convergence criterion as

$$\frac{\left|\sum\left(p_{i,j}\right)_{N-1} - \sum\left(p_{i,j}\right)_N\right|}{\sum\left(p_{i,j}\right)_N} \leq \text{a very small quantity, where } N \text{ is the number of iterations.}$$

The allowable error is to be kept to a very small fraction of 1%. Once the pressure distribution is obtained, the load capacity, volume rate flow, coefficient of friction can be calculated numerically. Usually the

Simpson's 1/3 rule for numerical integration and three-point backward or forward difference rule is applied for differentiation. For details of such numerical method, refer to any book on numerical analysis.

The above numerical procedure can be easily applied to a plain slider bearing. The dimensions of the bearing in this case will be $L$ and $B$ and $h$ will be a function of $x$ only. The finite difference form of concerned differential equation can also be easily written.

The load capacity and coefficient of friction for a plain slider bearing can be calculated numerically in the similar way. In Table 5.1, the dimensionless load capacity and friction force, etc. are given for $n = 2$ and for various $L/B$ ratios. The coefficient of friction from these results can be easily calculated.

In Table 5.2, the performance characteristics in terms of Sommerfeld number, friction variable, and attitude angle of full (360°) journal bearings having $L/D = \infty, 1, \frac{1}{2}, \frac{1}{4}$ are shown (Pinkus 1958).

### 5.3.2 | Error Analysis

The above equation is elliptical in nature. From the finite difference equations it is seen that the mesh size plays a significant role in error analysis. Hence the order of error is $O(\Delta x)^2$ or $O(\Delta y)^2$. If we solve a Laplace equation with given boundary conditions using mesh size $\Delta x = \Delta y = \Delta$ and $\Delta x = \Delta y = \Delta/2$, the error will be reduced significantly. For details, refer to Sarkar (2004).

### 5.3.3 | Finite Element Method *

The finite element method was incepted in the early 1950s as a method to primarily solve structural problems. Later due to the efforts of Zienkiwickz and others (1990), the method was applied to different problems and gained acceptance as a general technique to solve engineering problems and in particular field problems which can be formulated as an extremum or stationary value problem (Reddy 1993).

**Table 5.1** | Performance of Plane Slider Bearings

| *L/B* | ∞ | 2 | 1.5 | 1.0 | 0.75 | 0.5 |
|---|---|---|---|---|---|---|
| $\bar{W} = \dfrac{Wh_2^2}{\eta UB^2L}$ | 0.1589 | 0.1096 | 0.0946 | 0.0689 | 0.0504 | 0.0289 |
| $\bar{F} = \dfrac{Fh_2}{\eta UBL}$ | 0.7726 | 0.7480 | 0.7404 | 0.7276 | 0.7183 | 0.7076 |
| $\bar{Q} = \dfrac{Q}{Uh_2B}$ | 0 | 0.2589 | 0.2586 | 0.2462 | 0.2229 | 0.1788 |
| $x_1/B$ | 0.5687 | 0.5730 | 0.5756 | 0.5818 | 0.5838 | 0.6005 |

*This section is contributed by Dr Ram Turaga and this is highly appreciated by the authors.

Table 5.2 | Full Journal Bearings

| L/D | ε | S | μ (R/C) | φ° |
|---|---|---|---|---|
| ∞ | 0.1 | 0.247 | — | 69 |
| | 0.2 | 0.123 | 2.57 | 67 |
| | 0.3 | 0.0823 | 1.90 | 64 |
| | 0.4 | 0.0628 | 1.53 | 62 |
| | 0.5 | 0.0483 | 1.32 | 58 |
| | 0.6 | 0.0389 | 1.20 | 54 |
| | 0.7 | 0.0297 | 1.10 | 49 |
| | 0.8 | 0.0211 | 0.962 | 42 |
| | 0.9 | 0.00114 | 0.721 | 32 |
| 1 | 0.1 | 1.35 | — | 79 |
| | 0.2 | 0.632 | 12.9 | 74 |
| | 0.3 | 0.382 | 8.04 | 68 |
| | 0.4 | 0.261 | 5.80 | 62 |
| | 0.5 | 0.179 | 4.31 | 56 |
| | 0.6 | 0.120 | 3.21 | 50 |
| | 0.7 | 0.0765 | 2.36 | 43 |
| | 0.8 | 0.0448 | 1.71 | 36 |
| | 0.9 | 0.0191 | 1.06 | 25 |
| $\frac{1}{2}$ | 0.1 | 4.30 | — | 81 |
| | 0.2 | 2.01 | 40.9 | 75 |
| | 0.3 | 1.235 | 25.7 | 68 |
| | 0.4 | 0.785 | 17.11 | 62 |
| | 0.5 | 0.497 | 11.95 | 55 |
| | 0.6 | 0.320 | 8.08 | 48 |
| | 0.7 | 0.185 | 5.48 | 41 |
| | 0.8 | 0.092 | 3.25 | 33 |
| | 0.9 | 0.031 | 1.59 | 23 |
| $\frac{1}{4}$ | 0.1 | 15.9 | — | 83 |
| | 0.2 | 7.58 | 153 | 75 |
| | 0.3 | 4.69 | 98.5 | 68 |
| | 0.4 | 2.85 | 61.4 | 61 |
| | 0.5 | 1.78 | 40.0 | 54 |
| | 0.6 | 1.07 | 26.7 | 47 |
| | 0.7 | 0.591 | 16.6 | 39 |
| | 0.8 | 0.266 | 8.93 | 31 |
| | 0.9 | 0.0738 | 3.49 | 8 |

For solving lubrication problems, the finite element method offers several advantages:

- It offers complete generality with regard to geometry and field property variation.
- Prescription of boundary conditions in terms of pressure or flow.

- Prescription of symmetry conditions which are due to symmetry of equations generated by this method result in banded, definitive matrices which can be solved with minimum computational storage and time.

This technique has since been widely used in the solution of lubrication problems and a summary of some of the early contributions is given in Reddy (1969), Booker and Huebner (1972), and Huebner and Thornto (1982).

In the finite element method, we reduce a field problem described by an extremum principle to one where there are a finite number of unknowns. To achieve this, a number of localized functions called interpolation functions, each valid for a small subregion of the domain of interest are used. Thus, the overall state of the unknown field variable is represented piecewise, with continuity of the variables across the boundaries of the subregions or elements being assured by suitable requirements placed on the choice of the interpolation functions. Further the finite element method assumes that the state of field variable within the subregion or element is described completely by values of the unknown variable at the finite number of points or nodes located on the boundary of the element.

The main steps in the solution of a problem using the finite element method and computer implementation are given in Fig. 5.2.

As the finite element method is a technique for constructing approximate function required in an element-wise application of any variation method, the weak formulation facilitates the classification of the boundary conditions into natural and essential boundary conditions which play a crucial role in the derivation of the approximate functions and the selection of the nodal degrees of freedom in the finite element model. The primary purpose of developing a weighted integral statement of a differential equation is to have the means to obtain $N$ linearly independent algebraic relations which is accomplished by choosing $N$ linearly independent weight functions in the integral statement.

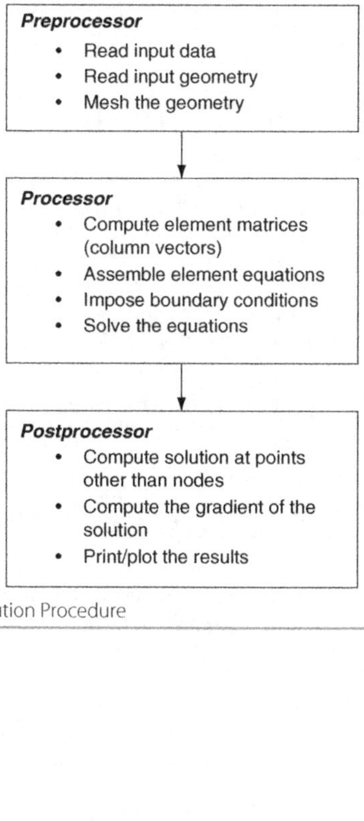


**Figure 5.2** | Main Steps in the Solution Procedure

Starting with the governing equation:

$$\frac{\partial}{\partial x}\left(h^3\frac{\partial p}{\partial x}\right)+\frac{\partial}{\partial y}\left(h^3\frac{\partial p}{\partial y}\right)=6\eta r\left(\omega-2\frac{\partial\phi}{\partial t}\right)\frac{\partial h}{\partial x}+12\eta\frac{\partial h}{\partial t} \quad (5.19)$$

### *Weak Form*

$$0=\int_{\Omega} w\left(\frac{\partial}{\partial x}\left(h^3\frac{\partial p}{\partial x}\right)+\frac{\partial}{\partial y}\left(h^3\frac{\partial p}{\partial y}\right)-6\eta r\left(\omega-2\frac{\partial\phi}{\partial t}\right)\frac{\partial h}{\partial x}-12\eta\frac{\partial h}{\partial t}\right)dA \quad (5.20)$$

Using the following expressions

$$-w\frac{\partial F_1}{\partial x}=F_1\frac{\partial w}{\partial x}-\frac{\partial}{\partial x}(wF_1)$$

$$-\int_{\Omega} w\frac{\partial F_1}{\partial x}=\int_{\Omega} F_1\frac{\partial w}{\partial x}-\int_{\Omega}\frac{\partial}{\partial x}(wF_1)$$

Using Greens theorem

$$\int_{\Omega}\frac{\partial wF_1}{\partial x}dA=\oint_{\Gamma} wF_1\, nds$$

$$\int_{\Omega} w\left[\frac{\partial}{\partial x}\left(h^3\frac{\partial p}{\partial x}\right)\right]dA=\int_{\Omega}\left(h^3\frac{\partial p}{\partial x}\frac{\partial w}{\partial x}-\frac{\partial}{\partial x}\left(wh^3\frac{\partial p}{\partial x}\right)\right)\Bigg]dA$$

$$=\int_{\Omega}\left(h^3\frac{\partial p}{\partial x}\frac{\partial w}{\partial x}\right)dA-\oint_{\Gamma} wh^3\frac{\partial p}{\partial x}nds$$

$$\int_{\Omega} w\left(\frac{\partial}{\partial y}\left(h^3\frac{\partial p}{\partial y}\right)\right)dA=\int_{\Omega}\left(h^3\frac{\partial p}{\partial y}\frac{\partial w}{\partial y}-\frac{\partial}{\partial y}\left(wh^3\frac{\partial p}{\partial y}\right)\right)dA$$

$$=\int_{\Omega}\left(h^3\frac{\partial p}{\partial y}\frac{\partial w}{\partial y}\right)dA-\oint_{\Gamma} wh^3\frac{\partial p}{\partial y}nds$$

$$\int_{\Omega}\left(\left(h^3\frac{\partial p}{\partial x}\frac{\partial w}{\partial y}\right)+\left(h^3\frac{\partial p}{\partial y}\frac{\partial w}{\partial y}\right)-\left(6\eta r\left(\omega-2\frac{\partial\phi}{\partial t}\right)h\frac{dw}{dx}-12\frac{dh}{dt}w\right)\right)dA$$

$$-\oint_{\Gamma} wh^3\frac{\partial p}{\partial x}n_x ds-\oint_{\Gamma} wh^3\frac{\partial p}{\partial y}n_y ds-\oint_{\Gamma} w6\eta r\left(\omega-2\frac{\partial\phi}{\partial t}\right)hn_x ds=0 \quad (5.21)$$

The above Equation (5.21) is called the weak form of the differential Equation (5.19). The term 'Weak' refers to the reduced continuity of 'p', which is required to be only once differentiable.

This can be expressed as,

$$0 = B(w,p) - L(w) \tag{5.22}$$

where

$$B(w,p) = \int_{\Omega} \left( \begin{array}{c} \left( h^3 \dfrac{\partial p}{\partial x}\dfrac{\partial w}{\partial y} \right) + \left( h^3 \dfrac{\partial p}{\partial y}\dfrac{\partial w}{\partial y} \right) \\ -\left( 6\eta r \left( \omega - 2\dfrac{\partial \phi}{\partial t} \right) h \dfrac{dw}{dx} - 12w\dfrac{dh}{dt} \right) \end{array} \right) dA \tag{5.23}$$

$$\begin{aligned} L(w) &= \int_{\Omega} \left( w12\frac{dh}{dt} \right) dA - \oint_{\Gamma} wh^3 \frac{\partial p}{\partial x} n_x ds \\ &- \oint_{\Gamma} wh^3 \frac{\partial p}{\partial y} n_y ds \oint_{\Gamma} w6\eta r \left( \omega - 2\frac{\partial \phi}{\partial t} \right) hn_x ds \end{aligned} \tag{5.24}$$

Using the variational

$$B(w,p) = L(w)$$

This holds for any $w$ that satisfies the homogeneous form of the specified essential boundary conditions and continuity conditions implied by the weak form.

The weak form or the variational form (used interchangeably) is equivalent to the original differential equation and the specific natural boundary conditions of the problem. The weak form exists for all differential equation and when the equation is linear and of even order, the resulting weak form will have a symmetric bilinear form in the dependent variable p and weight function $w$.

Thus, when $B(w,p)$ is bilinear and symmetric and $L(w)$ is linear the associated quadratic functional can be derived as follows.

The function $w$ can be viewed as a variation of the actual solution $p^*$.

$$p = p^* + w$$

$$w = \delta p$$

$$0 = B(\delta p, p) - l(\delta p)$$

As $B(\bullet,\bullet)$ is symmetric

$$0 = \delta\left(\frac{1}{2}B(p,p)\right) - \delta l(p)$$

$$= \delta I(p)$$

where

$$I(p) = \frac{1}{2}B(p,p) - l(p) \tag{5.25}$$

In the derivation of the functional $I(p)$ from the weak form, the linearity and symmetry of the bilinear form $B(p,p)$ is essential.

The functional $I(p)$ can be written as

$$
\begin{aligned}
I(p) = \frac{1}{2}\int_{\Omega}\left[[(h^3\nabla p) - 6\eta U(1-\frac{2}{\omega}\frac{\partial\phi}{\partial t})]\bullet\nabla p - 12\frac{\partial h}{\partial t}p\right]dA \\
-\oint_{\Gamma} n\bullet[6\eta U(1-\frac{2}{\omega}\frac{\partial\phi}{\partial t})h - h^3\nabla p]ds
\end{aligned}
\tag{5.26}
$$

## *Functional*

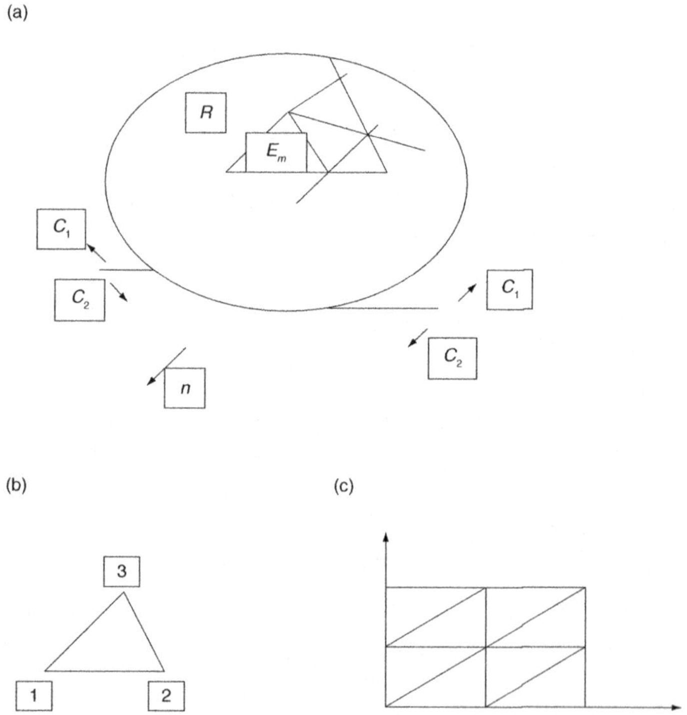


**Figure 5.3** | Shape Functions

This can be simplified and written as

$$I(p)=\frac{1}{2}\int_{\Omega}\left[[(h^3\nabla p)-(6\eta U(1-\frac{2}{\omega}\frac{\partial\phi}{\partial t})]\cdot\nabla p-12\frac{\partial h}{\partial t}p\right]dA+\oint_{\Gamma}Qpds \tag{5.27}$$

where $Q=q\cdot n=h\vec{u}\cdot\vec{n}=h(6\eta U(1-\frac{2}{\omega}\frac{\partial\phi}{\partial t})-h^2\nabla p)\cdot\vec{n}$

The boundary conditions are

$$\begin{aligned}&1.\ p=p(\bar{x},y)\ on C_1\\&2.\ q=h(6\eta U(1-\frac{2}{\omega}\frac{\partial\phi}{\partial t})-h^2\nabla p)\cdot\vec{n}\ on C_2\end{aligned} \tag{5.28}$$

Equation (5.27) after substituting for shape functions can be restated to the form

$$I(p)=(P^TK_p-K_u\{U\}+K_h\{\dot{h}\}+\{Q\})P \tag{5.29}$$

The first variation of the functional with respect to the nodal pressures yields

$$\begin{aligned}&\delta I(p)=\sum_{i=1}^{N}\frac{\partial I(p)}{\partial p_i}\partial p_i\\&\frac{\partial I(p)}{\partial p_i}=2K_p-K_u\{U\}+K_h\{\dot{h}\}+\{Q\}=0\end{aligned} \tag{5.30}$$

Neglecting the squeeze term we get

$$2[K_p]\{P\}=[K_u]\{U\}-\{Q\} \tag{5.31}$$

## Element Equations

Using the area co-ordinates $L_1, L_2, L_3$ for the triangle 1, 2, 3 (Fig. 5.3b), the relation between these and Cartesian system is (1990):

$$\begin{aligned}x&=L_1x_1+L_2x_2+L_3x_3\\y&=L_1y_1+L_2y_2+L_3y_3\\1&=L_1+L_2+L_3\end{aligned}$$

and in terms of $L_1, L_2, L_3$

$$L_1 = (a_1 + b_1 x + c_1 y)/2\Delta$$

$$L_2 = (a_2 + b_2 x + c_2 y)/2\Delta$$

$$L_3 = (a_3 + b_3 x + c_3 y)/2\Delta \quad \text{and}$$

$$2\Delta = area123 \quad \text{and}$$

$$\begin{aligned} a_1 &= x_2 y_3 - x_3 y_2 \\ b_1 &= y_2 - y_3 \\ c_1 &= x_2 - x_3 \end{aligned}$$

The shape functions are defined for a triangular element as shown in Fig. 5.3b.

$$N_1 = L_1, N_2 = L_2, N_3 = L_3$$

### 5.3.4 | Finite Bearings

The final expressions neglecting the squeeze term for the deterministic case using 3-node linear triangular elements are:

$$\begin{aligned} 2[K_p]\{P\} &= [K_u]\{U\} - \{Q\} \\ [K_p] &= \sum_{m=1}^{N} \left( \frac{b_1 b_2 + c_i c_j}{48\eta\Delta^2} \right) \iint h_m^3 dA_m \\ [K_u] &= \sum_{m=1}^{N} \left( \frac{b_1 + c_i}{2\Delta} \right) \iint h_m M_j dA_m \\ [Q] &= \sum_{m=1}^{N} \int q N_m ds \end{aligned} \tag{5.32}$$

The bearing surface is developed as shown in Fig. 5.3c. For a finite slider bearing the length and width of the developed surface are equal to the length and width of the bearing dimensions. For the journal bearing, the length is equal to the developed cylindrical surface. One can solve either for the full developed surface or half surface by using the geometric symmetry and applying the appropriate boundary conditions.

The boundary conditions for the prescribed pressure are applied by changing the appropriate rows in the matrix form for the assembly of elements. This means $p = 0$, when $\theta = 0, 2\pi$, also on the other two sides. To account for cavitation the pressure is made equal to zero whenever $p < 0$.

The assembled $K_p$ matrix is banded, symmetric, and sparse. These properties can be used in the optimum storage on the computer and solution of the linear equations.

## Steady–State Characteristics

Steady-state characteristics for finite slider bearings are calculated using finite element method.

- Load carrying capacity

$$W = \iint p\,dA$$

$$= \sum_{m=1}^{N} \iint \left(N_i p_i + N_j p_j + N_k p_k\right) dA_m \tag{5.33}$$

- Friction force

$$F = \iint \tau\,dA$$

$$\tau = \frac{H}{2}\frac{\partial p}{\partial x} + \eta \frac{U}{H}$$

$$\iint \eta \frac{U}{h} dA = \sum_{m=1}^{N} \eta \frac{U A_m}{h_m} \tag{5.34}$$

where $h_m$ is the average film thickness for the element m.

- Co-efficient of friction

$$\mu = \frac{F}{W} \tag{5.35}$$

- Flow in the direction of motion:

$$q_x = \int \left( h\frac{U}{2} - \frac{1}{12\eta}\left(\frac{\partial p}{\partial x}\right) h^3 \right) dy$$

$$q_x = \int h \frac{U}{2} dy - \left(\frac{1}{12\eta}\right)\left(\frac{b_i p_i + b_j p_j + b_k p_k}{2\Delta}\right) \int h^3 dy \tag{5.36}$$

Mean flow is calculated by integrating Equation (5.36) after substituting for $h$ and $h^3$. Similarly, the side flow can be calculated from

$$q_x = \int \left( \frac{1}{12\eta}\left(\frac{\partial p}{\partial y}\right) h^3 \right) dx$$

$$q_x = \left(\frac{1}{12\eta}\right)\left(\frac{c_i p_i + c_j p_j + c_k p_k}{2\Delta}\right) \int \left(h^3\right) dy \tag{5.37}$$

For a finite journal bearing, similar procedure as shown above is adopted.

## Solution of Dynamic Equations

The dynamic equations can be solved on principles developed in the earlier sections. Since these have already been developed and reported in literature (Majumdar 1986), we can proceed to use the same with necessary modifications. The real and imaginary parts of the equations are formulated separately.

Taking the pressure distribution in the interior of the element as

$$p_{ei} = \sum_{j=1}^{n} N_{ij}(\theta, z)\, p_{ij} \tag{5.38}$$

Introducing Equation (5.38) in the variational, we get the following

$$\frac{\partial I(p_i)_j}{\partial \bar{p}_{ij}} = -\int_A (-h_0^3\left(\frac{\partial N_{ij}}{\partial \theta}\sum_{k=1}^{n}\frac{\partial N_{ik}}{\partial \theta}p_{ik} + \left(\frac{d}{b}\right)^2\frac{\partial N_{ij}}{\partial y}\sum_{k=1}^{n}\frac{\partial N_{ik}}{\partial y}p_{ik} + 6h_0\frac{\partial N_j}{\partial \theta}\right)dA = 0 \tag{5.39}$$

Here $n$ is the number of element nodes, index $j$ indicates node number and

$$p_i = p_0$$

$$\frac{\partial I(p_i)_j}{\partial \bar{p}_{ij}} = -\int_A (-h_0^3\left(\frac{\partial N_{ij}}{\partial \theta}\sum_{k=1}^{n}\frac{\partial N_{ik}}{\partial \theta}p_{ik} + \left(\frac{d}{b}\right)^2\frac{\partial N_{ij}}{\partial y}\sum_{k=1}^{n}\frac{\partial N_{ik}}{\partial y}p_{ik}\right)$$

$$-\left(3\bar{h}_0^2\cos\theta\frac{\partial p_0}{\partial \theta} - 6\cos\theta\right)\frac{\partial N_j}{\partial \theta} - \left(3\left(\frac{d}{b}\right)^2\bar{h}_0^2\cos\theta\frac{\partial p_0}{\partial y}\right)\frac{\partial N_j}{\partial y})dA = 0 \tag{5.40}$$

In Equation (5.40) $p_i = p_1$

$$\frac{\partial I(p_i)_j}{\partial \bar{p}_{ij}} = -\int_A (-\bar{h}_0^3\left(\frac{\partial N_{ij}}{\partial \theta}\sum_{k=1}^{n}\frac{\partial N_{ik}}{\partial \theta}p_{ik} + \left(\frac{d}{b}\right)^2\frac{\partial N_{ij}}{\partial y}\sum_{k=1}^{n}\frac{\partial N_{ik}}{\partial y}p_{ik}\right)$$

$$-\left(3\bar{h}_0^2\cos\theta\frac{\partial p_0}{\partial \theta} - 6\cos\theta\right)\frac{\partial N_j}{\partial \theta} - \left(3\left(\frac{d}{b}\right)^2\bar{h}_0^2\cos\theta\frac{\partial p_0}{\partial y}\right)\frac{\partial N_j}{\partial y})dA = 0 \tag{5.41}$$

In Equation (5.41) $p_i = p_2$

$$\frac{\partial I(p_i)_j}{\partial \bar{p}_{ij}} = -\int_A (-\bar{h}_0^3\left(\frac{\partial N_{ij}}{\partial \theta}\sum_{k=1}^{n}\frac{\partial N_{ik}}{\partial \theta}p_{ik} + \left(\frac{d}{b}\right)^2\frac{\partial N_{ij}}{\partial y}\sum_{k=1}^{n}\frac{\partial N_{ik}}{\partial y}p_{ik}\right)$$

$$-12\lambda\sin\theta\frac{\partial N_j}{\partial \theta})dA = 0 \tag{5.42}$$

In Equation (5.42) $p_i = p_1$ imaginary component

$$\frac{\partial I(p_i)_j}{\partial \bar{p}_{ij}} = -\int_A (-\bar{h}_0^3 \left( \frac{\partial N_{ij}}{\partial \theta} \sum_{k=1}^{n} \frac{\partial N_{ik}}{\partial \theta} p_{ik} + \left(\frac{d}{b}\right)^2 \frac{\partial N_{ij}}{\partial y} \sum_{k=1}^{n} \frac{\partial N_{ik}}{\partial y} p_{ik} \right)$$

$$-24\lambda \cos\theta \frac{\partial N_j}{\partial \theta})dA = 0 \tag{5.43}$$

In Equation (5.43) $p_i = p_2$ imaginary component. The above equations reduce to the form

$$\{K_p\}\{\bar{p}_i\} = \{K_{ui}\} \tag{5.44}$$

where $i = 0, 1, 2$

The boundary conditions are—in Equation (5.39) $p_0 = 0$, when $\theta = 0, 2\pi$, also on the other two sides, and also when $p_0 < 0$. In Equations (5.40 to 5.43) $p_1$ and $p_2$ are made equal to zero at the same $\theta$ and nodes at which $p_0$ was equal to zero.

From Equation (5.39) we get $p_0$ which on integration gives steady-state load (non-dimensional). On solving Equations (5.40 to 5.43), we can obtain $p_1$ and $p_2$ which on integrations give us 4 stiffness and 4 damping coefficients, respectively.

## Examples

***E.5.1*** Calculation of Load Carrying Capacity of a Finite Plane Slider Bearing:
A plane slider bearing with a fixed shoe is operating under the following conditions (Majumdar 1986):
Width of the bearing B = 50 mm
Length to width ratio (L/B) = 1.0
Sliding velocity U = 5 m/s
Minimum Film thickness $h_2 = 0.02$ mm
Absolute viscosity of the lubricant $\eta = 0.02$ Pa.s
Find the load carrying capacity and friction force

***Solution:***

The steady-state solution of the slider bearing problem is obtained by solving the governing Reynolds Equation (5.19). In this equation, the time derivative components are ignored.

1. The key preprocessing steps are (a) read the data given in the problem, (b) form the geometry of the slide bearing, and (c) mesh it using 3 node triangular elements as shown in Figs 5.3b and c.
2. In the processor stage, (a) compute the element matrices using Equations 5.32, (b) assemble the element matrices, (c) apply boundary conditions, and (d) solve the assembled equations to get the pressures at each node.
3. In postprocessor, compute (a) load carrying capacity by integrating the nodal pressures (Equation 5.33), (b) compute gradients to calculate the friction force and coefficient of friction (Equation 5.34). We can compute other quantities of interest also.

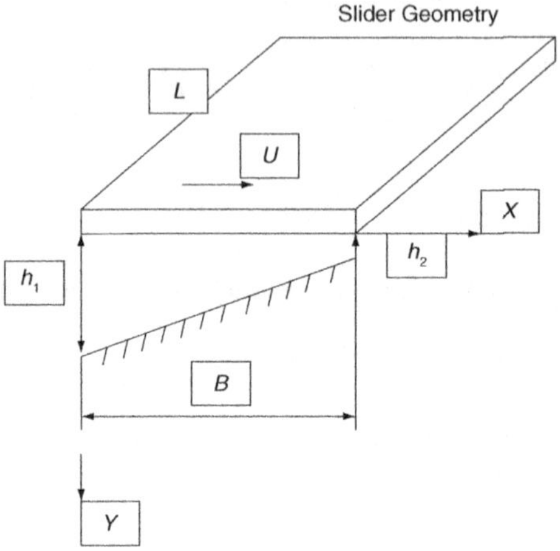


**Figure 5.4** | A Plane Slider Bearing

### *Results*

The results obtained by the finite element method are given in Table 5.3. They are compared to results obtained by finite difference method (shown in bracket) as given in Majumdar (1986).

The results show an excellent correlation for load carrying capacity and are fairly close while calculating derived quantities such as friction force.

***E.5.2*** Calculation of Load Carrying Capacity of a Finite Journal Bearing:
A full journal bearing is having the following specifications:
Journal Diameter D = 100 mm
Length to Diameter ratio L/D = 1.0
Radial Clearance C = 0.025 mm
Journal Speed N = 3000 rpm
Operating Eccentricity $\varepsilon = 0.6$
Average Viscosity of Lubricant $\eta = 0.02$ Pa.s
Calculate the Load Carrying Capacity and Friction Force of the Bearing

**Table 5.3** | Plane Slider Bearings

| *L/B* | 2 | 1.5 | 1.0 | 0.75 | 0.5 |
|---|---|---|---|---|---|
| $\bar{w} = \dfrac{Wh_2^2}{\eta ULB^2}$ | 0.109<br>(0.109) | 0.09455<br>(0.0946) | 0.06887<br>(0.06889) | 0.0503<br>(0.0504) | 0.02898<br>(0.0289) |
| $\bar{F} = \dfrac{Fh_2}{\eta ULB}$ | 0.8576<br>(0.748) | 0.8349<br>(0.740) | 0.7964<br>(0.727) | 0.7687<br>(0.718) | 0.7366<br>(0.707) |

**Table 5.4** | Finite Journal Bearings

| $L/D = 1$ | $\varepsilon$ | $S$ | $\left(\frac{R}{c}\right)\mu$ | $\phi°$ |
|---|---|---|---|---|
| | 0.1 | 1.404<br>(1.35) | 27.85 | 84.93<br>(79) |
| | 0.2 | 0.676<br>(0.632) | 13.62<br>(12.9) | 79.80<br>(74) |
| | 0.3 | 0.4228<br>(0.3282) | 8.744<br>(8.04) | 74.56<br>(68) |
| | 0.4 | 0.288<br>(0.261) | 6.198<br>(5.8) | 69.118<br>(62) |
| | 0.5 | 0.2012<br>(0.179) | 4.582<br>(4.31) | 63.368<br>(56) |
| | 0.6 | 0.1389<br>(0.12) | 3.425<br>(3.21) | 57.157<br>(50) |
| | 0.7 | 0.09135<br>(0.0765) | 2.52<br>(2.36) | 50.224<br>(43) |
| | 0.8 | 0.0535<br>(0.0448) | 1.758<br>(1.71) | 42.048<br>(36) |
| | 0.9 | 0.0233<br>(0.0191) | 1.0533<br>(1.06) | 31.251<br>(25) |

***Results***

The results obtained by the finite element method are given in Table 5.4. They are compared with results obtained by finite difference method (shown in bracket) as given in Majumdar (1986). The results show a good correlation for load carrying capacity and friction force.

## 5.4 | Cavitation and Cavitation Boundary Conditions

In oil film bearing usually there are two types of cavitation (Taylor 1974). These are gaseous cavitation and vapour cavitations. The first type is observed when a gas dissolved in oil that is emitted from the solution as the pressure falls to the saturation pressure of the gas in the liquid. The second one is found to occur in situations where the liquid is subjected to rapid changes in pressure as in dynamically loaded bearings and hydraulic machinery. Here the vapour bubbles collapse on to the boundary surfaces that can cause 'pitting'.

The location of the film-cavity interface is of interest to us as such information is important in estimating proper boundary conditions for the Reynolds equation. The Swift–Stieber (1931, 1933) conditions, presently used, ends the film where both pressure and pressure gradient become zero or equal to cavitation pressure. These conditions are also known as Reynolds boundary conditions. The Swift–Stieber conditions were derived independently in the 1930s. The former used stability consideration, but the latter employed the concept of flow continuity.

In the recent years, Floberg (1961) proposed a boundary condition for lightly loaded bearings having the familiar finger-like cavities. His boundary condition is that no fluid flows past the cavity boundary and the fluid pressure is same as the cavity pressure. Coyne and Elrod (1968) gave an analysis which took into account not only the liquid-gas interface which developed in practice but also the other effects like inertia, gravity, and surface tension.

In short, there are three realistic boundary conditions:

1. The Swift–Stieber or continuity conditions
2. The separation conditions
3. The Floberg conditions are well known

One of the advantages of the Swift–Stieber (sometimes Reynolds) boundary conditions is easy to apply. Numerical solution of Reynolds equation with this condition is straightforward for both one-dimensional and two-dimensional flow. Although the separation condition has often been suggested for lightly loaded bearings, not many design information are available. For an infinitely long bearing it is shown that $\frac{\partial^2 p}{\partial x^2} = 0$ at the cavitated region. The Floberg boundary conditions are not easy to apply numerically. However, there is an experimental evidence (Cole 1956), which leads strong support to Floberg's approach. Cavitation boundary conditions have been discussed in Chapter 4. For exact boundary condition, one can look into details given there.

It has been shown by Christopherson (1942) that in the process of iteration whenever the pressure becomes negative at any mesh point (in case of a numerical solution), it should be made equal to zero or the cavitation pressure and then further iteration can be carried out. This process is likely to satisfy the boundary conditions of both $p = 0$ and $\nabla p = 0$ at the cavitation boundary. The method shown by Christopherson has been used by Raimodi and Boyd (1958). Majumdar and Hamrock (1981) have also found that it satisfied the Swift–Stieber condition for infinitely long and finite oil journal bearings.

In recent years, Kicinski (1986) proposed a cavitation model taking into account the oil film history (Fig. 5.4). The following general observation should be considered to formulate exact mathematical model of a cavitation model:

- The process occurring in the cavitation zone is described on the assumption that the shape of the bubbles and the flows are determined by the flow continuity equation. This assumption allows neglecting the processes induced by surface tension, diffusion, bubble implosion, and it suggests that the large cavitation bubbles are generated mainly by air which is sucked in across the bearing lateral edges.
- The boundaries of the cavitation zone can be determined under the condition that positive pressure occurs only when the oil fills the oil gap completely. The time and space varying shape of the flows illustrates the flow prehistory under consideration. The Reynolds equation is integrated within boundaries determined from the continuity equation.
- The above procedure gives a unique solution which does not depend on the initial condition, except for periodically varying load.

For the cavitation zone, it is assumed that

- The flow within the real cavitation zone, which includes several more or less regular 'finger-like' bubbles, corresponds to the flow in the theoretical zone with only one 'analytical' bubble.

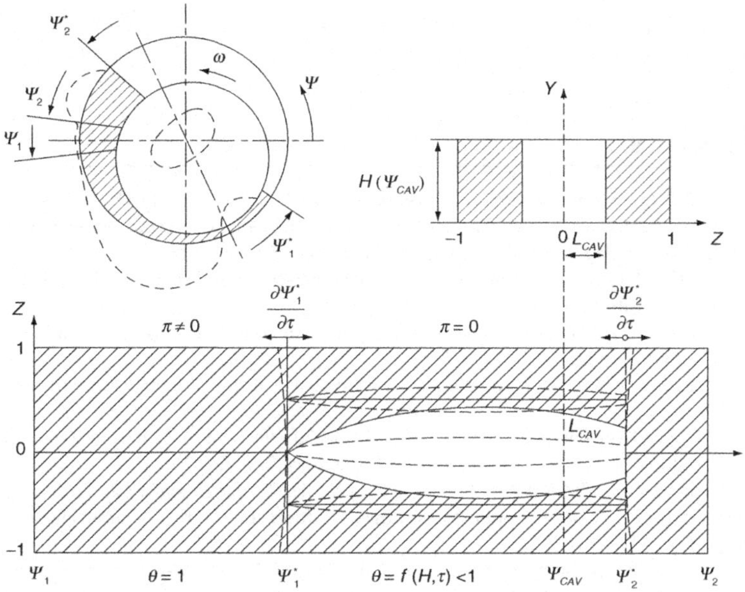


**Figure 5.4** | Theoretical Model of Cavitation Zone

- All oil flows around the analytical bubble which has a flat surface in the direction of the film thickness and therefore the shape of the analytical bubble, resulting from the continuity equation will be defined by its width.
- The flow within the cavitated zone is gradient less and the pressure is equal to atmosphere pressure.

In the figure, $\psi_{10}^*$ and $\psi_{20}^*$ are the arbitrary initial boundaries, $L_{CAV}$ is the dimensionless width of the analytical bubble, $\theta$ is the oil gap filling coefficient, $H$ is the oil gap height, $\pi$ is the dimensionless pressure, $\tau$ is the dimensionless time.

## Examples

The following examples show the results for slider and journal bearings using finite difference method.

***E.5.1 (repeated)*** For a slider bearing, $L = B = 50$ mm, $U = 5$ m/s, $h_2 = 0.02$ mm, $n = 2.0$ and $\eta = 0.02$ Pas. From Table 5.1, we get for $L/B = 1.0$, $n = 2$, the dimensional load capacity, friction force, and end flow as 0.0689, 0.7276, and 0.2462, respectively. The absolute value of these parameters can be found as 2.153 kN. 9.035 N and $1.231 \times 10^{-3}$ $m^3/s$. The coefficient of friction is $4.197 \times 10^{-3}$.

***E.5.2 (repeated)*** For a full finite journal bearing having $L/D = 1.0$, $D = 100$ mm, $C = 0.025$ mm, $N =$ 3000 rpm, $\varepsilon = 0.6$ and $\eta = 0.02$ Pas, we get $W = 333.33$ kN, $\mu = 1.615 \times 10^{-3}$ (Refer to Table 5.2 for $L/D = 1.0, \varepsilon = 0.6$).

As expected, actual load and coefficient of friction are lower and higher, respectively, than the infinitely long journal bearing solution. This is also revealed from the experimental findings (1930).

## REFERENCES

*Boegli, C.P.* (1947), The hydrodynamic lubrication of finite sliders, *J. Appl. Phys.* **18**, 482–488.

*Booker, J.F.* and *Huebner, K.H.* (1972), Application of finite element methods to lubrication: an engineering approach, *Trans ASME Journal of Lubrication Technology*, **94**, 313–323.

*Christopherson, D.G.* (1942), A new mathematical model for solution of film lubrication problems, *Proc. Instn. of Mech. Engrs.*, **146**, 126–135.

*Cole, J.*A. and *Hughes, C.J.* (1956), Oil flow and film extent in complete journal bearings, *Proc. Instn, Mech. Engrs.*, **170,** 4999–4510.

*Coyne, J.*C. and *Elrod, H.G.* (1968), An exact asymptotic solution for a separating fluid film, *ASME Paper D*-8-70.

*Floberg, L.* (1961). On hydrodynamic lubrication with special reference to cavitation in bearings, Chalmeres Tenkniska Hogskola Doctora vahandlinger, Gothenburg, Sweden, No. 30.

*Huebner, K.H.* and *Thornto, E.A.* (1982), *The Finite Element Method for Engineers*, 2nd Edition, John Wiley.

*Kicinski, J.* (1986), Influence of flow prehistory in the cavitation zone in the dynamic characteristics of solid bearings, *Wear*, **111**, 289–311.

*Kingsbury, A.* (1930), On problem in the theory of fluid film lubrication with an experimental method of solution, *ASME paper APM*. 53–5.

*Klit, P.* and *Lund, J.W.* (1986), Calculation of the dynamic coefficients of a journal bearing, using a variational approach, *Trans ASME Journal of Tribology*, **108**, 421–425.

*Majumdar, B.C.* (1986), *Introduction to Tribology of Bearings*, A.H. Wheeler & Co. Pvt. Ltd., Allahabad.

*Majumdar, B.C. and Hamrock, B.J.* (1981), Effect of surface roughness on hydrodynamic bearings, *NASA, TM* 81771.

*Pinkus, O.* (1958), Solution of Reynolds equation for finite journal bearing. *Trans, ASME*, **80**, 858–864.

*Raimondi, A.A.* and *Boyd, J.* (1958), A solution for the finite journal bearing and its application to analysis and design, I, II and III, *ASLE Trans*, **1**, 159–209.

*Reddy, J.N.* (1969), Finite element solution of the incompressible lubrication problem, *Trans ASME Journal of Lubrication Technology*, **91**, 524–533.

*Reddy, J.N.* (1993), *The Finite Element Method*, 3rd Edition, McGraw Hill International Edition.

*Sarkar, R.*K. (2004) *Numerical Methods for Science and Engineers* P. 347–348. Publisher Eswar Press, Chennai.

*Stieber, W.* (1933), Das Schwimn larger, Krayan VDI, Berlin.

*Swift, H.W.* (1931), The stability of lubrication films in journal bearings, *Proc. Inst. Civil Eng., London*, **233,**267–288.

*Taylor, C.*M. (1974), Separation cavitation in lightly loaded fluid film bearings with both surfaces in motion, *J. Mech. Engg. Sci., I. Mech. E.*, **16,** 147–155.

*Zienkiewicz, O.C.* (1990), *The Finite Element Method*, 3rd Edition, Tata McGraw Hill Co. Ltd.

Chapter 6

# Thermohydrodynamic Analysis of Fluid Film Bearings

## 6.1 | Introduction

Thermohydrodynamic analysis of fluid film lubrication of mechanical components, viz., bearings, gears, and seals is basically a heat transfer analysis wherein heat generated due to viscous shear of the lubricant is dissipated through convection, conduction and to some extent by radiation is balanced. However, generally the dominant mode of heat dissipation in fluid film lubrication is due to convection, i.e., heat is carried away by the fluid flowing out of boundaries open to atmosphere and into the sump to collect the fluid. In some situations, conduction to solid bodies surrounding the fluid will also carry away significant amount of heat generated. Generally, radiation is negligible. Therefore, in the analysis heat transfer due to convection and conduction is considered to seek balance between the heat generated and heat dissipated to achieve an equilibrium condition. This helps in determining the temperature rise in the fluid film and the surrounding solids. The following sections will describe the methods to take into account the thermal effect and to perform thermohydrodynamic analysis of bearings. The procedures discussed are general and can be applied to any fluid film lubrication problem.

## 6.2 | Thermal Analysis of Sector-Shaped Tilting Pad Thrust Bearings

In large turbo generator bearings, significant amount of heat is generated due to viscous shear and churning of oil. These bearings can operate either in laminar or turbulent regime in case of large-sized thrust bearings depending on the rotor speed. Higher speeds usually increase power losses and temperature rise, whereas the oil flow is reduced. Power consumed in these bearings and their supporting systems becomes a substantial drain on the output of the system. The most important parameter for thrust bearings for evaluating their power efficiency is to evaluate the ratio of the load carried to power consumed. Since the power consumed is dependent on the load bearing area, this requires raising the average loading at which the bearing operates in order to reduce its gross area. In general, this reduces the minimum film thickness in the bearing, so that various design alternatives can be studied based on a safe and reliable minimum film thickness. Also white metal or Babbitt metal temperature rise is an important consideration. Therefore, a thermal analysis of the entire system is essential to the solution of the problem. In the case of pivoted pad thrust bearings which are the most widely used configuration, it is also necessary to estimate the thermoelastic distortion of the pad to accurately predict the film shape and minimum film thickness under its actual operating conditions.

A generalized analysis for sector-shaped pivoted pad thrust bearings developed by Ettles (1976) would be discussed. This methodology is an effective design tool and can also be used for trouble shooting in industries. Heat transfer analysis details have been dealt with by Vohr (1981).

Sector-shaped thrust bearing and pad are shown in Figs 6.1 and 6.2, respectively.

The analysis incorporates several factors, for example:

1. Simultaneous solutions of coupled Reynolds equation having viscosity variation with temperature incorporated into it and energy equation for pad temperature distribution.
2. Complete heat transfer analysis including conduction to bearing surface and convection due to side leakage, conduction to the runner and heat transferred to the cold oil in the groove between pads. A detailed investigation to account for the complete heat transfer presented by Vohr (1981) is also included in the analysis.
3. Elastic distortion of pad in bending and shear and thermo elastic distortion of pad is incorporated.
4. Hot oil carryover from one pad to the other is accounted for.

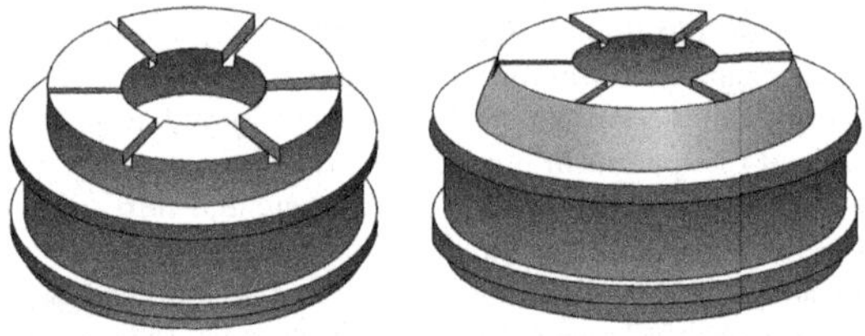

**Figure 6.1** | Fixed or Tapered Land Thrust Bearing

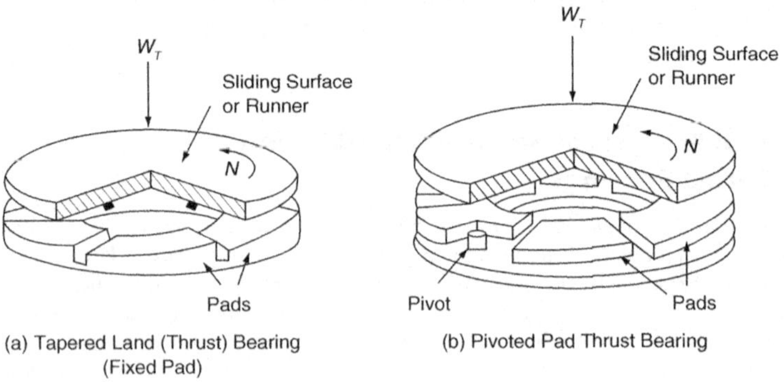


**Figure 6.2** | Thrust Pad Bearings

The Reynolds equation for a sector pad in cylindrical coordinates ($r$, $\theta$) is written for an incompressible fluid as:

$$\frac{1}{r}\frac{\partial}{\partial \theta}\left(\frac{h^3}{\eta}\frac{\partial p}{\partial \theta}\right)+\frac{\partial}{\partial r}\left\{\frac{rh^3}{\eta}\frac{\partial p}{\partial r}\right\}=6\omega r\frac{\partial h}{\partial \theta} \tag{6.1}$$

where both film pressure $p$ and viscosity of the oil, $\eta$, are function of ($r$, $\theta$).

Equation (6.1) is subject to boundary conditions such that $p = 0$ everywhere at the boundary of the pad and cavitation boundary condition, i.e., $p=\frac{dp}{d\theta}=0$ at the locations where the pressure in the lubricant film falls below ambient pressure.

The energy equation for temperature distribution in the lubricant film with the assumption that temperature across the film remains constant is a function of $r$, $\theta$ only. It is expressed in cylindrical coordinates as:

$$q_\theta\frac{\partial T}{\partial \theta}+q_r\frac{\partial T}{\partial r}=\frac{1}{\rho c_p}\left[\frac{\eta(r\omega)^2}{h}+\frac{h^3}{12\eta}\left\{\left(\frac{1}{r}\frac{\partial p}{\partial \theta}\right)^2+\left(\frac{\partial p}{\partial r}\right)^2\right\}\right]$$
$$+H_u\left(T_u-T\right)+H_l\left(T_b-T\right) \tag{6.2}$$

where $\rho$ and $c_p$ are density and specific heat of the lubricant at constant pressure, respectively, and $T$ is lubricant film temperature.

Equation (6.2) is basically a heat balance equation. Lubricant flow rates $q_\theta$ and $q_r$ in $r$ and $\theta$ directions, respectively, are written as:

$$q_\theta=-\frac{h^3}{12\eta}\frac{1}{r}\frac{\partial p}{\partial \theta}\omega r\frac{h}{2} \tag{6.3}$$

$$q_r=-\frac{h^3}{12\eta}\frac{\partial p}{\partial r} \tag{6.4}$$

The viscosity $\eta$, density $\rho$ and $c_p$ of the oil are known a priori. The viscosity of the lubricant is a function of temperature $T$, whereas density $\rho$ and $c_p$ are considered constant, i.e., independent of temperature. Pressure dependence of $\eta$, $\rho$, and $c_p$ is ignored.

$T_u$ is the temperature of upper surface, i.e., runner; $T_b$ is the temperature of the bottom surface of the pad. $H_u$ and $H_l$ are heat transfer coefficients at the upper and lower surface of the film, respectively. The last two terms on the right side of the energy equation represent heat flow due to conduction from the fluid film to the runner and lower side of the pad. It is again presumed that the heat transfer coefficients are either known or can be determined.

The expression for film thickness $h$ in case of a sector-shaped pivoted bearing can be expressed as:

$$h=h_p+\alpha_t\left\{r_p\sin\left(\theta_p-\beta/2\right)-r\sin\left(\theta-\beta/2\right)\right\}$$
$$-\alpha_r\left\{r_p\cos\left(\theta_p-\beta/2\right)-r\cos\left(\theta-\beta/2\right)\right\}+w \tag{6.5}$$

where $w$ includes elastic and thermal distortion of the pad, $\alpha_t$, and $\alpha_r$ are pad inclination in $\theta$ and $r$ directions, respectively. $h_p$ is the film thickness at the pivot location.

Elastic bending due to pressure of the film and thermal distortion of the pad due to temperature gradient can be calculated following Ritz procedure (1961) as shown by Sternlicht *et al.* (1961).

According to the procedure outlined by Sternlicht *et al.*, the pad deformation due to bending of the pad as a result of pressure load and thermal gradient through the pad ($w$) is expressed as:

$$w = \sum_{i=1}^{6} A_i f_i = A_1 f_1 + A_2 f_2 + A_3 f_3 + A_4 f_4 + A_5 f_5 + A_6 f_6 \tag{6.6}$$

Coefficients $A_1$ to $A_6$ are determined by solving the following matrix equation

$$\left[I_{i,j}\right]\left\{A_i\right\} + \left\{k_j\right\} = 0 \tag{6.7}$$

where $k_j$ are determined from pressure and temperature distribution in the bearing film. $I_{i,j}$ are calculated using mode shape functions as given below:

$$f_1 = \left(r_c / r_b\right)^2 - 1$$

$$f_2 = \left(r_c / r_b\right)^4 - 1$$

$$f_3 = \left(r_c / r_b\right)^2 \sin 2\phi_c$$

$$f_5 = \left(\frac{r_c}{r_b}\right)^2 \left[\left(\frac{r_c}{r_b}\right)^2 - 1\right] \sin 2\phi_c$$

$$f_6 = \left(\frac{r_c}{r_b}\right)^2 \left[\left(\frac{r_c}{r_b}\right)^2 - 1\right] \cos 2\phi_c \tag{6.8}$$

where $\phi_c$ is the angle shown in Fig. 6.3 of sector pad.

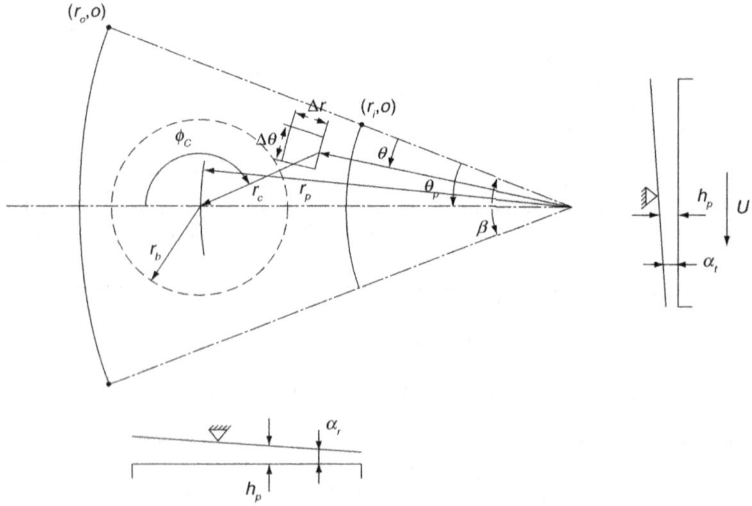


**Figure 6.3** | Coordinate System for Bearing Pad [Vohr, ASME JOLT, 1980]

Ettles and Cameron (1976) developed a simplified analysis to evaluate the distortion of the pad due to pressure of lubricant film and thermal effects.

Boundary conditions for the Reynolds Equation (6.1) are written as:

$$p(0,r) = p(\beta,r) = 0$$
$$p(\theta,r_i) = p(\theta,r_0) = 0 \tag{6.9}$$

At the cavitation boundary both pressure and pressure gradient in θ direction would be zero, i.e.,

$$p(\theta,r) = \frac{dp}{d\theta}(\theta,r) = 0 \tag{6.10}$$

Boundary conditions for the energy Equation (6.2) are:

$$T(0,r) = 0 \text{ and } \frac{\partial T}{\partial r} = 0 \text{ at } r = r_i \text{ and } r_o \tag{6.11}$$

Both Reynolds and energy equation are discretized using finite difference method and solved by an iterative procedure described briefly as follows:

- Initial input data are read, viz., $T_{in}$ (oil inlet temperature), film thickness at the pivot $h_p$, initial pressure and temperature distributions over the pad, pad inclinations $\alpha_r$, $\alpha_t$ and initial pad deformation $w$ is calculated using these data.
- Determine pressure and temperature distributions by solving Equations (6.1) and (6.2) satisfying appropriate boundary conditions and pad deformation, $w$ is calculated using new pressure and temperature distributions. Load carrying capacity of the pad, $W$ is also calculated.
- Determine the film thickness at the pivot $h_p$ and pad inclinations by satisfying moment balance equation at the pivot.

Steps 1 and 2 must be repeated until a preassigned convergence criterion is satisfied. Step 3 is to be repeated also until the correct film thickness at the pivot is obtained which ensures load convergence. The foregoing process must be repeated till the solution to the problem is obtained.

The load capacity of the pad is determined as:

$$W = \int_0^\beta \int_{r_i}^{r_0} prd\theta dr \tag{6.12}$$

The friction coefficient due to torque is defined as:

$$f = \frac{M_r}{Wr} \tag{6.13}$$

where $M_r$ is the frictional torque and is found from:

$$M_r = \int_0^\beta \int_{r_i}^{r_o} \left( \frac{\text{rh}}{2}\frac{\partial p}{\partial \theta} + \frac{\eta r^3}{h} \right) dr \; d\theta \tag{6.14}$$

The load and torque are determined using a suitable numerical integration method.

To develop complete heat transfer analysis the following steps have to be done after the above procedure is completed without accounting for heat transfer from the liquid film to bearing pad and runner surface. The procedure to account for the heat transfer from the fluid film to the pad and to the rotor surface has been described in detail by Vohr (1981) as given below.

### 6.2.1 | Heat Transfer Modes

The modes of heat transfer in the thrust bearing as described by Vohr (1981) are:

- Heat removal by the lubricant side leakage from the pad edges.
- Heat conduction from the film to runner and from runner oil to the bearing tub oil.
- Heat conduction from the bearing film to the pad and to the tub oil.
- Heat transferred to the cold oil in the groove between pads by convection along the runner.

Ettles and Cameron (1968, 1970) investigated the aspects of groove flow and developed solutions for groove flow. Hot oil carryover aspect in thrust bearings was also investigated. According to this analysis, a hot oil carry over factor $k$ is defined as:

$$k = \frac{T_{in} - T_s}{T_r - T_s} \tag{6.15}$$

where $T_s$ – oil supply temperature in the groove, $T_r$ – bearing runner temperature which is assumed as average of the pad inlet and outlet temperatures, i.e., $T_{in}$ and $T_o$,

thus,

$$T_r = \frac{T_{in} + T_o}{2} \tag{6.16}$$

Pad inlet temperature is thus given by:

$$T_{in} = \frac{2(1-k)}{(2-k)} T_s + \frac{k}{2-k} T_o \tag{6.17}$$

According to Ettles (1970), hot oil carryover factor varies between 0.71 and 0.93.
Heat transfer by various modes mentioned above can be estimated in following way (Fig. 6.4):
1. Heat transferred by the side leakage flow ($Q_l$) is given by

$$Q_l = n\rho c_p \begin{bmatrix} \int_o^\beta q_r(r_o,\theta)\left[T(r_o,\theta) - T_b\right] r_o d\theta \\ -\int_o^\beta q_r(r_i,\theta)\left[T(r_i,\theta) - T_b\right] r_i d\theta \end{bmatrix} \tag{6.18}$$

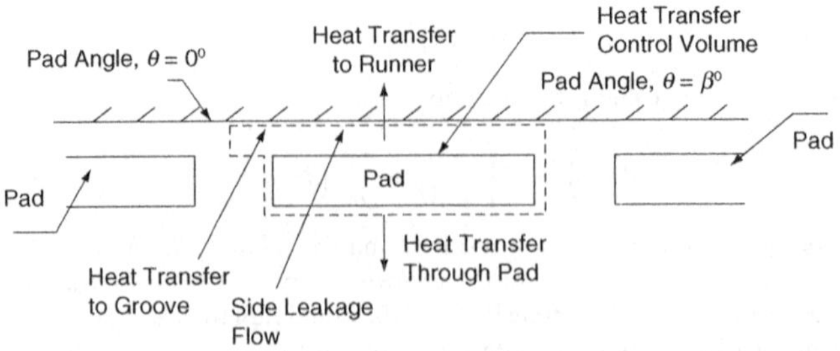


**Figure 6.4** | Schematic Diagram of Heat Transfer Analysis [Vohr, ASME JOLT, 1980]

2. Heat conduction from the fluid film to the bearing pad through the runner assuming the runner temperature to be constant at $T_r$, i.e., for isothermal condition is given as:

$$Q_r = H_u A_s \left(T_f - T_r\right) = U_r \left(T_r - T_b\right) \tag{6.19}$$

where $A_s = \pi\left(r_o^2 - r_i^2\right)$ is the surface area of the pad, $T_f$ is the average pad film temperature. Heat transfer coefficient $H_u$ is determined by the assumption that the mean Nusselt number of fluid film $N_u = 2H_u h_p / k$ where thermal conductivity of the lubricant is $k$. It has been shown by Kays (1980) that an approximate value of $N_u$ may be taken as 7.2 based on convective heat transfer solutions for laminar flow between parallel plates. $U_r$ is the overall heat transfer coefficient through the runner and is given by $U_r = 466$ R, W/°C, R is the radius of the runner in meter.

3. Heat conduction from the lubricant film to the bearing pad can be determined by solving heat conduction equation for the pad which is expressed as:

$$\frac{1}{r}\frac{\partial}{\partial r}\left[r\frac{\partial T_1}{\partial r}\right] + \frac{1}{r^2}\frac{\partial^2 T_1}{\partial \theta^2} + \frac{\partial^2 T_1}{\partial z^2} = 0 \tag{6.20}$$

where $T_1$ is bearing pad temperature, $z$ is the coordinate normal to the pad surface, i.e., across the pad thickness. Equation (6.20) must be solved along with the energy Equation (6.2). This is computationally intensive and also prone to numerical instability. Therefore, generally heat flow in the plane of the pad is neglected, i.e., $\frac{\partial T_1}{\partial r}$ and $\frac{\partial T_1}{\partial \theta}$ are taken to be zero. This approach has been adopted by Safar and Szeri (1974), Suganami and Szeri (1979). In that case, Equation (6.20) reduces to

$$\frac{\partial^2 T_1}{\partial z^2} = 0 \tag{6.21}$$

For temperature continuity at the interface of bearing pad and lubricant film it is required that $T_1 = T$ at $z = 0$, i.e., at the upper surface of the pad.

At the bottom surface of the pad

$$-k_p \frac{\partial T_1}{\partial z}\bigg|_{z=t} = H_p\left(T_l - T_b\right) \tag{6.22}$$

where $k_p$ is thermal conductivity of the pad material and $t$ its thickness. $H_p$ is the heat transfer coefficient at the pad bottom surface. Thus, overall heat transfer coefficient ($H_l$) in Equation (6.2) can be written as:

$$\frac{1}{H_l} = \frac{1}{H_p} + \frac{t}{k_p}$$

or

$$H_l = \left(\frac{H_p k_p}{H_p t + k_p}\right) \tag{6.23}$$

Heat transfer on the bottom of the pad varies approximately linearly with runner peripheral velocity, $R\omega$ as has been reported by many researchers. For large pads of approximately 0.8 m long radial the values of $H_p$ can lie between 260 to 450 W/m$^2$°C. A suitable value can be chosen for $H_p$ depending on size of the pad and speed of the runner. Heat transferred through the bearing pads to the tub oil is given by:

$$Q_p = H_l\left(T_f - T_b\right) A_p \tag{6.24}$$

where $A_p$ is the total pad surface area.

4. Heat transferred to the cold oil in the groove between pads $Q_g$ is given by:

$$Q_g = H_g \left(T_r - T_b\right)\left(A_p - A_p\right) \tag{6.25}$$

where $H_g$ – groove heat transfer coefficient.

Ettles (1968) studied the problem of groove heat transfer. It was concluded that the characteristic Nusselt number lie in the range of 55–60. $H_g$ for a groove dimension of 0.025 m worked out as 340 W/m² –°C which is very low. Test program undertaken by Vohr (1981) yielded the value of $H_g$ in the range of 2550–3670. For details, one can look into the work of Vohr (1981). The algorithm followed by Vohr is shown in Fig. 6.5.

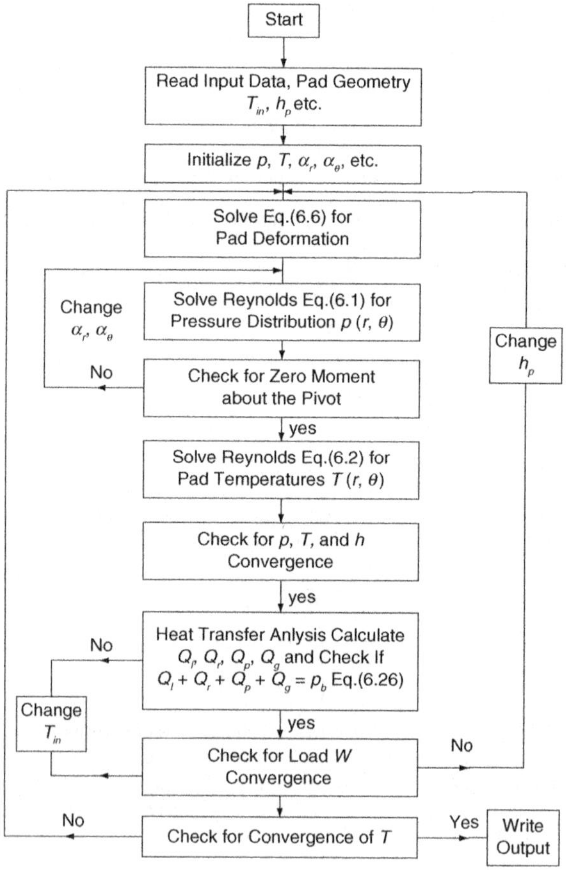


**Figure 6.5** | Flowchart According to Vohr, ASME JOLT, 1980

Therefore, to determine pad inlet temperature heat transferred from the bearing must be equal to the power dissipated in the lubricant film $P_b$. Thus,

$$Q_l + Q_r + Q_p + Q_g = P_b \tag{6.26}$$

Once all the performance characteristics are determined for the initialized $T_{in}$ and $h_p$, heat balance is done evaluating all heat transfer quantities. If the heat balance is satisfied, the solution is accepted or else $T_{in}$ is adjusted and the process is repeated. Thereafter, load convergence is checked. If the calculated load does not match the desired input load, $h_p$ is adjusted and the process is repeated until the $T_{in}$, $h_p$ and load convergence is obtained. Vohr (1981) observed that entire iteration scheme converged quickly. This entire approach is very vital for successful design of pivoted pad thrust bearings for vertical hydro turbo generators. Thermal analysis of sector pad thrust bearings including elasticity effect was developed by Castelli and Malanoski (1969). It has been seen by several investigators that high operating speeds can result in turbulence. Power loss in the bearing increases and pad temperature also increases. Gregory (1974), Capitao (1974), Huebner (1974), and Hashimoto and Wada (1985) have investigated thermohydrodynamic lubrication problem of thrust bearings in the turbulent flow regime. Frictional losses in hydrodynamic thrust bearings of high speed turbines in power plants can be as high as 0.2% of total output. In a 500 MW station with a turbine speed of 3600 rpm these losses are about 1000 kW in thrust bearings only as reported by Pinkus *et al.* (1977). One method of reducing this power loss is to adopt directed lubrication as shown by New (1974) and Mikula and Gregory (1983). Directed lubrication is also referred as leading edge lubrication in which lubricant is supplied to each pad over a confined region of its leading edge which eliminates churning losses in the bearing casing. It can reduce friction loss by almost 50%. However, it can also result in incomplete film and reduce load capacity of the bearing by reducing lubricated area of the pad as has been analyzed by Etsion and Barkan (1981).

## 6.3 | Thermohydrodynamic Analysis of Journal Bearings

Thermal effect in journal bearings is generally evaluated by calculating an effective temperature and the corresponding effective viscosity of the lubricant through a heat balance between heat generated and heat dissipated using isothermal analysis. However, in high speed lubrication temperature variations are significant and important. Therefore, variation of fluid properties due to temperature, especially viscosity is very important because of strong dependence of lubricant viscosity on temperature. Thermal analysis of journal bearings must take into account the viscosity variation due to temperature. Experimental results of Dowson *et al.* (1966), Tonnesen and Hansen (1981), and De Choudhary and Barth (1981) lend strong support to the need for a rigorous analytical approach to handle the thermal effect problem in journal bearings. Dowson (1962) developed generalized Reynolds equation which accounts for viscosity as well as density variation of lubricant with temperature not only in the flow direction but also across the film thickness. In Chapter 3, a generalized Reynolds equation has been developed. Since the development of generalized Reynolds equation several thermohydrodynamic analyses of journal bearings have been presented. A review of these methods has been reported by Pinkus and Wilcock (1980). Briefly stating there are mainly three different approaches to solve the thermohydrodynamic (THD) lubrication problems in journal bearings.

1. Adiabatic solutions in which viscosity variations across the film thickness are neglected. Heat generated is removed totally by convection due to flow of the fluid.
2. THD lubrication solution with more accurate heat transfer considerations which take into account heat flux continuity conditions at fluid solid boundaries.
3. THD lubrication analysis including thermoelastic deformation and other considerations such as reverse flow in the inlet and cold oil mixing at the inlet groove, etc.

A general analysis of THD lubrication of journal bearings is being discussed now following Ferron *et al.* (1983), Gethin (1987), and Singh and Majumdar (2005). These analyses account for thermoelastic deformation of the bush. Laminar flow regime is assumed to prevail.

A developed view of the journal bearing and the coordinate system is shown in Fig. 6.6 and Fig. 6.7 respectively.

The generalized thermal Reynolds equation as derived by Dowson (1962) can be written for steady-state incompressible Newtonian lubricant as:

$$\frac{\partial}{\partial\theta}\left(\bar{h}^3 F_2 \frac{\partial\bar{p}}{\partial\theta}\right)+\left(\frac{R}{L}\right)^2 \frac{\partial}{\partial\bar{y}}\left(\bar{h}^3 F_2 \frac{\partial\bar{p}}{\partial\theta}\right)=\frac{\partial}{\partial\theta}\left(\bar{h}-\bar{h}\frac{F_1}{F_o}\right) \tag{6.27}$$

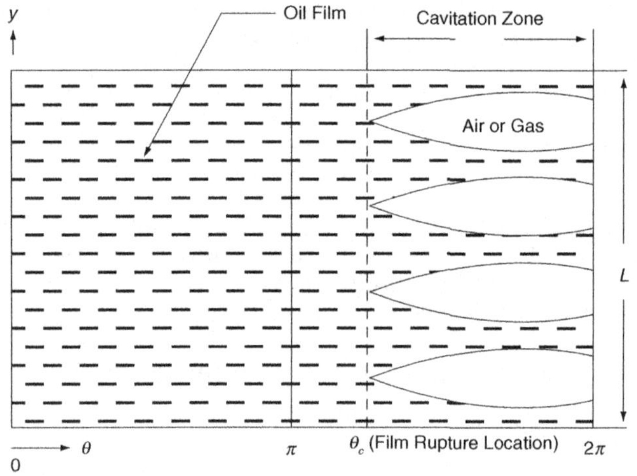


**Figure 6.6** | Developed View of Bearing Surface Showing Cavitation

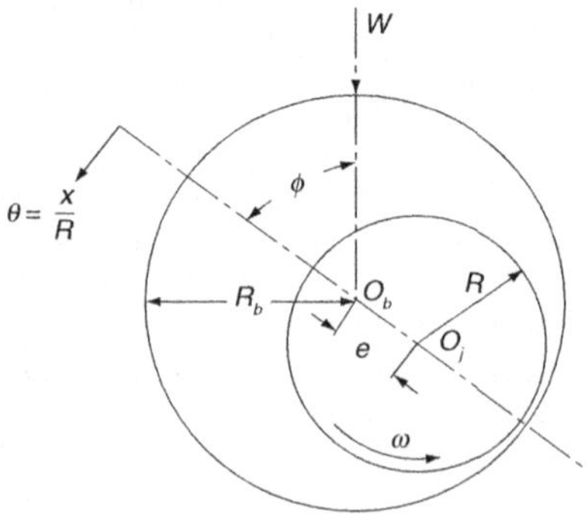


**Figure 6.7** | Coordinate System

where

$$F_0 = \int_0^1 \frac{d\bar{z}}{\bar{\eta}}$$

$$F_1 = \int_0^1 \frac{\bar{z}d\bar{z}}{\bar{\eta}}$$

$$F_2 = \int_0^1 \frac{\bar{z}}{\bar{\eta}}\left(\bar{z} - \frac{F_1}{F_0}\right) d\bar{z}$$

Density of the lubricant has been assumed to remain constant.

Dimensionless variables are defined as:

$$\bar{p} = pC^2 / \eta_0 \omega R^2, \; \bar{h} = h / C, \; \bar{\eta} = \eta / \eta_0, \; \theta = x/R, \; \bar{z} = z / h; \; \bar{y} = y / L$$

Energy equation for the lubricant flow in the bearing can be expressed as:

$$\rho c_p \left( u\frac{\partial T}{\partial x} + v\frac{\partial T}{\partial y} + w\frac{\partial T}{\partial z} \right) = \eta \left\{ \left(\frac{\partial u}{\partial z}\right)^2 + \left(\frac{\partial v}{\partial z}\right)^2 + \left(k_f \frac{\partial T}{\partial z}\right) \right\} \tag{6.28}$$

where $u$, $v$, $w$ are flow velocity components in x, y, and z directions, $\rho$ and $c_p$ are density and specific heat of the lubricant, $k_f$ – lubricant thermal conductivity. It is assumed that conductivity of lubricant does not influence the heat taken away by the lubricant. Heat generation is due to viscous shear and heat dissipation is through convection by the lubricant in the flow directions and conduction across the film thickness.

Energy Equation (6.28) can be written in the dimensionless form as:

$$P_e \left( \bar{u}\frac{\partial \bar{T}}{\partial \theta^*} + \frac{R}{L}\cdot \bar{v}\frac{\partial \bar{T}}{\partial \bar{y}} + \frac{\bar{w}}{\bar{h}}\frac{\partial \bar{T}}{\partial \bar{z}} \right) = \frac{\alpha\bar{\eta}}{\bar{h}^2}\left[ \left(\frac{\partial \bar{u}}{\partial \bar{z}}\right)^2 + \left(\frac{\partial \bar{v}}{\partial \bar{z}}\right)^2 \right] + \frac{1}{\bar{h}^2}\frac{\partial^2 \bar{T}}{\partial \bar{z}^2} \tag{6.29}$$

where dimensionless variables are defined as:

$$\bar{u} = u / U, \bar{v} = v / U; \quad \bar{w} = \frac{wR}{CU}, P_e = \rho c_p UC^2 / Rk_f; \quad \bar{T} = \frac{T}{T_i};$$

$U = R\omega$, journal speed

$P_e$ = Peclet number of the fluid

$\alpha = \eta_0 U^2 / k_f T_i$ dissipation number

$T_i$ = inlet oil temperature

$T$ = lubricant film temperature

The operator $\frac{\partial}{\partial \theta^*} = \frac{\partial}{\partial \theta} - \frac{\bar{z}}{\bar{h}}\frac{d\bar{h}}{d\theta}\frac{\partial}{\partial \bar{z}}$ is introduced to allow for change of shape of film from cylindrical to rectangular coordinate. In the inactive or cavitation zone, there is no pressure developed so the flow is uniform in the axial direction. Therefore, temperature is also constant in the axial direction.

Viscosity–temperature relationship can be given as:

$$\bar{\eta} = \eta / \eta_0 = k_0 - k_1\bar{T} + k_2\bar{T}^2 \tag{6.30}$$

where $k_0$, $k_1$, $k_2$ are constants which characterize the viscosity–temperature relationship of the lubricant.

Alternatively, Barus or Roelands relationship given by Equations (2.18) and (2.20) can be used also.

In addition to the above equations, heat transfer in the bearing is determined from the solution of following heat conduction equation in the solid bush neglecting convection as:

$$\frac{\partial^2 \overline{T}}{\partial \overline{r}^2}+\frac{1}{r}\frac{\partial T}{\partial r}+\frac{1}{\overline{r}^2}\frac{\partial^2 \overline{T}}{\partial \theta^2}+(R/L)^2\frac{\partial^2 \overline{T}}{\partial \overline{y}^2}=0 \tag{6.31}$$

where $\overline{r}=r/R$

Same equation can be used for heat conduction in the shaft with the assumption that the temperature is independent of $\theta$. Alternatively adiabatic boundary condition can also be assumed and shaft temperature can be taken as constant and equal to the lubricant temperature on the shaft surface.

Heat transfer in the shaft assuming that the temperature of the shaft is independent of $\theta$, is given by neglecting convection as:

$$\frac{\partial^2 \overline{T}}{\partial \overline{r}^2}+\frac{1}{\overline{r}}\frac{\partial \overline{T}}{\partial \overline{r}}+\left(\frac{R}{L}\right)^2\frac{\partial^2 \overline{T}}{\partial \overline{y}^2}=0 \tag{6.32}$$

Boundary conditions for the Equations (6.27), (6.29), (6.31), and (6.32), are given as:

1. Boundary conditions for Reynolds equation are:

$$\overline{p}(\theta,\pm 1/2)=0 \text{ for } 0\leq\theta\geq 2\pi$$

$$\overline{p}=0 \text{ and } \frac{\partial \overline{p}}{\partial \theta}=0 \text{ at } \theta=\theta_{cav}, \text{ film rupture boundary condition} \tag{6.33}$$

$$\left(\frac{\partial \overline{p}}{\partial y}\right)(\theta,0)=0 \text{ for } 0\leq\theta\geq 2\pi$$

Reynolds boundary conditions have been assumed for film rupture in the divergent portion of the film.

2. Temperature boundary condition are as follows:

(a) At the ends of the shaft, i.e., for $\overline{y}=\pm \frac{1}{2}$, a free convection is assumed to prevail which gives

$$\left.\frac{\partial \overline{T}}{\partial \overline{y}}\right|_{\overline{y}=\pm\frac{1}{2}}=-B_{is}\left(\left.\overline{T}\right|_{\overline{y}=\pm\frac{1}{2}}-1\right) \tag{6.34}$$

$B_{is}$ is the Biot number for the shaft and

$$B_{is}=\frac{h_s L}{k_s} \tag{6.35}$$

(b) For the outer surface of the bush, free convection and radiation modes of heat transfer is assumed which gives

$$\left.\frac{\partial \overline{T}}{\partial \overline{r}}\right|_{\overline{r}=\overline{R}_2}=-B_{ib}\left(\left.\overline{T}\right|_{\overline{r}=\overline{R}_2}-1\right) \tag{6.36}$$

where $B_{ib}$ is Biot number for the bush and

$$B_{ib}=\frac{h_b R_b}{k_b} \tag{6.37}$$

$R_b$ is outside radius of the bush

(c) On the surface between the bush and the fluid, the temperature is given by

$$\left.\frac{\partial \bar{T}}{\partial \bar{r}}\right|_{\bar{r}=1} = -\frac{k(\theta)}{k_b}\frac{R}{C}\frac{1}{\bar{h}}\left.\frac{\partial \bar{T}}{\partial \bar{y}}\right|_{\bar{y}=0} \tag{6.38}$$

where $k(\theta)$ is the thermal conductivity of the lubricant in the active zone and is equal to $k_f$ and $k_b$ is thermal conductivity of the bush.

In the inactive zone

$$k(\theta) = k_a - \frac{L'(\theta)}{L}\left(k_a - k_f\right)$$

$k_a$ is the thermal conductivity of the air and $L'(\theta)$ is defined as the apparent length of the liquid zone in the cavitations region (Fig. 6.6).

(d) For the lateral surface of the bush

$$\left.\frac{\partial \bar{T}}{\partial \bar{y}}\right|_{\bar{y}=\pm 1/2} = -\frac{L}{R}B_{ib}\left(\left.\bar{T}\right|_{\bar{y}=\pm 1/2} - 1\right) \tag{6.39}$$

(e) At shaft fluid interface, the shaft temperature is assumed to be independent of $\theta$, heat flux continuity condition gives:

$$\left.\frac{\partial \bar{T}}{\partial \bar{r}}\right|_{\bar{r}=1} = -\frac{1}{2\pi}\frac{k_o}{k_s}\frac{R}{C}\int_0^{2\pi}\frac{1}{\bar{h}}\left.\frac{\partial \bar{T}}{\partial \bar{z}}\right|_{\bar{z}=1} d\theta \tag{6.40}$$

In the boundary condition defined in (a) and (b), $h_s$ and $h_b$ are heat transfer coefficients of shaft and bush, respectively.

## 6.4 | Solution Procedure

Solution of thermohydrodynamic lubrication problem in journal bearings with heat transfer requires simultaneous solutions of Reynolds Equation (6.27) satisfying proper boundary conditions given by Equation (6.33) and energy Equation (6.29) and heat conduction equations for the bush and shaft given by Equations (6.31) and (6.32) satisfying appropriate boundary conditions given through Equations (6.34) to (6.40) following an iterative numerical technique. Finite difference and Gauss-Seidel methods with an over-relaxation factor are generally adopted to solve the equations.

The iterative procedure begins with an initial pressure and temperature field prescribed to determine the integrals in the generalized Reynolds equation. Reynolds equation is then solved for pressure distribution in the bearing. Reynolds boundary condition is satisfied by following Christopherson algorithm, i.e., by setting negative pressure equal to zero as and when it appears in the process of iteration. Energy equation and heat transfer equations in the solids are then solved simultaneously to determine new temperature distribution. The process is repeated with the new temperature distribution to determine the new pressure distribution by solving Reynolds equation. This process is continued until the converged solution is obtained, i.e., differences between two successive iterations converge within a prescribed error limit. This procedure has been followed by most researchers, for example, Dowson and March (1966), Mitsui (1978), Ferron *et al.* (1985), and Singh and Majumdar (2005) adopted similar procedure. Most of investigators observed that the iterative procedure converged fairly quickly. The algorithm followed by Singh and Majumdar is given in Fig. 6.8. Sample of results of above analysis in terms of pressure, load, temperature, etc. as obtained are shown in Figs 6.9 and 6.10 for a load of 6000 N at a journal speed of 4000 rpm. Bearing geometrical dimensions and fluid properties, etc. are given in Table 6.1. It is worth mentioning and necessary too that thermo hydrodynamic analysis of bearings is very bearing specific since the heat transfer analysis, etc. are dependent on the size of the bearing and are also lubricant and material specific due its dependence on thermo-physical properties. Therefore, generalization of results in the form of design charts or design curves is extremely difficult and no such design charts are available. Singh and Majumdar (2005) have developed empirical relationships for load capacity, oil flow rate, and friction factor based on numerical calculations which are given below:

Dimensionless load:

$$\bar{W} = a_1\left[a_2\left(L/D\right)^b + e^{c\left(L/D\right)^d}\right]\left(a_3\varepsilon_0 + \varepsilon_0{}^f\right)\left(L/D\right)^2 \tag{6.41}$$

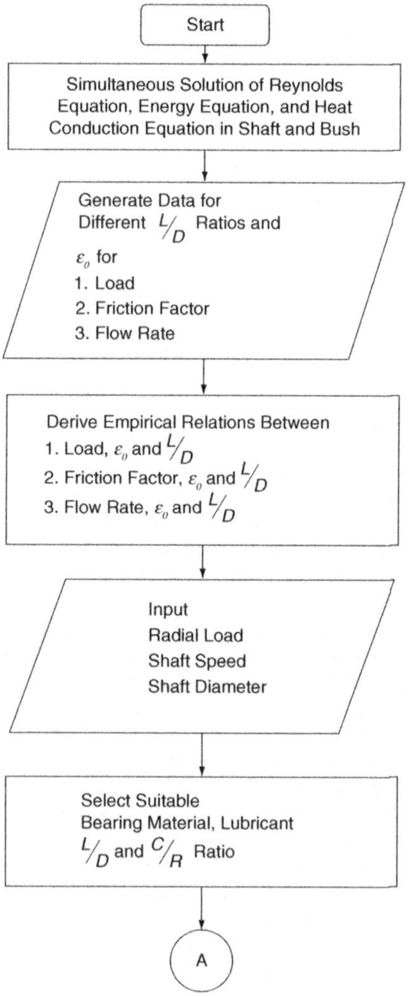

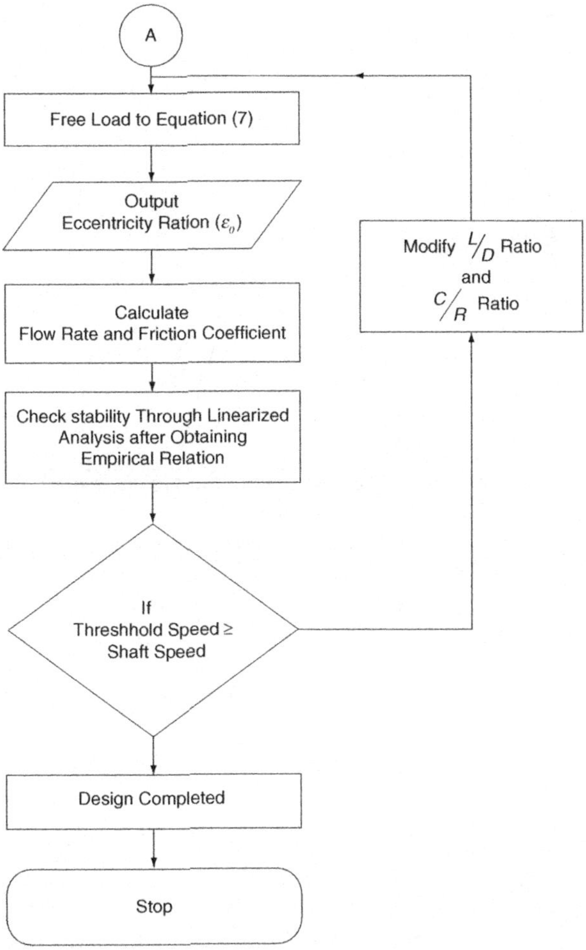


**Figure 6.8** | Design Flowchart

where

$$a_1 = 0.3137,\ a_2 = 3.4377,\ b = -1.8138,\ c = 2.0499,\ d = 0.5372,$$
$$a_3 = 0.6492,\ f = 4.0854$$

$$\bar{W} = \frac{WC^2}{\eta_0 \omega R^3 L}$$

$W$ is load acting on the bearing.

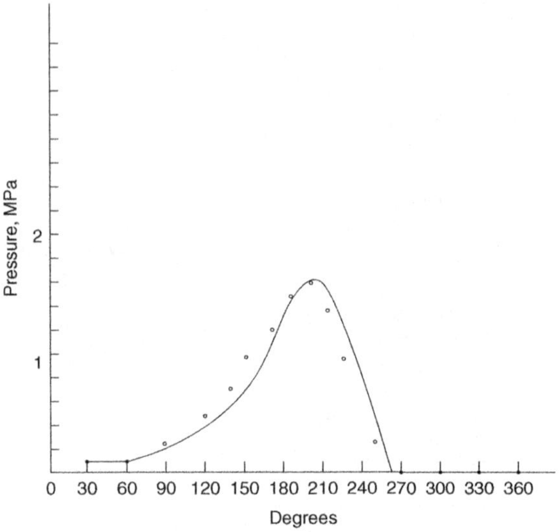


**Figure 6.9** | Typical Pressure Distribution in Circumferential Direction [Ferron *et al.*, ASME JOLT, 1983]

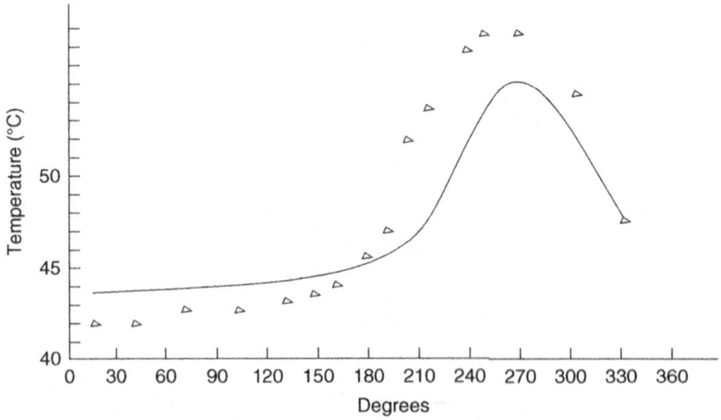


**Figure 6.10** | Typical Mid-Film Temperature Distribution [Ferron *et al.*, ASME JOLT, 1983]

**Table 6.1** | Bearing Geometrical and Material Properties

| **Material and Oil Properties** | |
|---|---|
| Inlet temperature | $T_i = 40°C$ |
| Lubricant viscosity at 40°C | $\eta_0 = 0.0277$ Pa s |
| Viscosity coefficients | $k_0 = 3.287$, $k_1 = 3.064$, $k_2 = 0.777$ |
| Lubricant density at 40°C | $\rho_f = 860$ kg/m³ |
| Bush material density | $\rho_b = 8522$ kg/m³ |
| Shaft material density | $\rho_s = 7753$ kg/m³ |
| Lubricant specific heat capacity | $c_r = 2000$ J/kg °C |
| Bush material specific heat capacity | $c_b = 385$ J/kg °C |
| Shaft material specific heat capacity | $c_s = 486$ J/kg °C |
| Lubricant thermal conductivity | $k_f = 0.13$ W/m °C |
| Air thermal conductivity | $k_a = 0.25$ W/m °C |
| Bush thermal conductivity | $k_b = 111$ W/m°C |
| Shaft thermal conductivity | $k_s = 36$ W/m°C |
| Convection heat transfer coefficient | $h_b = 80$ W/m°C, $h_s = 100$ W/m°C |

Fiction factor:

$$\left(\frac{R}{C}\right)\mu = \left(a_1 + e^b + e^c\right)\left[a_6\left(L/D\right)^{a_7} + a_8\right] \tag{6.42}$$

where

$$a_1 = -0.0264,\ b = a_2\varepsilon_0^{\ a_3},\ a_2 = -3.6662, a_3 = 0.4695,\ c = a_4\varepsilon_0^{\ a_5},$$
$$a_4 = -34.5477, a_5 = 1.0892, a_6 = 88.4788, a_7 = -2.0146, a_8 = 28.2124$$

$\mu$ is coefficient of friction.

Flow rate

$$\bar{Q} = \left[a_1\varepsilon_0^{\ a_2} + a_3\varepsilon_0^{\ a_4}\cos\left(\varepsilon_0 a_5\right)\right]\left(L/D\right)^{a_6} \tag{6.43}$$

where

$$a_1 = 0.5509, a_2 = 0.9468, a_3 = 0.2233, a_4 = 7.2743, a_5 = -3.2285, a_6 = 1.4227$$

$$\bar{Q} = \frac{QL}{2R^3\omega C},$$

$Q$ is the flow rate.

Equations (6.41), (6.42), and (6.43) relate to dimensionless load, friction factor, and flow rate with L/D and static eccentricity ratio. Figs 6.11, 6.12, and 6.13 present the variation of nondimensional load, flow rate, and friction factor against eccentricity ratio, respectively, and compare the theoretical values with values

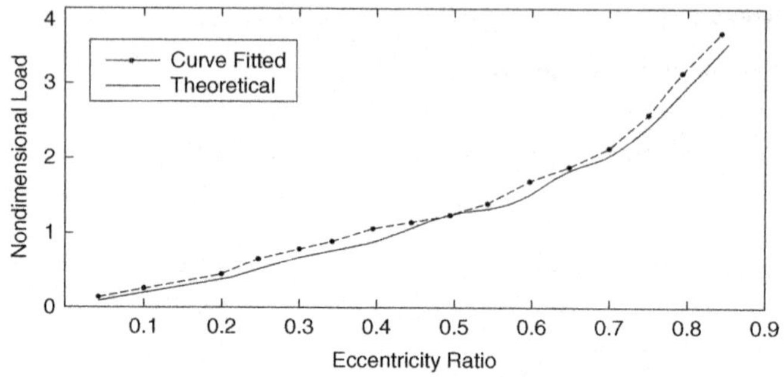


**Figure 6.11** | Curve Fitted Load Capacity for L/D = 1 [Singh and Majumdar, IMechE, UK, 2005]

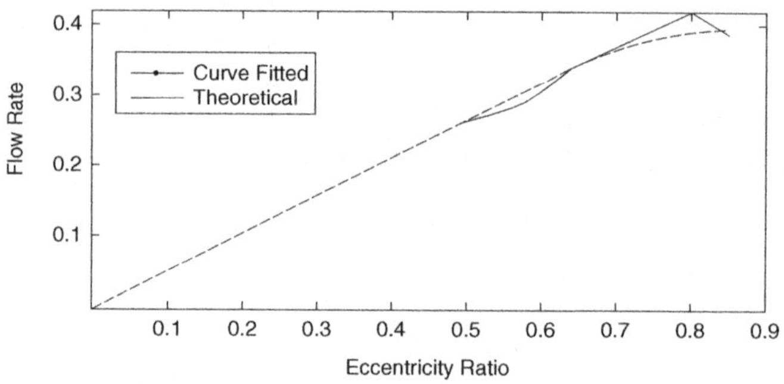


**Figure 6.12** | Curve Fitted Flow Rate for L/D [Singh and Majumdar, IMechE, UK, 2005]

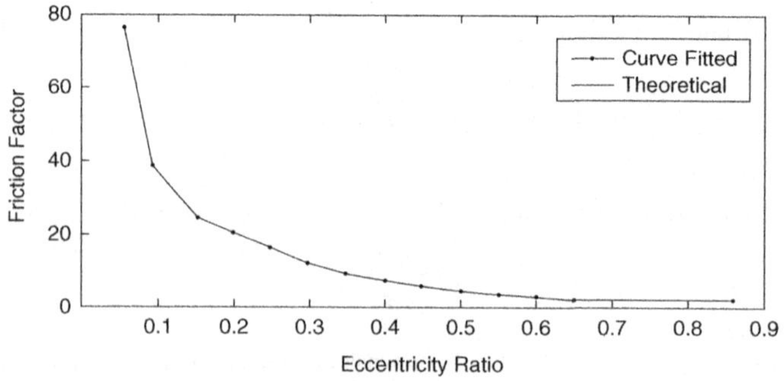


**Figure 6.13** | Curve Fitted Friction Factor for L/D [Singh and Majumdar, IMechE, UK, 2005]

obtained from empirical equations. Therefore, the empirical equations generated give very accurate results and can be used to design a journal bearing.

It has also been found that in high speed bearings temperature rise in the bush, etc. is significant and thermal distortions of the shaft and bush significantly alter the film shape in the bearing. In other words, pressure distribution and temperature distribution also depend on the thermal distortion particularly at high operating eccentricity ratios when both pressure and temperature generated could be high. Gethin (1985, 1987) solved the problem including thermoelastic deformation of the bush. Finite element method was employed by them to analyze the thermo hydrodynamic lubrication problem. Oil is generally supplied to the bearing through an oil groove. Mixing of cold and hot oil takes place in groove due to recirculation of oil in the groove. This affects the temperature of the oil entering the bearing film. Heshmat and Pinkus (1985) have investigated this problem theoretically.

## 6.5 | Thermoelastic Deformation of Shaft–Bush System Using Finite Element Method

The distortions of the bush and shaft due to imposed temperature and pressure field can be determined using equation of solid mechanics and finite element method. Gethin (1987) used finite element method with the assumption that the axial temperature variation is small and for moderate eccentricity ratios the pressure field is not too severe. Therefore, distortion of the bush at the center line was only calculated. Ghosh and Brewe (1993) also used finite element method to calculate thermoelastic deformations due to known temperature and pressure field in a bearing system. The method is general and can be used in any tribo system.

Thermoelastic stresses and deformation in an elastic isotropic solid are due to mechanical loading by surface and body forces and thermal state given by the temperature field in the solid.

Following Hooke's law, stress–strain equations for an isotropic material including thermal effects are given as:

$$\{\varepsilon\} = [D]\{\sigma\} + \{\varepsilon_0\} \tag{6.44}$$

$\varepsilon_0$ – initial strain due to non-uniform temperature distribution

$\sigma$, $\varepsilon$ are stress and strain, respectively, and $[D]$ is material flexibility matrix.

In an isotropic solid, the strains due to thermal expansion are uniform in all directions, hence, only normal strains are altered, shear strains are unaltered. In a three dimensional state, the thermal strains are given by:

$$[\varepsilon] = \left[\alpha\left(T - T_i\right), \alpha\left(T - T_i\right), \alpha\left(T - T_i\right), 0, 0, 0\right] \tag{6.45}$$

Stresses in the solid may be computed by solving equation (6.44) for the stress vector. The stress-strain law may then be written as:

$$\{\sigma\} = [C]\{\varepsilon\} - [C]\{\varepsilon_0\} \tag{6.46}$$

Utilizing the principle of minimum potential energy which states that among all the displacement fields of an admissible form, one which satisfies the equilibrium conditions makes the potential energy of a deformed body minimum.

Thus, for minimum potential energy of a deformed body an expression of the following form can be written following Zienkiewicz (1977).

$$\oiiint_v [B]^T [C][B]\{\delta\}\, dv = \oiiint_v p\, dv - \oiint_s q\, ds \tag{6.47}$$

or

$$[K]\{\delta\}=\{F\}$$

is obtained which can be solved for $\{\delta\}$, the displacement in the domain.

In the above equation, the integral over the volume domain of the solid includes thermal loading due to temperature while the surface integral over surface domain, contains contributions due to imposed pressure that is at the elemental level.

Thus,

$$\oint_{v} q dv = \oint \alpha f(T) dv \tag{6.48}$$

where $f(T)$ is the temperature field over the element and

$$\int_{s} = \oint_{s} f(\bar{p}) ds \tag{6.49}$$

where $f(\bar{p})$ is the pressure field over the element edge.

The strain–displacement relationship is written as

$$\{\varepsilon\}=[B]\{\delta\}=[B]\begin{Bmatrix} u \\ v \\ w \end{Bmatrix} \tag{6.50}$$

coordinates $u$, $v$, $w$ are displacement vectors in Cartesian coordinates.

The matrix $[C]$ is the material stiffness matrix and is written for an isotropic elastic solid as:

$$[C]=\frac{E}{(1+v)(1-v)}\begin{bmatrix} (1-v) & 0 & 0 & 0 & 0 & 0 \\ 0 & (1-v) & 0 & 0 & 0 & 0 \\ 0 & 0 & (1-v) & 0 & 0 & 0 \\ 0 & 0 & 0 & \left(\frac{1-2v}{2}\right) & 0 & 0 \\ 0 & 0 & 0 & 0 & \left(\frac{1-2v}{2}\right) & 0 \\ 0 & 0 & 0 & 0 & 0 & \left(\frac{1-2v}{2}\right) \end{bmatrix} \tag{6.51}$$

and matrix [B] is written as:

$$[B]=\begin{bmatrix} \frac{\partial}{\partial x} & 0 & 0 \\ 0 & \frac{\partial}{\partial y} & 0 \\ 0 & 0 & \partial/\partial z \\ \frac{\partial}{\partial y} & \frac{\partial}{\partial y} & 0 \\ 0 & \frac{\partial}{\partial y} & \frac{\partial}{\partial y} \\ \frac{\partial}{\partial z} & 0 & \partial/\partial x \end{bmatrix} \tag{6.52}$$

The displacement vector for an element with r nodes is

$$\{\delta\}^{(e)} = \begin{Bmatrix} u_1 \\ v_1 \\ w_1 \\ \cdot \\ \cdot \\ \cdot \\ u_r \\ v_r \\ w_r \end{Bmatrix} \tag{6.53}$$

The displacement field is approximately related to its nodal values by r interpolating function $N_i\,(x, y, z)$ so that displacement field is expressed as

$$\{\delta\}^{(e)} = \begin{Bmatrix} \sum_{i=1}^{r} N_i(x, y, z,)\; u_i \\ \sum_{i=1}^{r} N_i(x, y, z,)\; v_i \\ \sum_{i=1}^{r} N_i(x, y, z,)\; w_i \end{Bmatrix} = [N]\{\delta\}^{(e)} \tag{6.54}$$

where $[N]$ is the interpolation function matrix.

Therefore, within an element the matrix $[B]$ relating to strain and displacement at node q takes the form:

$$\left[B_q\right]^{(e)} = \begin{bmatrix} \frac{\partial N}{\partial x} & 0 & 0 \\ 0 & \frac{\partial N}{\partial y} & 0 \\ 0 & 0 & \frac{\partial N}{\partial z} \\ \frac{\partial N}{\partial y} & \frac{\partial N}{\partial x} & 0 \\ 0 & \frac{\partial N}{\partial z} & \frac{\partial N}{\partial y} \\ \frac{\partial N}{\partial z} & 0 & \frac{\partial N}{\partial x} \end{bmatrix} \quad \text{For } q = 1, 2 \ldots r \tag{6.55}$$

Using Equations (6.54) and (6.55), Equation (6.47) can be reduced to the form as written below:

$$[K]\{\delta\} = \{F\} \tag{6.56}$$

Displacement matrix $\{\delta\}$ can be determined from the above relationship. This is a general procedure and can be adopted for all types of thermo mechanical stress and deformation analysis including bearing deformation in thermohydrodynamic lubrication.

The steps involved in thermohydrodynamic lubrication including thermo elastic deformation are:

1. Solve the thermohydrodynamic lubrication and energy equation in the film and heat transfer equations in the solid.
2. Solve for the stresses and deformation in the bush and shaft using the film pressure and temperature field.

3. Update the film thickness profile, i.e., $h' = h - \delta_B - \delta_S$, $\delta_B$ and $\delta_S$ are bush and shaft deformations respectively.

This process is repeated until convergence is obtained within preassigned error limit.

## 6.6 | Adiabatic Solution for Thermohydrodynamic Lubrication Problem in Journal Bearing

The geometry of the journal bearing is shown in Fig. 6.14.

In adiabatic solution for thermohydrodynamic lubrication problems, it is assumed that viscosity of the lubricant remains constant across the film and therefore temperature across the film is constant. Heat generated due to viscous shear heating is fully removed by the lubricant flow from the bearing film and there is no other mode of heat transfer. This decouples the energy equation from the Reynolds equation if the pressure gradient terms in the energy equation are neglected as have been shown by McCallion *et al.* (1970) in their analysis. This assumption does not introduce any significant error in bearing performance over the full range of possible eccentricity ratios.

With the above assumptions, the energy equation is written in the reduced form as:

$$\frac{\partial T}{\partial \theta} = \frac{\mathrm{E}\ \bar{\eta}}{\beta\, H^2} \tag{6.57}$$

where

$$\bar{\eta} = \frac{\eta}{\eta_0} = \exp\left[-\beta\left(T - T_i\right)\right] \tag{6.58}$$

$T$ is the film temperature and $\beta$ is temperature–viscosity coefficient of the lubricant

$\eta_0$ is lubricant viscosity at temperature $T_i$, the inlet temperature of the lubricant

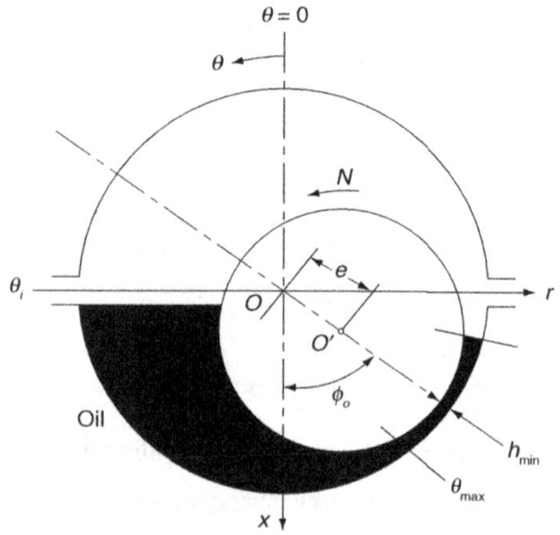


**Figure 6.14** | Hydrodynamic Journal Bearing

Shear stress $\tau = \eta\left(\frac{\partial u}{\partial z}\right)$ and shear strain rate $\frac{\partial u}{\partial z} = \left(\frac{U}{h}\right)$

where $U$ is the journal surface velocity and $h$ is the film thickness

$$\mathrm{E} = \frac{2\omega(R/C)^2}{\left(\rho c_p g/\beta\eta_0\right)}; \quad U = R\omega$$

Combining Equations (6.57) and (6.58), the energy equation takes the form of

$$\frac{d}{d\theta}\left(\frac{1}{\bar{\eta}}\right) = \frac{\mathrm{E}}{H^2}; \quad H = h/C \tag{6.59}$$

Equation (6.59) is integrated to give

$$\frac{1}{\bar{\eta}} = 1 + \mathrm{E}\left[I(\theta') - I(\theta_i')\right] \quad \text{where } \theta' = \theta - \phi_0 \tag{6.60}$$

The integral

$$\left(I(\theta) = \int \frac{d\theta}{H^2(\theta)}\right)$$

$$I(\theta) = \frac{1}{\left(1-\varepsilon^2\right)}\left[-\frac{\varepsilon\sin\theta}{1+\varepsilon\cos\theta} + \frac{\delta}{\left(1-\varepsilon^2\right)^{1/2}}\cos^{-1}\left(\frac{\varepsilon+\cos\theta}{1+\varepsilon\cos\theta}\right)\right] \tag{6.61}$$

where $\delta = 1$ for $\sin\theta > 0$ and $\delta = -1$ for $\sin\theta < 0$.

Adiabatic Reynolds equation for an incompressible lubricant is given by

$$\frac{\partial}{\partial\theta}\left(\frac{H^3}{\bar{\eta}}\frac{\partial\bar{p}}{\partial\theta}\right) + \frac{\partial}{\partial\bar{y}}\left(\frac{H^3}{\bar{\eta}}\frac{\partial\bar{p}}{\partial\bar{y}}\right) = \frac{\partial H}{\partial\theta} \tag{6.62}$$

where

$$\bar{p} = \frac{p}{12\pi\eta_0 N}\left(\frac{C}{R}\right)^2, \; p \text{ is film pressure and } \bar{y} = \frac{y}{R};$$

$R$ is the radius of the journal.

Substituting $\psi = G^{3/2}$

where

$$G = H/\bar{\eta}^{1/3} \tag{6.63}$$

The Reynolds equation becomes

$$\frac{\partial^2\psi}{\partial\theta^2} + \frac{\partial^2\psi}{\partial y^2} + f(\theta)\psi = g(\theta) \tag{6.64}$$

where

$$f(\theta) = -\frac{3}{4}\frac{1}{G^2}\left\{\left(\frac{\partial G}{\partial\theta}\right)^2 + 2G\frac{\partial^2 G}{\partial\theta^2}\right\}$$

and

$$g(\theta) = \frac{1}{G^{3/2}} \frac{\partial H}{\partial \theta}$$

A finite difference approach is generally used to solve Equation (6.61) which reduces to

$$\psi_{i,j-1} + a_j \psi_{i,j} + \psi_{i,j+1} + b\psi_{i-1,j} + b\psi_{i+1,j} = (\Delta\theta)^2 g_i \tag{6.65}$$

$$a_j = -2(1+b) + (\Delta\theta)^2 f_j$$
$$b = \left(\frac{\Delta\theta}{\Delta\bar{y}}\right)^2$$

A Gaussian elimination technique is used to get the solution of Equation (6.65) satisfying following which boundary condition:

$$\bar{p} = 0 \text{ at } y = \pm = L/2$$
$$\bar{p} = 0 \text{ at } \theta = \theta_i$$
$$\bar{p} = \frac{\partial \bar{p}}{\partial \theta} = 0 \text{ at } \theta = \theta_2 \text{ cavitation boundary.}$$

The above procedure was developed by Pinkus and Bupara (1979) to analyze thermohydrodynamic lubrication of finite journal bearings. The procedure is general and can be adopted to get solution of thermo hydrodynamic lubrication problems in bearings.

## 6.7 | Thermohydrodynamic Analysis Using Lobatto Quadrature Method

Lobatto quadrature technique to solve thermohydrodynamic lubrication problems wherein viscosity variation across the film thickness is taken into account was developed by Elrod and Brewe (1986) and Elrod (1991). It is a computationally very efficient and accurate method in comparison to finite difference and finite element methods. This theory is now presented below.

The momentum equation for noninertial laminar lubricating films and the corresponding energy equation are, respectively, as follows:

$$\nabla p = \frac{\partial}{\partial z}\left(\eta \frac{\partial u}{\partial z}\right), \tag{6.66}$$

$$\rho c_p u \frac{\partial T}{\partial x} = \frac{\partial}{\partial z}\left(k \frac{\partial T}{\partial z}\right) + \phi, \tag{6.67}$$

where $\phi = \eta(\partial u/\partial z)^2$ is the viscous dissipation function.

Along with Equations (6.66) and (6.67), following mass continuity equation for an incompressible fluid must also be satisfied:

$$\nabla \cdot u = 0 \tag{6.68}$$

The numerical solution to the flow field is sought by sampling the velocities, pressures, and temperatures over chosen grid points and appropriate physical laws mentioned above are satisfied through an algorithm to which these values are interlinked. For a given configuration in Fig. 6.15 which represents an inclined pad hydrodynamic thrust bearing, the temperature variation across the film is represented by a Legendre polynomial of order $N$, $P_N(\zeta)$ and the sampling points then are $N$ Lobatto points. It can be shown that $N$ such internally selected points permit exact numerical integration of a polynomial of order $2N + 1$ over range $-1 < \zeta < 1$. Thus,

$$\int_{-1}^{1} T(\zeta)d\zeta = \sum w_k T_k$$

and for $N = 2$, the Lobatto location $\zeta_k$ and weight factor $w_k$ are as follows and includes end-point values as given in Table 6.2. Therefore, if end-point temperatures are known, then it requires only two interior Lobatto point temperatures to be determined. The fluidity $\xi = 1/\eta$ is also collocated to its Lobatto point values by the series $\zeta_k = \zeta(T_k)$. The Galerkin style analysis used here involves the expansion of the temperature in a truncated series of Legendre polynomials.

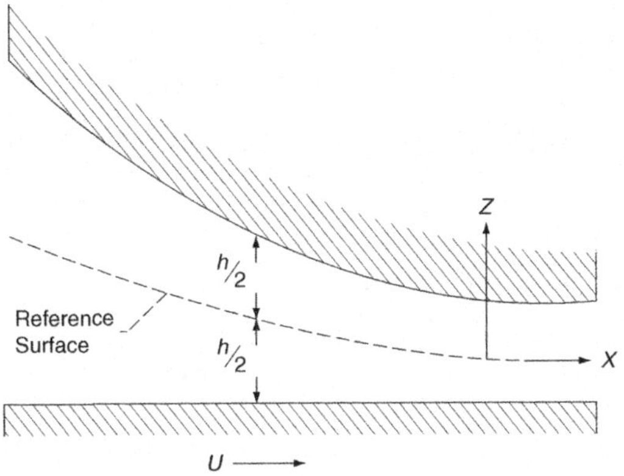


**Figure 6.15** | Coordinate Definition for Slider Bearing

**Table 6.2** | Lobatto Locations and Weight Factors for $N = 2$

| Location | Weight factor |
|---|---|
| $-1$ | 1/6 |
| $-1/\sqrt{5}$ | 5/6 |
| $1/\sqrt{5}$ | 5/6 |
| 1 | 1/6 |

Satisfaction is required of as many moments of the energy equation as there are unknowns in this series. The ensuing partial differential equations for the Legendre components are then solved. In the present solution, only two unknowns are used and for these it is feasible to carry out explicit integration as follows:

$$\int u\frac{\partial T}{\partial x}dz = k\left\{\left(\frac{\partial T}{\partial z}\right)_2 - \left(\frac{\partial T}{\partial z}\right)_{-2}\right\} + \frac{1}{\rho c_p}\int \phi dz,$$

or

$$\int_{-1}^{1}\frac{h}{2}u\frac{\partial T}{\partial x}d\zeta = k\left\{\left(\frac{\partial T}{\partial z}\right)_2 - \left(\frac{\partial T}{\partial z}\right)_{-2}\right\} + \frac{1}{\rho c_p}\int_{-1}^{1}\frac{h}{2}\phi d\zeta \tag{6.69}$$

and for the second moment

$$\int u\frac{\partial T}{\partial x}zdz = k\frac{h}{2}\left\{\left(\frac{\partial T}{\partial z}\right)_2 + \left(\frac{\partial T}{\partial z}\right)_{-2}\right\} + \frac{1}{\rho c_p}\int \phi z dz,$$

or

$$\int_{-1}^{1}\frac{h^2}{4}u\frac{\partial T}{\partial x}d\zeta = k\frac{h}{2}\left\{\left(\frac{\partial T}{\partial z}\right)_2 + \left(\frac{\partial T}{\partial z}\right)_{-2}\right\} + \frac{1}{\rho c_p}\int_{-1}^{1}\frac{h^2}{4}\phi\zeta d\zeta \tag{6.70}$$

The temperature distribution which passes through the Lobatto points expressed in Legendre polynomials as

$$T(\zeta) = \sum_{k=0}^{3}\tilde{T}_k P_k(\zeta), \tag{6.71}$$

then the Legendre coefficients are easily evaluated by integration

$$\int_{-1}^{1}T(\zeta)P_k(\zeta)d\zeta = \frac{2}{2k+1}\tilde{T}_{k,}$$

or

$$\tilde{T}_k = \frac{2k+1}{2}\sum_{i=0}^{3}w_i T_i P_k(\zeta), \tag{6.72}$$

The above linear set of equations can be solved for the $\tilde{T}_k$ For $N = 2$

$$\tilde{T}_0 = \frac{1}{12}\{T_{-2} + 5T_{-1} + 5T_1 + T_2\}, \tag{6.73a}$$

$$\tilde{T}_1 = \{T_2 - T_{-2} + \sqrt{5}\,(T_1 + T_{-1})\}/4 \tag{6.73b}$$

$$\tilde{T}_2 = \frac{5}{12}\{T_{-2} + T_{-2} - (T_1 + T_{-1})\}, \tag{6.73c}$$

$$\tilde{T}_3 = \{T_2 - T_{-2} + \sqrt{5}(T_1 + T_{-1})\}/4, \tag{6.73d}$$

The wall or surface temperatures $T_2$ and $T_{-2}$ are considered known for purposes of film calculation. Thus,

$$\tilde{T}_2 = \left(T_2 + T_{-2}\right)/2 - \tilde{T}_{0,} \tag{6.73e}$$

$$\tilde{T}_3 = \left(T_2 - T_{-2}\right)/2 - \tilde{T}_{1,} \tag{6.73f}$$

Similar expressions are obtained for fluidity with the Lobatto point temperatures. Thus,

$$\tilde{\xi}_0 = \frac{1}{12}\left\{\xi_{-2} + 5\xi_{-1} + \xi_1 + \xi_2\right\} \tag{6.74}$$

and fluidity distribution is

$$\xi = \sum_{k=0}^{3} \tilde{\xi}_k P_k(\zeta), \tag{6.75}$$

### 6.7.1 | Velocity Distribution and Mass Flux

A double integration of Equation (6.66) with $\xi = 1/\eta$ gives the tangential velocity vector

$$u = u_{-2} + A\int_{-1}^{\zeta} \xi d\zeta + B\int_{-1}^{\zeta} \xi\zeta d\zeta \tag{6.76}$$

where

$$A = \frac{u_2 - u_{-2} - B\int_{-1}^{1} \xi\zeta d\zeta}{\int_{-1} \xi d\zeta} \tag{6.77}$$

and

$$B = \left(\frac{h}{2}\right)^2 \nabla p$$

The linear mass flux is obtained as

$$\frac{\dot{m}}{\rho} = \int_{-1}^{1} \frac{h}{2} u d\zeta, \tag{6.78}$$

$$\frac{\dot{m}}{\rho} = \left(u_2 + u_{-2}\right)\frac{h}{2} - \frac{2}{3}\tilde{\xi}_1 A \frac{h}{2} - \frac{2}{3} B\left(\tilde{\xi}_0 + \frac{2}{5}\tilde{\xi}_2\right)\frac{h}{2}. \tag{6.79a}$$

For symmetric cross-film temperature distribution arithmetic averaging of fluidities at Lobatto points can be done and therefore, mass flux is given by

$$\frac{\dot{m}}{\rho} = \left(u_2 + u_{-2}\right)\frac{h}{2} - \frac{h^3}{12}\nabla p\left\{\frac{3}{2}\int_{-1}^{1} \xi\zeta^2 d\zeta\right\}. \tag{6.79b}$$

The cross-film temperature distribution would be symmetric when the surface velocities $u_2$ and $u_{-2}$ are equal and so also the surface temperatures $T_2$ and $T_{-2}$.

Mass continuity given by Equation (6.68) when applied to mass flux leads to the generalized Reynolds equation, as follows:

$$\nabla\cdot\frac{\dot{m}}{\rho}=\frac{d}{dx}\left\{\left(u_2+u_{-2}\right)\right\}\frac{h}{2}-\frac{d}{dx}\left\{\frac{2}{3}\tilde{\xi}_1 A\frac{h}{2}\right\}-\frac{d}{dx}\left\{\frac{2}{3}B\left(\tilde{\xi}_0+\frac{2}{5}\tilde{\xi}_2\right)\frac{h}{2}\right\}=0 \tag{6.80}$$

### 6.7.2 | Temperature Equation

With the aid of the Legendre series for temperature and fluidity integrals in the zeroth and first moment of energy, Equations (6.69) and (6.70) can be evaluated as follows:

Equation (6.69) becomes

$$\frac{12k}{h}\tilde{T}_0+\frac{h}{4}\alpha_7\frac{d}{dx}\left(\tilde{T}_0\right)+0.\tilde{T}_1+\frac{h}{4}\alpha_8\frac{d}{dx}\left(\tilde{T}_1\right)=\frac{12k}{h}\left\{\frac{T_2+T_{-2}}{2}\right\}+\frac{1}{\rho c_p}\int_{-h/2}^{h/2}\phi dz \tag{6.81}$$

The temperature $\tilde{T}_0$ is $\zeta-$space mean temperature where the integral of the dissipation function is:

$$\int_{-h/2}^{h/2}\phi dz=\frac{2}{h}\left\{2\tilde{\xi}_0 A^2+\frac{2}{3}\tilde{\xi}_1\left(2AB\right)+\frac{2}{3}\left(\tilde{\xi}_0+\frac{2}{5}\tilde{\xi}_2\right)B^2\right\} \tag{6.82}$$

and Equation (6.70) becomes:

$$0.\tilde{T}_0+\frac{h^2}{8}\beta_1\frac{d}{dx}\left(\tilde{T}_0\right)+10k\tilde{T}_1+\frac{h^2}{8}\beta_4\frac{d}{dx}\left(\tilde{T}_1\right)=10k\left\{\frac{T_2-T_{-2}}{2}\right\}+\frac{1}{\rho c_p}\int_{-1}^{1}\left(\frac{h^2}{4}\phi\zeta d\zeta\right) \tag{6.83}$$

where the moment of the dissipation function is:

$$\int_{-1}^{1}\phi\zeta d\zeta=\frac{2}{3}\tilde{\xi}_1 A^2+\frac{2}{3}\left(\tilde{\xi}_0+\frac{2}{5}\tilde{\xi}_2\right)\left(2A\cdot B\right)+\frac{2}{5}\left(\tilde{\xi}_1+\frac{2}{7}\tilde{\xi}_3\right)B^2 \tag{6.84}$$

The constants of Equations (6.81) and (6.83) are given in the Appendix.

Two simultaneous partial differential Equations (6.81) and (6.83) with two variables $\tilde{T}_0$ and $\tilde{T}_1$ are obtained by eliminating $\tilde{T}_2$ and $\tilde{T}_3$ via Equations (6.73e) and (6.73f) coupled with the generalized Reynolds Equation (6.80) provide the solution to the thermohydrodynamic lubrication problem for laminar films.

### 6.7.3 | Computational Procedure

The solution to the thermohydrodynamic lubrication problems begins with the known pressure distribution within the lubricant film as obtained from Equation (6.80) for isothermal films assume that the temperature in the entire fluid film is equal to the inlet oil temperature at 311 K.

Fluid film is discretized in the flow direction and the terms $\left(d\tilde{T}_0/dx\right)$ and $\left(d\tilde{T}_1/dx\right)$ in Equations (6.81) and (6.83) are expressed in finite-difference form with backward differencing. Separating the terms, simultaneous algebraic equations in terms of two unknowns $\tilde{T}_0$ and $\tilde{T}_1$ are obtained which are solved to determine them. This is done starting from the inlet to the outlet of the film in a forward marching manner. Iterations are done to obtain a converged solution. For known surface temperatures $T_2$ and $T_{-2}$ equal to the inlet oil temperature 311° K, the temperature distribution within the film region $T_1$ and $T_{-1}$ are thus obtained. Reverse flow situations, whenever occur, are handled by resorting to upwind differencing for the terms in Equations (6.81) and (6.83).

Once the Lobatto point temperatures are known over the entire film domain, fluidity functions given in the appendix are evaluated afresh. Then the generalized Reynolds Equation (6.80) is solved iteratively for pressure distribution following the finite-difference method with appropriate boundary conditions satisfied at the inlet and exit of the film.

For the pressure distribution thus known, we return to determine the temperature distribution within the entire film region for the new pressures by solving Equations (6.81) and (6.83) for $\tilde{T}_0$ and $\tilde{T}_1$.

As discussed earlier, this process is repeated until both pressure and temperature converge simultaneously. Pressure and temperature are treated as converged when the following convergence criteria are satisfied:

For pressure

$$\frac{\left|\left(\Sigma p_i\right)_{N-1} - \left(\Sigma p_i\right)_N\right|}{\left|\left(\Sigma p_i\right)_N\right|} \leq 0.0001.$$

For temperature

$$\frac{\left|\left(\Sigma T_i\right)_{N-1} - \left(\Sigma T_i\right)_N\right|}{\left|\left(\Sigma T_i\right)_N\right|} \leq 0.0001.$$

where $N$ is the number of iterations. However, convergence is quick and is obtained in only four or five iterations.

### 6.7.4 | Load Capacity and Friction Force

Load capacity per unit width of the pad is determined as:

$$\frac{W}{L} = \int_{x_i}^{x_e} p\,dx \tag{6.85}$$

and the friction force per unit width of the pad is expressed as:

$$\frac{F}{L} = \int_{x_i}^{x_e} \eta \left(\frac{\partial u}{\partial z}\right)_{0,h} dx. \tag{6.86}$$

when expressed in dimensionless form these are:

**Load Capacity**

$$\bar{W} = \frac{W h_{\min}}{24 L R u\, \eta_0} \tag{6.87}$$

**Friction Force**

$$\bar{F} = \frac{F h_{min}}{24 L R u\, \eta_0} \tag{6.88}$$

and the coefficient of friction $\mu = \frac{\bar{F}}{\bar{W}}$.

Results of temperature distribution for thermohydrodynamic lubrication of an infinitely wide inclined pad thrust bearing using this method are shown in Fig. 6.16. The geometry and fluid properties are given below.

Density: $1.7577 \times 10^6$ kg/m$^3$
Thermal diffusivity: $7.306 \times 10^{-8}$ m$^2$/s
Viscosity of lubricant: $0.13885 \times \exp(-0.045(T-311.11))$ PaS
Lubricant entrance temperature: 311.11 K
Solid surface temperatures: Uniformly at 311.11 K
Runner velocity: 31.946 m/s
Bearing length: 0.18288 m
Minimum gap: 0.00009144 m

The computations involved are much less in comparison to finite difference method.

This method has been extensively used in recent years by several researchers. Sharma and Pandey (2006) presented results of temperature contours of an inclined pad hydrodynamic thrust bearing as shown in Fig. 6.17 and 6.18, and also investigated the accuracy of the presented method in comparison to parabolic approximation of temperature variation across the film.

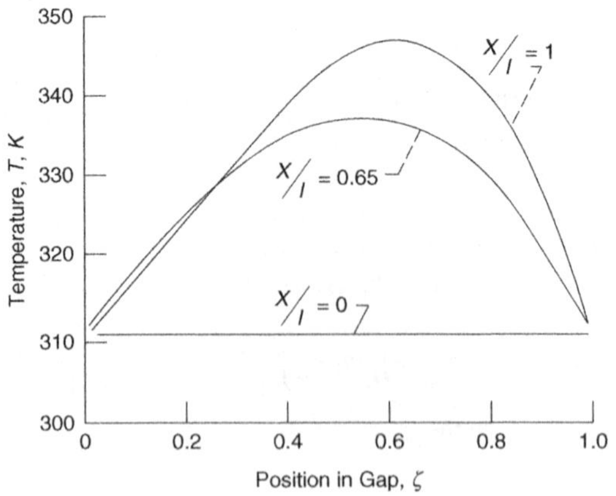


**Figure 6.16** | Temperature Versus Film Position in the Gap

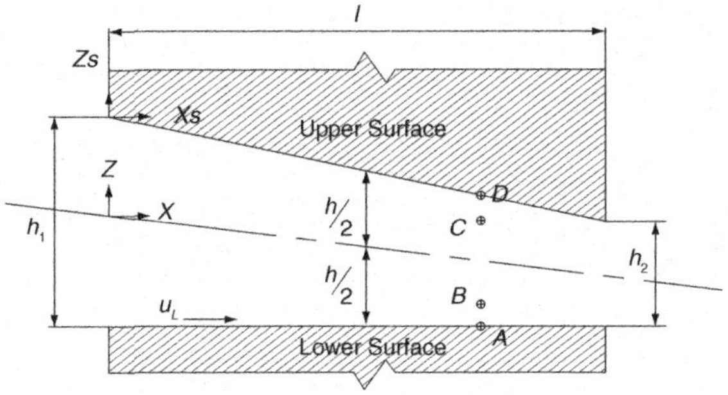


**Figure 6.17(a)** | Coordinate System for Legendre Polynomial Temperature Profile [Sharma and Pandey, Tribology Online, JST, 2006]

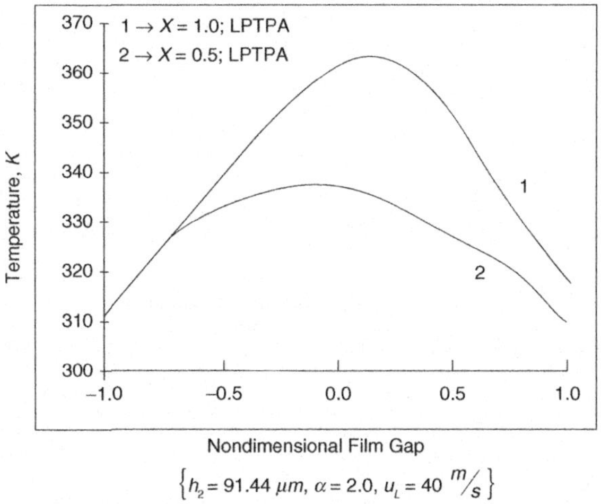


**Figure 6.17(b)** | Temperature Variation Across the Sliding Direction Computed with LPTPA [Sharma and Pandey, Tribology Online, JST, 2006]

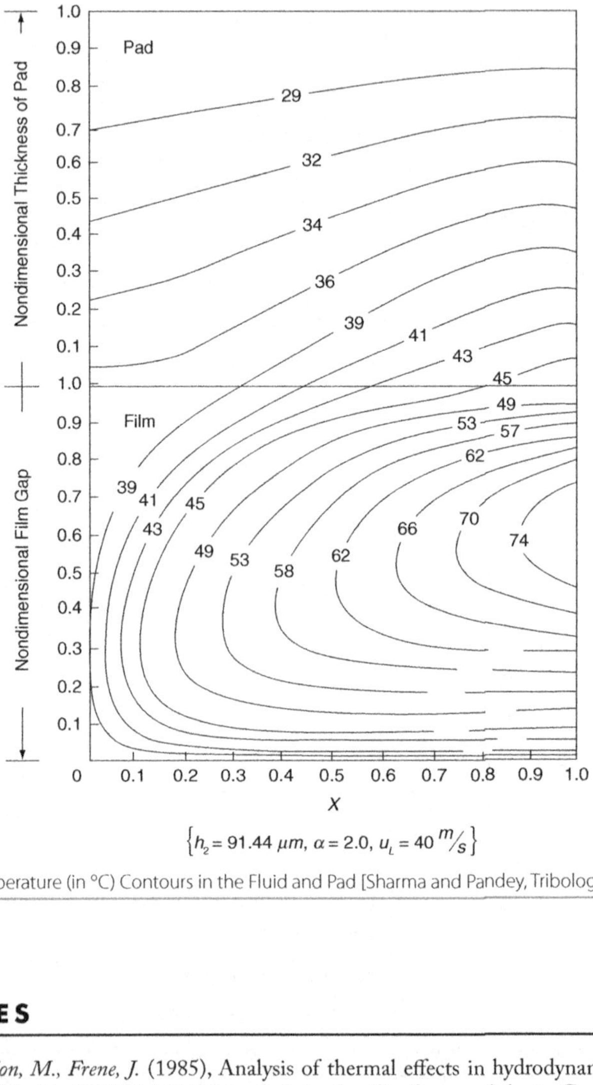


**Figure 6.18** | Temperature (in °C) Contours in the Fluid and Pad [Sharma and Pandey, Tribology Online, JST, 2006]

## REFERENCES


*Boncompain, R., Fillon, M., Frene, J.* (1985), Analysis of thermal effects in hydrodynamic bearings, *ASME*, Paper no. 85-Trib-21, ASME/ ASLE Joint Lubrication Conference, Atlanta, Ga.

*Capitao, J.W.* (1984), Influence of turbulence on performance characteristics of the tilting pad thrust bearings. *ASME, J. Lub. Tech.*, **96**, 110–117.

*Castelli, V. and Malanoski S.B.* (1979), Method for solution of lubrication problems with temperature and elasticity effects-application to sector, tilting pad bearings, *ASME, J. Lub. Tech.*, **91**, 634–640.

*De Choudhary, P. and Barth, E.W.* (1981), A comparison of film temperature and oil discharge temperature of a tilting pad journal bearing, ASME, *J. Lub. Tech.*, **103**, 115–119.

*Dowson, D.* (1962), A generalized Reynolds' equation for film lubrication, *Int. J. Mech. Sci.*, Pergamon Press Ltd., **4**, 159–170.

*Dowson, D. and March, C.N.* (1966), A thermo hydrodynamic analysis of journal bearings, *Proc., I. Mech. E*, London, U.K. **181**, 117–126.

*Dowson, D., Hundson, J.D., Hunter, C.N. and March C.N.* (1966), An experimental investigation of thermal equilibrium of steadily loaded journal bearings, *Proc. Inst. Mech. Engrs.*, London, U.K., **181**, Part 3B, 70–80.

*Elrod, H.G.* (1991), Efficient numerical method for computation of the thermohydrodynamics of laminar lubricating films, *Trans ASME, Journal of Tribology*, **113**, 506–511.

*Elrod, H.G. and Brewe, D.E.* (1986), Thermohydrodynamic analysis for laminar films, *NASA Technical Memorandum* 88845.

*Etsion, I. and Barkan, I.* (1981), Analysis of a hydrodynamic thrust bearings with incomplete film, *ASME, J. Lub. Tech.*, **103**, 355–360.

*Ettles, C. and Cameron, A.* (1969–70), Consideration of flow across a bearing groove, *ASME, J. Lub. Tech.*, Vol. 90, N1, p. 312, 1968.

*Ettles, C.* (1968), Solutions for flow in a bearings groove, Tribology Convention, *Proc., I. Mech. E*, London, U.K, **184**, 120.

*Ettles, C.* (1976), The development of a generalized computer analysis for sector shaped tilting pad thrust bearings, *ASLE Transaction*, **19**, 153–163.

*Ettles, C.*, Hot oil carry over in thrust bearings, *Proc. I. Mech. E*, London U.K., **184**, 75.

*Ferron, J., Frene, J, Boncompain, R.* (1983), A study of the thermo hydrodynamic performance of a plain journal bearing: comparison between theory and experiments, *ASME, J. Lub. Tech.*, **105**, 422–428.

*Gethin, D.T.* (1985), An investigation into plain journal bearing behavior including thermo elastic deformation of the bush, *Proc. Inst. Mech. Engrs.*, London, U.K., **199**,215–223.

*Gethin, D.T.* (1987), An application of finite element method to the thermo hydrodynamic analysis of a thin film cylindrical bore bearing running at high sliding speed, *ASME, Journal of Tribology*, **109**, 283–289.

*Ghosh, M.K. and Brewe, D.E.* (1993), Thermo-mechanical analysis of a dry shaft bush tribo system using the finite element method, *Computers and Structures*, **49**, 207–218.

*Gregory, R.S.* (1974), Performance of thrust bearings at high operating speeds, *ASME, J. Lub. Tech.*, **96**, 7–14.

*Gregory, R.S.* (1979), Factors Influencing Power Loss of Tilting Pad Thrust Bearings, *ASME, J. Lub. Tech.*, **101**, 154–163.

*Hashimoto, H. and Wada, S.* (1985), Turbulent Lubrication of Tilting Pad Thrust Bearings with Thermal and Elastic Deformations, *ASME, Journal of Tribology*, **107**, 82–86.

*Heshmat, H. and Pinkus, O.* (1985), Mixing inlet temperatures in hydrodynamic bearings, *ASME*, Paper N. 85-Trib-26, ASME/ASLE Joint Lubrication Conference, Atlanta, Ga.

*Huebner, K.H.* (1974), Solution for the pressure and temperature in thrust bearings operating in the thermo hydrodynamic turbulent regime, ***ASME**, J. Lub. Tech.*, **96**, 58–68.

*Kays, W.M.* (1980), *Convective Heat and Mass Transfer*, McGraw-Hill, New York, N.Y.

*McCallion, M., Yousif, F. and Lloyd, T.* (1970), The analysis of thermal effect in full journal bearing, *ASME, JOLT*, **92**, 578–587.

*Mikula, A.M. and Gregory, R.S.* (1983), A comparison of tilting pad thrust bearing lubricant supply methods, *ASME, J. Lub. Tech.*, **105**, 39–47.

*Mitsui, J. and Yamada, T.* (1978), A study of lubricant film characteristics in journal bearings, Part 1, A thermo hydrodynamic analysis with particular reference to viscosity variation with in lubricating film, *Bull. JSME*, **22**, 1491–1498.

*New, N.H.* (1974), Experimental comparison of flooded, directed, and inlet orifice type of lubrication for a tilting pad thrust bearing, *ASME, J. Lub. Tech.*, **96**, 22–27.

*Pinkus, O. and Bupara, S.S.* (1979), Adiabatic solutions for finite journal bearings, *ASME J. Lub. Tech.*, **101**, 492–476.

*Pinkus, O. and Wilcock, D.J.* (1980), Thermal effects in fluid bearings, *Proc. of Sixth Leeds-Lyon Symposium on Tribology-Thermal Effects in Tribology*, Mechanical Engineering Publication, 3–23.

*Sharma, R.K. and Pandey, R.K.* (2006), An investigation into the validity of the temperature profile approximations across the film thickness in the THD analysis of infinitely wide slider bearing, *Tribology Online, Japanese Society of Tribologists*, **1**, 19–24.

*Singh, D.S. and Majumdar, B.C.* (2005), Computer aided design of hydrodynamic journal bearings considering thermal effects, *Proc. I. mech. E.*, U.K., *Journal of Engineering for Tribology*, **219**, 133–143.

*Sternlicht, B., Carter, G.K. and Arawas, E.B.* (1961), Adiabtic analysis of elastic, centrally pivoted sector thrust bearing pads, *ASME, Journal* of *Applied Mechanics*, **28**, 179–187.

*Tonnesen, J. and Hansen, P.K.* (1981), Some experiments on the steady state characteristics of a cylindrical fluid film bearing considering thermal effects, *ASME, Journal of Lub. Tech.*, **103**, 107–114.

*Vohr, J.* (1981), Prediction of the operating temperature of thrust bearings, *ASME, J. Lub. Tech.*, **103**, 97–106.

## APPENDIX 6.1

$$\tilde{\xi}_1 = \tilde{\xi}(T_1); \tilde{\xi}_0 = \left(\xi_2 + 5\tilde{\xi}_1 + 5\tilde{\xi}_{-1} + \tilde{\xi}_{-2}\right)/12$$

$$\tilde{\xi}_2 = \tilde{\xi}(T_2); \tilde{\xi}_1 = \left[\xi_2 - \xi_{-2} + \sqrt{5}\left(\tilde{\xi}_1 - \xi_{-1}\right)\right]/4$$

$$\bar{\xi}_{-1} = \tilde{\xi}(T_{-1}); \tilde{\xi}_2 = 5\left(\tilde{\xi}_2 + \tilde{\xi}_{-2} - \xi_1 - \bar{\xi}_{-1}\right)/12$$

$$\tilde{\xi}_{-2} = \xi(T_{-2}); \tilde{\xi}_3 = \left[\tilde{\xi}_2 - \tilde{\xi}_{-2} - \sqrt{5}\left(\tilde{\xi}_1 - \xi_{-1}\right)\right]/4$$

$$\alpha_1 = \tilde{\xi}_0 - \xi_1/2 + \tilde{\xi}_3/8$$

$$\alpha_2 = -\tilde{\xi}_0/2 + \tilde{\xi}_1/3 - \xi_2/8$$

$$\alpha_3 = 2\xi_1/15 - \frac{9}{70}\tilde{\xi}_3 + \frac{4}{3}\alpha_1$$

$$\alpha_4 = \frac{2}{15}\tilde{\xi}_0 - \frac{1}{42}\tilde{\xi}_2 + \frac{4}{3}\alpha_2$$

$$\alpha_5 = \frac{4}{15}\tilde{\xi}_0 - \frac{8}{108}\tilde{\xi}_2$$

$$\alpha_6 = \frac{4}{105}\tilde{\xi}_1 - \frac{8}{315}\tilde{\xi}_3$$

$$\alpha_7 = 4\bar{u}_{-2} + 3\alpha_3\bar{A} + 3\alpha_4\bar{B}$$

$$\alpha_8 = 5\alpha_5\bar{A} + 5\alpha_6\bar{B}$$

$$\beta_1 = \frac{3}{5}\alpha_8$$

$$\beta_2 = \frac{2}{35}\tilde{\xi}_1 = \frac{29}{630}\tilde{\xi}_3 + \frac{4}{15}\alpha_1$$

$$\beta_3 = \frac{2}{35}\tilde{\xi}_0 = \frac{1}{210}\tilde{\xi}_3 + \frac{4}{15}\alpha_2$$

$$\beta_4 = \frac{4}{3}\bar{u}_{-2} + 5\beta_2\bar{A} + 5\beta_3\bar{B}$$

Chapter 7

# Design of Hydrodynamic Bearings

## 7.1 | Introduction

The theory of hydrodynamic oil bearings is given in Chapters 4 and 5. The three important parameters concerning bearing design are

1. Load capacity
2. Coefficient of friction
3. Oil flow

A bearing designed should be able to support the applied thrust or radial load calculated from other design considerations. The coefficient of fluid friction upon which depends the frictional power loss should be minimum. Finally, the flow rate of lubricant through the bearing is to be determined correctly so that the bearing does not starve and consequently oil film which supports the load does not break. Furthermore, an adequate oil flow is required for dissipating the heat generated due to fluid friction.

## 7.2 | Practical Considerations

Theoretical load carrying capacity derived in Chapter 5 shows that it increases with increase in the eccentricity ratio. Thus, a reduction in minimum film thickness gives to higher load capacity. If one can make an infinitesimal minimum film thickness, an infinite load can be achieved. However, the value of minimum oil film thickness depends on several practical considerations, like surface finish, rigidity of shaft, geometry of bearing surface, etc. The choice of bearing materials which plays a vital role is briefly outlined. For details refer to Radzimovsky (1959).

A few practical considerations will now be discussed (Etchells, 1942). Most of these are based on experience. These may help in finding the initial bearing dimensions. Amongst the various practical considerations, length to diameter ($L/D$) ratio in the case of journal bearing (width to length ratio in slider bearing), radial clearance, $C$ (sometimes minimum film thickness), and bearing pressure, $P$ (unit load) are most important. In the following sections, factors which play important role on $L/D$ ratio, $C$ and $P$ are discussed.

### 7.2.1 | Length to Diameter Ratio

Theoretical analysis of load capacity shows that it can be increased with an increase in bearing length because of large projected area. It has also been found that rate of flow of oil decreases with increase in the length of bearing. Thus, a longer bearing gives higher load and less flow, whereas the reverse is true for a shorter bearing. A shorter bearing is preferable when the problem of heat dissipation in the bearing is critical.

Rigidity of the shaft and the bearing is an important factor for deciding a correct $L/D$ ratio. If the bearing is rigidly supported and cannot deflect with the shaft, the journal may be damaged at the ends due to surface contact. When the bearing is relatively long and the bearing material has poor conformability, this problem is severe.

The space sometimes puts a limitation on $L/D$ ratio. A smaller $L/D$ is generally preferred in aircraft engine bearings because of space limitation.

There are many such practical aspects which determine the possible $L/D$ ratio for longer service life. But there is no general rule for the best $L/D$ ratio, since both large and small $L/D$ ratios have their advantages and disadvantages. Therefore, a designer has to rely mostly on the operating conditions. It is usually the practice that $L/D$ ratio should be taken close to 1, with flexibilities to increase when alignment and cooling the bearing do not pose a serious problem; to decrease when the flow rate is a primary consideration or when the bearing material has a poor conformability.

### 7.2.2 | Radial Clearance

Under given operating conditions, the clearance in a journal bearing has a significant effect on load capacity and flow rate. For a given operating condition, an increase in radial clearance decreases the load capacity and increases the oil flow rate. As the oil flow rate is directly proportional to cube of radial clearance, a slight increase in clearance will increase the oil flow rate substantially thereby reducing the temperature of bearing surface.

Some practical considerations, such as properties of bearing alloy, geometry of bearing ($L$ and $D$), and nature and magnitude of load and speed are important in deciding initial bearing clearance. Table 7.1 may give a preliminary idea for selection of radial clearance.

It has also been recommended (Shaw 1949) to use a slightly higher $C/R$ for small bearings and lower $C/R$ for large bearings.

**Table 7.1** | Recommended Radial Clearance Ratio for Different Bearing Alloys

| Alloys | C/R ratio |
|---|---|
| Tin-base Babbitt | 0.0005 |
| Cadmium-silver | 0.0008 |
| Copper-lead | 0.0010 |
| Sliver-lead-indium | 0.0010 |
| Aluminum | 0.0010 |

Table 7.2 | Maximum Unit Load for Various Bearing Metals

| Bearing metals | Maximum $P$ in kN/m$^2$ |
|---|---|
| Lead-base Babbitt | 4200–5600 |
| Tin-base Babbitt | 5600–7000 |
| Cadmium-base bearing metal | 8400–10 500 |
| Copper-lead (Pb 45%, Cu 55%) | 14 000–21 000 |
| Copper-lead (Pb 25%, Sn 3%, Cu 72%) | 21 000–28 000 |
| Silver (lead-indium overlay) | 35 000 and up |
| Bronzes | 70 000 |

### 7.2.3 | Bearing Unit Load

The unit load P is defined as the load per unit projected area of bearing. This unit load is a function of Sommerfeld number. If for a bearing, *C/R*, $\eta$, and *N* are kept constant, Sommerfeld number is inversely proportional to the unit load. Therefore, for successful hydrodynamic lubrication the bearing load should be such that the Sommerfeld number does not become critical $(S \geq 0.04)$. This is from the theoretical point of view. When the active oil film is short and the attitude is relatively high, the maximum hydrodynamic pressure $p_{max}$ in the oil film is several times higher than the unit load *P*. The bearing may fail by fatigue if the maximum pressure in the oil film exceeds the fatigue strength of the bearing material. A number of factors which influence the fatigue life of a bearing are the fatigue strength, the operating temperature, and the character and the frequency of load application. It is also quite difficult to have correlation of fatigue life and the unit load of a particular bearing. However, as a guide to preliminary selection of unit load the data (1942) given in Table 7.2 may be recommended.

It is expected that life of bearing used in various applications is from 1000 to 20 000 hours. The shortest life may be expected in aircraft engines, whereas the largest one in stationary heavy engines. For automobile engine bearing, the average life expectancy is usually between 2000 and 4000 hours.

An accurate design procedure recently published (2005) is given in Section 7.4.1.

## 7.3 | Bearing Materials

### 7.3.1 | Introduction

When fluid film bearings are operated under hydrodynamic lubrication (i.e., no metal-to-metal contact), any pair of materials can be used as bearing materials provided they have enough strength to support the applied load and their ability to be machined. Unfortunately, fluid film bearings do not always operate under this hydrodynamic lubrication. There will always be some amount of metal-to-metal contact between the surfaces during starting and stopping when adequate fluid film pressure will not be generated. In some situations bearing operates under a heavy load or with too low a speed to develop pressures that are not adequate to support applied load avoiding metal-to-metal contact. In addition to

- The shaft and bearing surfaces are not always smooth enough to avoid contact.
- There may be misalignment between the journal and bearing.
- The lubricant supply may be inadequate for forming required fluid film.

Due to the above reasons, there is a possibility of contact between the surfaces having relative motion under load. In the previous chapters, it was assumed that the applied load is constant and unidirectional. Many bearings, such as connecting rod and main bearings in I.C. engines and in piston pumps and compressors operate where the load is fluctuating during loading cycle.

These situations place additional requirements on choice of bearing materials. As bearings operate under various conditions, a particular material will never satisfy all the requirements of bearing material. For example, a bearing material having good deformability has low compressive and fatigue strength. Hence, the selection of a proper bearing material is a matter of judgment.

### 7.3.2 | Properties

A good bearing material is required to have the properties, like

- Compatibility
- Conformability
- Embedability
- Compressive strength
- Fatigue strength
- Thermal conductivity
- Machinability, and so on.

The above properties are well known and do not require further explanation.

The bearings are likely to adhere or weld under heavy load to the shaft if there is a contact between them. This situation is observed in the case of boundary lubrication. The measure of anti-weld or anti-scoring property of a bearing material operating particularly with a journal bearing material is called compatibility. Amongst the available bearing materials, tin-base and lead-base alloys are common where score resistance is a primary requirement. Nylon and zinc alloy give the best and the worst anti-scoring properties, respectively. Normally bearing alloys having the best anti-scoring resistance have low hardness and strength.

The conformability of a bearing material is its ability to yield to deformation under running condition. The situation is highly critical if deflection of shaft occurs, thereby giving a possibility of having good conformability adjusts itself by wearing without disturbing the normal operation. White metal alloys, particularly tin-base Babbitt have the ability to conform other geometric errors.

Embedability is the ability to absorb the foreign particles present in the lubricant to avoid scoring and to some extent wear. Bearing materials with good conformability and embedability have low hardness, fatigue strength and Young's modulus. Babbitts have good embedability and conformability but have less hardness and poor fatigue strength.

A bearing material having a good compressive strength resists pressure without plastic deformation. The compressive strength is calculated as load per unit projected area of the bearing. A material with good compressive strength is having poor conformability and embedability.

The fatigue strength of a bearing material depends on the range and number of stress cycles to which it is subjected to. This is important when the load fluctuates and changes in direction. Such a load is applied to reciprocating I.C. engines and piston compressors. Under the action of such load application fatigue cracks and surface pits may develop. Bronzes have the high fatigue strength, whereas lead- and tin-base Babbitts are relatively weak in fatigue strength. A thin layer of Babbitt is bonded on a backing material (steel) having higher strength and rigidity.

Lubricants sometimes tend to form organic acid and oxidize with time and temperature. This is usually seen on the piston and cylinder walls of an I.C. engine. These give rise to corrosion and as a result insoluble organic soaps are deposited in the system. Alloys having lead, cadmium, zinc, copper, and silver are subjected to corrosion. Tin-base Babbitt is the most corrosion-resistant alloy. Antimony, arsenic, or indium when added to lead-base Babbitt may increase the corrosion resistance. It is advisable to replace the used oil periodically by fresh one to avoid corrosion.

### 7.3.3 | Materials

The most common bearing materials found in use are alloys of tin, lead, and copper as Babbitts or white metals. A white metal usually comprises nearly 90% tin, 9% antimony, and 1% copper. Babbitts are classified as tin-base and lead-base alloys. In the first category, tin-base alloys have the same structure of the original Babbitt. The lead-base alloys have pure lead and compounds of tin and antimony. These alloys are having low hardness and excellent anti-scoring properties. These are quite good as far as the conformability and embedability are concerned. However, these alloys have low fatigue strength that decreases rapidly with increase in the temperature. Hence, these alloys are used at low temperatures. To improve the fatigue strength, a thin layer of Babbitt is used as an intermediate layer of high fatigue-strength material and bonded to a steel backing. These bearing materials can support high load and have anti-scoring resistance and embedability. Tin-base Babbitt has better corrosion resistance and is harder than lead-base one at room temperature. They can be bonded to steel backings fairly easily. However, the tin-base Babbitts are expensive.

If small amount of calcium, tin, or mercury is added to a lead-base Babbitt, the hardness of the lead increases to a great extent. These are called alkali-hardened lead and can be used at a moderately high temperature. These alloys have poor corrosion resistance properties.

Cadmium-base alloys (alloys of cadmium contain nearly 1% nickel, 0.5% copper, and 1% silver) have good conformability and embedability. These have better hardness than Babbitts and can be used at higher temperatures. They have higher strength than Babbitts. However, these alloys do not have good corrosion resistance.

A commonly used bearing material is bronze. Bronzes are fairly strong so that these can be used without any backing. The fatigue strength for bronze is high. Bronzes can be quite easily cast and machined. The amount of lead and tin present in the alloy is vital as it can make bronze softer or harder. A typical leaded bronze has nearly 10% to 25% lead and 2% to 10% tin. Lead provides anti-scoring properties whereas tin increases the hardness and fatigue strength. If more than 20% tin and about 10% lead are present, these are called tin bronze. Although these are quite hard, they have low conformability and embedability.

Some alloys, such as copper-lead alloys, aluminum alloys, silver sintered metals and cast iron, are used as bearing materials.

Sintered bearings which are used in some specialized applications are made from two powder metals. These are mixed and pressed in dies and then sintered by heating at a high temperature. As these bearings can be mass produced, their cost is relatively less. A typical composition of such a bearing is about 10% of tin and the rest copper. These bearings are some times called porous bearings that impregnated in oil before use. The oil is supplied through the interconnected pores. Hence, porous bearings are self-lubricating.

## 7.4 | Bearing Design

### 7.4.1 | Design Procedure

In the previous chapters, the steady state performance characteristics of journal bearings are found considering constant viscosity of oil. As already mentioned, for an accurate solution the Reynolds equation is to be solved simultaneously with the energy and the heat conduction equations. The solution can be obtained numerically using inverse method satisfying the appropriate pressure and temperature boundary conditions using finite difference method with successive over-relaxation scheme under steady state condition. The details of the solution procedure are available (2005).

The data bank for load, flow, and coefficient of friction is created from a large number of *L/D* and eccentricity ratios for given lubricant and material properties. These data are curve fitted to frame the relations dealing with various design parameters. The relations for load, flow, and coefficient of friction are given as:

### *For Load*

$$\bar{W} = a_1\left(a_2(L/D)^{a_3} + e^{a_4(L/D)^{a_5}}\right)\left(a_6\varepsilon_0 + \varepsilon_0^{\,a_7}\right)(L/D)^2 \tag{7.1}$$

where $\bar{W} = \dfrac{WC^2}{\eta\omega R^3 L}$, $\omega$ is the angular velocity of journal

$$a_1 = 0.3137,\ a_2 = 3.4377,\ a_3 = -1.8138,\ a_4 = 2.049?$$
$$a_5 = 0.5372,\ a_6 = 0.6492,\ a_7 = 4.0854$$

### *Flow Rate*

$$\bar{Q} = \left(a_1\varepsilon_0^{\,a_2} + a_3\varepsilon_0^{\,a_4}\cos\left(\varepsilon_0 a_5\right)\right)(L/D)^{a_6} \tag{7.2}$$

where $\bar{Q} = \dfrac{LQ}{2UR^2C}$

$$a_1 = 0.5509,\ a_2 = 0.9468,\ a_3 = 0.2233$$
$$a_4 = 7.2743,\ a_5 = -3.2285,\ a_6 = 1.4227$$

### *Friction Factor*

$$\left(R/C\right)\mu = \left(a_1 + e^{a_2\varepsilon_0^{a_3}} + e^{a_4\varepsilon_0^{a_5}}\right)\left(a_6(L/D)^{a_7} + a_8\right) \tag{7.3}$$

where $\bar{\mu} = \dfrac{R}{C}\mu$

$$a_1 = -0.0264,\ a_2 = -3.6662,\ a_3 = 0.4695,\ a_4 = -34.5477$$
$$a_5 = 1.0892,\ a_6 = 88.4788,\ a_7 = -2.0146,\ a_8 = 28.2124$$

These equations are found iteratively using least square optimization technique. The values obtained from the above expressions for load, flow, and coefficient of friction agrees very well with the theoretical prediction. The relations given in Equations (7.1), (7.2), and (7.3) are used to design a journal bearing.

Knowing the specific application of the bearing, assume *L/D* ratio. The bearing length *L* is then found. Choose a suitable bearing material and accordingly assume the clearance *C*. The operating eccentricity ratio $\varepsilon_0$ can be determined from Equation (7.1) for the applied load and the chosen *L/D* ratio. Then Equations (7.2) and (7.3) are used to find flow rate and coefficient of friction, respectively.

## 7.4.2 | Design Example

A full journal is operating under the following conditions:

- Shaft (journal) diameter = 100 mm
- Shaft speed *N* = 3500 rpm
- Radial load *W* = 250 kN

It is assumed that inlet oil temperature is 40$^0$C and the corresponding viscosity of oil used is 0.018 Pa-s. From the consideration of space and other limitations let the *L*/*D* be 1. Hence $L = 100$ mm. From the consideration of surface finish of the shaft and bush, the clearance ratio *C/R* be taken as 0.0005. The radial clearance *C* is then equal to 0.025 mm. The dimensionless load is $\bar{W} = 1.228$. From Equation (7.1), the eccentricity ratio $\varepsilon_0$ is obtained as 0.469. The volume rate of flow and coefficient of friction then determined from Equations (7.2) and (7.3) are $6.165 \times 10^{-6}$ m$^3$/s and 0.0029, respectively.

The conventional design of such a bearing deals with the design data of journal bearings under steady state condition. Heat balance, i.e., heat generated, $H_g$ is made equal to heat dissipated $H_d$ is satisfied using Cameron and Woods' so–called 'two-thirds' temperature rise rule. The average equilibrium temperature, (hence average viscosity) is then determined for a given oil.

In the forgoing study, a rigorous analysis considering energy and heat conduction equations are solved simultaneously with Reynolds equation. The temperature distribution (hence viscosity) in the film is known. To carry out heat balance, the following steps are suggested:

1. Calculate $H_g$ from coefficient of friction determined above.
2. Calculate heat dissipated $H_{d1}$, due to convection from the classical equation: $\rho QS\Delta T$, where $\rho$ = mass density of oil, $Q$ = volume rate of flow, $S$ = specific heat of oil and $\Delta T$ = temperature rise. $\Delta T$ can be approximately estimated as $\Delta T = (T_{max} - T_i)$, where $T_{max}$ = maximum oil temperature and $T_i$ = inlet oil temperature.
3. Heat dissipation $H_{d2}$ due to conduction from oil film to shaft and that to bush after knowing the temperature distributions in the shaft and the bush.
4. For heat balance the Equation (7.4) is to be satisfied.

$$H_g = H_{d1} + H_{d2} \tag{7.4}$$

5. If Equation (7.4) is not satisfied in the first attempt, choose different *L*/*D*, and *C*/*R* ratios and other parameters relating to properties of oil and bearing materials given in reference Singh and Majumdar (2005). It may sometimes be necessary to generate a new set of data with the above variables.

Once Equation (7.4) is satisfied, the design is all right.

Using the above method, preliminary design dimensions are found. Then the journal speed is checked for the given load and see if the bearing is operating stably for complete design.

## REFERENCES

*Etchells, E.B. and Underwood, A.F.* (1942), Practical aspects of bearing design, *Machine Design*, **9**, 85.

*Radzimovsky, E.I.* (1959), *Lubrication of Bearings*, The Ronald Press Co., 279–295.

*Shaw, M.C. and Macks, E.F.* (1949), *Analysis and Lubrication of Bearings*, McGraw-Hill Book Co. Inc., New York.

*Singh, D.S. and Majumdar, B.C.* (2005), Computer-aided design of hydrodynamic bearings considering thermal effects. *J. Engg. Trib, Proce. IMechE*, **219**, 133–143.

Chapter 8

# Dynamics of Fluid Film Bearings

## 8.1 | Introduction

In the previous chapters, analysis for load capacity, oil flow rate, power loss, and heat balance has been presented for hydrodynamic bearings under steady load. However, in reality, the bearings are invariably subjected to time dependent dynamic loads. For example, rotors of turbines, pumps and compressors are subjected to dynamic load due to rotor unbalance which becomes very high at high speeds. Dynamic force also rotates at the same frequency as that of rotational speed of the journal. Similarly rotors/crank shafts of reciprocating machines, e.g., diesel and petrol engines, reciprocating pumps, and compressors are also subjected to large variations in magnitude and direction of the load. Fluid film bearings of these machines are therefore subjected to severe dynamic loading. Maintaining a minimum value of film thickness is a challenging task for the designers of the fluid film bearings. Ensuring stability of the rotors supported on hydrodynamic bearings is of utmost importance. Rotors are subjected to self excited vibrations and instability may occur due to unsteady fluid film forces. Journal bearing stability is one of the most important and vital aspects of design of rotor-bearing systems. Besides, there are numerous applications where relative sliding velocity is very small or zero but the surfaces approach each other during the course of action. In other words, there exists a normal velocity of approach. In the absence of lubrication, surfaces would impact each other and would get damaged. However, if the gap between the surfaces is filled with a viscous lubricant, then a positive pressure would be generated due to squeezing out of the lubricant from the gap and thereby a load carrying capacity would be generated. This mode of lubrication is generally referred to as squeeze film lubrication. It was recognized immediately after the Second World War that squeeze film lubrication plays an important role in the lubrication of piston- cylinder contact, piston pin, and connecting rod lubrication of internal combustion engines. It has also been recognized that mode of lubrication in human joints is also due to squeeze action when synovial fluid in the joints is squeezed out due to movement of limbs. This chapter deals with the issues related to squeeze film lubrication, dynamic loading, and stability aspects of fluid film bearings.

## 8.2 | Derivation of Reynolds Equation for Journal Bearing under Dynamic Condition

The Reynolds equation for incompressible, isoviscous lubrication has been derived in Chapter 3 and is rewritten as

$$\frac{\partial}{\partial x}\left(h^3\frac{\partial p}{\partial x}\right)+\frac{\partial}{\partial y}\left(h^3\frac{\partial p}{\partial y}\right)=6\eta U\frac{\partial h}{\partial x}+12\eta\frac{\partial h}{\partial t} \tag{8.1}$$

where $U$ is the surface speed of the journal and $U = R\omega$. $\frac{\partial h}{\partial t}$, is the velocity of approach of journal surface toward the bearing surface and is usually referred to as squeeze velocity. Figure 8.1 shows the journal bearing under dynamic loading condition. It shows the journal center velocities in radial and tangential directions. In this situation, attitude angle $\phi$ changes with time, i.e., attitude line has an angular velocity $\dot{\phi}$. It is also assumed that the load magnitude and direction can change. In other words, load line can also rotate with an angular velocity $\dot{\psi}$ while its magnitude may also change. This situation is generally encountered in the main bearing of internal combustion engines. Turbo-generator rotors are usually very heavy and can weigh few tons in the case of 500–2000 MW power plant. Rotors are generally balanced precisely but in spite of that a small amount of unbalance exists which would introduce large centrifugal force. Therefore, the journal center undergoes an orbital motion around its steady state equilibrium position.

The velocity of the journal center $\frac{\partial h}{\partial t}$ can be determined by differentiating film thickness expression with respect to time.

Film thickness expression is written as

$$h = C\left[1+\varepsilon\cos\left\{\alpha-(\phi+\psi)\right\}\right] \tag{8.2}$$

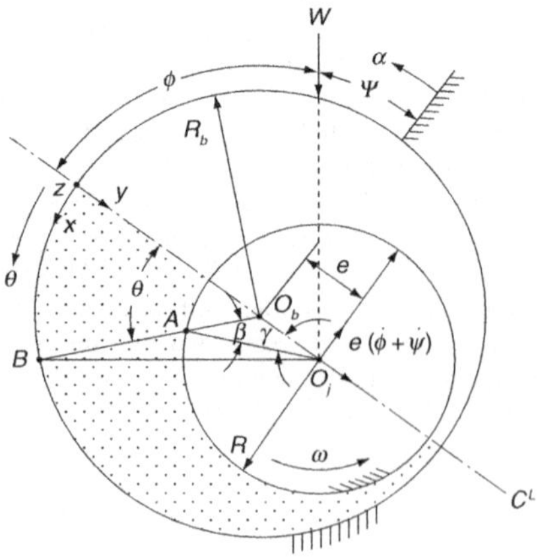


**Figure 8.1** | Journal Bearing Geometry and Nomenclature

or $h = C(1+\varepsilon\cos\theta)$ where $\theta = \alpha - (\phi + \psi)$ and for $\alpha = 0$, i.e., a fixed position we get

$$\frac{\partial h}{\partial t} = C\left\{\dot{\varepsilon}\cos\theta + \varepsilon(\dot{\varphi} + \dot{\Psi})\sin\theta\right\} \tag{8.3}$$

Equation (8.1) with the substitution of Equation (8.3) becomes:

$$\frac{\partial}{\partial x}\left(h^3\frac{\partial p}{\partial x}\right) + \frac{\partial}{\partial y}\left(h^3\frac{\partial p}{\partial y}\right) = 6\eta\omega\frac{\partial h}{\partial\theta} + 12\eta C\left\{\dot{\varepsilon}\cos\theta + \varepsilon(\dot{\varphi} + \dot{\psi})\sin\theta\right\} \tag{8.4}$$

Since $\bar{h} = \dfrac{h}{C} = 1 + \varepsilon\cos\theta$ and $\theta = x/R$

Equation (8.4) can be written as

$$\begin{aligned}\frac{\partial}{\partial\theta}\left(\bar{h}^3\frac{\partial p}{\partial\theta}\right) + R^2\frac{\partial}{\partial y}\left(\bar{h}^3\frac{\partial p}{\partial y}\right) &= 6\eta\left(\frac{R}{C}\right)^2\omega\frac{\partial\bar{h}}{\partial\theta} \\ &\quad + 12\eta\left(\frac{R}{C}\right)^2\dot{\varepsilon}\cos\theta - 12\eta\left(\frac{R}{C}\right)^2\frac{\partial\bar{h}}{\partial\theta}(\dot{\varphi} + \dot{\psi})\end{aligned} \tag{8.5}$$

or

$$\begin{aligned}\frac{\partial}{\partial\theta}\left(\bar{h}^3\frac{\partial p}{\partial\theta}\right) + R^2\frac{\partial}{\partial y}\left(\bar{h}^3\frac{\partial p}{\partial y}\right) &= 6\eta(R/C)^2\left\{\omega - 2(\dot{\varphi} + \dot{\psi})\right\}\frac{\partial\bar{h}}{\partial\theta} \\ &\quad + 12\eta(R/C)^2\dot{\varepsilon}\cos\theta\end{aligned} \tag{8.6}$$

## 8.3 | Dynamics of Rotor-Bearings Systems

Rotor-bearing systems of high speed turbo-machinery are susceptible to severe vibrations and instabilities. Unbalance causes severe dynamic loading while the rotor moves beyond its first or second critical speed. Hydrodynamic bearings affect the dynamics of a rotor in a far more complex way than does a rolling-element bearing. Fluid film bearings can induce self-excited vibrations in a rotor-bearing system which may damage the rotor and the bearing. In liquid lubricated bearings, this phenomenon is known as oil whirl, resonant whirl, or half frequency whirl in case of lightly loaded rotors. The onset of self-excited vibrations is caused by the fact that rotor-bearing system becomes unstable. Because, the instabilities occur more readily as the dynamic effects become stronger, for instance, at higher speeds one may say that the application of fluid film bearings to support rotor literally stands or falls with the possibilities of controlling the dynamics of the system in such a way as to avoid instabilities.

Rotor-bearing stability study can be outlined as follows:

- To evaluate the rotor dynamic coefficients, i.e., stiffness and damping coefficients of fluid film of a bearing using lubrication theory and linear perturbation theory for small amplitude vibrations. Rotor dynamic coefficients thus determined are used to find the stability of a rotor-bearing system.
- To study bearing configurations which are known to have favorable influence on rotor stability, viz., tilting pad bearings, noncircular bearings, pressurized bearings, support flexibility, etc. Stiffness and damping coefficients of these configurations along with support flexibility are known to influence the stability of the rotor favorably.

- To determine critical speeds of large flexible rotors and balancing of such rotors to minimize vibrations and support reactions. Some of these aspects will be dealt with in the following sections.

## 8.4 | Stiffness and Damping Coefficients

### 8.4.1 | Analytical Approach for Short and Long Bearings

Stiffness and damping coefficients of plain journal bearings can be determined using short and long bearing assumptions from the dynamic film forces obtained from the solution of dynamic Reynolds equation with appropriate assumptions. Holmes (1960) derived expressions for rotor dynamic coefficients using short bearing assumptions. Lund (1966), Lund and Saibel (1967), Kirk and Gunter (1976), Badgley and Booker (1969), Reddy and Trumpler (1962), and Muszynska (1986) also investigated various aspects of rotor dynamics including oil whirl using short and long bearing theories.

#### Short Bearing Approximation

Integrating Reynolds Equation (8.6) by substituting $\frac{\partial p}{\partial \theta} = 0$ and $\dot{\psi} = 0$ for nonrotating load, subject to appropriate boundary conditions would yield the pressure distribution as

$$p(y) = \frac{3\eta\left(y^2 - \frac{L^2}{4}\right)}{C^2}\left\{\left[\frac{2\dot{\varepsilon}\cos\theta - (\omega - 2\dot{\varphi})\varepsilon_0\sin\theta}{(2+\varepsilon_0^{\,2})(1+\varepsilon_0\cos\theta)^2}\right] + \frac{\dot{\varepsilon}}{\varepsilon_0}\left[\frac{1}{(1+\varepsilon_0\cos\theta)^2} - \frac{1}{(1+\varepsilon_0)^2}\right]\right\} \tag{8.7}$$

The boundary conditions are given as

$$p\left(\theta,\ \pm\frac{L}{2}\right) = 0 \tag{8.8}$$

The pressure distribution given by Equation (8.7) can be used to determine the dynamic film forces of the bearing for $\pi$ film extent in the circumferential direction (see Fig. 8.2), which can be used to find stiffness and damping coefficients of the bearing.

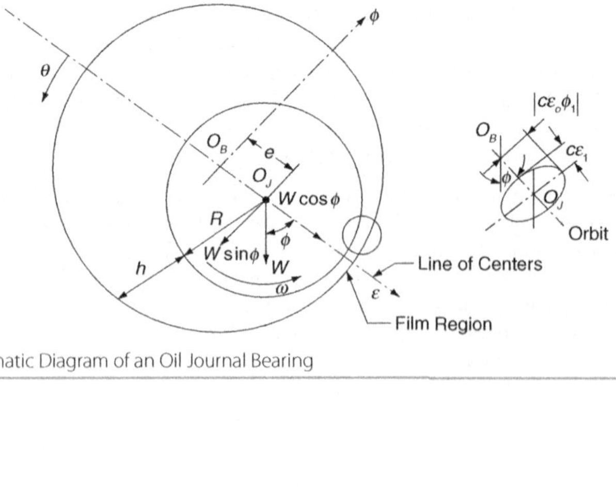


**Figure 8.2** | Schematic Diagram of an Oil Journal Bearing

Dynamic film force components along the line of centers and normal to it are written as:

$$F_{\varepsilon} = -\eta RL\left(\frac{L}{C}\right)^2\left[\left|\omega - 2\dot{\phi}\right|\frac{\varepsilon^2}{\left(1-\varepsilon^2\right)^2} + \frac{\pi}{2}\frac{\left(1+2\varepsilon^2\right)\dot{\varepsilon}}{\left(1-\varepsilon^2\right)^{5/2}}\right] \tag{8.9}$$

$$F_{\phi} = \eta RL\left(\frac{L}{C}\right)^2\left[\left(\omega - 2\dot{\phi}\right)\frac{\pi\varepsilon}{4\left(1-\varepsilon^2\right)^{3/2}} + \frac{2\varepsilon\dot{\varepsilon}}{\left(1-\varepsilon^2\right)^2}\right] \tag{8.10}$$

For short bearing $\pi$ film solution, modified Sommerfeld number is expressed as

$$S' = \pi S\left(\frac{L}{D}\right)^2 = \frac{\eta\omega}{8}\left(\frac{L}{C}\right)^2\frac{DL}{W}$$

and at steady state position of the journal

$$S' = \left(1-\varepsilon_0^2\right)^2 / \varepsilon_0\left[16\varepsilon_0^2 + \pi^2\left(1-\varepsilon_0^2\right)\right]^{1/2} \tag{8.11}$$

Attitude angle

$$\phi_0 = \tan^{-1}\left[\frac{\left(1-\varepsilon_0^{\,2}\right)^{1/2}}{4\varepsilon_0}\right]$$

Dimensionless stiffness and damping coefficients are given in two perpendicular directions as

### ***Stiffness Coefficients***

$$k_{\varepsilon\varepsilon} = \frac{8\left(1+\varepsilon_0^2\right)}{\left(1-\varepsilon_0^2\right)}f\left(\varepsilon_0\right)$$

$$k_{\varepsilon\phi} = \frac{\pi\left(1-\varepsilon_0^2\right)^{1/2}}{\varepsilon_0}f\left(\varepsilon_0\right) \tag{8.12}$$

$$k_{\phi\varepsilon} = \frac{\pi\left(1+2\varepsilon_0^2\right)}{\varepsilon_0\left(1-\varepsilon_0^2\right)^{1/2}}f\left(\varepsilon_0\right)$$

$$k_{\phi\phi} = 4f\left(\varepsilon_0\right)$$

***Damping Coefficients***

$$b_{\varepsilon\varepsilon} = \frac{2}{\omega} k_{\phi\varepsilon}$$

$$b_{\varepsilon\varphi} = b_{\varphi\varepsilon} = \frac{2}{\omega} k_{\varphi\varphi} \tag{8.13}$$

$$b_{\varphi\varphi} = \frac{2}{\omega} k_{\varepsilon\varphi}$$

where

$$f(\varepsilon_0) = \frac{1}{\left\{\pi^2 + (16 - \pi^2)\varepsilon_0^{\,2}\right\}^{1/2}} \tag{8.14}$$

Stiffness and damping coefficients in $x, y$ directions can also be determined after obtaining fluid film forces.

Dimensionless stiffness and damping coefficients in $x, y$ coordinates may be expressed as

***Stiffness Coefficients***

$$k_{xx} = 4\left\{2\pi^2 + (16 - \pi^2)\varepsilon_0^{\,2}\right\} f(\varepsilon_0)$$

$$k_{xy} = \frac{\pi\left\{-\pi^2 + 2\pi^2\varepsilon_0^{\,2} + (16 - \pi^2)\varepsilon_0^{\,4}\right\} f(\varepsilon_0)}{\varepsilon_0 (1 - \varepsilon_0^{\,2})^{1/2}}$$

$$k_{yx} = \frac{\pi\left\{\pi^2 + (32 + \pi^2)\varepsilon_0^{\,2} + 2(16 - \varepsilon_0^{\,2})\varepsilon_0^{\,4}\right\} f(\varepsilon_0)}{\varepsilon_0 (1 - \varepsilon_0^{\,2})^{1/2}} \tag{8.15}$$

$$k_{yy} = \frac{4\left\{\pi^2 + (32 + \pi^2)\varepsilon_0^{\,2} + 2(16 - \pi^2)\varepsilon_0^{\,4}\right\} f(\varepsilon_0)}{(1 - \varepsilon_0^{\,2})}$$

***Damping Coefficients***

$$b_{xx} = \frac{2\pi(1 - \varepsilon_0^{\,2})^{1/2}\left\{\pi^2 + 2(\pi^2 - 8)\varepsilon_0^{\,2}\right\} f(\varepsilon_0)}{\varepsilon}$$

$$b_{xy} = 8\left\{\pi^2 + 2(\pi^2 - 8)\varepsilon_0^{\,2}\right\} f(\varepsilon_0)$$

$$b_{yx} = b_{xy} \tag{8.16}$$

$$b_{yy} = \frac{2\pi\left\{\pi^2 + 2(24 - \pi^2)\varepsilon_0^{\,2} + \pi^2\varepsilon^4\right\}}{\varepsilon(1 - \varepsilon^2)^{1/2}}$$

where

$$f(\varepsilon_0) = \frac{1}{\left\{\pi^2 + (16 - \pi^2)\varepsilon_0^{\,2}\right\}^{3/2}} \tag{8.17}$$

In Equations (8.11) to (8.16), the dimensional stiffness and damping coefficients can be obtained from their respective dimensionless quantities as per the relationships given below:

$$k_{ij} = \frac{K_{ij}C}{W},\; b_{ij} = \frac{B_{ij}C\omega}{W} \qquad i = x, j = y \tag{8.18}$$

where $W = \frac{\eta\omega}{4}\left(\frac{L}{C}\right)^2 LR \frac{\left(1-\varepsilon_0\right)^2}{\varepsilon_0\left[16\varepsilon_0^{\,2}+\pi^2\left(1-\varepsilon_0^{\,2}\right)^2\right]^{1/2}}$

Figures 8.3 and 8.4 show the plots of variation of rotor dynamic coefficients with eccentricity ratio as obtained from the above equations for short bearings. Similar plots of rotor dynamic coefficients of short journal bearings were also presented by Holmes (1960). These are very useful in determining stability and small amplitude unbalance vibration response of the journal center.

Pivoted pad journal and thrust bearings are also very widely used bearing configurations. Lund (1964), Nicholas *et al.* (1979), Allaire *et al.* (1981), and Raimondi and Szeri (1984) investigated stiffness and damping coefficients of pivoted pad journal and thrust bearings.

## Long Bearing Approximation

Pressure distribution in the lubricant film can be determined using long bearing assumption by integrating Equation (8.6) after substituting $\frac{\partial p}{\partial y} = 0$ and $\dot{\psi} = 0$ for nonrotating load satisfying appropriate boundary conditions given by Equation (8.8) as

$$p(\theta) = \frac{6\eta R^2}{C^2}\left\{\frac{(\omega-2\dot{\varphi})\varepsilon\sin\theta(2+\varepsilon\cos\theta)}{(2+\varepsilon^2)(1+\varepsilon\cos\theta)^2} + \frac{\dot{\varepsilon}}{\varepsilon}\left[\frac{1}{(1+\varepsilon\cos\theta)^2} - \frac{1}{(1+\varepsilon)^2}\right]\right\} \tag{8.19}$$

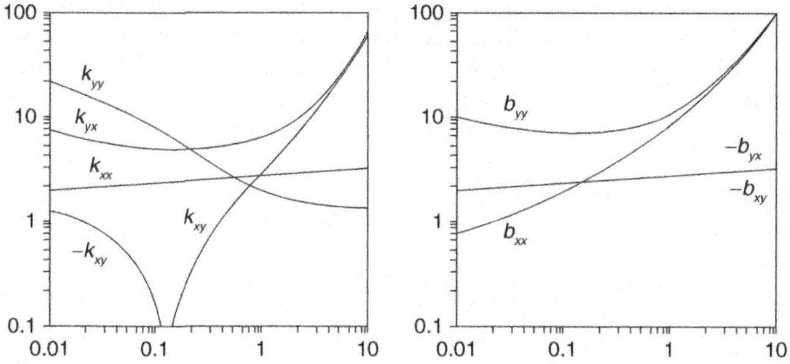


**Figure 8.3** | Stiffness and Damping Coefficients of Short Bearing Versus Sommerfeld Number [Lund, 1966]

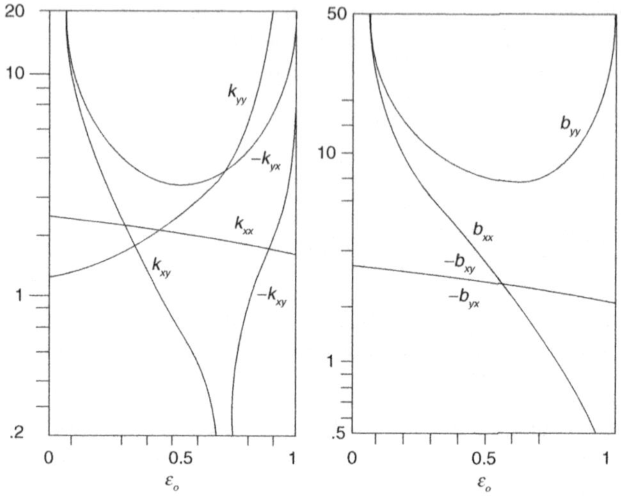


**Figure 8.4** | Stiffness and Damping Coefficients of Short Bearing Versus Eccentricity Ratio [Lund, 1966]

Dynamic fluid film forces can be determined using the pressure distribution as given above. Stiffness and damping coefficients can be calculated in a similar manner as has been done for short bearing approximation.

### 8.4.2 | Finite Length Bearings

It is assumed that the journal center undergoes small amplitude vibration around its steady state eccentric position under dynamic loading. Further, that the dynamic load responsible for vibration is not large in comparison to steady state load. The frequency of dynamic load may be same as that speed of rotation of the journal as in the case of unbalance load or may be different if the loading is from an external source. To determine the rotor dynamic coefficients, the steady state journal center position is given small perturbations, i.e., displacements and velocities to estimate the perturbed fluid film forces which are used to determine the rotor-dynamic coefficients.

The Reynolds equation is written for nonrotating load, i.e., for $\dot{\psi} = 0$ as

$$\frac{\partial}{\partial\theta}\left(\bar{h}^3\frac{\partial p}{\partial\theta}\right) + R^2\frac{\partial}{\partial y^2}\left(\bar{h}^3\frac{\partial p}{\partial y}\right) = 6\eta(R/C)^2(\omega - 2\dot{\varphi}) + 12\eta(R/C)^2\,\dot{\varepsilon}\cos\theta \tag{8.20}$$

Fluid film forces are calculated from the pressure distribution after solving Reynolds equation for a known equilibrium position $(\varepsilon_0, \varphi_0)$ of the journal center satisfying appropriate boundary conditions.

In the case of small amplitude orbital motion of the journal center, linear vibration theory can be used and perturbation theory can be employed to determine the characteristics of the fluid film in terms of its stiffness and damping coefficients. Components of the dynamic fluid film forces can be determined by integrating the pressure over the bearing area for a given equilibrium position as [Refer to Fig. 8.2].

Force along the line of centers

$$F_\varepsilon = -\oiint p(\theta, y)\cos\theta R d\theta dy \tag{8.21}$$

Force perpendicular to the line of centers

$$F_\phi = \oiint p(\theta, y)\sin\theta R d\theta dy$$

Attitude angle $\phi = \tan^{-1}\left(\frac{F_\phi}{F_\varepsilon}\right)$

These forces when resolved in the Cartesian coordinates are

$$\begin{aligned} F_x &= F_\varepsilon \cos\varphi - F_\varphi \sin\varphi \\ F_y &= F_\varepsilon \sin\varphi + F_\varphi \cos\varphi \end{aligned} \tag{8.22}$$

Under dynamic condition lubricant pressure, $p$, depends only on the position of the journal center or film thickness, $h$ and its time derivative, $\frac{\partial h}{\partial t}$.

If it is further assumed that the displacements of the journal center from its equilibrium position $(\varepsilon_0, \phi_0)$ or $(x_0, y_0)$ are so small that the linearization of additional reaction forces due to fluid film pressure is permissible and that the effects of misalignment is negligible, one can write with the assumptions that $\varepsilon = \varepsilon_0 + d\varepsilon$ and $\phi = \phi_0 + d\phi$. Thus, film forces can be expressed as

$$F_\varepsilon(\varepsilon, \phi, \dot{\varepsilon}, \dot{\phi}) = F_\varepsilon(\varepsilon_0, \phi_0, 0, 0) + \Delta F_\varepsilon \tag{8.23}$$

where

$$\Delta F_\varepsilon = -K_{\varepsilon\varepsilon}(C d\varepsilon) - K_{\varepsilon\phi}(C\varepsilon_0 d\phi) - B_{\varepsilon\varepsilon}(C d\dot{\varepsilon}) - B_{\varepsilon\phi}(C\varepsilon_0 d\dot{\phi})$$

and

$$F_\phi(\varepsilon, \phi, \dot{\varepsilon}, \dot{\phi}) = F_\phi(\varepsilon_0, \phi_0, 0, 0) + \Delta F_\phi \tag{8.24}$$

where

$$\Delta F_\phi = -K_{\phi\varepsilon}(C d\varepsilon) - K_{\phi\phi}(C\varepsilon_0 d\phi) - B_{\phi\varepsilon}(C d\dot{\varepsilon}) - B_{\phi\phi}(C\varepsilon_0 d\dot{\phi})$$

In the above expressions, all higher order derivatives have been neglected. Thus, stiffness and damping coefficients of the fluid film are defined as

### *Stiffness Coefficients*

$$\begin{aligned} K_{\varepsilon\varepsilon} &= -\left(\frac{\partial F_\varepsilon}{C\partial\varepsilon}\right); \quad K_{\varepsilon\varphi} = -\left(\frac{\partial F_\varepsilon}{C\varepsilon_0 \partial\varphi}\right); \\ K_{\varphi\varepsilon} &= -\left(\frac{\partial F_\varphi}{C\partial\varepsilon}\right); \quad K_{\varphi\varphi} = -\left(\frac{\partial F_\varphi}{C\varepsilon_0 \partial\varphi}\right); \end{aligned} \tag{8.25}$$

***Damping Coefficients***

$$B_{\varepsilon\varepsilon} = -\left(\frac{\partial F_{\varepsilon}}{C\partial\dot{\varepsilon}}\right); \quad B_{\varepsilon\varphi} = -\left(\frac{\partial F_{\varepsilon}}{C\varepsilon_0\partial\dot{\varphi}}\right)$$
$$B_{\varphi\varepsilon} = -\left(\frac{\partial F_{\varphi}}{C\partial\dot{\varepsilon}}\right); \quad B_{\varphi\varphi} = -\left(\frac{\partial F_{\varphi}}{C\varepsilon_0\partial\dot{\varphi}}\right) \tag{8.26}$$

$K_{\varepsilon\varepsilon}$, $K_{\varphi\varphi}$ are called direct stiffness coefficients, whereas $K_{\varepsilon\varphi}$, $K_{\varphi\varepsilon}$ are cross stiffness coefficients. Similarly, $B_{\varepsilon\varepsilon}$, $B_{\varphi\varphi}$ are referred to as direct damping coefficients and $B_{\varepsilon\varphi}$, $B_{\varphi\varepsilon}$ are cross damping coefficients.

The above partial derivatives of the fluid film forces are obtained at the equilibrium position of the journal center given by $\varepsilon_0$, $\varphi_0$. Perturbed forces are then determined by giving small displacement and velocity to the journal center from its equilibrium position in two directions alternately. Derivatives of the force components are determined from the change in the film forces in two directions obtained by perturbing the equilibrium position of the journal center. To determine damping coefficients, journal center equilibrium position is subjected to small velocities to determine change in the fluid film forces in two directions, respectively.

In Cartesian coordinates, dynamic fluid film forces are expressed as

$$F_x = F_{x0} + \Delta F_x$$
$$F_y = F_{y0} + \Delta F_y$$

$F_{x0}$ and $F_{y0}$, represent fluid film forces that correspond to the journal center equilibrium position $(x_0, y_0)$ and $\Delta F_x$ and $\Delta F_y$ are perturbed fluid film forces.

The perturbed fluid film forces in Cartesian coordinates are

$$\Delta F_x = \Delta F_{\varepsilon} \cos\varphi - \Delta F_{\varphi} \sin\varphi$$
$$\Delta F_y = \Delta F_{\varepsilon} \sin\varphi + \Delta F_{\varphi} \cos\varphi \tag{8.27}$$
$$\Delta F_x = -K_{xx}x - K_{xy}y - B_{xx}\dot{x} - B_{xy}\dot{y}$$
$$\Delta F_y = -K_{yy}y - K_{yx}x - B_{yx}\dot{x} - B_{yy}\dot{y} \tag{8.28}$$

***Stiffness Coefficients***

$$K_{xx} = -\left(\frac{\partial F_x}{\partial x}\right); \quad K_{xy} = -\left(\frac{\partial F_x}{\partial y}\right)$$
$$K_{yx} = -\left(\frac{\partial F_y}{\partial x}\right); \quad K_{yy} = -\left(\frac{\partial F_x}{\partial y}\right) \tag{8.29}$$

***Damping Coefficients***

$$B_{xx} = -\left(\frac{\partial F_x}{\partial \dot{x}}\right); \quad B_{xy} = -\left(\frac{\partial F_x}{\partial \dot{y}}\right)$$
$$B_{yx} = -\left(\frac{\partial F_y}{\partial \dot{x}}\right); \quad B_{yy} = -\left(\frac{\partial F_x}{\partial \dot{y}}\right) \tag{8.30}$$

$\Delta F_x$ and $\Delta F_y$, are to be determined by perturbing the equilibrium position of the journal center $x_0, y_0$ by giving small displacement $x, y$, and velocities $(\dot{x}, \dot{y})$, respectively, about the equilibrium position.

An alternative approach to determine the rotor dynamic coefficients is to determine the dynamic fluid film forces by solving the perturbed dynamic Reynolds equations. To illustrate this, it is assumed that the journal center undergoes small amplitude harmonic oscillations about its steady state equilibrium position $\varepsilon_0, \phi_0$ and therefore dynamic position of the journal center is written as

$$\varepsilon = \varepsilon_0 + \varepsilon_1 e^{i\nu t} \text{and } \phi = \phi_0 + \phi_1 e^{i\nu t} \tag{8.31}$$

where $\varepsilon_1, \phi_1$ are dynamic amplitudes of oscillations and $\nu$ is the frequency of oscillation.

Thus, dynamic pressure and film thickness can be expressed as

$$\begin{aligned} p &= p_0 + \varepsilon_1 e^{i\nu t} p_1 + \varepsilon_0 \phi_1 e^{i\nu t} p_2 \\ \bar{h} &= \bar{h}_0 + \varepsilon_1 e^{i\nu t} \cos\theta + \varepsilon_0 \phi_1 e^{i\nu t} \sin\theta \end{aligned} \tag{8.32}$$

where $p_1$ and $p_2$ are dynamic film pressures and $\bar{h}_0 = (1 + \varepsilon_0 \cos\theta)$ is steady state film thickness.

Substituting the above equations into dynamic Reynolds Equation (8.20) and collecting only zeroth and first order terms, neglecting all higher order terms one gets:

Zeroth order terms

$$\frac{\partial}{\partial\theta}\left(\bar{h}_0^{3} \frac{\partial p_0}{\partial\theta}\right) + R^2 \frac{\partial}{\partial y}\left(\bar{h}_0^{3} \frac{\partial p_0}{\partial y}\right) = 6\eta\omega\left(\frac{R}{C}\right)^2 \frac{\partial \bar{h}_0}{\partial\theta} \tag{8.33}$$

First order terms

$$\begin{aligned} &\frac{\partial}{\partial\theta}\left(\bar{h}_0^{3} \frac{\partial p_1}{\partial\theta}\right) + \frac{\partial}{\partial\theta}\left(3\bar{h}_0^{3} \cos\theta \frac{\partial p_0}{\partial\theta}\right) + R^2 \frac{\partial}{\partial y}\left(\bar{h}_0^{3} \frac{\partial p_1}{\partial y}\right) \\ &+ R^2 \frac{\partial}{\partial y}\left(3\bar{h}_0^{2} \cos\theta \frac{\partial p_1}{\partial y}\right) = i12\eta\left(\frac{R}{C}\right)^2 \gamma\omega \cos\theta - 6\eta\omega\left(\frac{R}{C}\right)^2 \sin\theta \end{aligned} \tag{8.34}$$

$\varepsilon_0 \phi_1 e^{i\tau}$ terms

$$\begin{aligned} &\frac{\partial}{\partial\theta}\left(\bar{h}_0^{3} \frac{\partial p_2}{\partial\theta}\right) + \frac{\partial}{\partial\theta}\left(3\bar{h}_0^{3} \sin\theta \frac{\partial p_0}{\partial\theta}\right) + R^2 \frac{\partial}{\partial y}\left(\bar{h}_0^{3} \frac{\partial p_2}{\partial y}\right) \\ &+ R^2 \frac{\partial}{\partial y}\left(3\bar{h}_0^{2} \sin\theta \frac{\partial p_0}{\partial y}\right) = i12\eta\left(\frac{R}{C}\right)^2 \gamma\omega \sin\theta + 6\eta\omega\left(\frac{R}{C}\right)^2 \cos\theta \end{aligned} \tag{8.35}$$

where whirl frequency ratio '$\gamma$' is defined as the ratio of the frequency of oscillation to the angular frequency of rotation of the journal. Thus,

$$\gamma = \frac{\nu}{\omega} \tag{8.36}$$

In synchronous whirl when the frequency of oscillation is same as the frequency of rotation of the journal, i.e., $\gamma = 1.0$. Synchronous vibration occurs when the vibration is caused by the unbalance.

Stiffness and damping coefficients can be used to determine either the unbalance response of the rotor or the stability of the rotor. Therefore, dynamic fluid film forces can be determined for the synchronous condition of the vibration of the rotor, i.e., by substituting $\nu = \omega$ in the dynamic Reynolds Equations (8.20) and (8.21). To determine stiffness and damping coefficients it is necessary to determine steady state pressure distribution, $p_0$ for the equilibrium position of the journal center $(\varepsilon_0, \phi_0)$. Perturbed dynamic pressure distributions $p_1$ and $p_2$ are determined from the solution of dynamic Reynolds equations (8.34) and (8.35) satisfying appropriate boundary conditions at inlet, cavitations boundary and bearing sides.

### 8.4.3 | Stiffness and Damping Coefficients

The pressures $p_1$ and $p_2$ in the film region are developed due to the dynamic displacements of the journal center $\mathrm{Re}(C\varepsilon_1 e^{i\tau})$ parallel to and $\mathrm{Re}(C\varepsilon_0 \phi_1 e^{i\tau})$ perpendicular to the line of centers. The components of the load due to pressure $p_1$ along and perpendicular to the line of centers can be written as

$$(W_1)_\varepsilon = -2\int_0^{L/2}\int_{\theta_1}^{\theta_2} p_1 R\cos\theta \, d\theta \, dy \tag{8.37a}$$

$$(W_1)_\varphi = 2\int_0^{L/2}\int_{\theta_1}^{\theta_2} p_1 R\sin\theta \, d\theta \, dy \tag{8.37b}$$

Since the pressure $p_1$ is complex, the load components will consists of two parts, i.e., real and imaginary parts.

The fluid film which supports the rotor is treated as equivalent to a spring and a dashpot system. Since the journal executes small harmonic oscillation about its steady state position, the dynamic load carrying capacity can be expressed as a spring and a viscous damping force as given below

$$-(W_1)_\varepsilon \varepsilon_1 e^{i\tau} = K_{\varepsilon\varepsilon} X + B_{\varepsilon\varepsilon}\frac{dX}{dt} \tag{8.38a}$$

$$-(W_1)_\phi \varepsilon_1 e^{i\tau} = K_{\phi\varepsilon} X + B_{\phi\varepsilon}\frac{dX}{dt} \tag{8.38b}$$

where the position of the journal center is given by $X = C\varepsilon_1 e^{i\tau}$, $\tau = \omega t$.

Now,

$$-(\bar{W}_1)_\varepsilon = -\frac{(W_1)_\varepsilon}{W} = \frac{K_{\varepsilon\varepsilon}C}{W} + i\frac{B_{\varepsilon\varepsilon}C\omega}{W} \tag{8.39a}$$

$$-(\bar{W}_1)_\phi = -\frac{(W_1)_\phi}{W} = \frac{K_{\phi\varepsilon}C}{W} + i\frac{B_{\phi\varepsilon}C\omega}{W} \tag{8.39b}$$

Since $p_1$ is complex, the dynamic load $W_1$ is also complex and can be expressed in terms of real and imaginary parts as

$$W_1 = \mathrm{Re}(W_1) + \mathrm{Im}(W_1) \tag{8.40}$$

Thus, stiffness and damping coefficients can be expressed in dimensionless form as

$$k_{\varepsilon\varepsilon} = -\mathrm{Re}\left(\bar{W}_1\right)_\varepsilon = \frac{K_{\varepsilon\varepsilon}C}{W} \; ; k_{\phi\varepsilon} = -\mathrm{Re}\left(\bar{W}_1\right)_\phi = \frac{K_{\phi\varepsilon}C}{W} \tag{8.41a}$$

$$b_{\varepsilon\varepsilon} = -\mathrm{Im}\left(\bar{W}_1\right)_\varepsilon = \frac{B_{\varepsilon\varepsilon}C\omega}{W} \; ; b_{\phi\varepsilon} = -\mathrm{Im}\left(\bar{W}_1\right)_\phi = \frac{B_{\phi\varepsilon}C\omega}{W} \tag{8.41b}$$

Similarly, considering the dynamic displacement of the journal center along $\phi$-direction as $Y = C\varepsilon_0\phi_1 e^{i\tau}$ we can get another set of equations given by

$$\left(W_2\right)_\varepsilon = -2\int_0^{L/2}\int_{\theta_1}^{\theta_2} p_2 R\cos\theta \, d\theta \, dy \tag{8.42a}$$

$$\left(W_2\right)_\phi = 2\int_0^{L/2}\int_{\theta_1}^{\theta_2} p_2 R\sin\theta \, d\theta \, dy \tag{8.42b}$$

Since $p_2$ is complex, the dynamic load $W_2$ is also complex and can be expressed in terms of real and imaginary parts as

$$W_2 = \mathrm{Re}\left(W_2\right) + \mathrm{Im}\left(W_2\right)$$

$$-\left(\bar{W}_2\right)_\varepsilon = -\frac{\left(W_2\right)_\varepsilon}{W} = \frac{K_{\varepsilon\phi}C}{W} + i\frac{B_{\varepsilon\phi}C\omega}{W} \tag{8.43a}$$

$$-\left(\bar{W}_2\right)_\phi = -\frac{\left(W_2\right)_\phi}{W} = \frac{K_{\phi}C}{W} + i\frac{B_{\phi\phi}C\omega}{W} \tag{8.43b}$$

$$k_{\varepsilon\phi} = -\mathrm{Re}\left(\bar{W}_2\right)_\varepsilon = \frac{K_{\varepsilon\phi}C}{W} \; ; k_{\phi\phi} = -\mathrm{Re}\left(\bar{W}_2\right)_\phi = \frac{K_{\phi\phi}C}{W} \tag{8.44a}$$

$$b_{\varepsilon\phi} = -\mathrm{Im}\left(\bar{W}_2\right)_\varepsilon = \frac{B_{\varepsilon\phi}C\omega}{W} \; ; b_{\phi\phi} = -\mathrm{Im}\left(\bar{W}_2\right)_\phi = \frac{B_{\phi\phi}C\omega}{W} \tag{8.44b}$$

## 8.5 | Stability of Rigid Rotors Supported on Fluid Film Bearings

Stability of rigid rotors supported on plain journal bearings can be analyzed in two different ways. The first approach investigates the stability of small amplitude vibration of a rigid rotor about its steady equilibrium position. This approach uses the rotor dynamic coefficients of journal bearings to write down the equations of motion of the rotor. Concept of rotor dynamic coefficients was first postulated by Stodola (1927) and was later evaluated by Pestel (1954). Newkirk and Taylor(1925) discovered the phenomenon of bearing induced vibration which was originally called oil whip and later generalized as half frequency whirl. The second approach uses the nonlinear dynamic equations of motion of the rotor to predict the onset of instability of the rotor equilibrium position which result in orbital motion of the journal center around its equilibrium position. Hori (1959) first developed the theory of oil whip subsequent to initial efforts by Hagg (1946). Tondl (1965) and Holmes *et al.* (1965–66) investigated the vibration of a rigid rotor on short journal bearings and oil whirl of a rigid rotor, respectively. Reddi and Trumpler (1962) found the stability of high speed journal bearing under steady load for incompressible lubricant film. However, pioneering work on rotor dynamics of rigid and flexible rotors was done by Lund (1965) and that was followed by further work in subsequent years. Antiwhirl bearing configurations, e.g., multilobe, noncircular, pivoted pad, and pressure dam journal bearings have also been investigated by several researchers to enhance the stability of rotors supported on fluid film journal bearings to suppress oil whirl.

A symmetric rotor bearing system is shown in Fig. 8.5. The rotor can execute harmonic vibrations in either translational mode or conical mode. In translational mode, rotor axis undergoes mere translational motion with rotor and bearing axes parallel to each other. In conical mode centre of gravity of the rotor remains stationary. Equations of motion for translational and conical mode of vibration of the rotor are given in Cartesian coordinates as

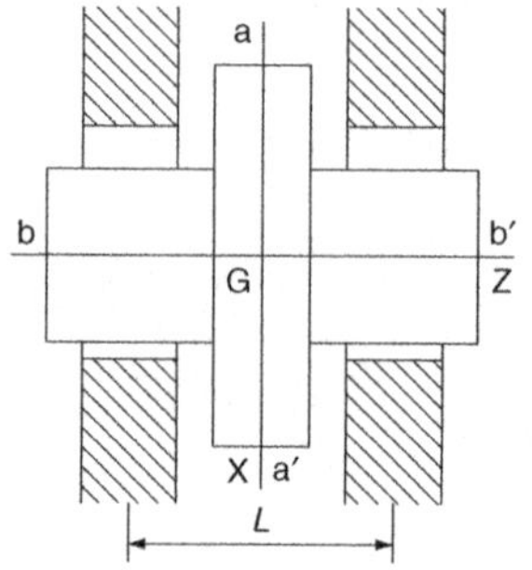


**Figure 8.5(a)** | Symmetric Rotor and Bearing System

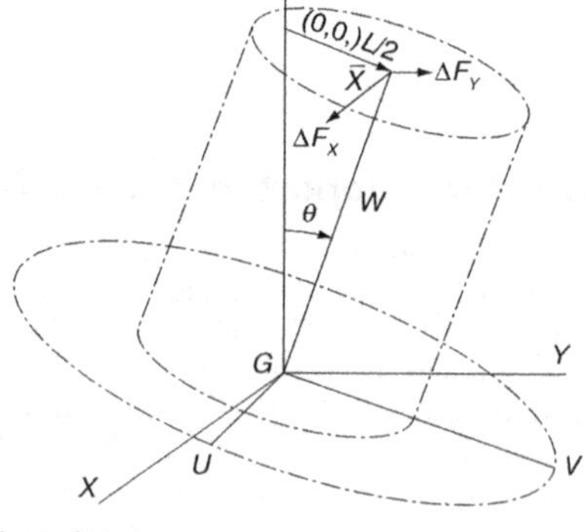


**Figure 8.5(b)** | Coordinates of Conical Motion

### 8.5.1 | Translational Mode

$$M\frac{d^2x}{dt^2} = \Delta F_x \tag{8.45}$$

$$M\frac{d^2y}{dt^2} = \Delta F_y \tag{8.46}$$

where $\Delta F_x$, $\Delta F_y$ are components of the dynamic fluid film forces in $x$, $y$ directions, respectively. $X$ and $Y$ are vibration amplitudes of the rotor in $x$, $y$ directions, respectively. $M$ is the mass of rotor per bearing.

### 8.5.2 | Conical Mode

It can be shown that equations of motion for conical mode are as follows

$$\frac{2J}{L^2}\frac{d^2x}{dt^2} + \frac{2\omega}{L^2}I\frac{dy}{dt} = \Delta F_x \tag{8.47}$$

$$\frac{2J}{L^2}\frac{d^2y}{dt^2} + \frac{2\omega}{L^2}I\frac{dx}{dt} = \Delta F_y \tag{8.48}$$

where $\omega$ Angular speed of rotation of the journal
$J$ Moment of inertia of the rotor about $x$ and $y$ axes
$I$ Moment of inertia of the rotor about $z$ axis

The above equations can be expressed in dimensionless form as

$$\bar{M}\left(\ddot{\bar{x}}\gamma^2 - \gamma\lambda\dot{\bar{y}}\right) = \Delta f_x \tag{8.49}$$

$$\bar{M}\left(\ddot{\bar{y}}\gamma^2 - \gamma\lambda\dot{\bar{x}}\right) = \Delta f_y \tag{8.50}$$

$\lambda$ is the gyroscopic parameter, $\lambda = 0$ for translational mode and $\lambda = \frac{I}{J}$ for conical mode

$\Delta f_x = \frac{\Delta F_x}{F_0}$; $\Delta f_y = \frac{\Delta F_y}{F_0}$; $\bar{M} = \frac{MC\omega^2}{F_0}$ for translational mode and for conical mode $\bar{M} = \frac{2JC\omega^2}{F_0L^2}$

$F_0$ – Reference force and usually $F_0 = W$, i.e., rotor weight on the bearing

$\bar{x} = \frac{x}{C}, \bar{y} = \frac{y}{C}$, where $C$ is the radial clearance

$\tau = \nu t$; $\nu$ is the angular frequency of vibration of the journal

$\gamma = \frac{\nu}{\omega}$, is defined as whirl frequency ratio

Derivatives of $x$ and $y$ are with respect to $\tau$.

$\Delta f_x$ and $\Delta f_y$ can be expressed in the dimensionless form using rotor dynamic coefficients as

$$\Delta f_x = -k_{xx}x - k_{xy}y - b_{xx}\dot{x} - b_{xy}\dot{y} \tag{8.51}$$

$$\Delta f_y = -k_{yx}x - k_{yy}y - b_{yx}\dot{x} - b_{yy}\dot{y} \tag{8.52}$$

In the above equations, the rotor dynamic coefficients are expressed as

$$k_{xx} = \frac{K_{xx}C}{F_0}; k_{xy} = \frac{K_{xy}C}{F_0}$$

$$k_{yy} = \frac{K_{yy}C}{F_0}; k_{yx} = \frac{K_{yx}C}{F_0}$$

$$b_{xx} = \frac{C\omega B_{xx}}{F_0}; b_{xy} = \frac{C\omega B_{xy}}{F_0} \tag{8.53}$$

$$b_{yy} = \frac{C\omega B_{yy}}{F_0}; b_{yx} = \frac{C\omega B_{yx}}{F_0}$$

Rotor dynamic coefficients are evaluated for a known journal speed, i.e., $\omega$ rad/sec at its equilibrium position due to steady load $W$. Equations (8.49) and (8.50) can be expressed in terms of dimensionless rotor dynamic coefficients as

$$\bar{M}\gamma^2\ddot{\bar{x}} + k_{xx}\bar{x} + \gamma b_{xx}\dot{\bar{x}} + \left(\gamma b_{xy} + \bar{M}\gamma\lambda\right)\dot{\bar{y}} + k_{xy}\bar{y} = 0 \tag{8.54}$$

$$\bar{M}\gamma^2\ddot{\bar{y}} + k_{yy}\bar{y} + \gamma b_{yy}\dot{\bar{y}} + \left(\gamma b_{yx} - \bar{M}\gamma\lambda\right)\dot{\bar{x}} + k_{yx}\bar{x} = 0 \tag{8.55}$$

Substituting $\bar{x} = \bar{x}_0 e^{i\tau}$ and $\bar{y} = \bar{y}_0 e^{i\tau}$, Equations (8.54) and (8.55) are reduced to

$$\begin{bmatrix} \left(-\bar{M}\gamma^2 + k_{xx} + i\gamma b_{xx}\right) & i\left(\gamma b_{xy} + \bar{M}\gamma\lambda\right) + k_{xy} \\ k_{yx} + i\left(\gamma b_{yx} - \bar{M}\gamma\lambda\right) & \left(-\bar{M}\gamma^2 + k_{yy} + i\gamma b_{yy}\right) \end{bmatrix} \begin{pmatrix} \bar{x}_0 \\ \bar{y}_0 \end{pmatrix} = \begin{pmatrix} 0 \\ 0 \end{pmatrix} \tag{8.56}$$

The characteristic equation for the threshold of stability is given as

$$\begin{vmatrix} \left(-\bar{M}\gamma^2 + k_{xx} + i\gamma b_{xx}\right) & k_{xy} + i\left(\gamma b_{xy} + \bar{M}\gamma\lambda\right) \\ k_{yx} + i\left(\gamma b_{yx} - \bar{M}\gamma\lambda\right) & \left(-\bar{M}\gamma^2 + k_{yy} + i\gamma b_{yy}\right) \end{vmatrix} = 0 \tag{8.57}$$

At the stability threshold, the above equation must be satisfied. To satisfy the above equation, real and imaginary part on the left hand side must become zero. The stability parameters, viz., critical mass of the rotor $\bar{M}_{cr}$ and whirl frequency ratio $\gamma$ can be determined in terms of rotor dynamic coefficients of the bearing from the above determinant as:

From the imaginary part being set to zero yields for translational mode, i.e., $\lambda = 0$

$$C_1 = \bar{M}_{cr}\gamma^2 = \frac{\left(b_{xx}k_{yy} + b_{yy}k_{xx}\right) - \left(b_{xy}k_{yx} + b_{yx}k_{yx}\right)}{\left(b_{xx} + b_{yy}\right)} \tag{8.58}$$

Similarly, setting real part equal to zero, yields for λ= 0

$$C_2 = \gamma^2 = \frac{\left(\bar{M}_{cr}\gamma^2 - k_{xx}\right)\left(\bar{M}_{cr}\gamma^2 - k_{yy}\right) - k_{xy}k_{yx}}{\left(b_{xx}b_{yy} - b_{xy}b_{yx}\right)} \tag{8.59}$$

Therefore,

Whirl frequency ratio, $\gamma = \sqrt{C_2}$ (8.60)

Critical Mass, $\bar{M}_{cr} = C_1 / C_2$ (8.61)

Similarly, both critical mass and whirl ratio can be determined for conical mode also. Plots of critical mass and whirl frequency ratio for plain journal bearings are shown in Figs 8.6 and 8.7 for various eccentricity ratios. If the rotor mass $\bar{M} < \bar{M}_{cr}$ the system is stable, whereas for $\bar{M} > \bar{M}_{cr}$ the system would become unstable. Figure 8.6 shows that at high eccentricity ratios, i.e., $\varepsilon_0 > 0.7$ the rotor is inherently stable. On the other hand, at low eccentricity ratios, the rotor critical mass for the stability threshold is very low and the rotors are likely to become unstable.

Whirl frequency ratio at $\varepsilon_0 \cong 0.0$ is 0.5, which indicates that very light rotors or rotors of vertical machines are prone to half frequency whirl oscillations. At higher eccentricity ratios, whirl frequency ratio drops and is far less than 0.5, indicating more stable operation. Muszynska (1986) has shown that in the case of flexible vertical rotors, half frequency whirl occurs at an angular speed of rotor twice that of the natural

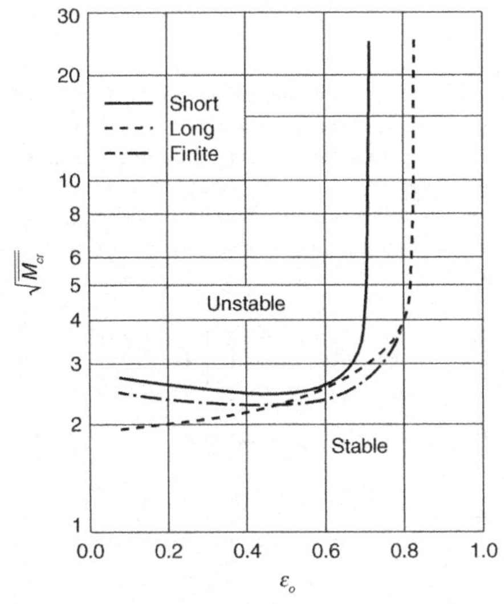


**Figure 8.6** | Instability Threshold Curves Indicating Bearing Assumptions After Small Initial Velocity Disturbances [ASME, JOLT, 1969]

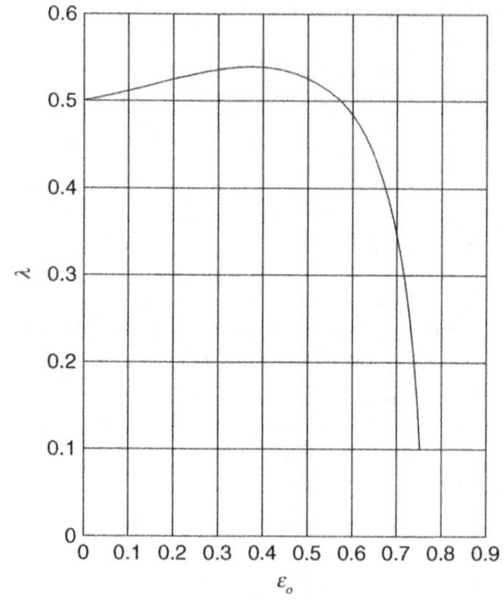


**Figure 8.7** | Whirl Frequency Ratio ($\lambda$) versus Equilibrium Eccentricity Ratio for the Short-$\pi$ Bearing [Lund, 1966]

frequency of the rotor and grows further with increasing speed of the rotor until it is twice the natural frequency or critical speed of the rotor. Further increase in speed shows that vibration grows further with frequency remaining the same as the natural frequency of the rotor. This phenomenon is generally described as 'oil whip', and can be very damaging.

Recently, Singh and Majumdar (2005) presented an empirical equation to determine the critical mass of a rigid rotor supported by journal bearings for its stable operation in terms of a known $L/D$ ratio and eccentricity ratio as given below.

$$\bar{M}\bar{W} = a_1\varepsilon_0 \exp(a_2(1-\varepsilon_0)^{a_3}\left(a_4 + \left(\frac{L}{D}\right)^{a_5}\right) \quad (8.62)$$

where

$$a_1 = 0.6755,\ a_2 = 4.2711,\ a_3 = -0.1730,\ a_4 = -0.7484,\ a_5 = 0.2768$$

$\bar{M} = \dfrac{MC\omega^2}{W}$ and $\bar{W} = \dfrac{WC^2}{\eta_0 \omega R^3 L}$, $M$ and $W$ are rotor mass per bearing and load acting on the bearing, respectively.

Based on the results of numerical experiments carried out by Sarangi and Majumdar, the following empirical relationships have been developed to evaluate dimensionless stiffness and damping coefficients of journal bearings.

### *Stiffness Coefficients*

$$k_{\varepsilon\varepsilon} = A + B\left(\frac{L}{D}\right)^{C}\left\{e^{\left(\varepsilon^{D}\right)}\right\}^{E}$$

where $A = -2.4213, B = 3.4615, C = 0.59607, D = 2.8659, E = 4.5796$

$$k_{\varepsilon\phi} = A + B\left(\frac{L}{D}\right)^{C}\left\{e^{\left(\varepsilon^{D}\right)}\right\}^{E}$$

where $A = -1.0852$ $B = 2.8685, C = 0.75742, D = 1.6876, E = 1.1125$

$$k_{\phi\varepsilon} = A + B\left(\frac{L}{D}\right)^{C}\left\{e^{\left(\varepsilon^{D}\right)}\right\}^{E}$$

where, $A = 2.4507, B = -5.0640, C = 0.57466, D = 3.6776, E = 2.8886$

$$k_{\phi\phi} = A + B\left(\frac{L}{D}\right)^{C}\left\{e^{\left(\varepsilon^{D}\right)}\right\}^{E}$$

where $A = -0.70359, B = 0.93196, C = 0.62544, D = 2.1563, E = 3.0589$

### *Damping Coefficients*

$$b_{\varepsilon\varepsilon} = A + B\left(\frac{L}{D}\right)^{C}\left\{e^{\left(\varepsilon^{D}\right)}\right\}^{E}$$

where $A = -2.3549, B = 6.3994, C = 0.74104, D = 2.4433, E = 3.3566$

$$b_{\varepsilon\phi} = A + B\left(\frac{L}{D}\right)^{C}\left\{e^{\left(\varepsilon^{D}\right)}\right\}^{E}$$

where $A = 2.4578, B = -4.6640, C = 0.76365, D = 1.9043, E = 2.6587$

$$b_{\phi\varepsilon} = A + B\left(\frac{L}{D}\right)^{C}\left\{e^{\left(\varepsilon^{D}\right)}\right\}^{E}$$

where $A = 1.5692, B = -2.7366, C = 0.71807, D = 2.0439, E = 2.4963$

$$b_{\phi\phi} = A + B\left(\frac{L}{D}\right)^{C}\left\{e^{\left(\varepsilon^{D}\right)}\right\}^{E}$$

where $A = -3.2643, B = 12.690, C = 0.87248, D = 3.1030, E = 0.96971$

Alternately, the equations of motion of the rotor for stability analysis can also be written in $\varepsilon$ and $\phi$ coordinates for translational mode as

$$M\left(C\ddot{\varepsilon}_{d} - C\varepsilon_{0}\dot{\phi}_{d}^{2}\right) = \Delta F_{\varepsilon} \tag{8.63}$$

$$M\left(C\varepsilon_0\ddot{\phi}_d + 2C\dot{\varepsilon}_0\dot{\phi}_d^2\right) = \Delta F_t \tag{8.64}$$

where derivatives are respect to time, $t$. $\Delta F_\varepsilon$ and $\Delta F_\phi$ can be written in terms of stiffness and damping coefficients of the bearing. Dimensionless form of equations of motion thus become

$$\bar{M}\gamma^2\left(\ddot{\varepsilon}_d - \varepsilon_0\dot{\phi}_d^2\right) + k_{\varepsilon\varepsilon}\varepsilon_d + k_{\varepsilon\phi}\varepsilon_d\phi_d + \gamma b_{\varepsilon\varepsilon}\dot{\varepsilon}_d + \gamma b_{\varepsilon\phi}\dot{\varepsilon}_d = 0 \tag{8.65}$$

$$\bar{M}\gamma^2\left(\varepsilon_0\ddot{\phi}_d + 2\dot{\varepsilon}_d\dot{\phi}_d\right) + k_{\phi\phi}\varepsilon_d + k_{\phi\varepsilon}\varepsilon_0\phi_d + \gamma b_{\phi\varepsilon}\dot{\varepsilon}_d + \gamma b_{\phi\phi}\varepsilon_0\dot{\phi}_d = 0 \tag{8.66}$$

where $\varepsilon_d$ and $\phi_d$ are dynamic eccentricity ratio and attitude angle, respectively, and for small amplitude oscillations around the equilibrium position of the journal it can be written as

$$\varepsilon_d = \varepsilon_0 + \varepsilon_1 e^{i\tau}, \phi_d = \phi_0 + \phi_1 e^{i\tau} \text{ and } \tau = \nu t = \gamma\omega t \tag{8.67}$$

$\nu$, is the frequency of vibration and $\gamma$ is the whirl frequency ratio. $\varepsilon_1$ and $\varepsilon_0\phi_1$ are the amplitudes of harmonic vibration of the journal center about its equilibrium position. $\bar{M}_{cr}$ and $\gamma$ can be determined in the same way as described earlier for the system equations in Cartesian coordinates, neglecting higher order terms since $\varepsilon_1 << \varepsilon_0$ and $\phi_1 << \phi_0$.

In the above equations, rotor dynamic coefficients are written following Equations (8.41) and (8.44) as

$$k_{..} = \frac{K..C}{Fo}, \quad \text{and} \quad b_{..} = \frac{B..C\omega}{F_0}$$

Critical mass of the rotor for stability $\bar{M}_{cr}$ and whirl frequency ratio, $\gamma$ are thus expressed as

$$\bar{M}_{cr}\gamma^2 = \frac{\left(b_{\varepsilon\varepsilon}k_{\phi\phi} + b_{\phi\phi}k_{\varepsilon\varepsilon}\right) - \left(b_{\varepsilon\phi}k_{\phi\varepsilon} + b_{\phi\varepsilon}k_{\varepsilon\phi}\right)}{\left(b_{\varepsilon\varepsilon} + b_{\phi\phi}\right)} \tag{8.68}$$

and

$$\gamma^2 = \frac{\left(\bar{M}_{cr}\gamma^2 - k_{\varepsilon\varepsilon}\right)\left(\bar{M}_{cr}\gamma^2 - k_{\phi\phi}\right) - k_{\varepsilon\phi}k_{\phi\varepsilon}}{\left(b_{\varepsilon\varepsilon}b_{\phi\phi} - b_{\varepsilon\phi}b_{\phi\varepsilon}\right)} \tag{8.69}$$

Plain journal bearings are susceptible to instability, and therefore to improve stability noncircular geometries have been studied. Li *et al.* (1980) investigated stability of rigid rotors supported on multilobe bearings. Two lobe (or elliptical), three lobe, four lobe, and offset cylindrical configurations shown in Fig. 8.8 have been investigated to ascertain their stability characteristics.

Plots of critical mass parameter and whirl ratio against Sommerfeld number are shown in Figs 8.9 and 8.10, respectively, as determined by Li *et al.*

Higher critical mass parameter and lower whirl ratio confirm that these configurations are more stable than plain circular bearing. Tilting pad bearings shown in Fig. 8.11 completely eliminate rotor instability and are widely used in power plant turbo machinery.

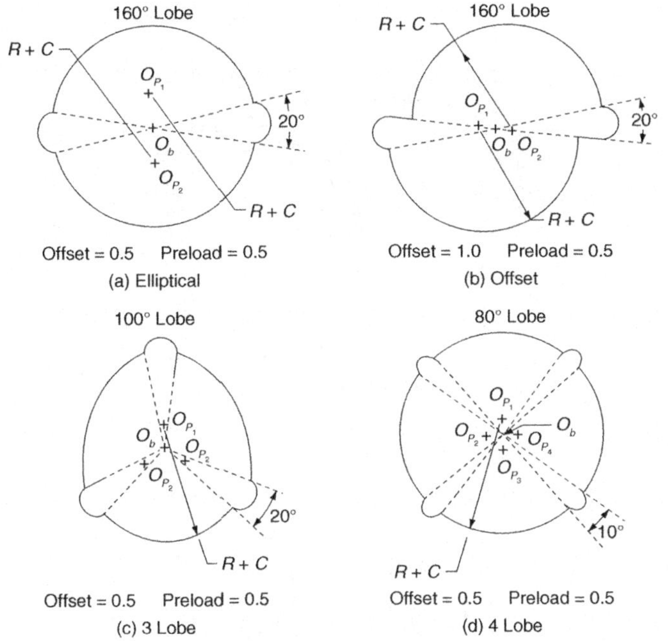


**Figure 8.8** | Multilobe Bearing Geometry [Li *et al.*, ASME, JOLT, 1980]

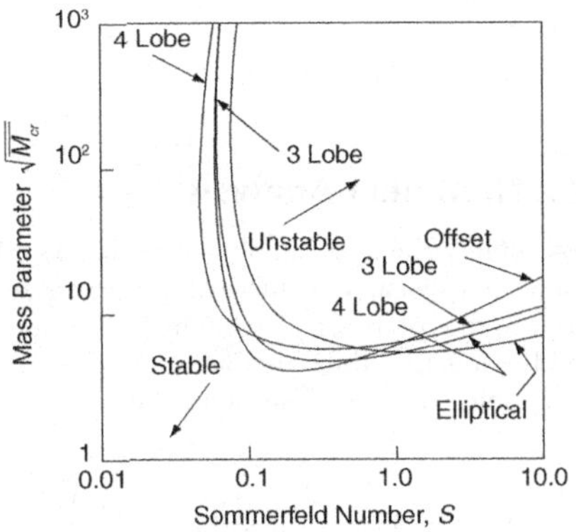


**Figure 8.9** | Linearized Bearing Stability for a Rigid Rotor [Li *et al.*, ASME, JOLT, 1980]

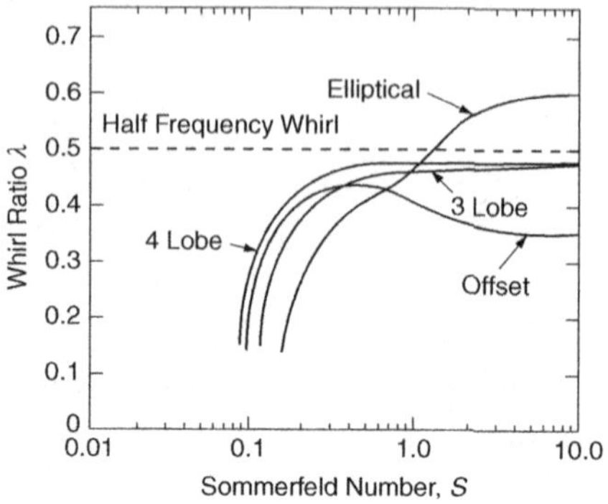


**Figure 8.10** | Whirl Speed Ratio versus Sommerfeld Number [Li *et al.*, ASME, JOLT, 1980]

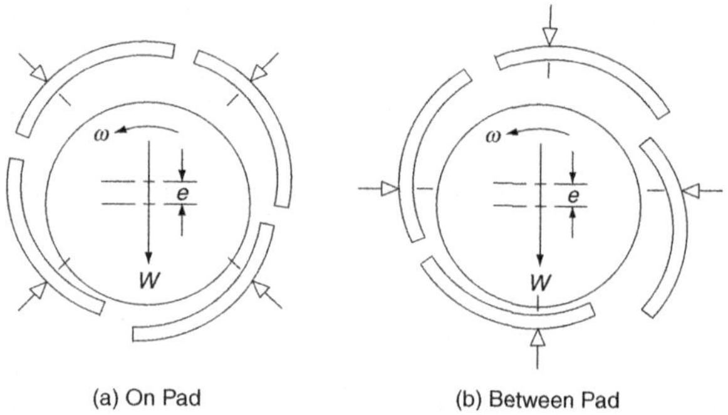


**Figure 8.11** | Loading for Four Pad Tilting Pad Bearing

## 8.6 | Rotor Instability: Nonlinear Analysis

We have seen that threshold of stability and small amplitude oscillation orbits of rotors can be investigated using linear vibration theory and rotor dynamic coefficients. However, large unbalance response representing some emergency situations in turbo rotors, e.g., large initial velocities due to blade loss are observed to produce unstable vibrations leading to large orbital motion of the journal center. Nonlinear analysis is used to investigate such situations. Assuming that only plane journal motion occurs, this motion can be described by the following equations:

$$\frac{d^2\varepsilon}{d\tau^2} = \frac{F_\varepsilon}{\omega^2 MC} + \frac{W_0}{\omega^2 MC}\cos\phi + \varepsilon\left(\frac{d\phi}{d\tau}\right)^2 \tag{8.70}$$

$$\frac{d^2\phi}{d\tau^2} = \frac{F_\phi}{\varepsilon\omega^2 MC} + \frac{W_o \sin\phi}{\varepsilon\omega^2 MC} - \frac{2}{\varepsilon}\left(\frac{d\varepsilon}{d\tau}\right)\left(\frac{d\phi}{d\tau}\right) \tag{8.71}$$

where $\omega$ = angular velocity of the rotor and $\tau = \omega t$

$W_0$ = static load and usually $W_0 = Mg$

$M$ = rotor mass per bearing

The accelerations $\frac{d^2\varepsilon}{d\tau^2}$ and $\frac{d^2\phi}{d\tau^2}$ can be evaluated using equations for the given initial values of $\varepsilon$, $\phi$, $\frac{d\varepsilon}{d\tau}$ and $\frac{d\phi}{d\tau}$. If these four initial values are specified, the resulting initial value problem may be solved by standard numerical marching techniques such as Runge–Kutta extrapolation technique or Newton's predictor-corrector method. Results of such solutions are shown in Fig. 8.12. It can be seen that initial transients, e.g., small initial velocity given to the journal center from its equilibrium position simulate the condition of an impact or an initial displacement given to the journal center from its equilibrium position result into the kind of response shown in Fig. 8.13. While determining the response, fluid film forces $F_\varepsilon$ and $F_\phi$ have to be evaluated at each time step by solving Reynolds equation based on short, long, or finite bearings as the case may be. Similar approach is also adopted to determine journal center orbit for dynamically loaded bearings.

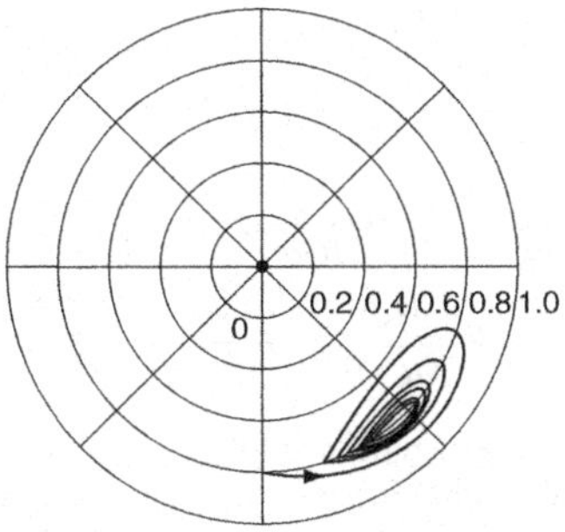


**Figure 8.12** | Journal Center Trajectory for a Unidirectional Constant Load ($L/D$ = 1.0, $\varepsilon_0 = 0.8$, $\bar{M} = 5$, $\Omega = 0.5$)

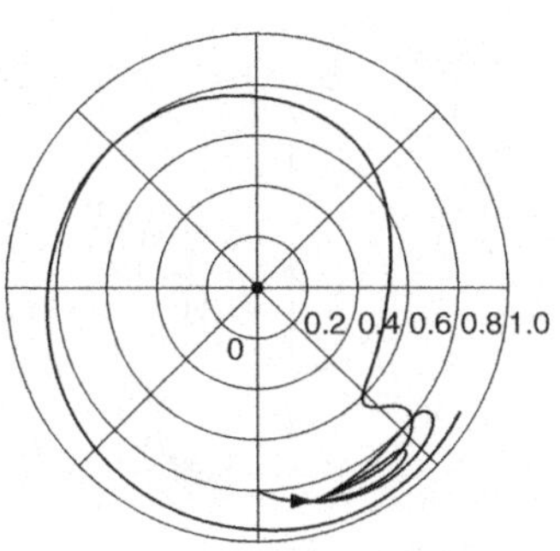


**Figure 8.13** | Journal Center Trajectory for a Unidirectional Periodic Load ($L/D$ = 1.0, $\varepsilon_0 = 0.8$, $\bar{M} = 5$, $\Omega = 0.5$)

## 8.7 | Dynamically Loaded Bearings: Nonlinear Analysis

Response of rotors can be determined using the equations of motion based on small amplitude vibration theory written in terms of stiffness and damping coefficients of the bearing at its equilibrium position.

However, equations of motion can be expressed in general case of an unbalanced dynamically loaded rotor in Cartesian coordinates as

$$M\ddot{x} = W_x(t) + F_x + Me_u\omega^2 \cos \omega t + W_0 \tag{8.72}$$

$$M\ddot{y} = W_y(t) + F_y + Me_u\omega^2 \sin \omega t \tag{8.73}$$

where $W_x$ and $W_y$ are dynamic load components, $F_x$ and $F_y$ are fluid film forces, $e_u$ is unbalance of the rotor, i.e., distance of the journal center from its geometric center.

Differentiating with respect to dimensionless time, $\tau$ we get the dimensionless form of the above equations as:

$$\ddot{\bar{x}} = \bar{W}_x(\tau) + f_x + \varepsilon_u \cos \tau + \bar{W}_0 \tag{8.74}$$

$$\ddot{\bar{y}} = \bar{W}_y(\tau) + f_y + \varepsilon_u \sin \tau \tag{8.75}$$

where

$$\bar{W}_x = \frac{W_x}{MC\omega^2}; \quad f_x = \frac{F_x}{MC\omega^2}, \quad \bar{x} = \frac{x}{C}, \quad \bar{W}_0 = \frac{W_0}{MC\omega^2}$$

$$\bar{y} = \frac{y}{C}, \quad \bar{W}_y = \frac{W_y}{MC\omega^2}, \quad f_y = \frac{F_y}{MC\omega^2}, \quad \varepsilon_u = e_u / C, \quad \tau = \omega t$$

Nonlinear response of the rotor can be determined following Runge–Kutta method in a time marching solution starting from a known initial condition. This approach is adopted to determine orbital motion of rotors of I.C. engines, pumps and compressors of reciprocating type, etc. A few typical response of a rigid rotor subject to dynamic loading of the bearing is shown in Fig. 8.14 (Majumdar *et al.*, 1987).

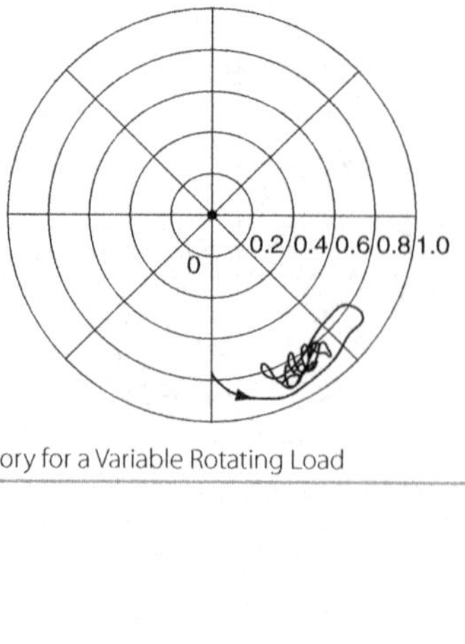


**Figure 8.14** | Journal Center Trajectory for a Variable Rotating Load

## 8.8 | Squeeze Film Lubrication

In this section squeeze film lubrication of mechanical components which is a prevalent mode of lubrication when surfaces approach each other under the action of dynamic load will be dealt. Positive pressure is generated due to squeezing of lubricant held in the gap between the surfaces, which is responsible for supporting load and thus prevent direct contact between the surfaces. We now discuss squeeze film lubrication in various conformal contacts.

### 8.8.1 | Squeeze Film Lubrication between Parallel Surfaces

The bearing configuration is shown in Fig. 8.15. Assuming that the bearing pad is infinitely wide in $y$-direction, the Reynolds equation for parallel surfaces approaching each other with a velocity $V$ reduces to

$$\frac{d^2 p}{dx^2} = \frac{12\eta V}{h^3} \tag{8.76}$$

Integrating Equation (8.76) twice with respect to x gives the expression for pressure generated due to squeeze action as

$$p = -\frac{6\eta V}{h^3}x^2 + C_1 x + C_2 \tag{8.77}$$

Using the boundary conditions for pressure as $p = 0$ at $x = \pm\frac{B}{2}$, one obtains $C_1 = 0$ and $C_2 = \frac{3}{2}\frac{\eta V B^2}{h^3}$

Therefore,

$$p = \frac{3\eta V}{2h^3}\left(B^2 - 4x^2\right) \tag{8.78}$$

The pressure distribution in dimensionless form can be written as

$$\bar{p} = \frac{3}{2}\left(1 - 4\bar{x}^2\right) \tag{8.79}$$

where $\bar{p} = \frac{p h^3}{\eta V B^2}$ and $\bar{x} = x / B$

Since, the pressure distribution is symmetrical about the center of the bearing, the maximum pressure will occur at $\bar{x} = 0$

Hence, $\bar{p}_{\max} = \frac{3}{2}$ (8.80)

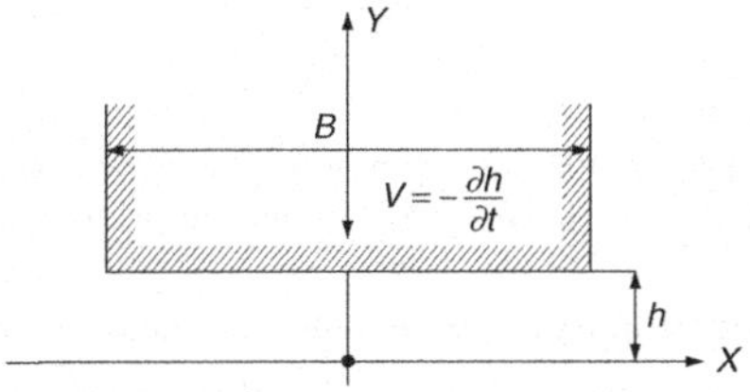


**Figure 8.15** | Squeeze Film Between Parallel Surfaces

The squeeze load capacity $W$ is determined as

$$W = L\int_{-B/2}^{B/2} p\,dx \tag{8.81}$$

where $L$ is the length of bearing pad

Substituting for $p$ given by Equation (8.78) into Equation (8.81) and integrating we get

$$W = \frac{\eta VLB^3}{h^3} \tag{8.82}$$

The volume rate of flow of lubricant is evaluated as

$$Q = -\frac{Lh^3}{12\eta}\frac{dp}{dx} \tag{8.83}$$

Using Equation (8.78) in Equation (8.83), the flow rate is obtained as

$$Q = LxV \tag{8.84}$$

The flow rate increases from zero at the center of the bearing to a maximum value of $\frac{1}{2}LBV$ at the edge of the bearing.

The time of approach can be calculated in the following way

Since, $V = -\frac{\partial h}{\partial t}$, one can write

$$W = -\frac{\eta LB^3}{h^3}\frac{\partial h}{\partial t} \tag{8.85}$$

or

$$-\frac{W}{\eta LB^3}\int_{t_1}^{t_2} dt = \int_{h_1}^{h_2}\frac{dh}{h^3}$$

or,

$$\Delta t = t_2 - t_1 = \frac{\eta LB^3}{2W}\left[\frac{1}{h_2^2} - \frac{1}{h_1^2}\right] \tag{8.86}$$

The final film thickness $h_2$ can be expressed in terms of the initial film thickness $h_1$ and the time interval $\Delta t$ as

$$h_2 = \frac{h_1}{\sqrt{\left(1 + \frac{2W\Delta t h_1^2}{\eta LB^3}\right)}} \tag{8.87}$$

Using Equations (8.86) and (8.87) time of approach of the surfaces from gap $h_1$ to $h_2$ or otherwise the final film thickness after a lapse of time $\Delta t$ can be determined, respectively.

### 8.8.2 | Squeeze Film Lubrication between Parallel Circular Plates

Consider a circular plate of radius $R$ approaching a parallel plane surface as shown in Fig. 8.16.

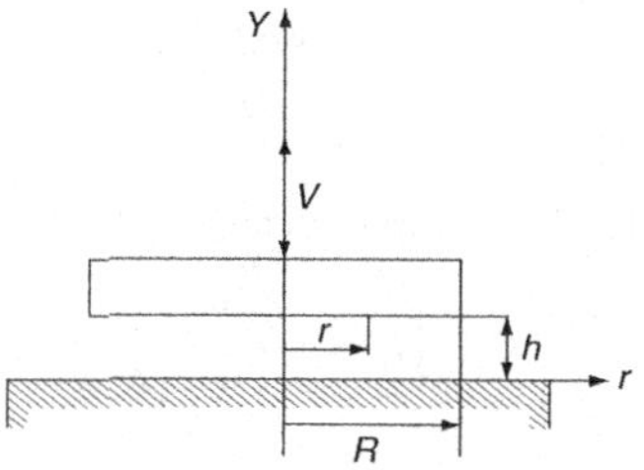


**Figure 8.16** | Squeeze Film Between Parallel Circular Plates

For an axisymmetric case and using polar coordinates, Reynolds equation can be written as

$$\frac{1}{r}\frac{d}{dr}\left(rh^3\frac{dp}{dr}\right) = -12\eta V \tag{8.88}$$

Integrating Equation (8.88), we get

$$\frac{dp}{dr} = -\frac{6\eta r V}{h^3} + \frac{C_1}{rh^3} \tag{8.89}$$

Since $\frac{dp}{dr}$ will not be infinity at $r = 0$, $C_1 = 0$

Hence,
$$\frac{dp}{dr} = -\frac{6\eta r V}{h^3} \tag{8.90}$$

Integration of Equation (8.90) with respect to $r$ yields

$$p = -\frac{3\eta r^2 V}{h^3} + C_2 \tag{8.91}$$

The constant $C_2$ can be evaluated from the boundary condition that $p = 0$ at $R = 0$. Thus,

$$C_2 = \frac{3\eta R^2 V}{h^3}$$

The pressure distribution is now given by

$$p = \frac{3\eta V}{h^3}\left(R^2 - r^2\right) \tag{8.92}$$

The squeeze film load capacity $W$ is

$$W = \int_0^R 2\pi r p dr = \frac{6\pi\eta V}{h^3}\left[R^2\frac{r^2}{2} - \frac{r^4}{4}\right]\Bigg|_0^R$$

or

$$W = \frac{3\pi\eta R^4 V}{2h^3} \tag{8.93}$$

Since $V = \frac{dh}{dt}$, the time of approach can thus be determined as

$$\int dt = -\frac{3\eta R^4}{2W}\int \frac{dh}{h^3}$$

or

$$\Delta t = \frac{3\pi\eta R^4}{4W}\left(\frac{1}{h_2^2} - \frac{1}{h_1^2}\right) \tag{8.94}$$

### 8.8.3 | Squeeze Film Lubrication between a Sphere and a Spherical Seat

Figure 8.17 shows a sphere of Radius $R$ on a hemispherical seat.
The amount of lubricant flow from the control volume due to Poiseuille and Couette flow can be written as

$$-\frac{\pi}{6\eta}\frac{d}{dx}\left(rh^3\frac{dp}{dx}\right)\delta x - \pi V\frac{d}{dx}\left(rh\sin\theta\right)\delta x \tag{8.95}$$

The rate of reduction of flow due to squeeze action is $2\pi rV\cos\theta\delta x$. To satisfy flow continuity relationship this is equated to terms in Equation (8.95). Thus, we get Reynolds equation as

$$\frac{d}{dx}\left(rh^3\frac{dp}{dx}\right) + 6\eta V\frac{d}{dx}\left(rh\sin\theta\right) = -12\,\eta Vr\cos\theta \tag{8.96}$$

Since, $6\eta V\frac{d}{dx}\left(rh\sin\theta\right)$ is smaller than the other terms, it can be neglected.
Again, since $x = R\theta$ and $r = R\sin\theta$ we get

$$\frac{d}{d\theta}\left(\sin\theta\, h^3\frac{dp}{d\theta}\right) = -6\eta R^2 V\sin 2\theta \tag{8.97}$$

Integrating Equation (8.97)

$$\frac{dp}{d\theta} = -6\eta R^2 V\frac{\sin\theta}{h^3} + \frac{C_1}{h^3\sin\theta} \tag{8.98}$$

where $C_1$ is a constant and

$$h = C\left(1 - \varepsilon\cos\theta\right) \tag{8.99}$$

For the configuration shown in Fig. 8.17 the boundary conditions are

$$\text{(i)}\quad \frac{dp}{d\theta} = 0 \quad \text{when } \theta = 0$$

$$\text{(ii)}\quad p = 0 \quad \text{when } \theta = \pi/2 \tag{8.100}$$

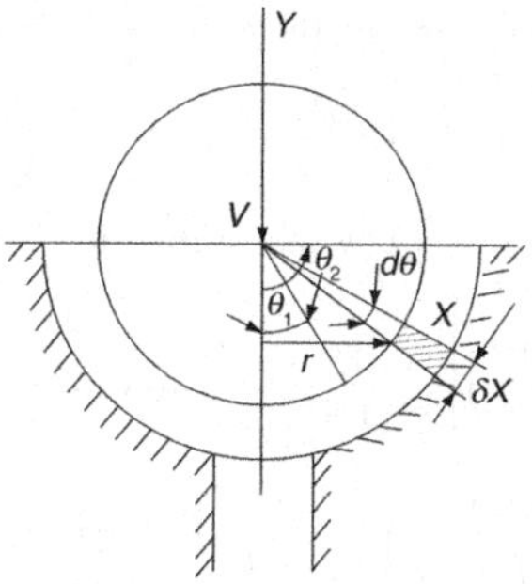


**Figure 8.17** | Squeeze Film Between a Sphere on a Spherical Seat

Using condition (i) of Equation (8.100) in Equation (8.98), we get $C_1 = 0$

Therefore,
$$\frac{dp}{d\theta} = -6\eta R^2 V \frac{\sin\theta}{h^3} \tag{8.101}$$

Integration of Equation (8.101) gives expression of pressure as

$$p = \frac{3\eta R^2 V}{\varepsilon C^3}\left[\frac{1}{(1-\varepsilon\cos\theta)^2}\right] + C_2 \tag{8.102}$$

Using second boundary condition of Equation (8.100), we get

$$C_2 = -\frac{3\eta R^2 V}{\varepsilon C^3}$$

Introducing this in Equation (8.102), the pressure distribution becomes

$$p = \frac{3\eta R^2 V}{\varepsilon C^3}\left[\frac{1}{(1-\varepsilon\cos\theta)^2} - 1\right] \tag{8.103}$$

The squeeze load capacity $W$ is determined as
$W = \int_0^{\pi/2} 2\pi R p R \cos\theta d\theta$. This can be rewritten as

$$W = 2\pi R^2 \int_0^{\pi/2} p \sin\theta \cos d\theta \tag{8.104}$$

Substituting $p$ in equation (5.98) and integrating, one gets

$$W = \frac{6\pi\eta R^4 V}{\varepsilon C^3} \int_0^{\pi/2} \left[\frac{\sin\theta\cos\theta}{(1-\varepsilon\cos\theta)^2} - \sin\theta\cos\theta\right] d\theta \tag{8.105}$$

Thus, squeeze load capacity is obtained on integration as

$$W = \frac{6\pi\eta R^4 V}{C^3}\left[\varepsilon^2 \frac{1}{(1-\varepsilon)} + \frac{1}{\varepsilon^3}\ln(1-\varepsilon) - \frac{1}{2\varepsilon}\right] \tag{8.106}$$

The time of approach may be evaluated using the relationship given below.

$$V\cos\theta = -\frac{dh}{dt} = -C\cos\theta\frac{d\varepsilon}{dt} \quad \text{or} \quad V = C\frac{d\varepsilon}{dt}$$

Thus, one obtains from Equation (8.106) by substituting for $V$ from above the expression for time of approach as

$$\int dt = \frac{6\pi\eta R^4}{C^2W}\int\left[\frac{1}{\varepsilon^2(1-\varepsilon)} + \frac{1}{\varepsilon^3}\ln(1-\varepsilon) - \frac{1}{2\varepsilon}\right]dt$$

Thus,

$$\Delta t = \frac{6\pi\eta R^4}{C^2W}\left[\frac{1}{\varepsilon} + \left(\frac{\varepsilon^2+1}{\varepsilon^2}\right)\ln(1-\varepsilon)\right]_{\varepsilon_1}^{\varepsilon_2} \quad (8.107)$$

### 8.8.4 | Squeeze Film Lubrication of Journal Bearing

In squeeze film lubrication, pressure generation is due to dynamic motion of the nonrotating journal as shown in Fig. 8.18.

In the absence of journal rotation, i.e., for normal squeeze velocity of the journal center, the Reynolds equation reduces to

$$\frac{\partial}{\partial\theta}\left(\frac{h^3}{\eta}\frac{\partial p}{\partial\theta}\right) + R^2\frac{\partial}{\partial y}\left(\frac{h^3}{\eta}\frac{\partial p}{\partial y}\right) = 12R^2\frac{\partial h}{\partial t} \quad (8.108)$$

where $\frac{\partial h}{\partial t}$ is the normal squeeze velocity, the film shape is given as

$$h = C(1-\varepsilon\cos\theta)$$

where $\theta$ is measured from the location of minimum film thickness.

For long bearing approximation, $\frac{\partial p}{\partial y} = 0$ and the Reynolds equation becomes

$$\frac{\partial}{\partial\theta}\left(\frac{h^3}{\eta}\frac{\partial p}{\partial\theta}\right) = -12R^2C\dot{\varepsilon}\cos\theta \quad (8.109)$$

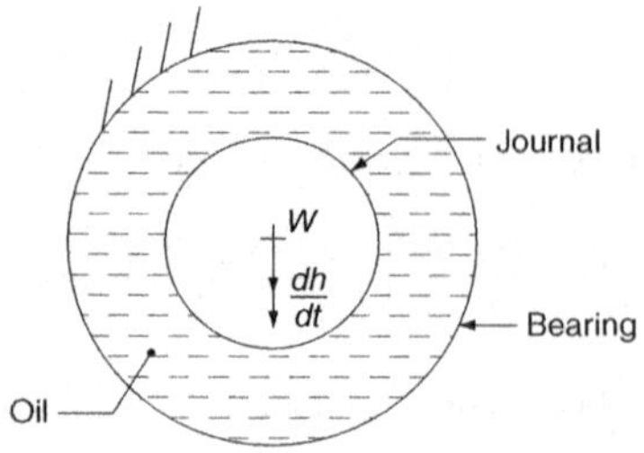


**Figure 8.18** | Squeeze Film Journal Bearing

Integrating with respect to $\theta$, we get

$$\frac{dp}{d\theta} = -\frac{12\eta R^2 \dot{\varepsilon}\sin\theta}{C^2(1-\varepsilon\cos\theta)^3} + \frac{A}{C^3(1-\varepsilon\cos\theta)^3} \tag{8.110}$$

For symmetry about the line of centers

$\frac{dp}{d\theta} = 0$ at $\theta = \pi$, when substituted in the above equation, gives the constant of integration $A = 0$.

Therefore,

$$\frac{dp}{d\theta} = -12\eta\left(\frac{R}{C}\right)^2 \frac{\dot{\varepsilon}\sin\theta}{(1-\varepsilon\cos\theta)^3} \tag{8.111}$$

Integrating further gives pressure distribution as

$$p = -6\eta\left(\frac{R}{C}\right)^2 \frac{\dot{\varepsilon}}{\varepsilon(1-\varepsilon\cos\theta)^2} + B \tag{8.112}$$

$p = 0$ at $\theta = \pi/2$ and $3\pi/2$. Since $p$ is also zero at $\theta = 0$, this condition makes it easier to compute the constant $B$. Then,

$$B = -6\eta\left(\frac{R}{C}\right)^2 \frac{\dot{\varepsilon}}{(1-\varepsilon)^2}$$

Thus,

$$p = 6\eta\left(\frac{R}{C}\right)^2 \frac{\dot{\varepsilon}}{\varepsilon}\left\{\frac{1}{(1-\varepsilon)^2} - \frac{1}{(1-\varepsilon\cos\theta)^2}\right\} \tag{8.113}$$

The normal load carrying capacity can be determined as

$$W_s = 2L\int_0^\pi pR\cos\theta d\theta$$

Substitution of $p$ from Equation (8.113) and integrating, we get

$$W_s = 12\pi\eta L\frac{R^3}{C^2}\frac{\dot{\varepsilon}}{(1-\varepsilon^2)^{3/2}} \tag{8.114}$$

$W_s$ is the squeeze film load capacity of the journal bearing.

In a similar way, squeeze film load capacity of the journal bearing can also be determined for short bearing assumption.

The lubrication of connecting rod bearings of I.C. engines, reciprocating pumps, and compressors are due to squeeze film action. Besides many other engineering manifestations of squeeze film action in real life, for example, lubrication of human joints and animal joints is also possible due to squeeze action of synovial fluid trapped in the joints. Engineering manifestations include failure of action in clutches and brakes due to oil or water entrapment and also skidding of vehicles due to water layer between the tire and wet road.

## 8.9 | Squeeze Film Damper

Rolling element bearings are used to support the rotors of modern aircraft engines due to high stiffness and high reliability. However, these bearings provide almost negligible damping to vibrations due to unbalance. To provide additional damping squeeze film dampers are used in an arrangement shown in Fig. 8.19. It is a journal bearing in which the journal i.e. outer race of rolling element bearing and housing are constrained to move. Short bearing approximation is generally used since most dampers have a low *L*/*D* ratio.

The Reynolds equation for short journal bearings can be written in Cartesian coordinates as

$$\frac{\partial}{\partial y}\left(\frac{h^3}{\eta}\frac{\partial p}{\partial y}\right)=12\frac{\partial h}{\partial t} \tag{8.115}$$

where $h = C - x\cos\theta - y\sin\theta$

$\theta$, is measured from the x axis as shown in Fig. 8.19.

The pressure distribution is obtained by integrating Equation (5.110) as

$$p(\theta, y)=\frac{3\eta}{h^3}\left(y^2-Ly\right)\cdot 2\left(\frac{\partial h}{\partial t}\right) \tag{8.116}$$

Bearing forces can be expressed in Cartesian coordinates as

$$\begin{Bmatrix} F_x \\ F_y \end{Bmatrix} = -\frac{\eta R L^3}{2}\int_0^{2\pi}\frac{2\left(\dot{x}\cos\theta+\dot{y}\sin\theta\right)}{\left(C-x\cos\theta-y\sin\theta\right)^3}\begin{Bmatrix}\cos\theta \\ \sin\theta\end{Bmatrix}d\theta \tag{8.117}$$

Since full film condition has been assumed, integration is performed between the limits 0 to $2\pi$. However, cavitations occur and the film is ruptured.

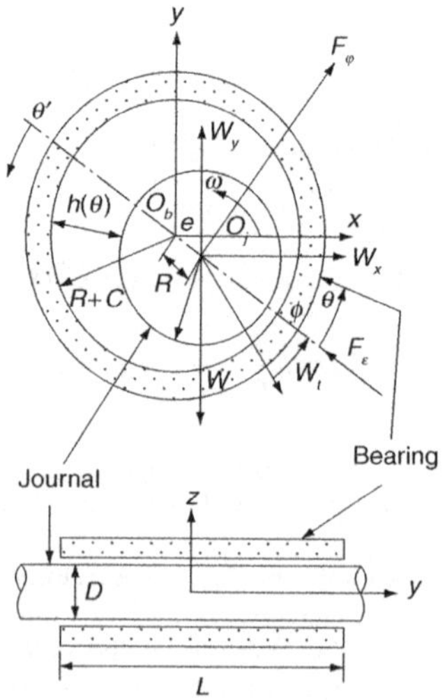


**Figure 8.19** | Schematic Diagram Geometry of Squeeze Film Damper [Gunter *et al.*, ASME, JOLT, 1977]

Assuming steady state precession of the journal center, the Reynolds equation for short journal bearing for nonrotating journal is written as

$$\frac{\partial}{\partial y}\left(\frac{h^3}{\eta}\frac{\partial p}{\partial y}\right) = 12C\left(\dot{\varepsilon}\cos\theta + \varepsilon\dot{\phi}\sin\theta\right) \tag{8.118}$$

where $h = C(1-\varepsilon\cos\theta)$ angle $\theta$ being measured from the location of minimum film thickness.

The force components in $\varepsilon$, $\phi$ coordinates, i.e., along and normal to the line of centers are

$$\begin{Bmatrix} F_\varepsilon \\ F_\phi \end{Bmatrix} = -\frac{-\eta RL^3}{C^2}\int_{\theta_1}^{\theta_2} \frac{\dot{\phi}\varepsilon\sin\theta + \dot{\varepsilon}\cos\theta}{(1+\varepsilon\cos\theta)^3} \begin{Bmatrix} \cos\theta \\ \sin\theta \end{Bmatrix} d\theta \tag{8.119}$$

$\theta_1, \theta_2$ define the domain of positive film pressure. The film extent is usually taken as $\pi$. It is assumed that the damper exhibits circular precession about the bearing center, and therefore $\dot{\varepsilon} = 0; \dot{\phi} = \omega$

The force components $F_\varepsilon$ and $F_\phi$ are given as

$$F_\varepsilon = \frac{-2\eta RL^3 \varepsilon e\omega}{C^3\left(1-\varepsilon^2\right)^2}; \quad F_\phi = -\frac{\eta RL^3 \pi e\omega}{2C^3\left(1-\varepsilon^2\right)^{3/2}} \tag{8.120}$$

where $\omega =$ angular velocity of precession of journal center. Equivalent damper stiffness and damping coefficients can be determined as follows:

Damper stiffness coefficient $K_d$ is given as

$$K_d = -\frac{\partial F_\phi}{\partial e} = \frac{2\eta RL^3 \varepsilon\omega}{C^3\left(1-\varepsilon^2\right)^2} \tag{8.121}$$

Damper damping coefficient $B_d$ as

$$B_d = -\frac{\partial F_\phi}{\partial(e\omega)} = \frac{\pi\eta RL^3}{2C^3\left(1-\varepsilon^2\right)^{3/2}} \tag{8.122}$$

For full film condition, the stiffness and damping coefficients are determined from $F_\varepsilon$ and $F_\varphi$ which are obtained as

$$F_\varepsilon = 0; \quad F_\phi = -\frac{\eta RL^3 \pi e\omega}{C^3\left(1-\varepsilon^2\right)^{3/2}} \tag{8.123}$$

Thus, damper stiffness is zero and damping is obtained as

$$B_d = \pi\eta RL^3 / C^3\left(1-\varepsilon^2\right)^{3/2} \tag{8.124}$$

The above expressions are derived for a plain damper but can be used for other damper configurations also. Generally two other configurations are commonly employed, viz., damper with circumferential oil feed groove with ends open which permit full end leakage or with end seals to prevent end leakage. In case of damper without end seals, hydrodynamic forces and damper coefficients decrease by a factor of 4. For dampers with end seals hydrodynamic pressure distribution is similar to a plain land with pressure at center equal to zero as shown in Fig. 8.20.

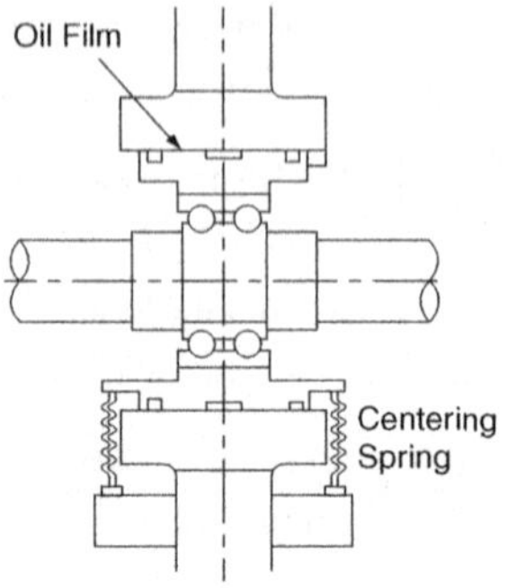


**Figure 8.20(a)** | Squeeze Film Damper Installation [Pan and Tonnesen, ASME, JOLT, 1978]

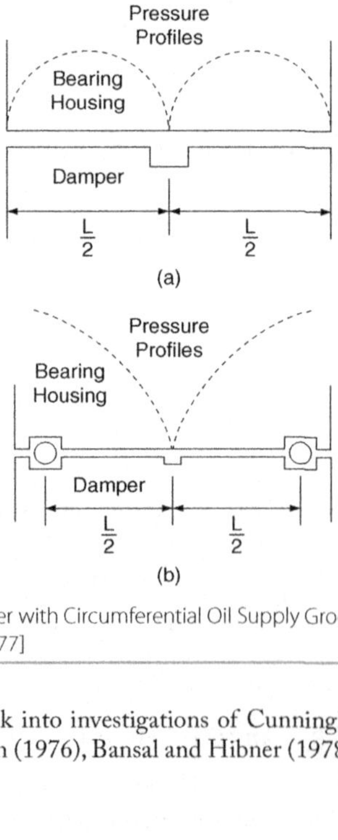


**Figure 8.20(b)** | Squeeze Film Damper with Circumferential Oil Supply Groove with and without End Seals [Gunter *et al.*, ASME, JOLT, 1977]

For further reading, one can look into investigations of Cunningham *et al.* (1975), Mohan and Hahn (1974), Gunter *et al.* (1977), Tonnesen (1976), Bansal and Hibner (1978), Lund *et al.* (1983), and Zeidan and Vance (1988,89).

## Problems

***P.8.1*** Neglecting side leakage obtain the expressions for pressure distribution, oil flow, and time taken for the film thickness to reduce by half in case of parallel surface squeeze film bearing of infinite width. Calculate the theoretical separation velocity required to reduce the oil film pressure between two parallel plates 0.025 m long and infinitely wide to a pressure of absolute zero. The film thickness separating the plates is 25 micrometers and the oil viscosity is 0.5 Pas. If the load per unit width of 20 KN/m is applied to the conditions as above, calculate the time required to reduce film thickness to 2.5 micro meters.

***P.8.2*** (a) Using long bearing theory, calculate the time taken for the journal center to move from an eccentricity ratio of 0.5 to 0.8 in a 8 cm diameter journal with a radial clearance of 40 $\mu$m. The viscosity of the oil is 1 poise when the load per unit length is 20 KN/m and $L/D = 1.5$. $L$ is width of the bearing and $D$ journal diameter.

(b) What difference will it make if $L/D = 0.5$? Use short bearing theory to do the calculations and derive the expressions used.

***P.8.3*** Use theory of narrow bearing, evaluate the stiffness and dynamic coefficients of a hydrodynamic journal of following specifications:

Journal diameter = 7.0 cm, Bearing length = 7.0 cm, Radial clearance = 0.02 mm, Journal angular speed = 1000 rpm, Mean viscosity of oil = 0.025 Pas, operating eccentricity ratio = 0.6. Determine the critical mass for stable operation of bearing at the operating eccentricity ratio.

## REFERENCES

*Allaire, P.E., Parsell, J.K. and Barrett, L.E.* (1981), A pad perturbation method for the dynamic coefficients of tilting pad journal bearings, *Wear*, **72**, 29–44.

*Badgley, R.H.* (1969) and Booker, J.F., Turbo rotor instability–effect of initial transients in plane motion, *ASME Journal of Lubrication Technology*, Vol. 91, pp. 625–633, 1969.

*Bansal. P. and Hibner, D.* (1978), Experimental and analytical investigation of squeeze-film damper forces induced by offset circular whirl orbits, *Journal of Mechanical Design, ASME Trans.*, 559–557.

*Booker, J.F.* (1977), Dynamically loaded journal bearing: Mobility method of solution, *ASME, Journal of Basic Engineering*, **87**, 537–546.

*Childs, D., Moes, H. and Leeuwen, H.* (1977), Journal bearing impedance descriptions for rotor dynamic applications, *ASME, Journal of Lubrication Tech.*, **99**, 198–219.

*Cunningham, R, Fleming, D. and Gunter, E.* (1975), Design of a squeeze-film damper for a multi mass flexible rotor, *ASME, Journal of Engineering for Industry*, 1383–1389.

*Goodwin, G. and Holmes, R.* (1977), Determination of oil film thickness in a crankshaft main bearing, *Journal of Automotive Engineering, Inst. Mech. Engineers*, **191**, 161–167.

*Gunter, E.J., Barrett, L.E. and Allaire, P.E.* (1977), Design of nonlinear squeeze-film dampers for aircraft engines, *ASME, Journal of Lubrication Technology*, **99**, 57–64.

*Hagg, A.C.* (1946), Influence of oil film journal bearings on the stability of rotating machines, *ASME Trans. Journal of Applied Mechanics*, **68**, 211.

*Holmes, R.* (1960), The vibration of a rigid shaft on short sleeve bearings, *J. Mech. Eng. Sci.*, Institution of Mechanical Engineers, U.K., **2**, 337.

*Holmes, R., Mitchell, Jr. and Byrne, J.* (1965–66), Oil whirl of a rigid rotor in 360° journal bearings: Further characteristics, *Proc. Inst. Mech. Engrs.*, U.K., **180**, 593–609.

*Hori, Y.* (1959), A theory of oil whip, *Trans. ASME Journal of Applied Mechanics*, **26**, 189.

*Li, D., Choy, K. and Allaire, P.* (1980), Stability and transient characteristic of four multi lobe journal bearings, *ASME Trans. Journal of Lubrication Technology*, **102**, 291–299.

*Lund, J.W.* (1964), Spring and damping coefficients for the tilting pad journal bearing, *ASLE Trans.*, **7**, 342–352.

*Lund, J.W.* (1966), Self-excited whirl orbits of a journal in a sleeve bearing, Ph. D thesis, Renseller Polytechnic Institute, Troy, N.Y., USA,

*Lund, J.W, Smalley, A.J., Tecza, J.A. and Walton, J.F.* (1983), Squeeze-film damper technology, Part 1—Prediction of finite length damper performance, *ASME*, paper No. 83-GT-247.

*Lund, J.W.* (1965), The stability of an elastic rotor in journal bearings with flexible, damped supports. *ASME, Journal of Applied Mechanics*, 911–920.

*Majumdar, B.C. and Brewe, D.E.* (1987), Stability of rigid rotor supported on oil-film journal bearings under dynamic load, *NASA Technical Memorandum 102309 and AVSCOM Technical Report*, 87-C-26.

*Mohan, S. and Hahn, E.J.* (1974), Design of squeeze film damper supports for rigid rotors, *ASME Journal of Engineering for Industry*, 976–982.

*Muszynska, A.* (1986), Whirl and whip-rotor/bearing stability problems, *Journal of Sound and Vibration*, **110**, 443–462.

*Newkirk, B.L. and Taylor, H.D.* (1925), Shaft whipping due to oil action in journal bearing, *Gen. El., Review*, **28**, 559.

*Nicholas, J. and Barrett, L.* (1985), The effect of support flexibility on critical speed prediction, *ASLE Trans.*, **29**, 329–338.

*Nicholas, J.C., Gunter, E.J. and Allaire, P. E.* (1979), Stiffness and damping properties for the five pad tilting pad bearing, *Trans. ASLE*, **22**, 113–224.

*Pestel, E.* (1954), Beitrag zur ermittlung der hydrodynamisches dampfungs und feder-eigenschaften von gleitlagern, *Ingenieu. Archive*, **XXII**, 147–155.

*Raimondi, A.A. and Szeri, A.Z.* (1984), Journal and thrust bearing, *CRC Handbook of Lubrication*, E.R. Booser (ed.), 413–462.

*Reddi, M.M and Trumpler, P.R.* (1962), Stability of the high speed journal bearing under steady load—I: incompressible film, ASME Trans., *Journal of Engineering for Industry*, **84**.

*Singh, D.S. and Majumdar, B.C.* (2005), Computer aided design of hydrodynamic journal bearings considering thermal effects, *Proc. I. mech. E.*, U.K., *Journal of Engineering for Tribology*, **219**, 133–143.

*Stodola, A.* (1927), *Steam and Gas Turbines*, McGraw Hill, New York, USA.

*Tondl, A.* (1965), Some problems of rotor dynamics, Publishing House of Czechoslovakian Academy of Science, Prague.

*Tonnesen, J.* (1976), Experimental parametric study of a squeeze film damper for a multi mass flexible rotor, *ASME, Journal of Lubrication Tech.*, 206–213.

*Zeidan, F. and Vance, J.* (1988, 1989), Cavitations regimes in squeeze film dampers and their effect on the pressure distribution, STLE Paper No. 89-AM-48-1.

Chapter 9

# Externally Pressurized Lubrication

## 9.1 | Introduction

It has been seen that hydrodynamic bearings can exhibit high load carrying capacity due to pressure developed by wedge action or self-action at high shaft speeds. However, hydrodynamic bearings are not good in generating sufficient load carrying capacity at low speeds and also for low viscosity fluids. Since high operating speed is essential to generate hydrodynamic fluid film by self-action, hydrodynamic bearings are mostly used in high speed machinery such as turbines, pumps, turbo generators, and turbo pumps. On the contrary, in externally pressurized bearings, the fluid film is generated by supplying high pressure fluid from an external source. Therefore, these bearings can operate at very low speeds. In many applications, operating speed may not be high enough to generate adequate load capacity by hydrodynamic action alone. In this situation, a combination of hydrostatic (or externally pressurized) and hydrodynamic lubrication have to be resorted to. These bearings are usually known as hybrid bearings.

Externally pressurized lubrication is extensively used now-a-days in aerospace and machine tool industries mainly due to high load capacity, high stiffness, and low frictional resistance. Multirecess externally pressurized journal and thrust bearings are often used in machine tools. The load capacity can be attributed to the presence of deep recesses which constitute large areas of uniform pressure, and high stiffness is generally attributed to the use of flow restrictors, e.g., capillary, orifice, constant flow control valve, and diaphragm type variable flow restrictor in between the recesses and the supply line.

The bearing configuration can either have axial drain grooves in between the recesses or can be without drain grooves. The bearings without axial grooves have better characteristics due to interaction of fluid between the recesses and have become popular. Bearings employing flow control valve or diaphragm type variable flow restrictors usually operate on a bearing pressure feedback system and can be designed for high stiffness. However, capillary and orifice restrictors being simpler in construction, more reliable in operation, and easier in maintenance are commonly used.

The initial cost of these bearings is high because of additional requirement of pump and its accessories to supply high pressure fluid, but this disadvantage is offset in most cases since the hydraulic system is already available in the machine for other functions and is also used for supplying oil to the bearing. Bearings are usually designed either for maximum load or maximum stiffness.

## 9.2 | Circular Step Externally Pressurized Thrust Bearing

A circular step pressurized thrust bearing system is shown in Fig. 9.1. Oil is supplied to the recess through a flow restrictor either a capillary or an orifice.

The Reynolds equation for the flow of lubricant is written in polar coordinates, i.e., $r, \theta$ coordinates as

$$\frac{\partial}{\partial r}\left(\frac{rh^3}{\eta}\frac{\partial p}{\partial r}\right)+\frac{1}{r}\frac{\partial}{\partial \theta}\left(\frac{h^3}{\eta}\frac{\partial p}{\partial \theta}\right)=6r\omega\frac{\partial h}{\partial \theta} \tag{9.1}$$

In case of parallel film, i.e., constant film thickness, pressure gradient in the circumferential direction would be zero, i.e., $\frac{\partial p}{\partial \theta}=0$ and $h=\text{constant}$. Since viscosity is also assumed to be constant, the Reynolds equation thus reduces to for $\omega=0$ as given below:

$$\frac{\partial}{\partial r}\left(r\frac{\partial p}{\partial r}\right)=0 \tag{9.2}$$

The boundary conditions are

$$\begin{aligned} p &= p_r \text{ at } r = r_i \\ p &= 0 \text{ at } r = r_o \end{aligned} \tag{9.3}$$

Integrating Equation (9.2) with respect to $r$ and substituting boundary conditions given by Equation (9.3), pressure distribution over bearing land area is obtained as

$$p = p_r \frac{\ln\left(\frac{r}{r_o}\right)}{\ln\left(\frac{r_i}{r_o}\right)} \tag{9.4}$$

Bearing performance parameters, viz., load capacity, lubricant flow rate, and power loss due to viscous friction, etc. can be evaluated using the above expression of pressure distribution.

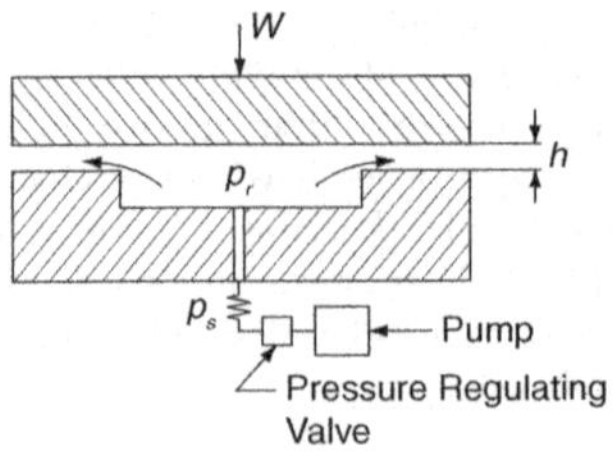


**Figure 9.1** | Single Recess Externally Pressurized Bearing

The expression for load capacity $W$ is written as

$$W = \pi r_i^2 p_r + 2\pi \int_{r_i}^{r_o} prdr \tag{9.5}$$

using Equation (9.4) for $p$ and integrating, load capacity $W$ is obtained as

$$W = \pi r_i^2 p_r + \int_{r_i}^{r_o} p_r \frac{\ln\left(\dfrac{r}{r_o}\right)}{\ln\left(\dfrac{r_i}{r_o}\right)} 2\pi r\, dr \tag{9.6}$$

which reduces to on integration as

$$W = \frac{\pi p_r \left(r_o^2 - r_i^2\right)}{2\ln\left(\dfrac{r_o}{r_i}\right)} = \frac{A p_r \left\{1 - \left(r_i/r_o\right)^2\right\}}{2\ln\left(\dfrac{r_o}{r_i}\right)} \tag{9.7}$$

where $A = \pi r_o^2$, bearing pad area.

While determining the expression for pressure distribution over the bearing pad area, it has been assumed that the depth of recess is very large in comparison to the film thickness, and therefore pressure in the recess area is constant at $p_r$.

The flow rate of the lubricant through the periphery of the bearing at radius $r = r_o$ can be determined in following manner.

The radial velocity of the lubricant

$$v_r = \frac{1}{2\eta}\frac{\partial p}{\partial r} z(z - h) \tag{9.8}$$

Outflow rate of the lubricant is given by

$$Q = \int_0^h 2\pi r v_r dz = \int_0^h 2\pi r \frac{1}{2\eta}\frac{\partial p}{\partial r} z(z - h)\, dz \tag{9.9}$$

Substituting for $\dfrac{\partial p}{\partial r}$ and integrating with respect to $z$, we get

$$Q = \frac{\pi}{6\ln\left(r_o/r_i\right)} \frac{h^3 p_r}{\eta} \tag{9.10}$$

Pumping power necessary to pump the above volume of lubricant into the bearing is given by:

$$\text{Pumping power, } P_p = p_s Q$$

where $p_s$ is supply pressure of the lubricant into the recess.

Power required to pump the fluid into the recess is thus determined as

$$P_p = \frac{\pi h^3}{6\ln\left(\dfrac{r_o}{r_i}\right)} \frac{p_s p_r}{\eta} \tag{9.11}$$

Frictional power loss due to rotation of the runner at an angular speed of $\omega$, can be determined as

$$P_f = \oint_A \tau . r\omega . dA = \int_{r_i}^{r_o} \tau . 2\pi r^2 \omega . dr \tag{9.12}$$

Shear stress $\tau$ due to rotation of the journal is given as: $\tau = \eta r\omega / h$. Therefore,

$$P_f = 2\pi\eta\omega^2 \int_{r_i}^{r_o} r^3 dr = 2\pi\eta\omega^2 \frac{\left(r_o^4 - r_i^4\right)}{4} \tag{9.13}$$

or

$$P_f = \frac{1 - \left(\frac{r_i}{r_o}\right)^4}{2} \frac{\eta r_o^2 \omega^2 A}{h} \tag{9.14}$$

where $A = \pi r_o^2$, total pad area including the recess.

Following coefficients, viz., load, flow, and frictional power coefficients which are purely dependent on the geometry of pad, i.e., recess and bearing radii are introduced and defined as

$$\begin{aligned} \text{Load coefficient } a_f &= \frac{1 - (r_i / r_o)^2}{2 \ln\left(\frac{r_o}{r_i}\right)} \\ \text{Flow coefficient } q_f &= \frac{\pi}{6 \ln\left(\frac{r_o}{r_i}\right)} \\ \text{Frictional power loss coefficient } h_f &= \frac{1 - \left(\frac{r_i}{r_o}\right)^4}{2} \end{aligned} \tag{9.15}$$

Thus, load capacity, oil flow rate through the bearing, pumping, and frictional power loss can be expressed in terms of the above coefficients as

$$\begin{aligned} W &= a_f A p_r \\ Q &= q_f \left(\frac{h^3 p_r}{\eta}\right) \\ P_p &= q_f \frac{h^3 p_r p_s}{\eta} \\ P_f &= h_f \eta \frac{r_o^2 \omega^2 A}{h} \end{aligned} \tag{9.16}$$

Therefore, one can say that for a given geometry of the bearing, the load capacity is not dependent directly on the film thickness $h$ or viscosity of the oil $\eta$, whereas pumping power and friction power loss are dependent on film thickness and viscosity of oil. In the case of uncompensated bearing when the flow restrictor is not put between the recess and supply line, the recess pressure $p_r$ would be equal to oil supply pressure $p_s$, and the bearing would always give a constant load capacity irrespective of the film thickness. Thus, stiffness of the bearing defined as $\left(-\frac{dW}{dh}\right)$ would be zero. Restrictors, e.g., orifice or capillary are therefore required to obtain variable recess pressure $p_r$ with the variation in the film thickness $h$ and provide necessary stiffness to the bearing.

Recess pressure is determined from the balance of flow of lubricant into recess from supply line through the restrictor and the flow out of the bearing.

Thus, recess flow continuity equation can be written for capillary and orifice compensation as:

### For Capillary Compensation

Flow through the capillary into the recess = flow out of the bearing periphery. Thus,

$$\frac{\pi d_c^4}{128\eta}\frac{(p_s - p_r)}{l_c} = q_f \frac{h^3}{\eta} p_r \tag{9.17}$$

Capillary design parameter $k_c$, a dimensionless parameter is defined as

$$k_c = \frac{\pi d_c^4}{128\eta l_c} \tag{9.18}$$

$d_c$ = diameter of the capillary tube
$l_c$ = length of the capillary tube
$h$ = film thickness
$k_c$ = capillary parameter

For capillary restrictor, recess flow continuity equation reduces to

$$k_c p_s (1-\beta) = B p_s h^3 \beta \tag{9.19}$$

where

$$\beta = p_r / p_s,\ B = \frac{q_f}{\eta}$$

### For Orifice Compensation

Similarly for orifice compensation, recess flow continuity equation can be written as

$$\frac{\pi}{4} d_o^2 c_d \sqrt{\frac{2}{\rho} p_s (1-\beta)} = q_f \frac{h^3}{\eta} p_r \tag{9.20}$$

In the dimensionless form, the above equation becomes

$$k_0 \sqrt{p_s (1-\beta)} = B h^3 p_s \beta \tag{9.21}$$

where

$$k_0 = \frac{\pi d_o^2 c_d}{4}\sqrt{\frac{2}{\rho}} \tag{9.22}$$

$d_0$ = orifice diameter
$c_d$ = discharge coefficient of orifice
$\rho$ = density of lubricant
$k_0$ = orifice parameter

Total power required to operate the bearing system is $P_t$ which can be determined as

$$P_t = P_p + P_f$$

$$P_t = q_f \frac{h^3}{\eta} p_r p_s + h_f \frac{\eta U_0^2 A}{h} \tag{9.23}$$

where $U_0 = r_0 \omega$.

The procedure to design the bearing is to optimize or minimize the total power required to operate the system. Besides, the bearing geometry and a chosen value of recess to supply pressure ratio $\beta$, the power loss is dependent on film thickness $h$ and viscosity of the oil $\eta$. Therefore, optimization can be done with respect to either film thickness for a given viscosity of the oil or with respect to viscosity for a given film thickness of lubricant film.

For optimum film thickness $\frac{\partial P_t}{\partial h} = 0$ when viscosity of oil $\eta$ is known. Total power given by Equation (9.23) is differentiated with respect to film thickness and set equal to zero for minimum power, i.e., $\frac{\partial P_t}{\partial h} = 0$ which gives the optimum film thickness as:

$$h_{opt} = \left( \frac{h_f \eta^2 U_0^2 A}{3 q_f p_r p_s} \right)^{1/4} \tag{9.24}$$

Substituting $h = h_{opt}$ in Equation (9.23), minimum power consumption is determined as:

$$P_t^{opt} = \frac{4}{3^{3/4}} \left( q_f h_f^3 p_r p_s \eta^2 U_0^6 A^3 \right)^{1/4} \tag{9.25}$$

This gives

$$\frac{P_f}{P_p} = 3 \text{ for } h = h_{opt}$$

Similarly for a given film thickness $h$, optimum viscosity can be determined as

$$\frac{\partial P_t}{\partial \eta} = 0 \text{ gives } \eta_{opt} = \left( \frac{q_f h^4 p_r p_s}{h_f U_0^2 A} \right) \tag{9.26a}$$

and

$$P_t^{opt} = (2\sqrt{q_f h_f A p_r p_s}) h U_0$$

This gives

$$\frac{P_f}{P_p} = 1 \text{ for } \eta = \eta_{opt} \tag{9.26b}$$

## 9.2.1 | Stiffness of the Bearing

Load capacity and flow of the lubricant of the bearing given by Equation (9.16) can be expressed as

$$W = A_{eff} p_r = A_{eff} \beta p_s; \tag{9.27a}$$

$$A_{eff} = a_f A$$

$$Q = q_f \frac{h^3 p_r}{\eta} = \left( q_f \frac{\beta p_s}{\eta} \right) h^3 \tag{9.27b}$$

Stiffness, $K$ of the bearing is defined as

$$K = -\frac{dW}{dh} = -A_{eff}\, p_s \frac{d\beta}{dh} \tag{9.28}$$

Using Equations (9.27a) and (9.27b), it can be shown that for a constant flow rate the stiffness of the bearing can be obtained as:

$$K = \frac{3W}{h} \tag{9.29}$$

Thus, dimensionless stiffness

$$\overline{K} = \frac{Kh}{W} = 3$$

### 9.2.2 | Optimum Design of the Bearing

The approach generally adopted in the design of externally pressurized bearings is to either optimize the load capacity or the stiffness of the bearing. Ling (1962) has shown that for a given film thickness bearing stiffness can be optimized with respect to recess to supply pressure ratio $\beta$.

The procedure to optimize either load or stiffness is as follows.

Recess flow continuity Equations (9.19) and (9.21) for capillary and orifice compensations, respectively, can be used to determine optimum stiffness of the bearing. From these it is clear that

$$h = h\left(B, p_s, k, \beta\right) \tag{9.30a}$$

and

$$k = k\left(B, p_s, h, \beta\right) \tag{9.30b}$$

Therefore, to determine optimum stiffness it is necessary to differentiate stiffness $K$ with respect to $\beta$ and equate to zero. Thus,

$$\frac{dK}{d\beta} = \frac{\partial K}{\partial k}\frac{\partial k}{\partial \beta} + \frac{\partial K}{\partial h}\frac{\partial h}{\partial \beta} = 0 \tag{9.31}$$

For a given restrictor parameter $k$ or a given film thickness $h$ above equation reduces, respectively, to

$$\frac{\partial K}{\partial h}\frac{\partial h}{\partial \beta} = 0 \text{ and } \frac{\partial K}{\partial k}\frac{\partial k}{\partial \beta} = 0 \tag{9.32}$$

Satisfying the above conditions, one obtains values of $\beta$ for which maximum stiffness would be obtained as:

$$\left.\begin{array}{l}\beta = 0.5 \text{ for } h = \text{constant} \\ \beta = 0.667 \text{ for } k_c \text{ constant}\end{array}\right\} \text{capillary restrictor}$$

and

$$\left.\begin{array}{l}\beta = 0.586 \text{ for } h = \text{constant} \\ \beta = 0.691 \text{ for } k_0 \text{ constant}\end{array}\right\} \text{orfice restrictor} \tag{9.33}$$

Graphs of pad coefficients versus recess to bearing radius ratio, i.e., $(r_i/r_o)$ and stiffness of the bearing for capillary and orifice restrictors versus pressure ratio $\beta$ are shown in Figs 9.2 and 9.3, respectively. For a given

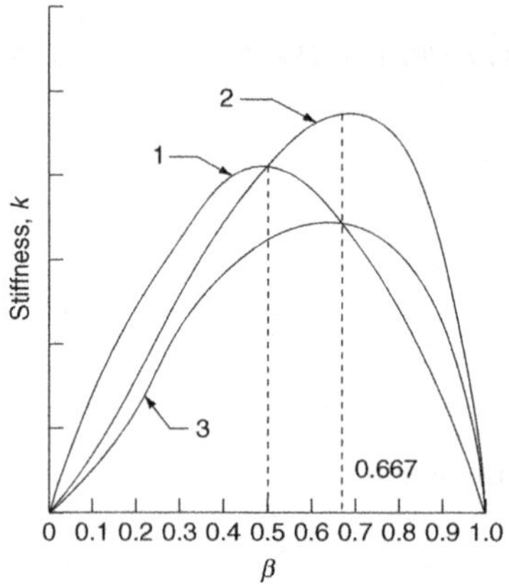


**Figure 9.2** | Capillary Compensated Thrust Bearing Stiffness versus Pressure Ratio $\beta$

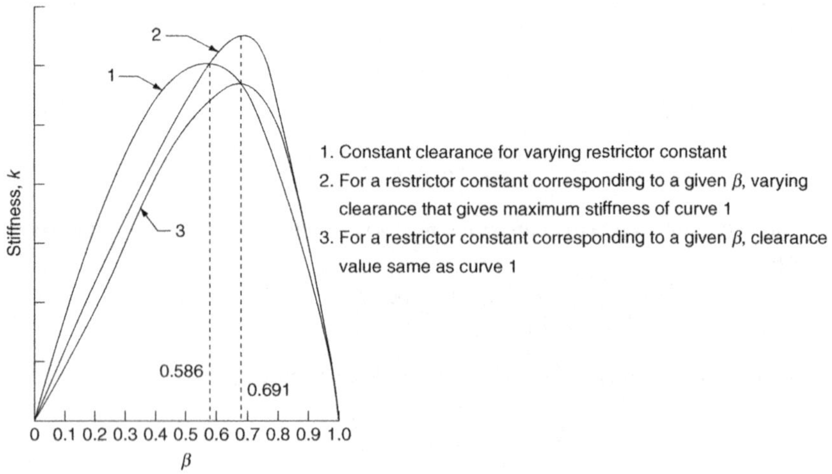


**Figure 9.3** | Orifice Compensated Thrust Bearing Stiffness versus Pressure Ratio $\beta$

film thickness, restrictor can be chosen to give either optimum load or stiffness for a given supply pressure $p_s$ and viscosity of oil $\eta$. For a given oil and film thickness, increase in supply pressure would increase the load capacity and stiffness of the bearing.

## 9.3 | Externally Pressurized Multirecess Journal Bearing with Short Sills

An excellent general method of analysis of multirecess externally pressurized journal bearing was developed by Davies (1969–70). The analysis is simple but very useful. It is valid for bearings with large recess area, i.e., when recess area is greater than 60% of the bearing area or in other words for bearings with short land/sill dimensions. Therefore, it assumes that the pressure drop across each axial and circumferential land is linear and pressure gradient is constant. It also assumes that variation of clearance across the circumferential land is negligible. A multirecess hydrostatic bearing system is shown in Fig. 9.4.

The position of the journal center is specified in this analysis by its eccentricity ratio, $\varepsilon$ and an attitude angle, $\alpha$, measured from the center line of a chosen circumferential land as shown in Fig. 9.5. Recess pressures are first determined and load, load angle, flow rate of the lubricant, power required, etc. are determined later.

To determine recess pressures for an operating eccentricity ratio, $\varepsilon$ and at a speed $N_s$ of the shaft for a bearing of diameter $D$, radial clearance $C$ for known orifice/capillary dimensions, viscosity of the oil $\eta$, etc, it is necessary to write down recess flow continuity equation.

In a $N$ recess bearing recess, pressures are assumed as $p_1, p_2, p_3 \ldots\ldots p_N$. For the $n$th recess pressure $p_n$ and adjacent recess pressures $p_{n-1}, p_{n+1}$ are assumed to be in increasing order, i.e., $p_{n-1} > p_n > p_{n+1}$.

Flow continuity equation for $n$th recess can be expressed as a flow balance relationship between the volume of the fluid entering into the recess from the restrictor and the fluid volume going out of the recess boundaries per unit time.

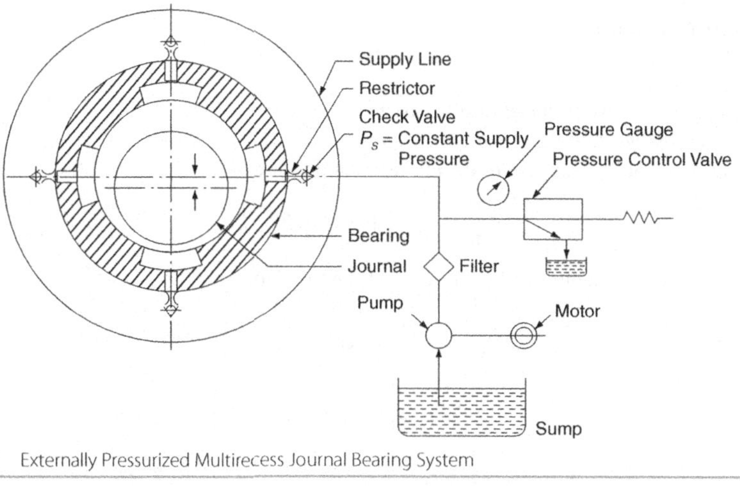


**Figure 9.4** | Externally Pressurized Multirecess Journal Bearing System

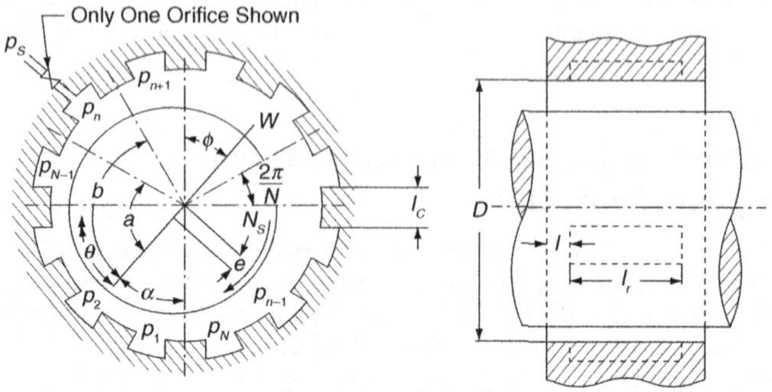


$$a = \frac{2\pi n - \alpha}{N} \quad b = \frac{2\pi(n-1) - \alpha}{N}$$

**Figure 9.5** | Schematic Diagram of the General Multirecess Hydrostatic Journal Bearing with $N$ Recesses [Davies, Proc. IMechE, UK, 1969–70]

1. For a supply pressure of lubricant at $p_s$, the flow rate entering the recess from the restriction in to nth recess is:

$$\frac{\pi}{4} c_d d^2 \sqrt{\frac{2}{\rho}(p_s - p_n)}$$

***For Orifice Restrictor***

$$\frac{\pi d^4}{128 l_t \eta}(p_s - p_n)$$

***For Capillary Restrictor***

$d$ = capillary or orifice diameter as the case may be
$l_t$ = length of the capillary tube
$\rho$, $\eta$ = density and viscosity of the lubricant respectively
$c_d$ = discharge coefficient of the orifice

Axial flow rate of lubricant out of the $n$th recess at pressure $p_n$ is

$$\frac{DC^3 p_n}{12\eta l} \int_{\frac{2\pi}{N}(n-1)-\alpha}^{\frac{2\pi}{N}n-\alpha} (1 - \varepsilon \cos\theta)^3 \, d\theta \tag{9.34}$$

2. Flow rate of lubricant out from $n$th recess into the $(n+1)$th recess is

$$\frac{l_r C^3}{12\eta l_c}(p_n - p_{n+1})\left\{1 - \varepsilon\cos\left(\frac{2\pi}{N}n - \alpha\right)\right\}^3 + \frac{\pi}{2} DCl_r N_s \left\{1 - \varepsilon\cos\left(\frac{2\pi}{N}n - \alpha\right)\right\} \tag{9.35}$$

The first term is due to pressure induced flow and second term is due to the shaft velocity induced flow.

3. Similarly, the flow into the $n$th recess form the ($n$-1)th recess is:

$$\frac{l_r C^3}{12\eta l_c}\left(p_{n-1}-p_n\right)\left\{1-\varepsilon\cos\left(\frac{2\pi n}{N}-\alpha\right)\right\}^3$$
$$+\frac{\pi}{2}DCl_r N_s\left\{1-\varepsilon\cos\left(\frac{2\pi(n-1)}{N}-\alpha\right)\right\} \tag{9.36}$$

Continuity of flow for the recess requires that lubricant flow rate into the recess from the restrictor be equated to the flow rate out of the recess which is sum of axial flow rate out of $n$th recess and flow rate out of the $n$th recess into ($n$+1)th recess minus the flow into $n$th recess from ($n$-1)th recess. The recess flow continuity equation for $n$th recess is written in dimensionless form as:

#### For Orifice Compensated Bearing

$$-C_{n-1}\bar{p}_{n-1}+\left(A_n+C_n+C_{n-1}\right)\bar{p}_n-C_{n-1}\bar{p}_{n+1}=\delta_0\sqrt{1-\bar{p}_n}-\omega S_n \tag{9.37}$$

#### For Capillary Compensated Bearing

$$-C_{n-1}\bar{p}_{n-1}+\left(A_n+C_n+C_{n-1}\right)\bar{p}_n-C_{n+1}\bar{p}_{n+1}=\delta_c\left(1-\bar{p}_n\right)-\omega S_n \tag{9.38}$$

where aspect ratio,

$$m=\frac{ll_r}{Dl_c};\ \bar{p}_n=p_n/p_s;\ \omega=\frac{6\pi\eta N_s ll_r}{C^2 p_s}$$

speed parameter,

$$A_n=\int_{\frac{2\pi}{N}(n-1)-\alpha}^{\frac{2\pi}{N}.n}\left(1-\varepsilon\cos\theta\right)^3 d\theta \tag{9.39}$$

$$C_n=m\left\{1-\varepsilon\cos\left(\frac{2\pi}{N}\eta-\alpha\right)\right\}^3$$

$$S_n=\varepsilon\left\{\cos\left(\frac{2\pi}{N}(\eta-1)-\alpha\right)-\cos\left(\frac{2\pi}{N}\eta-\alpha\right)\right\}$$

$\delta_0=\dfrac{3\sqrt{2}\pi c_d d^2\eta l}{C^3 D\sqrt{\rho ps}}$, dimensionless orifice design parameter

$\delta_c=\dfrac{3\pi d^4 l}{32 l_t DC^3}$, dimensionless capillary design parameter.

Equations (9.37) and (9.38) are general continuity equation for $n$th recess which would lead to $N$ equation for a bearing with $N$ recesses. To determine recess pressures $\bar{p}_1, \bar{p}_2, \ldots \bar{p}_N$ $N$ simultaneous nonlinear algebraic equations would have to solved in case of orifice compensated bearings. Similarly, $N$-simultaneous algebraic linear equations have to be solved using methods to solve simultaneous algebraic linear equations to determine recess pressures $\bar{p}_1, \bar{p}_2, \ldots \bar{p}_N$ in case of capillary compensated bearings. For example, equations for a 4-recess orifice compensated bearing the equations can be written in matrix notation as:

$$\begin{bmatrix} (A_1+C_1+C_4) & -C_1 & 0 & C_4 \\ -C_1 & (A_2+C_2+C_1) & -C_2 & 0 \\ 0 & -C_2 & (A_3+C_3+C_2) & -C_3 \\ -C_4 & 0 & -C_3 & (A_4+C_4+C_3) \end{bmatrix} \cdot \begin{bmatrix} \bar{p}_1 \\ \bar{p}_2 \\ \bar{p}_3 \\ \bar{p}_4 \end{bmatrix} = \delta_0 \begin{bmatrix} \sqrt{1-\bar{p}_1} \\ \sqrt{1-\bar{p}_2} \\ \sqrt{1-\bar{p}_3} \\ \sqrt{1-\bar{p}_4} \end{bmatrix} - \omega \begin{bmatrix} S_1 \\ S_2 \\ S_3 \\ S_4 \end{bmatrix} \tag{9.40}$$

For capillary compensated bearing this reduces to

$$\begin{bmatrix} (A_1+C_1+C_4+\delta_c) & -C_1 & 0 & -C_4 \\ -C_1 & (A_2+C_2+C_1+\delta_c) & -C_2 & 0 \\ 0 & -C_2 & (A_3+C_3+C_2+\delta_c) & -C_3 \\ -C_4 & 0 & -C_3 & (A_4+C_4+C_3+\delta_c) \end{bmatrix} \cdot \begin{bmatrix} \bar{p}_1 \\ \bar{p}_2 \\ \bar{p}_3 \\ \bar{p}_4 \end{bmatrix} = \begin{bmatrix} \delta_c - \omega S_1 \\ \delta_c - \omega S_2 \\ \delta_c - \omega S_3 \\ \delta_c - \omega S_4 \end{bmatrix} \tag{9.41}$$

Once the recess pressures are calculated for a given eccentricity ratio $\varepsilon$ and other bearing parameters using Equations (9.40) and (9.41) for orifice and capillary compensation, respectively, as the case may be, bearing performance can be evaluated in terms of load variable $(\bar{W})$, flow variable $(\bar{Q})$, and energy loss variable $(\bar{H})$. The orifice and capillary design parameters $\delta_0$ and $\delta_c$ can be expressed in terms of concentric pressure ratio $(\beta)$ following continuity equation.

For orifice compensated bearing Equation (9.40) gives for concentric position of the journal:

$$\frac{\beta}{\sqrt{1-\beta}} = \frac{N\delta_0}{2\pi} \tag{9.42}$$

Similarly for capillary compensated bearing Equation (9.41) gives

$$\frac{\beta}{1-\beta} = \frac{N\delta_c}{2\pi} \tag{9.43}$$

Load variable $\overline{W}$ is defined as load supported per unit bearing area at a supply pressure $p_s$. It is expressed as:

$$\overline{W} = \frac{W(\text{Total Load Carried})}{D(l_r + l)\,p_s}$$

$$\overline{W} = \left(\overline{W}_x^2 + \overline{W}_y^2\right)^{1/2} \tag{9.44}$$

where

$$\overline{W}_x = \sum_{n=1}^{N} \overline{p}_n \sin\frac{2\pi}{N}\left(n - \frac{1}{2}\right)\sin\frac{\pi}{N}$$

$$\overline{W}_y = \sum_{n=1}^{N} \overline{p}_n \cos\frac{2\pi}{N}\left(n - \frac{1}{2}\right)\sin\frac{\pi}{N}$$

Load angle

$$\phi = \tan^{-1}\frac{\overline{W}_x}{\overline{W}_y} \tag{9.45}$$

Flow variable $\overline{Q}$ is written as:

$$\overline{Q} = \frac{\text{Total lubricant flow rate through the bearing}}{2\dfrac{\pi D C^3 p_s}{12\eta l}}$$

and $\overline{Q} = \dfrac{\delta_0}{2\pi}\displaystyle\sum_{n=1}^{N}\sqrt{1 - \overline{p}_n}$ for orifice compensation

$$\overline{Q} = \frac{\delta_c}{2\pi}\sum_{\eta=1}^{N}\sqrt{1 - \overline{p}_n} \quad \text{for capillary compensation} \tag{9.46}$$

Energy loss variable $\overline{P}$ is determined as: $\overline{P} = \dfrac{\text{Total Energy Dissipation}}{\overline{Q}p_s\dfrac{2\pi D C^3 p_s}{12\eta l}}$

This assumes that frictional losses are significant only in the bearing land area. Thus,

$$\overline{P} = 1 + \frac{1}{6\pi\overline{Q}}\left(\frac{\omega}{mL}\right)^2\left(\frac{2\pi}{\sqrt{1-\varepsilon^2}} + mL^2\sum_{n=1}^{N}\frac{1}{1 - \varepsilon\cos\left(\dfrac{2\pi}{N}n - \alpha\right)}\right) \tag{9.47}$$

where $L = \dfrac{l_c}{l}$ is defined as land width ratio.

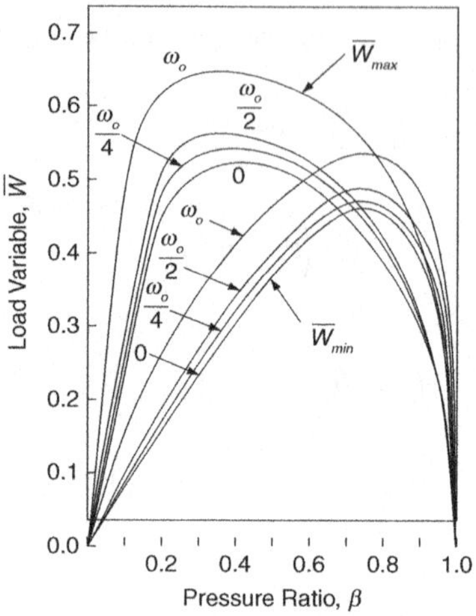


**Figure 9.6** | Variations of Maximum and Minimum Values of Load Variable as Functions of Pressure Ratio for an Eccentricity Ratio of $\varepsilon = 0.8$ and for Various Values of Speed [Davies, Proc. IMechE, UK 1969–70]

Optimum clearance $C_o$ can be found for known values of fluid properties and supply pressure by minimizing the energy loss variable $\overline{P}$. This can be found by equating $\frac{dP}{dC} = 0$ from the dimensional form of energy loss variable. Minimum value of $\overline{P}$ is obtained as 4. Alternately, optimum speed $N_s$ can also be determined for a known value of clearance $C$ by substituting $\overline{P} = 4$. Thus, optimum speed variable is found as:

$$\omega_0 = 3ml\left\{\frac{2\pi\beta}{2\pi + Nml^2}\right\}^{1/2} \tag{9.48}$$

Variation of maximum and minimum values of load variable as function of concentric pressure ratio $\beta$ for an eccentricity ratio of 0.8 and speed variable of 0, 0.25, 0.5, and 1 times the optimum speed variable $\omega_0$ is shown in Fig. 9. 6. for a 4-recess bearing. As can be seen in the figure, there is a maximum and minimum value of load variable, which depend on load angle $\phi$. For load angle, $\phi = 45^\circ$ maximum load is obtained, whereas for $\phi = 0^\circ$ minimum load is obtained. It is seen that load variable increases with increase in speed variable. For a known eccentricity ratio, there is a pressure ratio $\beta$ for which maximum load is obtained. This is dependent on speed variable and load angle.

## 9.4 | Multirecess Externally Pressurized Journal Bearings with Large Sill Dimensions

Multirecess externally pressurized journal bearings of large sill/land dimensions are very widely used as load support system in high speed turbo pumps and high speed machine tools. Reducing the recess size reduces the power required to pump large volume of lubricant in the case of bearings with large size recesses.

This also reduces the load capacity and stiffness when the bearing is used at low speed of operation. However, reducing the recess size or increasing bearing land area also results in additional load capacity due to hydrodynamic effect over the land area in case of high speeds of journal rotation.

Although pumping power to supply the lubricant is reduced, the frictional power loss over the land area increases at high speeds. Therefore, an optimization is done to minimize the power consumption and reduce the temperature rise.

The effect of shaft rotation on the steady state and dynamic behavior of multirecess externally pressurized bearings with large land area have been investigated by several researchers. O'Donoghue *et al.* (1970) and Cusano and Conry (1974) outlined a procedure for optimum design of multirecess hydrostatic journal bearings based on minimum power consumption and temperature rise. Optimum land width ratio and optimum speed parameters were determined. Optimum land width ratio was found to be 0.25 and it was also found that in optimized bearings cavitations of oil film do not occur at eccentricity ratios <0.6. However, at high eccentricity ratios (≥ 0.7) and at high speed parameters cavitations may occur. At high speeds, these bearings often operate in turbulent flow regime. Redcliffe and Vohr (1969) and Heller (1974) investigated the steady state and dynamic behavior in turbulent flow regime. Shinkle and Hornung (1965) and Ghigliazza and Michelini (1968) studied the frictional behavior of liquid hydrostatic bearings. Later on lot of studies were done by Ghosh and Majumdar (1978), Ghosh *et al.* (1979) on the steady state and dynamic behavior of multirecess externally pressurized oil journal bearings in the laminar flow regime. A methodology to determine recess pressures under eccentric condition was developed by Ghosh and Majumdar (1978). In bearings with finite sill dimensions, Reynolds equation for finite journal bearings has to be solved to determine the pressure distribution over the bearing lands.

Steady state analysis of multirecess hydrostatic journal bearings with shaft rotation is now presented for bearings with large sill dimensions.

### 9.4.1 | Steady State Analysis of Multirecess Journal Bearings with Large Sills

A multirecess externally pressurized journal bearing system is shown in Fig. 9.7. The developed view of the bearing surface is shown in Fig. 9.8. It is assumed that depth of the recesses is very large in comparison to the radial clearance between the journal and the bearing so that pressures in the recesses remain constant.

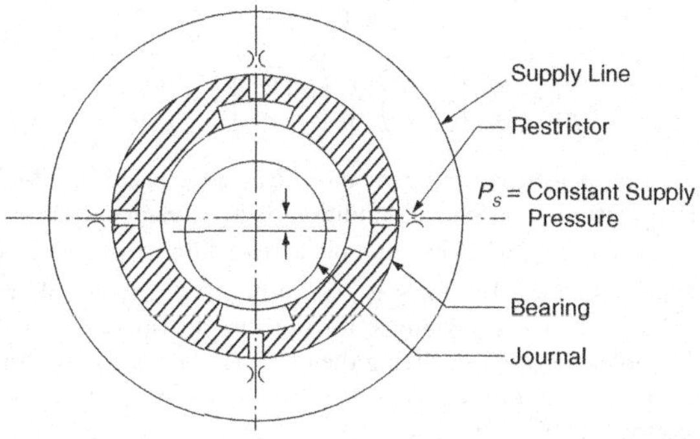


**Figure 9.7** | A Multirecess Hydrostatic Oil Journal Bearing

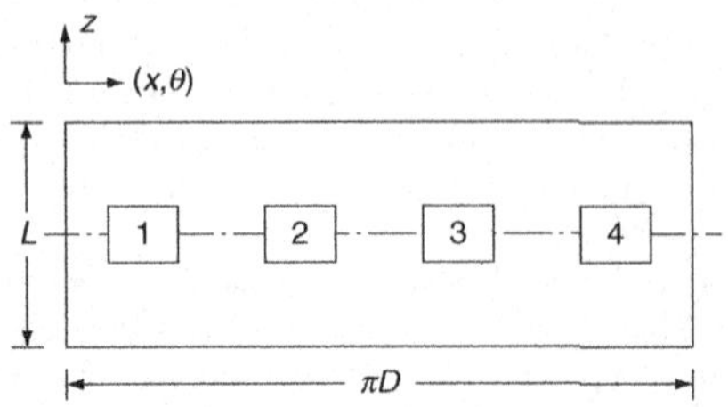


**Figure 9.8** | Developed View of the Bearing

The oil is supplied to the recesses from a constant pressure source at a supply pressure, $p_s$ and it enters into each recess through a restrictor either a capillary or an orifice. The journal position within the bearing depends on the steady load acting on the journal. For an $N$ recess bearing, recess pressures $p_1, p_2, p_3 \cdots\cdots p_N$ get adjusted in such a way as to carry the net load acting on the journal. Some of assumptions inherent in the geometrical configuration of the bearing and the compensating devices may be noted as follows:

- The recesses are symmetrically spaced and the recess depth is large in comparison to the radial clearance of the bearing to ensure approximately uniform pressure in the recesses.
- The flow in the capillary tube restrictor is laminar while the flow through the orifice restrictor is turbulent in the vicinity of the edge of the orifice and the discharge coefficient of the restrictors is constant and the same for all the restrictors.

The Reynolds equation which gives the pressure distribution of the oil film over the sill surface is given for an incompressible lubricant of constant viscosity as

$$\frac{\partial}{\partial x}\left(h^3\frac{\partial p}{\partial x}\right)+\frac{\partial}{\partial y}\left(h^3\frac{\partial p}{\partial y}\right)=6\eta U\frac{\partial h}{\partial x} \tag{9.49}$$

where $U = R\omega$, surface speed of the journal which rotates at an angular velocity $\omega$ rad/sec.

With the following substitution, the above equation is expressed in dimensionless form as:

$$\bar{h}=h/C, \bar{p}=p/p_s, \Lambda=\frac{6\eta\omega}{p_s(C/R)^2}; x=R\theta \text{ and } \bar{y}=y/L$$

$$\frac{\partial}{\partial\theta}\left(\bar{h}^3\frac{\partial\bar{p}}{\partial\theta}\right)+\left(\frac{R}{L}\right)^2\frac{\partial}{\partial\bar{y}}\left(\bar{h}^3\frac{\partial\bar{p}}{\partial\bar{y}}\right)=\Lambda\frac{\partial\bar{h}}{\partial\theta} \tag{9.50}$$

To determine the bearing performance parameters, i.e., load capacity, oil flow rate, friction coefficient, etc. for a given eccentricity ratio under loaded condition, it is necessary to determine the recess pressures $\bar{p}_1, \bar{p}_2 --- \bar{p}_N$. Ghosh and Majumdar (1978) adopted a procedure as described below:

The procedure adopted solves the Reynolds Equation (9.50) satisfying following boundary conditions by assigning a definite value to the recess pressures. The pressure distribution over the bearing surface is thus determined by solving Reynolds equation satisfying the necessary boundary conditions which is then used to determine the lubricant flow rate out of each recess from one recess into other.

Boundary conditions for Equation (9.50) are

1. $\bar{p}\left(\theta,-\tfrac{1}{2}\right)=\bar{p}\left(\theta,+\tfrac{1}{2}\right)=0$, for bearing ends open to ambient pressure

2. $\frac{\partial \bar{p}}{\partial \bar{y}}(\theta,0)=0$ at $\bar{y}=0$, at the center line of the bearing because of symmetry

3. $\bar{p}=\bar{p}_i, i=1,2,3.......N$, pressure at $i$th recess (assigned value)

4. $\bar{p}(\theta,\bar{y})=\frac{\partial \bar{p}}{\partial \theta}(\theta,\bar{y})=0$, Reynolds' boundary condition at film rupture boundary

The dimensionless recess pressure, say at the $i$th recess is assigned an arbitrary value equal to 1 while all the other recess pressures are assigned a value equal to zero. Equation (9.50) is solved with the boundary conditions mentioned using finite difference method and following Gauss–Seidel iterative procedure described in Chapter 5. The dimensionless volume rate of flow of lubricant from $i$th recess, i.e., $\bar{Q}_c(i,i)$ and lubricant flow rate into the other recesses due to pressure assigned at $i$th recess, i.e., $\bar{Q}_c(i,j)$ are evaluated using Equation (9.51) in the dimensionless form:

$$\bar{Q}_c(i,j)=-\bar{p}_r\int \bar{h}^3\frac{\partial \bar{p}}{\partial s}\hat{n}dl+\left(\frac{L}{R}\right)\oint_{\bar{y}_{r1}}^{\bar{y}_{r2}}\left(\Lambda\bar{h}-\bar{h}^3\frac{\partial \bar{p}}{\partial \theta}\right)_{\theta_{r1},\theta_{r2}}d\bar{y} \tag{9.51}$$

where $\bar{p}_r=p_r/p_s$, dimensionless recess pressure $r$ th recess, $r=i$ for the $i$th recess and $r=j$ when the flow is evaluated at the $j$th recess due to pressure assigned at $i$th recess. $\bar{y}_{r1}$ and $\bar{y}_{r2}$ are axial boundaries of $r$th recess, whereas $\theta_{r1}$ and $\theta_{r2}$ are the circumferential boundaries of the $r$th recess where

$dl$ = elementary length

$\frac{\partial \bar{p}}{\partial s}$ = pressure gradient in the direction perpendicular to $dl$

$\bar{n}$ = normal unit vector

The above integral is evaluated around all the four edges of the recesses for the first part and for the second part is to be evaluated for two sides of the recess only, i.e., edges at $\theta_{r1}$ and $\theta_{r2}$ for $r$th recess.

$$\bar{Q}_c(i,j)=Q_c(i,j)/(12\eta/C^3p_s)$$

This process is repeated by alternately assigning one of the recess pressures as 1 and keeping other recess pressures equal to 0. The flow coefficients $\bar{Q}_c(i,j)$ are thus evaluated for $N$ recesses and are expressed in the matrix form as:

$$\begin{bmatrix}\bar{Q}_c(1,1) & \bar{Q}_c(1,2) & \bar{Q}_c(1,N)\\ \bar{Q}_c(2,1) & \bar{Q}_c(2,2) & \bar{Q}_c(2,N)\\ \bar{Q}_c(N,1) & \bar{Q}_c(N,2) & \bar{Q}_c(N,N)\end{bmatrix} \tag{9.52}$$

Continuity equation for $i$th recess may be written as

Flow from the restrictor into the $i$th recess = flow out from $i$th recess – flow into $i$th recess from recesses adjacent to it.

The above relationship can be expressed as

### *For Capillary Compensated Bearings*

$$\delta_c(1-\bar{p}_i)=\bar{Q}_c(i,i)\bar{p}_i-\sum_{j=1}^{N}\delta_{ij}\bar{Q}_c(j,i)\bar{p}_j \tag{9.53}$$

***For Orifice Compensated Bearings***

$$\delta_0 (1-p_i)^{1/2} = \bar{Q}_c(i,i)\bar{p}_i - \sum_{j=1}^{N} \delta_{ij}\bar{Q}_c(j,i)\bar{p}_{ij} \tag{9.54}$$

where $\delta_c = \dfrac{3\pi}{32}\dfrac{d_c^4}{C^3 l_c}$, dimensionless parameter for capillary compensated bearing

$d_c$ = diameter of the capillary tube
$l_c$ = length of the capillary tube

$$\delta_0 = \frac{3\pi c_d d_0^2 \eta}{\left(\dfrac{\rho p_s}{2}\right)^{1/2} C^3},$$

dimensionless parameter for orifice compensated bearings and

$$\delta_{ij} = 1.0 \text{ for } i \neq j$$
$$\delta_{ij} = 0 \text{ for } i \neq j$$

For a bearing with $N$ number of recesses, the above equations can be expanded into $N$ simultaneous algebraic equations

***For Capillary Compensated Bearings***

$$\begin{aligned}
\delta_c(1-\bar{p}_1) &= \bar{Q}_c(1,1)\bar{p}_1 - \bar{Q}_c(2,1)\bar{p}_2 - \ldots\ldots\ldots\ldots - \bar{Q}_c(N,1)\bar{p}_N \\
\delta_c(1-\bar{p}_2) &= -\bar{Q}_c(1,2)\bar{p}_1 + \bar{Q}_c(2,2)\bar{p}_2 - \ldots\ldots\ldots\ldots - \bar{Q}_c(N,2)\bar{p}_N \\
&\ldots\ldots\ldots\ldots\ldots\ldots\ldots\ldots\ldots\ldots\ldots\ldots \\
&\ldots\ldots\ldots\ldots\ldots\ldots\ldots\ldots\ldots\ldots\ldots\ldots \\
\delta_c(1-\bar{p}_N) &= -\bar{Q}_c(1,N)\bar{p}_1 - \bar{Q}_c(2,N)\bar{p}_2 \ldots\ldots\ldots\ldots + \bar{Q}_c(N,N)\bar{p}_N
\end{aligned} \tag{9.55}$$

The above is a set of $N$ linear algebraic equations and can be solved to determine recess pressures $\bar{p}_1, \bar{p}_2, \ldots \bar{p}_N$ using Gauss elimination process.

***For Orifice Compensated Bearings***

$$\begin{aligned}
\delta_0(1-\bar{p}_1)^{1/2} &= \bar{Q}_c(1,1)\bar{p}_1 - \bar{Q}_c(2,1)\bar{p}_2 - \ldots\ldots\ldots\ldots - \bar{Q}_c(N,1)\bar{p}_N \\
\delta_0(1-\bar{p}_2)^{1/2} &= -\bar{Q}_c(1,2)\bar{p}_1 + \bar{Q}_c(2,2)\bar{p}_2 - \ldots\ldots\ldots\ldots - \bar{Q}_c(N,2)\bar{p}_N \\
&\ldots\ldots\ldots\ldots\ldots\ldots\ldots\ldots\ldots\ldots\ldots\ldots \\
&\ldots\ldots\ldots\ldots\ldots\ldots\ldots\ldots\ldots\ldots\ldots\ldots \\
\delta_0(1-\bar{p}_N)^{1/2} &= -\bar{Q}_c(1,N)\bar{p}_1 - \bar{Q}_c(2,N)\bar{p}_2 - \ldots\ldots\ldots\ldots + \bar{Q}_c(N,N)\bar{p}_N
\end{aligned} \tag{9.56}$$

This is a set of $N$ simultaneous nonlinear algebraic equations which can be solved using Newton–Raphson method to determine recess pressures $\bar{p}_1, \bar{p}_2, \ldots \bar{p}_N$.

Having determined the recess pressures for a known eccentricity ratio bearing performance parameters, viz., load capacity, oil flow rate, friction force, etc. can be evaluated.

Load components can be expressed as:

$$W_\varepsilon = -2\int_0^{L/2}\int_0^{2\pi} p\cos\theta \ \mathrm{R}d\theta \ dy \tag{9.57}$$

$$W_\phi = -2\int_0^{L/2}\int_0^{2\pi} p\sin\theta \ \mathrm{R}d\theta \ dy \tag{9.58}$$

Hence the load capacity of the bearing is given by

$$W = \left(W_\varepsilon^2 + W_\phi^2\right)^{1/2} \tag{9.59a}$$

and attitude angle $\phi$ is given as

$$\phi = \tan^{-1}\left(W_\phi / W_\varepsilon\right) \tag{9.59b}$$

The total friction force exerted on the moving journal due to shearing action of the viscous lubricant is given by

$$F_j = 2\int_0^{2\pi}\int_0^{L/2}\left(\frac{h}{2}\frac{\partial p}{\partial x} + \eta\frac{U}{h}\right) R \, d\theta \, dy \tag{9.60}$$

and hence, the coefficient of friction is given as

$$\mu = F_j / W \tag{9.61}$$

Friction parameter is defined as $\left(\frac{R}{C}\right)\mu$. The total oil flow rate from the bearing edges is determined as

$$Q = -2\int_0^{2\pi} \left.\frac{h^3}{12\eta}\frac{\partial p}{\partial y}\right|_{y=\pm L/2} Rd\theta \tag{9.62}$$

This flow rate must be equal to the sum of oil flow rate from all the recesses.

Heller (1974) adopted an iterative procedure to determine the recess pressures in loaded condition in a hybrid bearing.

In the iterative process which is followed, the dimensionless pressures in all the recesses are set equal to 1 and Reynolds Equation (9.50) is solved satisfying the boundary conditions to determine the pressure distribution in the bearing. Then flow of lubricant out of each recess is calculated using Equation (9.51) and recess pressures are evaluated by satisfying recess flow continuity equations as expressed below:

### For Capillary Compensation

$$\delta_c\left(1-\bar{p}_i\right) = \bar{Q}_c(i)\,\bar{p}_i \quad \text{for } i = 1,2,\ldots.N \tag{9.63}$$

### For Orifice Compensation

$$\delta_0\left(1-\bar{p}_i\right)^{1/2} = \bar{Q}_c(i)\,\bar{p}_i \quad \text{for } i = 1,2,\ldots..N \tag{9.64}$$

For the next iteration, recess pressures are set equal to the calculated values of the recess pressures from the previous iteration, i.e., $\bar{p}_i, i = 1, 2, .....N$ and Equation (9.50) is again solved satisfying proper boundary conditions. Flows through the recesses are again calculated and using Equations (9.63) and (9.64). New values of recess pressures are calculated using Equations (9.63) and (9.64). This process is repeated until the difference between two successive iterations in recess pressures is below a prescribed convergence value. This convergence is obtained in a few iterations only.

Results of load capacity, oil flow are presented for orifice and capillary compensated bearings for nonrotating journal against concentric pressure ratio $\beta$ for various eccentricity ratios for loading through the center line of a recess in Figs 9.9 to 9.14. Figure 9.9 shows that there is an optimum value of concentric pressure ratio $\beta$ at which maximum load capacity is obtained. Load capacity of orifice compensated bearing is higher than the load carrying capacity of capillary compensated bearing. Figure 9.10 shows that with increase in L/D ratio load capacity decreases, whereas the load capacity increases with increase in eccentricity ratio.

Similarly, results for load capacity are plotted against orifice and capillary design parameters in Figs 9.11 and 9.12.

Optimum values of $\delta_0$ and $\delta_c$ at which maximum load capacity is obtained can be determined for a given bearing geometry. Variation of lubricant flow rate against concentric pressure ratio $\beta$ and $L/D$ ratio are shown in Figs 9.13 and 9.14, respectively. Oil flow rate increases with increase in $\beta$ which is quite obvious and it also decreases with increase in $L/D$ ratio. In the case of nonrotating journal, the power consumption is due to requirement to pump the lubricant to the recesses at a supply pressure $p_s$. Hybrid bearings

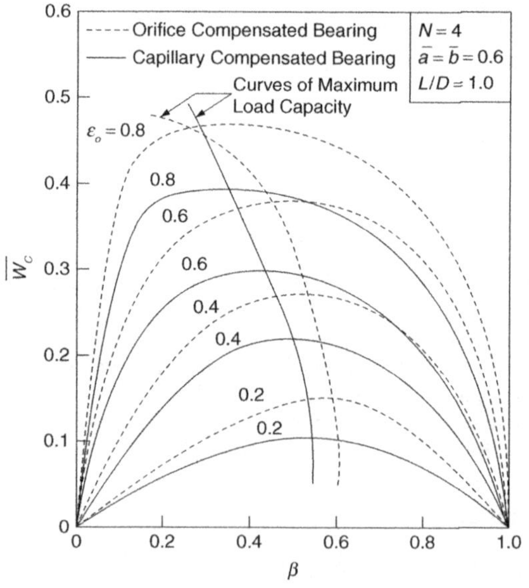


**Figure 9.9** | Load Capacity of Compensated Bearing Versus Concentric Pressure Ratio

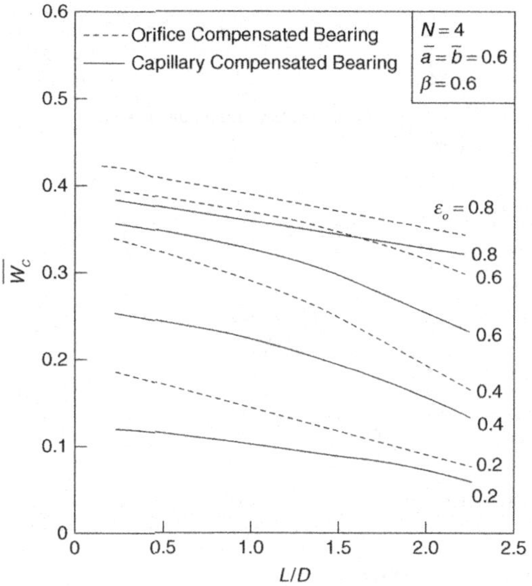


**Figure 9.10** | Load Capacity of Compensated Bearing Versus L/D Ratio

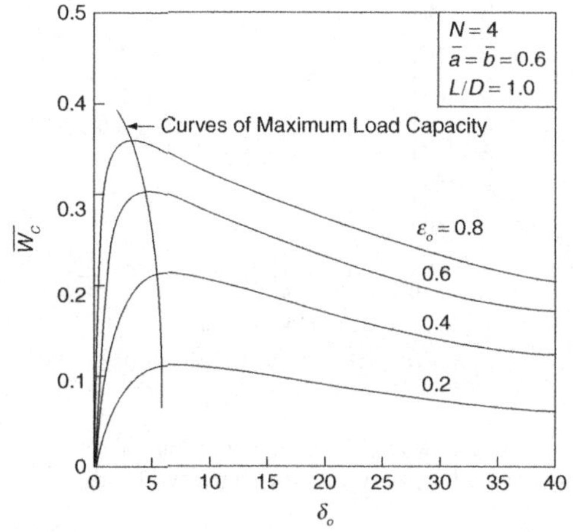


**Figure 9.11** | Load Capacity Versus Capillary Design Parameter

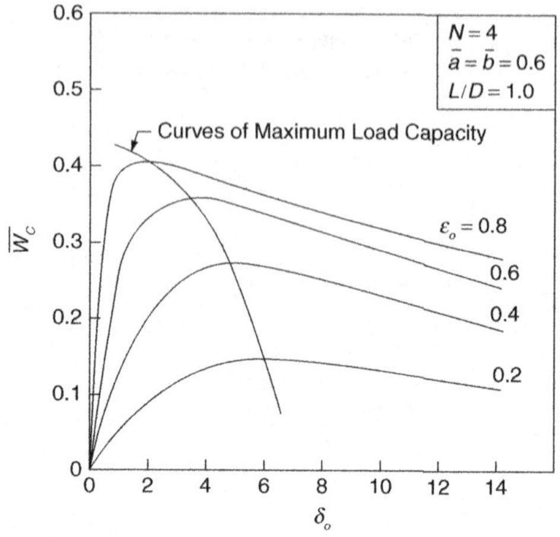


**Figure 9.12** | Load Capacity Versus Orifice Design Parameter

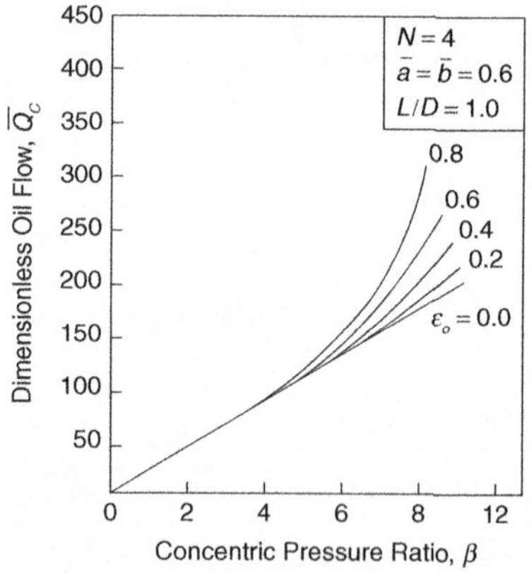


**Figure 9.13** | Oil Flow of Capillary Compensated Bearing Versus Pressure Ratio

(combined hydrostatic and hydrodynamic lubrication in which journal rotates) the results of recess pressures are given in Table 9.1. Results of load capacity, attitude angle, and friction factor are shown in Figs 9.15 to 9.17, respectively.

## 9.4.2 | Dynamic Behavior of Externally Pressurized Bearings

Dynamic characteristics of externally pressurized bearings have been the subject of several studies to determine their stiffness and damping characteristics. Dynamic response of rotors supported on externally pressurized

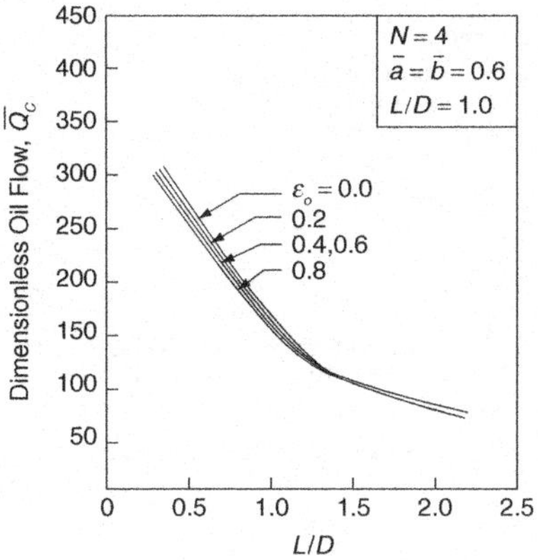


**Figure 9.14** | Oil Flow of Orifice Compensated Bearing Versus L/D Ratio

**Table 9.1** | Recess Pressures and Lubricant Flow for a Capillary Compensated Hybrid Bearing

$(N = 4,\ L/D = 1.0,\ \bar{a} = \bar{b} = 0.5,\ \beta = 0.6,\ \delta_c = 7.795)$

| $\varepsilon_0$ | $\Lambda$ | $\bar{p}_1$ | $\bar{p}_2$ | $\bar{p}_3$ | $\bar{p}_4$ | $\bar{Q}_b$ |
|---|---|---|---|---|---|---|
| 0.2 | 0.1 | 0.4934 | 0.5933 | 0.7187 | 0.5888 | 12.5656 |
| | 0.5 | 0.4928 | 0.6025 | 0.7187 | 0.5796 | 12.5603 |
| | 1.0 | 0.4927 | 0.6142 | 0.7187 | 0.5679 | 12.5587 |
| | 2.0 | 0.4927 | 0.6374 | 0.7187 | 0.5445 | 12.5581 |
| | 5.0 | 0.4929 | 0.7066 | 0.7187 | 0.4744 | 12.5573 |
| 0.5 | 0.1 | 0.3659 | 0.5543 | 0.8853 | 0.5435 | 12.9574 |
| | 0.5 | 0.3642 | 0.5773 | 0.8853 | 0.5219 | 12.9384 |
| | 1.0 | 0.3638 | 0.6051 | 0.8853 | 0.4939 | 12.9308 |
| | 2.0 | 0.3639 | 0.6603 | 0.8853 | 0.4380 | 12.9211 |
| | 5.0 | 0.3639 | 0.7483 | 0.8853 | 0.3557 | 12.9005 |

bearings depends very much on the dynamic properties of the bearings. Sternlicht and Elwell (1960), Brown (1961), Hunt (1964), Licht and Cooley (1964), Schwarzenback and Gill (1966), Rhode and Ezzat (1976), Ghosh (1978), Ghosh and Vishwanath (1987), Ghosh et al. (1989), and Ghosh and Majumdar (1982) investigated on the dynamic behavior of hydrostatic thrust and journal bearings. In the following sections, we will discuss the procedure to determine the dynamic behavior of hydrostatic bearings with orifice and capillary compensation.

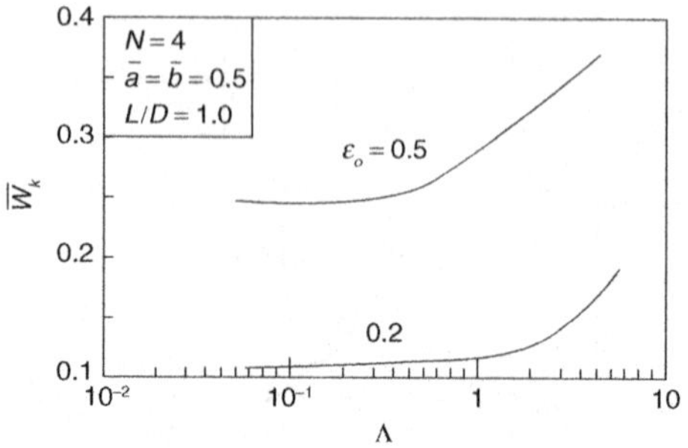


**Figure 9.15** | Load Capacity of Capillary Compensated Hybrid Bearing Versus Bearing Number

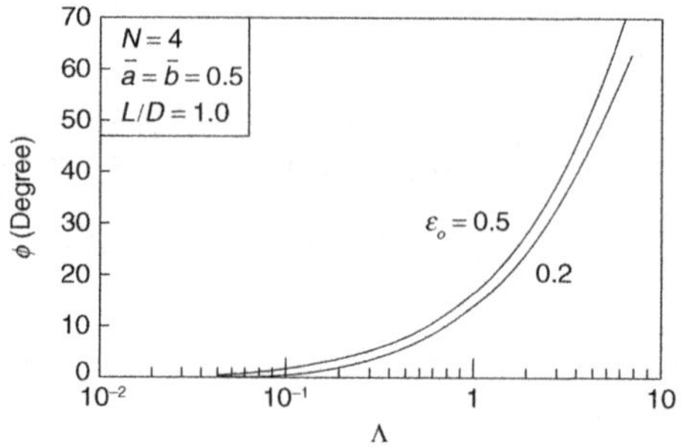


**Figure 9.16** | Attitude Angle of Capillary Compensated Hybrid Bearing Versus Bearing Number

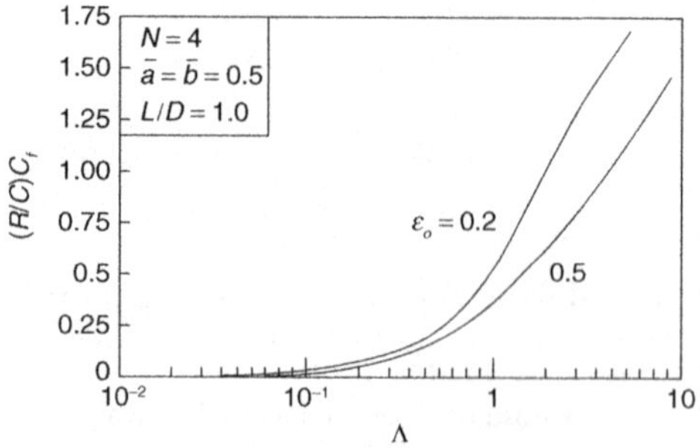


**Figure 9.17** | Friction Parameter of Capillary Compensated Hybrid Bearing Versus Bearing Number

### 9.4.3 | Dynamic Behavior of Compensated Hydrostatic Thrusts Bearings

A circular step thrust bearing is shown in Fig. 9.18. With the usual assumptions of an incompressible isothermal lubrication and following Dowson's (1961) approach which includes centrifugal fluid inertia, the momentum equations for the fluid film are written as

$$-\rho\frac{v_\theta^{\,2}}{r} = -\frac{\partial p}{\partial r} + \eta\frac{\partial^2 v_r}{\partial z^2} \tag{9.65}$$

$$\frac{\partial^2 v_\theta}{\partial z^2} = 0 \tag{9.66}$$

and the continuity equation is expressed as

$$\frac{1}{r}\frac{\partial}{\partial r}\left(rv_r\right) + \frac{\partial v_z}{\partial z} = 0 \tag{9.67}$$

$v_r$, $v_\theta$ and $v_z$; are flow velocity components in radial, circumferential, and normal directions, i.e., along $r$, $\theta$ and $z$ coordinates.

Integrating Equation (9.66) with following boundary conditions

$$z = 0,\ v_\theta = 0 \text{ and } z = h,\ v_\theta = r\Omega$$

where $\Omega$ is the angular speed of the runner in rad/sec, the tangential velocity, $v_\theta$ is given as

$$v_\theta = r\Omega z / h \tag{9.68}$$

Substituting Equation (9.68) into Equation (9.65) and integrating with the boundary conditions $v_r = 0$ at $z = 0$ and $z = h$, the radial velocity component of the fluid is obtained as

$$v_r = \frac{1}{2\eta}\frac{\partial p}{\partial r}\left(z^2 - zh\right) + \frac{\rho r\Omega^2}{12\eta h^2}\left(zh^3 - z^4\right) \tag{9.69}$$

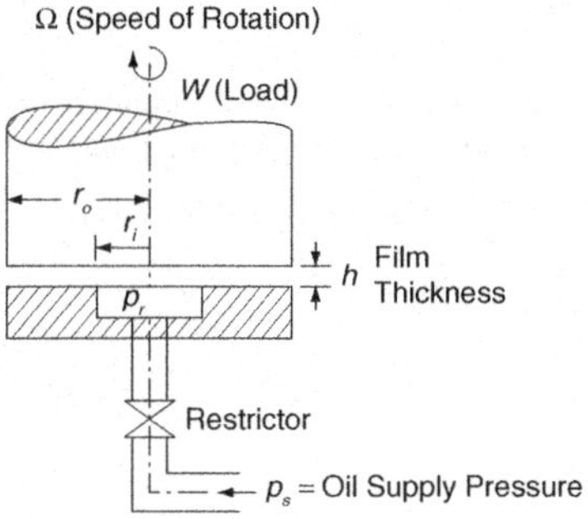


**Figure 9.18** | Hydrostatic Thrust Bearing System

Substituting Equations (9.69) into continuity Equation (9.67) and on integration with respect to $z$ within limits 0 to $h$ yields the Reynolds equation as given below

$$h^3 \frac{1}{r}\frac{\partial}{\partial r}\left(r\frac{\partial p}{\partial r}\right) = 0.6\rho\Omega^2 h^3 + 12\eta\frac{\partial h}{\partial t} \tag{9.70}$$

Parallel surfaces have been assumed with squeeze velocity $\frac{\partial h}{\partial t}$ for dynamic motion of the runner along the film thickness. Equation (9.70) is written in dimensionless form using the following substitutions

$$\bar{p} = p / p_s; \bar{r} = r / r_i; \bar{h} = h / h_0, \bar{p}_r = p_r / p_s \tag{9.71}$$

$h_0$, is steady state film thickness and $p_r$ recess pressure.

Equation (9.70) with above substitutions reduces to

$$\bar{h}^3 \frac{1}{\bar{r}}\frac{\partial}{\partial \bar{r}}\left(\bar{r}\frac{\partial \bar{p}}{\partial \bar{r}}\right) = S\bar{h}^3 + \sigma\frac{\partial \bar{h}}{\partial \tau} \tag{9.72}$$

where $S = 0.6\rho\Omega^2 r_i / p_s$, is defined as the speed parameter and

$$\sigma = 12\eta \upsilon r_i^2 / p_s h_0^2,$$

is defined as the squeeze number.

It has been assumed that the system undergoes small amplitude vibrations at a frequency $\upsilon$, and therefore film thickness and pressure can be expressed as

$$\bar{h} = 1 + \varepsilon e^{i\upsilon t}; \bar{p} = \bar{p}_0 + \varepsilon e^{i\upsilon t}\bar{p}_1; \tau = \upsilon t \tag{9.73}$$

where dimensionless steady state film pressure is $\bar{p}_0$ and $\bar{p}_1$ is the dimensionless dynamic film pressure.

Substituting Equation (9.73) into Equation (9.72) and retaining only up to first order terms, we get

$$\frac{1}{\bar{r}}\frac{\partial}{\partial \bar{r}}\left(\bar{r}\frac{\partial \bar{p}_0}{\partial r}\right) = S \tag{9.74}$$

$$\frac{1}{\bar{r}}\frac{\partial}{\partial \bar{r}}\left(\bar{r}\frac{\partial \bar{p}_1}{\partial \bar{r}}\right) = i\sigma \tag{9.75}$$

Integration of Equation (9.74) satisfying the following boundary conditions at the recess boundary and bearing periphery yields steady state film pressure distribution. Boundary conditions for Equation (9.74) are

$$\bar{p}_0 = \bar{p}_{r0} \text{ at } \bar{r} = 1; \bar{p}_0 = 0 \text{ at } \bar{r} = \bar{r}_o \tag{9.76}$$

Steady state pressure distribution is obtained as

$$\bar{p}_0 = \bar{p}_{r0} + \frac{S}{4}(\bar{r}^2 - 1) - \left[\frac{S}{4}(\bar{r}_o^2 - 1) + \bar{p}_{r0}\right]\frac{\ln \bar{r}}{\ln \bar{r}_o} \tag{9.77}$$

Steady state load capacity is obtained as

$$W = \bar{p}_{r0}\pi r_i^2 p_s + r_i^2 p_s \pi \int_1^{\bar{R}} \bar{p}_0 \bar{r} d\bar{r} \tag{9.78}$$

This is determined in dimensionless form as

$$\bar{W} = 0.5\left(\bar{r}_o - 1^2\right)\left\{\frac{\bar{p}_{r0}}{\ln \bar{r}_o} - \frac{S}{4}\left[\bar{r}_o^2 + 1 - \frac{\left(\bar{r}_o^2 - 1\right)}{\ln \bar{r}_o}\right]\right\} \tag{9.79}$$

where,

$$\bar{W} = W / \pi r_i^2 p_s$$

The oil flow from the recess boundary is determined as

$$Q_{r0} = -\int_0^{2\pi} \frac{h^3}{12\eta} \frac{\partial p_0}{\partial r}\bigg|_{r=r_i} r d\theta \tag{9.80}$$

It is expressed in the dimensionless form as

$$\bar{Q}_{r0} = -\frac{\partial \bar{p}_0}{\partial \bar{r}}\bigg|_{\bar{r}=1} = -\left[\frac{S}{2} - \frac{S}{4}\left(\bar{r}_o^2 - 1\right) / \ln \bar{r}_o\right] + \frac{\bar{p}_{r0}}{\ln \bar{r}_o} \tag{9.81}$$

where

$$\bar{Q}_{r0} = -\frac{6\eta Q_{r0}}{p_s h_0^3 \pi}$$

Friction torque can be determined as

$$M = \int_0^{r} \int_0^{2\pi} \frac{\eta r \Omega}{h} r^2 d\theta dr \tag{9.82}$$

$$M = \left(\frac{\pi \Omega r_i^4}{2h_0}\right) \bar{r}_o^4 \tag{9.83}$$

Equation (9.79), (9.81), and (9.83) give values of load capacity, oil flow rate, and frictional torque of the bearing at film thickness $h_0$.

Dynamic characteristics of the bearing, i.e., stiffness and damping coefficients can be determined from the solution of dynamic Reynolds Equation (9.75) satisfying following boundary conditions.

$$\bar{p}_1 = \bar{p}_{r1} \text{ at } \bar{r} = 1, \ \bar{p}_1 = 0 \text{ at } \bar{r} = \bar{r}_o \tag{9.84}$$

Dynamic pressure distribution $\bar{p}_1$ is obtained on integration of Equation (9.75) satisfying above boundary condition as

$$\bar{p}_1 = \bar{p}_{r1} + i\frac{\sigma}{4}\left(\bar{r}^2 - 1\right) - \left[i\frac{\sigma}{4}\left(\bar{r}_o^2 - 1\right) + \bar{p}_{r1}\right]\frac{\ln \bar{r}}{\ln \bar{r}_o} \tag{9.85}$$

Dynamic load capacity of the bearing can be determined as

$$W_d = \pi r_i^2 p_{r1} + 2\pi \int_{r_i}^{r_o} p_1 r dr \tag{9.86}$$

$$\overline{W}_d = \overline{p}_{r1} + 2\int_1^{\overline{r}_o} \overline{p}_1 \overline{r} d\overline{r} \tag{9.87}$$

This is determined in the dimensionless form from the above expression substituting for $\overline{p}_1$ from Equation (9.86) as

$$\overline{W}_d = 0.5\left(\overline{r}_o^2 - 1\right)\left\{\frac{\overline{p}_{r1}}{\ln \overline{r}_o} - i0.25\sigma\left[\overline{r}_o^2 + -\frac{\left(\overline{r}_o^2 - 1\right)}{\ln \overline{r}_o}\right]\right\} \tag{9.88}$$

where

$$\overline{W}_d = W_d / \pi r_i^2 p_s$$

Dynamic load capacity can be expressed in the form of stiffness and damping coefficients as

$$\varepsilon e^{i\upsilon t} W_d = -K_d \varepsilon h_0 e^{i\upsilon t} - i\upsilon B_d h_0 \varepsilon e^{i\upsilon t} \tag{9.89}$$

or

$$\overline{W}_d = -\left(\frac{K_d h_0}{\pi r_i^2 p_s}\right) - i\left(\frac{\upsilon B_d h_0}{\pi r_i^2 p_s}\right)$$

$K_d$, $B_d$ are stiffness and damping coefficients of the bearing, respectively, which can be written in the dimensionless form as

$$\begin{aligned}\overline{K}_d &= -\mathrm{Re}\left(\overline{W}_d\right) = \frac{K_d h_0}{\pi r_i^2 p_s} \\ \overline{B}_d &= -\frac{\mathrm{Im}\left(\overline{W}_d\right)}{\sigma} = \frac{\upsilon B_d h_0}{\left(\pi r_i^2 p_s\right)\sigma} = \left(\frac{B_d h_0^3}{12\eta\pi r_i^4}\right)\end{aligned} \tag{9.90}$$

The steady state recess pressure $\overline{p}_{r0}$ and dynamic recess pressure $\overline{p}_{r1}$ are determined from the recess flow continuity equation as described below:

Recess flow continuity equation can be written in the dimensionless form including the recess volume fluid compressibility effect as:

### *Capillary Compensation*

$$\delta_c\left(1 - \overline{p}_r\right) = \overline{Q}_r \overline{p}_r + \frac{12\eta}{h_0^3 p_s} A_r \frac{\partial h}{\partial t} + \frac{12\eta}{h_0^3 p_s}\frac{1}{\rho}\frac{\partial}{\partial t}\left(\rho V_0\right)$$

Or

$$\delta_c\left(1 - \overline{p}_r\right) = \overline{Q}_r \overline{p}_r + \pi\sigma\frac{\partial \overline{h}}{\partial \tau} + \pi\sigma\gamma\frac{\partial \overline{p}}{\partial \tau} \tag{9.91}$$

### Orifice Compensation

$$\delta_0\left(1-\bar{p}_r\right)^{0.5}=\bar{Q}_r\bar{p}_r+\frac{12\eta A_r}{h_0^3 p_s}\frac{\partial h}{\partial t}+\frac{12\eta}{h_0^3 p_s}\frac{1}{\rho}\frac{\partial}{\partial t}\left(\rho V_0\right)$$

or (9.92)

$$\delta_0\left(1-\bar{p}_r\right)^{0.5}=\bar{Q}_r\bar{p}_r+\pi\sigma\frac{\partial\bar{h}}{\partial\tau}+\pi\sigma\gamma\frac{\partial\bar{p}}{\partial\tau}$$

where $A_r=\pi r_i^2$ is recess area and $V_0=A_r d_r+V_1$ is volume of fluid contained between the recess and the restrictor. $d_r$ is recess depth, $V_1$ is the volume between the restrictor to the entrance of the recess. Fluid in the recess volume experiences compressibility effect under dynamic conditions when recess pressure is a harmonic function of time. Thus, $\gamma$ is defined as the recess volume compressibility parameter and is given as

$$\gamma=\bar{V}_0\beta p_s\ ;\ \bar{V}=V_0/A_r h_0 \text{ and } \beta=\frac{1}{\rho}\left(\frac{\partial\rho}{\partial p}\right);$$

$\beta$ is the bulk modulus of the fluid.

For small amplitude oscillations, first order perturbations are valid, therefore, one can express film thickness and recess pressure as

$$\bar{h}=1+\varepsilon e^{i\tau};\bar{p}_r=\bar{p}_{r0}+\varepsilon e^{i\tau}p_{r1} \tag{9.93}$$

Substituting Equation (9.93) into Equation (9.91) and (9.92), one can obtain:

### For Capillary Compensation

$$\bar{p}_{r0}=\delta_c/\lambda$$

where

$$\lambda=\left(\bar{Q}_{r0}+\delta_c\right)$$

### For Orifice Compensation

$$\bar{p}_{r0}=\frac{1}{2}\left[-\left(\frac{\delta_0}{Q_{r0}}\right)^2+\sqrt{\left(\frac{\delta_0}{Q_{r0}}\right)^4+4.0\left(\frac{\delta_0}{Q_{r0}}\right)^2}\right] \tag{9.94}$$

Dynamic recess pressure $\bar{p}_{r1}$ as

$$\bar{p}_{r1}=\frac{\left[\bar{Q}_{r1}\bar{p}_{r0}\lambda+(\pi\sigma)^2\gamma\right]+i\left\{\pi\sigma\lambda-Q_{r1}\bar{p}_{r0}\pi\sigma\gamma\right\}}{\lambda^2+(\pi\sigma\gamma)^2} \tag{9.95}$$

where $\bar{Q}_{r1}=\frac{\partial\bar{Q}_r}{\partial\bar{h}}\Big|_{\bar{h}=1}$ and $\lambda=\bar{Q}_{r0}+0.5\delta_0\left(1-\bar{p}_{r0}\right)^{0.5}$

Here, $\delta_c=\frac{3\pi d_c^4}{32h_0^3 l_c}$ and $\delta_0=\frac{3\pi c_d d_0^2\sqrt{2}}{h_0^3\left(\rho p_s\right)^{1/2}}$ are capillary and orifice design parameters defined earlier in Equations (9.91) and (9.92).

Results of this analysis for stiffness and damping coefficients are shown in Figs 9.19a and 9.19b, respectively. Stiffness coefficient $\bar{K}_d$ increases with squeeze parameter $\sigma$ for $\sigma > 50$ and is dependent on recess volume compressibility parameter $\gamma$. Stiffness increases with $\gamma$.

On the contrary, damping coefficient $\bar{B}_d$ decreases with frequency or squeeze parameter as well as with compressibility parameter $\gamma$. For $\beta = 0$, dimensionless stiffness and damping coefficients reduce to its static value and is independent of $\sigma$. Thus, fluid compressibility effect in the recess volume results in frequency dependence of both stiffness and damping coefficient of the bearing.

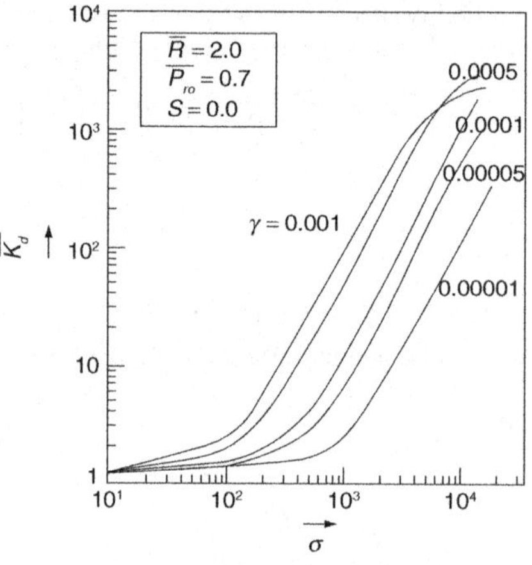


**Figure 9.19(a)** | Dimensionless Stiffness Coefficient Versus Squeeze Number for Various Recess Parameters for Capillary Compensated Bearings

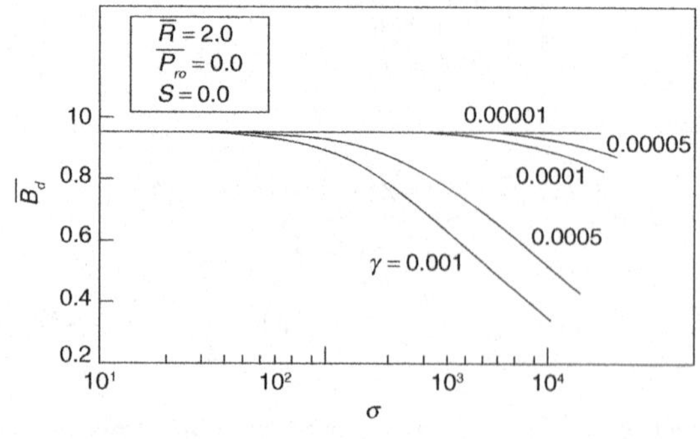


**Figure 9.19(b)** | Dimensionless Damping Coefficient Versus Squeeze Number for Various Recess Parameters for Orifice Compensated Bearings

## 9.5 | A General Analysis of Dynamic Characteristics of Multirecess Externally Pressurized Journal Bearings with Large Sills

Dynamic behavior of multirecess hydrostatic journal bearings has been investigated by several researchers. Preliminary investigations were made by Davies and Leonard (1970) and Leonard and Rowe (1973) for bearings with short lands. Later studies concentrated on the dynamic performance of bearings with finite sill dimensions. Most of investigations determined stiffness and damping coefficients of hydrostatic/hybrid bearings and the stability of the bearing. Noteworthy amongst the studies are those of Singh *et al.* (1979), Ghosh and Majumdar (1978), Ghosh *et al.* (1979), Ghosh and Vishwanath (1987), and Ghosh et al. (1989) in the laminar flow regime. Investigations were also made in the turbulent flow regime by Heller (1974), Artiles *et al.* (1982), Braun *et al.* (1985), and San Andres (1990).

A general procedure adopted by Ghosh *et al.* (1978, 1989) will be discussed here. The procedure is suitable for both hydrostatic and hybrid bearings in laminar flow.

The bearing configuration is shown in Fig. 9.7, whereas vibrating hybrid bearing configuration is shown in Fig. 9.20.

To determine the dynamic characteristics of a bearing, it is assumed that the journal rotates about its own axis and also undergoes small amplitude oscillations about its steady state equilibrium position given by $(\varepsilon_0, \phi_0)$.

For an isoviscous, incompressible lubricant, the Reynolds equation for flow of lubricant on the bearing lands is given as:

$$\frac{\partial}{\partial x}\left(\frac{h^3}{\eta}\frac{\partial p}{\partial x}\right)+\frac{\partial}{\partial y}\left(\frac{h^3}{\eta}\frac{\partial p}{\partial y}\right)=6U\frac{\partial h}{\partial x}+12\frac{\partial h}{\partial t} \tag{9.96}$$

The above equation is written in the dimensionless form as $\tau = vt$ with the following substitutions

$$\bar{h}=\frac{h}{C},\ \theta = x/R,\ U = R\omega$$
$$\bar{p} = p/p_s \text{ and } \tau = vt,\ \bar{y} = y/L/2$$

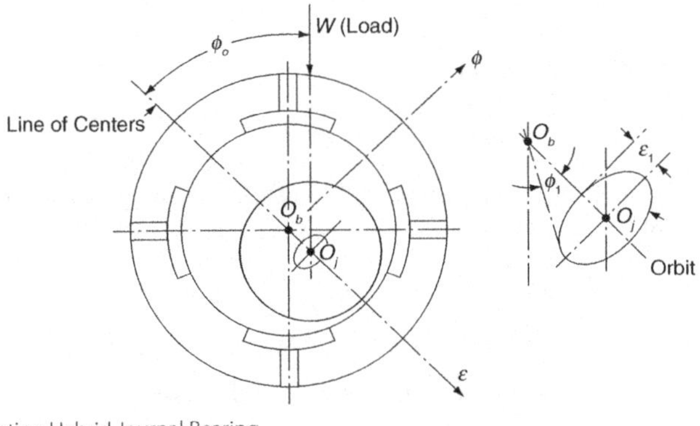


**Figure 9.20** | Vibrating Hybrid Journal Bearing

where $\upsilon$ is the frequency of vibration of the journal center about its equilibrium position.

Thus, Equation (9.96) becomes

$$\frac{\partial}{\partial\theta}\left(\bar{h}^3\frac{\partial\bar{p}}{\partial\theta}\right)+\left(\frac{D}{L}\right)^2\frac{\partial}{\partial\bar{y}}\left(\bar{h}^3\frac{\partial\bar{p}}{\partial\bar{y}}\right)=\Lambda\frac{\partial\bar{h}}{\partial\theta}+\sigma\frac{\partial\bar{h}}{\partial\theta} \tag{9.97}$$

where

$$\Lambda=\frac{6\eta\omega}{p_s(C/R)^2}$$

is defined as bearing number and

$$\sigma=12\eta\upsilon/p_s\left(\frac{C}{R}\right)^2$$

is defined as squeeze number or frequency parameter.

First order perturbations can be used to express dynamic pressures and film thickness as

$$\begin{aligned}\bar{p}&=\bar{p}_0+\varepsilon_1 e^{i\tau}\bar{p}_1+\varepsilon_0\phi_1 e^{i\tau}\bar{p}_2\\ \bar{h}&=\bar{h}_0+\varepsilon_1 e^{i\tau}\cos\theta+\varepsilon_0\phi_1 e^{i\tau}\sin\theta\end{aligned} \tag{9.98}$$

The above expressions are written with the assumption that

$$\begin{aligned}\varepsilon&=\varepsilon_0+\varepsilon_1 e^{i\tau}\\ \phi&=\phi_0+\phi_1 e^{i\tau}\end{aligned} \tag{9.99}$$

where

$$\varepsilon_1<<\varepsilon_0 \text{ and } \phi_1<<\phi_0$$

The steady state film thickness $\bar{h}_0$ is given as

$$\bar{h}_0=1+\varepsilon_0\cos\theta \tag{9.100}$$

Substituting Equation (9.98) into Equation (9.97) and collecting terms only up to first order, we obtain the following equations:

### *Zeroth Order*

$$\frac{\partial}{\partial\theta}\left(\bar{h}_0^{\,3}\frac{\partial\bar{p}_0}{\partial\theta}\right)+\left(\frac{D}{L}\right)^2\frac{\partial}{\partial\bar{y}}\left(\bar{h}_0^{\,3}\frac{\partial\bar{p}_0}{\partial\bar{y}}\right)=\Lambda\frac{\partial\bar{h}_0}{\partial\theta} \tag{9.101}$$

### First Order

$$\frac{\partial}{\partial\theta}\left(\bar{h}_0^3\frac{\partial\bar{p}_1}{\partial\theta}\right)+\frac{\partial}{\partial\theta}\left(3\bar{h}_0^3\cos\theta\frac{\partial\bar{p}_0}{\partial\theta}\right)+\left(\frac{D}{L}\right)^2\frac{\partial}{\partial\bar{y}}\left(\bar{h}_0^3\frac{\partial\bar{p}_1}{\partial\bar{y}}\right)$$
$$+\left(\frac{D}{L}\right)^2\frac{\partial}{\partial\bar{y}}\left(3\bar{h}_0^3\cos\theta\frac{\partial\bar{p}_0}{\partial\bar{y}}\right)=i\sigma\cos\theta-\Lambda\sin\theta \tag{9.102}$$

and

$$\frac{\partial}{\partial\theta}\left(\bar{h}_0^3\frac{\partial\bar{p}_2}{\partial\theta}\right)+\frac{\partial}{\partial\theta}\left(3\bar{h}_0^3\sin\theta\frac{\partial\bar{p}_0}{\partial\theta}\right)+\left(\frac{D}{L}\right)^2\frac{\partial}{\partial\bar{y}}\left(\bar{h}_0^3\frac{\partial\bar{p}_2}{\partial\bar{y}}\right)$$
$$+\left(\frac{D}{L}\right)^2\frac{\partial}{\partial\bar{y}}\left(3\bar{h}_0^3\sin\theta\frac{\partial\bar{p}_0}{\partial\bar{y}}\right)=i\sigma\sin\theta+\Lambda\cos\theta \tag{9.103}$$

The appropriate boundary conditions for the above equations are expressed as

$$\bar{p}_i(\theta,-1)=\bar{p}_i(\theta,+1)=0,\ \text{for ambient pressure at the bearing edges}$$
$$\frac{\partial\bar{p}_i}{\partial\bar{z}}(\theta,0)=0,\ \text{for symmetry of pressure about the bearing center line}$$
$$\bar{p}_i=\bar{p}_{ri},\ \text{at the } r\text{th recess} \tag{9.104}$$
$$\bar{p}_i(\theta,\bar{y})=\bar{p}_i(\theta+2\pi,\bar{y}),\ \text{for cyclic continuity}$$

where $i = 0, 1, 2$ for Equations (9.101), (9.102), and (9.103), respectively.

The above equations are written in the finite difference form and solved using Gauss–Seidel iterative procedure with an over relaxation factor in a high speed digital computer until convergence is obtained within a prescribed error criterion.

However, the solution of the above equations is dependent on the condition that recess pressures, i.e., steady state and dynamic pressures $\bar{p}_{r0}$, $\bar{p}_{r1}$ and $\bar{p}_{r2}$ are known for a prescribed steady state position of the journal center.

Methodology to determine the recess pressures $(\bar{p}_{r0})$ under steady state eccentric position of the journal center given by $\varepsilon_0, \phi_0$ has been described in detail in Section 9.4.1, and one of the procedures described there can be used to determine steady state recess pressures $p_{r0}$, $r = 1, 2, \ldots N$.

To determine dynamic recess pressures $\bar{p}_{r1}$ and $\bar{p}_{r2}$ recess flow continuity equation for capillary and orifice compensation developed by Ghosh *et al.* (1987 and 1989) is adopted as described below:

### For Capillary Compensation

$$\frac{\pi d_c^4}{128 l_c \eta}(p_s p_r)=Q_r+\frac{\partial}{\partial t}(A_r h_r)+\frac{1}{\rho}\frac{\partial}{\partial t}(\rho V_0) \tag{9.105}$$

where $A_r$ = recess area
$V_0$ = recess volume
$h_r$ = film thickness at the center of $r$th recess
$Q_r$ = flow from $r$th recess

The above equation is expressed in the dimensionless form by multiplying the equation by $12\eta C^3 / p_s$ as

$$\delta_c\left(1-\bar{p}_r\right)=\bar{Q}_r\bar{p}_r+\frac{\partial}{\partial\tau}\left(A_r\bar{h}_r\right)\frac{12\eta\upsilon}{C^2 p_s}+\frac{12\eta\upsilon}{C^3 p_s}\frac{1}{\rho}\frac{\partial}{\partial\tau}\left(\rho V_0\right) \tag{9.106}$$

Similarly, recess flow continuity equation for $r$th recess for orifice compensation is written as

$$\frac{\pi d_0^2}{2\sqrt{2\rho}}\left(p_s-p_r\right)^{0.5}=Q_r+\frac{\partial}{\partial t}\left(A_r h_r\right)+\frac{1}{\rho}\frac{\partial}{\partial t}\left(\rho V_0\right) \tag{9.107}$$

In the dimensionless form, it reduces to

$$\delta_0\left(1-\bar{p}_r\right)^{1/2}=\bar{Q}_r\bar{p}_r+\frac{\partial}{\partial\tau}\left(A_r\bar{h}_r\right)\frac{12\eta\upsilon}{C^2 p_s}+\frac{12\eta\upsilon}{C^3 p_s}\frac{1}{\rho}\frac{\partial}{\partial\tau}\left(\rho V_0\right) \tag{9.108}$$

In the above expressions, pressure drop due to fluid inertia effect at the recess edges have been neglected. It has been quite convincingly demonstrated by Heller (1974) that although fluid inertia effect at the recess edges grossly affect bulk flow rate at high Reynolds number flow regime, the bearing performance parameters are not affected significantly.

The dynamic recess pressures $\bar{p}_{r1}$ and $\bar{p}_{r2}$ can be evaluated following perturbation theory for small amplitude oscillations of $\bar{p}_r$, $\bar{Q}_r$ and $\bar{h}_r$ as given below

$$\begin{aligned}
\bar{p}_r &= \bar{p}_{r0}+\varepsilon_1 e^{i\tau}\bar{p}_{r1}+\varepsilon_0\phi_1 e^{i\tau}\bar{p}_2\\
\bar{Q}_r &= \bar{Q}_{r0}+\varepsilon_1 e^{i\tau}\bar{Q}_{r1}+\varepsilon_0\phi_1 e^{i\tau}Q_{r2}\\
\bar{h}_r &= \bar{h}_{r0}+\varepsilon_1 e^{i\tau}\cos\theta_r+\varepsilon_0\phi_1 e^{i\tau}\sin\theta_r
\end{aligned} \tag{9.109}$$

where

$$\bar{h}_{r0}=\left(1+\varepsilon_0\cos\theta_r\right),\ \bar{Q}_{r1}=\left.\frac{\partial\bar{Q}_r}{\partial\varepsilon}\right|_{\varepsilon=\varepsilon_0}\ \text{and}\ \bar{Q}_{r2}=\left.\frac{1}{\varepsilon_0}\frac{\partial\bar{Q}_r}{\partial\phi}\right|_{\phi=\phi_0}$$

The dynamic recess pressures $\bar{p}_{r1}$ and $\bar{p}_{r2}$ are obtained by substituting Equation (9.109) into Equations (9.106) and (9.108). For capillary and orifice compensation, respectively, $\bar{p}_{r1}$ and $\bar{p}_{r2}$ are obtained as

$$\begin{aligned}
\bar{p}_{r1} &= -\frac{\bar{Q}_{r1}\bar{p}_{r0}\lambda+\psi^2\gamma\cos\theta_r}{\left(\lambda^2+\psi^2\gamma^2\right)}-\\
&\quad i\frac{\psi\lambda\cos\theta_r-\bar{Q}_{r1}\bar{p}_{r0}\psi\gamma}{\left(\lambda^2+\psi^2\gamma^2\right)}
\end{aligned} \tag{9.110}$$

and

$$\begin{aligned}
\bar{p}_{r2} &= -\frac{\bar{Q}_{r2}\bar{p}_{r0}\lambda+\psi^2\gamma\sin\theta_r}{\left(\lambda^2+\psi^2\gamma^2\right)}-\\
&\quad i\frac{\psi\lambda\sin\theta_r-\bar{Q}_{r1}\bar{p}_{r0}\psi\gamma}{\left(\lambda^2+\psi^2\gamma^2\right)}
\end{aligned} \tag{9.111}$$

where

$$\delta_c = 3\pi d_c^4 / 32C^3 l_c$$

$$\delta_0 = 3\pi c_d d_0^2 \eta / (\rho p_s / 2)^{1/2}$$

$$\gamma = \bar{V}_0 \beta p_s,\ \bar{V}_0 = V_0 / A_r C$$

$$\psi = \frac{\sigma A_r}{R^2}, \sigma = 12\eta \upsilon / p_s (C/R)^2$$

$$\bar{Q}_r = \frac{12\eta Q_r}{C^3 p_r}, \bar{Q}_{r0} = \frac{12\eta Q_{r0}}{C^3 p_{r0}}$$

The steady state recess pressures are determined for capillary and orifice compensations, respectively, as follows:

**For Capillary Compensation**

$$\bar{p}_{r0} = \frac{\delta_c}{\lambda}$$

$$\lambda = \bar{Q}_{r0} + \delta_c \tag{9.112a}$$

**For Orifice Compensation**

$$\bar{p}_{r0} = -0.5\left(\frac{\delta_0}{\bar{Q}_{r0}}\right)^2 + 0.5\left[\left(\frac{\delta_0}{\bar{Q}_{r0}}\right)^4 + 4\left(\frac{\delta_0}{\bar{Q}_{r0}}\right)^2\right]^{1/2} \tag{9.112b}$$

$$\lambda = 0.5\delta_0 (1 - \bar{p}_{r0})^{-0.5} + \bar{Q}_{r0}$$

### 9.5.1 | Stiffness and Damping Characteristic of the Bearing

The components of the restoring dynamic load due to dynamic film pressure $\varepsilon_1 e^{i\tau} p_1$ can be determined along $\varepsilon$ and $\phi$ directions, i.e., along the line of centers and normal to it as

$$(W_1)_\varepsilon e^{i\tau} = -2\int_0^{L/2}\int_0^{2\pi} \varepsilon_1 e^{i\tau} p_1 R \cos\theta \, d\theta \, dy$$

$$(W_1)_\phi e^{i\tau} = -2\int_0^{L/2}\int_0^{2\pi} \varepsilon_1 e^{i\tau} p_1 R \sin\theta \, d\theta \, dy \tag{9.113}$$

The dynamic load resulting from small periodic pressure $\varepsilon_1 e^{i\tau} p_1$ can be expressed in terms of linear spring and damping coefficients as

$$(W_1)_\varepsilon e^{i\tau} = -K_{\varepsilon\varepsilon} y_\varepsilon - B_{\varepsilon\varepsilon} \frac{dy_\varepsilon}{dt}$$

$$(W_1)_\phi e^{i\tau} = -K_{\phi\varepsilon} y_\varepsilon - B_{\phi\varepsilon} \frac{dy_\varepsilon}{dt} \tag{9.114}$$

where the amplitude of oscillation of the journal center along the direction of eccentricity is given by

$$y_\varepsilon = C\varepsilon_1 e^{i\tau} \tag{9.115}$$

Substituting Equation (9.115) into Equations (9.114), we obtain

$$\begin{aligned}(\bar{W}_1)_\varepsilon &= \frac{(W_1)_\varepsilon}{LDp_s\varepsilon_1} = -\frac{K_{\varepsilon\varepsilon}C}{LDp_s} - i\frac{B_{\varepsilon\varepsilon}Cv}{LDp_s}\\ (\bar{W}_1)_\phi &= \frac{(W_1)_\phi}{LDp_s\varepsilon_1} = -\frac{K_{\phi\varepsilon}C}{LDp_s} - i\frac{B_{\phi\varepsilon}Cv}{LDp_s}\end{aligned} \tag{9.116}$$

$K_{\varepsilon\varepsilon}$, $K_{\phi\varepsilon}$ are direct and cross stiffness coefficients, respectively, and $B_{\varepsilon\varepsilon}$ and $B_{\phi\varepsilon}$ are direct and cross damping coefficients, respectively. Stiffness and damping can be expressed in dimensionless form as

$$\begin{aligned}k_{\varepsilon\varepsilon} &= -\mathrm{Re}\{(\bar{W}_1)_\varepsilon\} = \frac{K_{\varepsilon\varepsilon}C}{LDp_s}\\ k_{\phi\varepsilon} &= -\mathrm{Re}\{(\bar{W}_1)_\phi\} = \frac{K_{\phi\varepsilon}C}{LDp_s}\\ b_{\varepsilon\varepsilon} &= -\frac{\mathrm{Im}\{(\bar{W}_1)_\varepsilon\}}{\sigma} = \frac{B_{\varepsilon\varepsilon}}{24\eta L(R/C)^3}\\ b_{\phi\varepsilon} &= -\frac{\mathrm{Im}\{(\bar{W}_1)_\phi\}}{\sigma} = \frac{B_{\phi\varepsilon}}{24\eta L(R/C)^3}\end{aligned} \tag{9.117}$$

Similarly, the components of restoring dynamic load due to perturbed film pressure $\varepsilon_0\phi_1 e^{i\tau} p_2$ can be written as

$$\begin{aligned}(W_2)_\varepsilon e^{i\tau} &= -2\int_0^{L/2}\int_0^{2\pi} \varepsilon_0\phi_1 e^{i\tau} p_2 R \cos\theta d\theta dy\\ (W_2)_\phi e^{i\tau} &= 2\int_0^{L/2}\int_0^{2\pi} \varepsilon_0\phi_1 e^{i\tau} p_2 R \sin\theta d\theta dy\end{aligned} \tag{9.118}$$

The dynamic load can be expressed in terms of linear spring and damping coefficients as

$$\begin{aligned}(W_2)_\varepsilon e^{i\tau} &= -K_{\varepsilon\phi} y_\phi - B_{\varepsilon\phi}\frac{dy_\phi}{dt}\\ (W_2)_\phi e^{i\tau} &= -K_{\phi\phi} y_\phi - B_{\phi\phi}\frac{dy_\phi}{dt}\end{aligned} \tag{9.119}$$

The amplitude of oscillation of the journal center along the direction normal to the line of centers is given as

$$y_\phi = C\varepsilon_0\phi_1 e^{i\tau} \tag{9.120}$$

Substituting Equation (9.115) into Equation (9.114), we obtain

$$\begin{aligned}(\bar{W}_2)_\varepsilon &= \frac{(W_2)_\varepsilon}{LDp_s\varepsilon_0\phi_1} = -\frac{K_{\varepsilon\phi}C}{LDp_s} - i\frac{B_{\varepsilon\phi}Cv}{LDp_s}\\ (\bar{W}_2)_\phi &= \frac{(\bar{W}_2)_\phi}{LDp_s\varepsilon_0\phi_1} = -\frac{K_{\phi\phi}C}{LDp_s} - i\frac{B_{\phi\phi}Cv}{LDp_s}\end{aligned} \tag{9.121}$$

$K_{rt}$ and $K_{tt}$, are spring coefficients and $B_{rt}$ and $B_{tt}$ are damping coefficients of the fluid film which can be expressed in the dimensionless form as

$$\begin{aligned}k_{\varepsilon\phi} &= -\mathrm{Re}\{(\bar{W}_2)_\varepsilon\} = \frac{K_{\varepsilon\phi}C}{LDp_s}\\ k_{\phi\phi} &= -\mathrm{Re}\{(\bar{W}_2)_\phi\} = \frac{K_{\phi\phi}C}{LDp_s}\\ b_{\varepsilon\phi} &= -\frac{\mathrm{Im}\{(\bar{W}_2)_\varepsilon\}}{\sigma} = \frac{B_{\varepsilon\phi}}{24\eta L(R/C)^3}\\ b_{\phi\phi} &= -\frac{\mathrm{Im}\{(\bar{W}_2)_\phi\}}{\sigma} = \frac{B_{\phi\phi}}{24\eta L(R/C)^3}\end{aligned} \tag{9.122}$$

In case the journal is not rotating, the Reynolds Equation (9.97) can be rewritten as

$$\frac{\partial}{\partial\theta}\left(\bar{h}^3\frac{\partial\bar{p}}{\partial\theta}\right) + \left(\frac{D}{L}\right)^2\frac{\partial}{\partial\bar{y}}\left(\bar{h}^3\frac{\partial\bar{p}}{\partial\bar{y}}\right) = \sigma\frac{\partial\bar{h}}{\partial\tau} \tag{9.123}$$

The dynamic pressure and film thickness can be expressed in the dimensionless form assuming the oscillations to be harmonic and in one plane only as

$$\begin{aligned}&\bar{p} = \bar{p}_0 + \varepsilon_1 e^{i\tau}\bar{p}_1 \text{ and } \bar{h} = \bar{h}_0 + \varepsilon_1 e^{i\tau}\cos\theta\\ &\varepsilon = \varepsilon_0 + \varepsilon_1 e^{i\tau}\end{aligned} \tag{9.124}$$

Substituting the above equation into Equation (9.123) and collecting only zeroth and first order terms following equations can be obtained as:

$$\varepsilon_1^0 : \frac{\partial}{\partial\theta}\left(\bar{h}_0^3\frac{\partial\bar{p}}{\partial\theta}\right) + \left(\frac{D}{L}\right)^2\frac{\partial}{\partial\bar{y}}\left(\bar{h}_0^3\frac{\partial\bar{p}_0}{\partial\bar{y}}\right) = 0 \tag{9.125}$$

and

$$\begin{aligned}\varepsilon_1^1 : &\frac{\partial}{\partial\theta}\left(\bar{h}_0^3\frac{\partial\bar{p}_1}{\partial\theta}\right) + \frac{\partial}{\partial\theta}\left(3\bar{h}_0^3\cos\theta\frac{\partial\bar{p}_0}{\partial\bar{y}}\right) + \left(\frac{D}{L}\right)^2\frac{\partial}{\partial\bar{y}}\left(\bar{h}_0^3\frac{\partial\bar{p}_1}{\partial\bar{y}}\right)\\ &+ \left(\frac{D}{L}\right)^2\frac{\partial}{\partial\bar{y}}\left(3\bar{h}_0^3\cos\theta\frac{\partial\bar{p}_0}{\partial\bar{y}}\right) = i\sigma\cos\theta\end{aligned} \tag{9.126}$$

The dynamic recess pressure boundary condition as given by Equation (9.110) in terms of $\bar{p}_{r1}$ would describe the boundary condition for Equation (9.126). For Equation (9.125) the boundary conditions given by Equation (9.112) would be used. Stiffness and damping coefficients are determined from dynamic load given as

$$W_d = -2\int_0^{L/2} \int_0^{2\pi} p_1 R \cos\ \theta\ d\theta\, dy \tag{9.127}$$

or

$$\varepsilon_1 e^{i\tau} W_d = -Ky - B\frac{dy}{dt};\ \tau = \upsilon t \tag{9.128}$$

where

$$y = C\varepsilon_1 e^{i\tau} \tag{9.129}$$

Substituting Equation (9.129) into Equation (9.128), we get

$$W_d = -KC - iBC\upsilon$$

which can be expressed in as the dimensionless form

$$\bar{W}_d = \frac{W_d}{LDp_s} = -\frac{KC}{LDp_s} - i\frac{BC\upsilon}{LDp_s} \tag{9.130}$$

Dimensionless stiffness of the bearing $K_S$ is defined as:

$$K_S = -\text{Re}\left(\bar{W}_d\right) = \frac{KC}{LDp_s} \tag{9.131}$$

and dimensionless damping coefficient, $B_S$ as :

$$B_S = -\frac{\text{Im}\left(\bar{W}_d\right)}{\sigma} = \frac{B}{24\eta L(R/C)^3} \tag{9.132}$$

Neglecting fluid compressibility effect in the recess volume, i.e., for $\gamma = 0$, stiffness $K_S$ and damping $B_S$ characteristics of capillary and orifice compensated bearings are plotted against concentric pressure ratio $\beta$ and orifice and capillary design parameters $\delta_0$ and $\delta_c$ in Figs 9.21 to 9.26 for nonrotating journal.

For both capillary and orifice compensated bearings, an optimum value of concentric pressure ratio $\beta$ is observed at an eccentricity ratio where the bearing stiffness is maximum. Orifice compensated bearings exhibit higher stiffness than capillary compensated bearings. For a particular bearing, an optimum value of $\delta_0$ or $\delta_c$ is also observed at which stiffness is maximum. It is also observed that damping decreases considerably with increase in concentric pressure ratio $\beta$.

Damping is an important parameter which helps in attenuating vibration amplitude under dynamic conditions. From the static and dynamic characteristics, it is found that a bearing which has good damping characteristics will have poor load capacity and stiffness. Damping ratio is high at high eccentricity ratios.

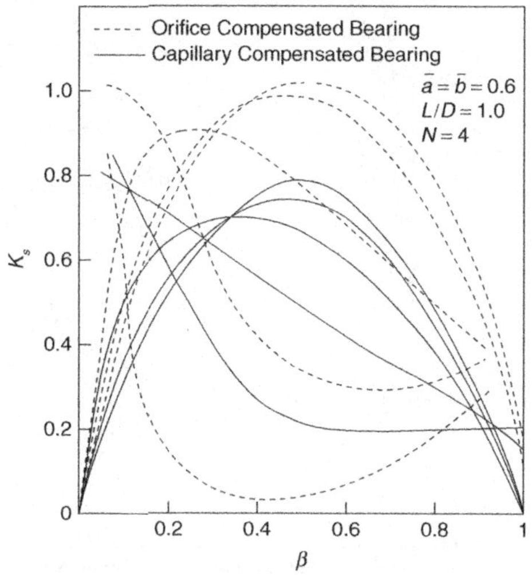


**Figure 9.21** | Stiffness Coefficient Versus Capillary Parameter

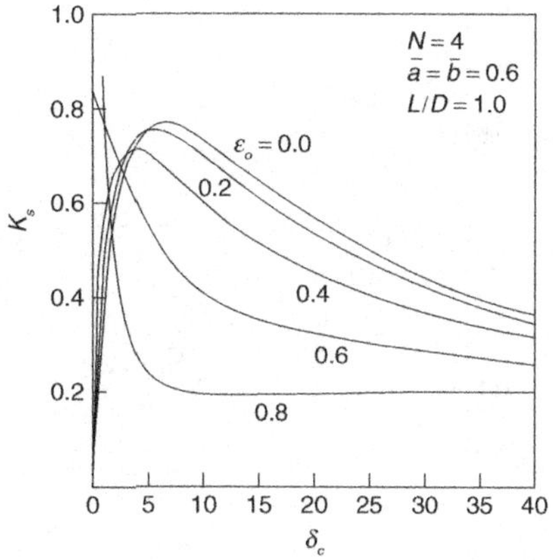


**Figure 9.22** | Stiffness Coefficient Versus Capillary Parameter

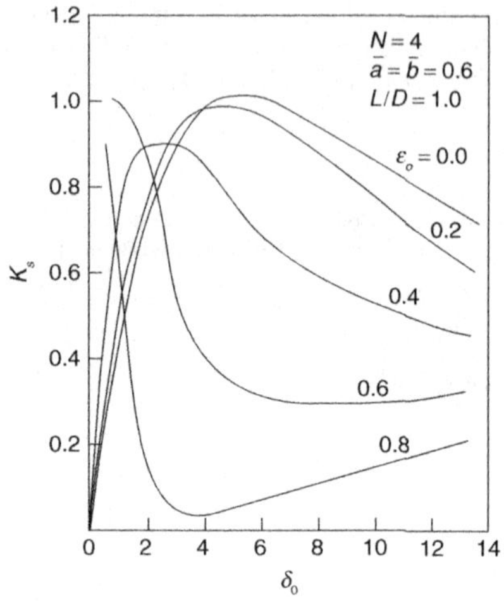


**Figure 9.23** | Stiffness Coefficient Versus Orifice Parameter

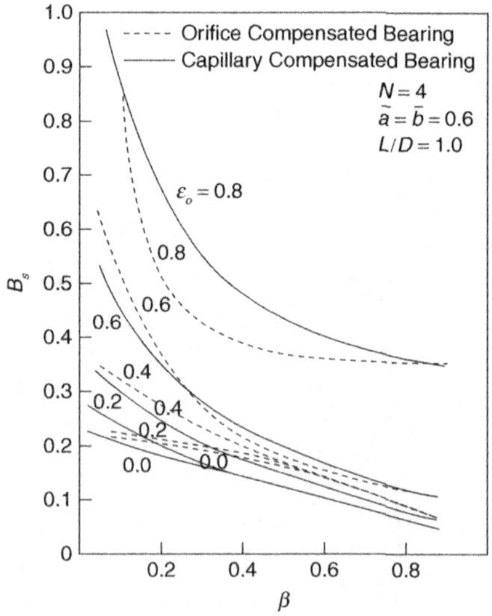


**Figure 9.24** | Damping Coefficient Versus Concentric Pressure Ratio

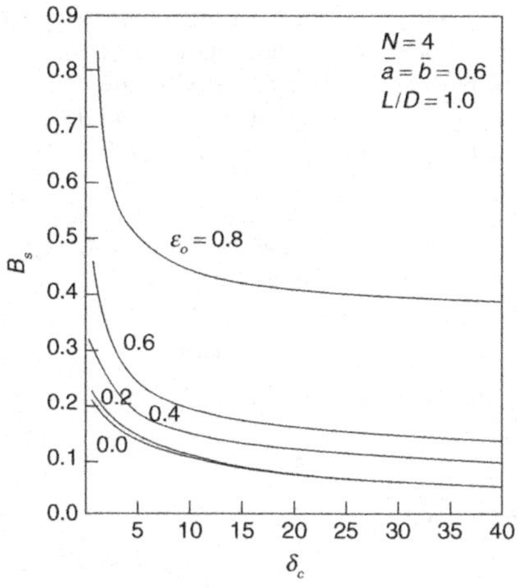


**Figure 9.25** | Damping Coefficient Versus Capillary Parameter

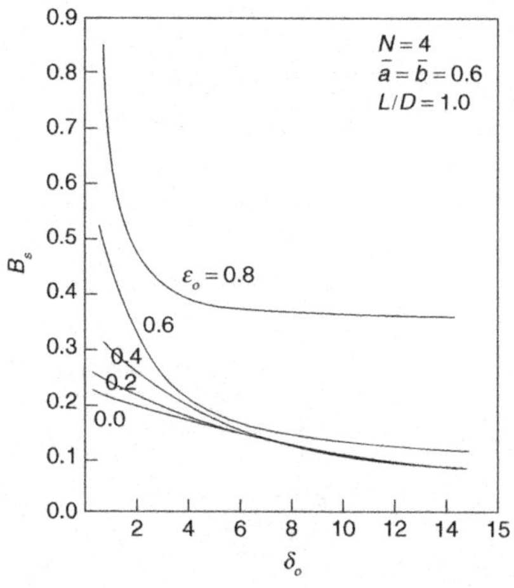


**Figure 9.26** | Damping Coefficient Versus Orifice Parameter

Hybrid bearing dynamic characteristics is shown in Figs 9.27 and 9.28 for a capillary compensated bearing. Dynamic behavior of hybrid bearing has been presented in terms of the critical mass of a rigid rotor for stability and whirl ratio, which has been determined following linear small amplitude vibration theory and using stiffness and damping coefficients of the bearing.

Dynamic characteristic of externally pressurized bearings considering recess volume fluid compressibility effect are presented through Figs 9.29 to 9.34 for a nonrotating journal.

Investigation carried out by Ghosh and Viswanath (1987) shows the recess volume fluid compressibility effect in terms of parameter '$\gamma$'. It has been found that stiffness and damping coefficients become frequency dependent. Stiffness increases with frequency parameter $\sigma$ and attains a high constant value at high

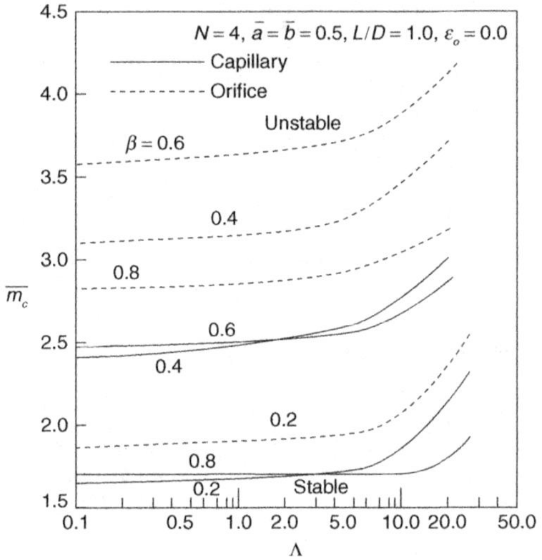


**Figure 9.27** | Stability Parameter Versus Bearing Number for Different Value of Concentric Pressure Ratios

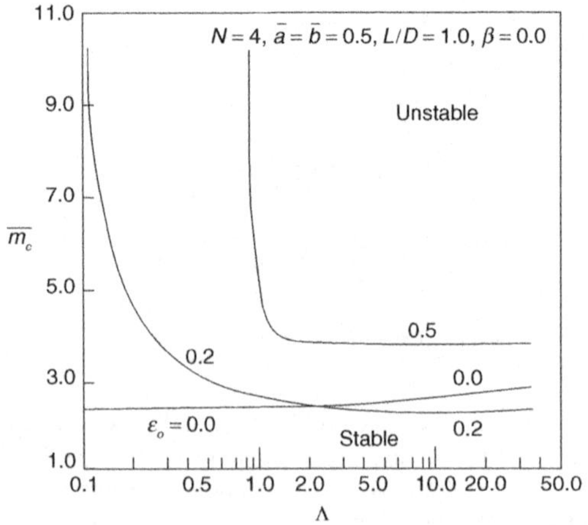


**Figure 9.28** | Stability Parameter Versus Bearing Number for Different Eccentricity Ratios

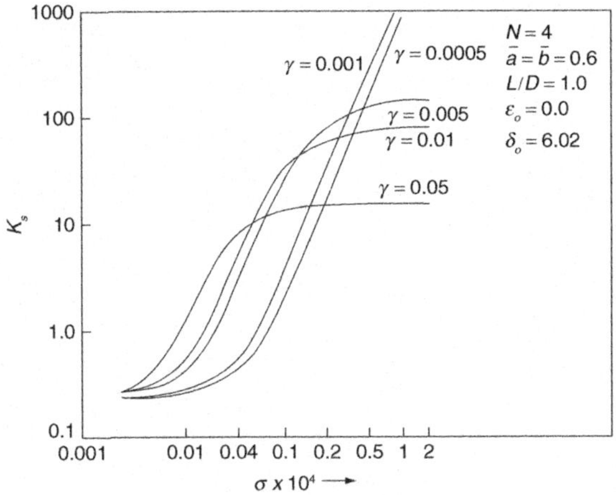


**Figure 9.29** | $K_S$ Versus $\sigma$ for Different Values of $\gamma$ of an Orifice-compensated Bearing

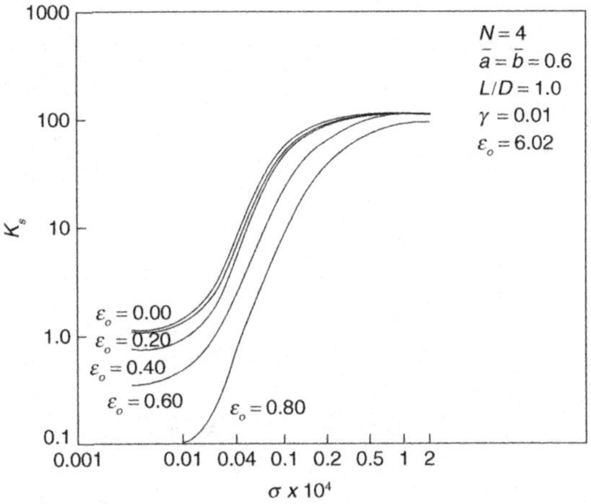


**Figure 9.30** | $K_S$ Versus $\sigma$ for Different Values of $\varepsilon_0$ of an Orifice-compensated Bearing

frequencies. The frequency at which it attains a constant value is referred to as the break frequency which decreases with increases in $\gamma$. On the other hand, damping coefficient decreases with increase in $\gamma$ and attains a reduced constant value at high frequencies. Thus, it is observed that recess volume compressibility parameter has a very significant influence on the dynamic coefficients. A proper value of recess depth or recess volume must be chosen at the design stage to achieve a suitable dynamic response.

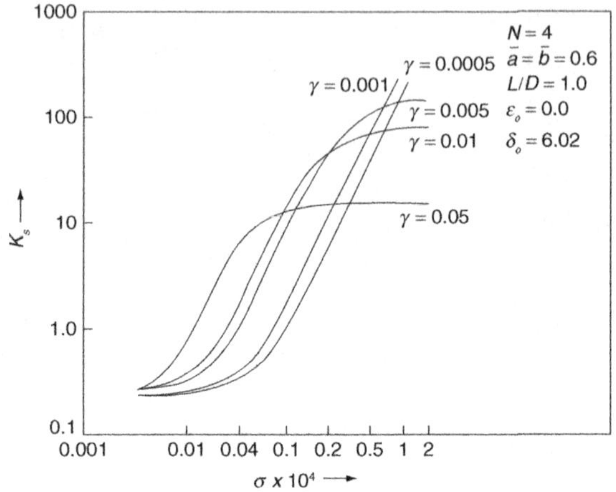


**Figure 9.31** | $K_s$ Versus $\sigma$ for Different Values of $\gamma$ of an Orifice-compensated Bearing

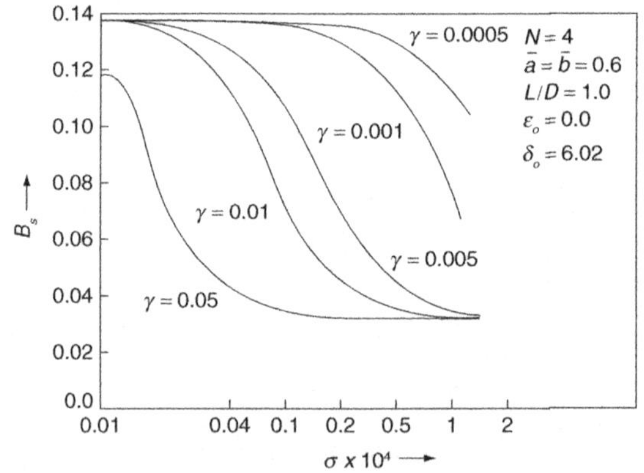


**Figure 9.32** | $B_s$ Versus $\sigma$ for Different Values of $\gamma$ of an Orifice-compensated Bearing

## 9.6 | Analysis of Fluid Seals

Seals are mechanical components generally used in fluid machinery to prevent leakage of high pressure fluid to low pressure region or to the surrounding medium from the casing of the machine. Two types of seals are very commonly employed, viz., mechanical face seals and annular pressure seals.

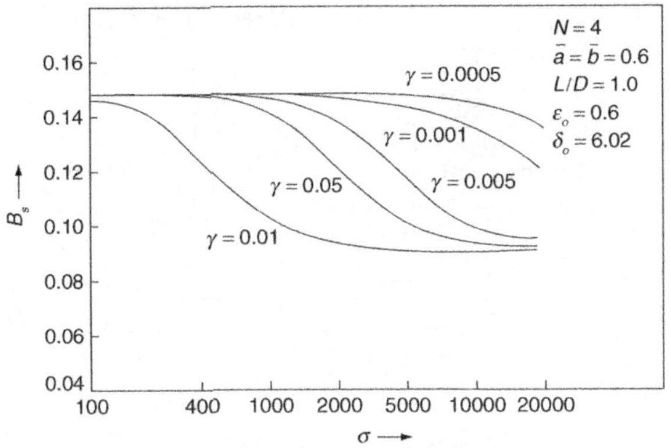


**Figure 9.33** | $B_s$ Versus $\sigma$ for Different Values of $\gamma$ of an Orifice-compensated Bearing

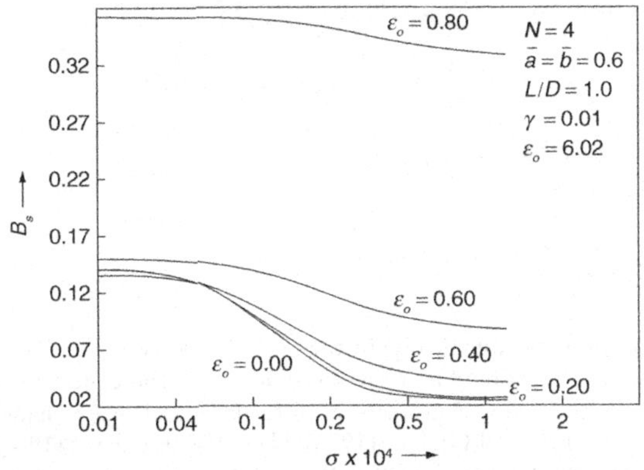


**Figure 9.34** | $B_s$ Versus $\sigma$ for Different Values of $\varepsilon_0$ of an Orifice-compensated Bearing

A typical mechanical face seal is shown schematically in Fig. 9.35. Prevention of leakage is the primary task of a face seal. Similarly, an annular pressure seal is shown schematically in Fig. 9.36. Its primary task is to reduce leakage of fluid between high pressure stages to an intermediate or low pressure stage in multistage fluid machinery. Fluid seals are generally similar to externally pressurized bearing, and therefore their analysis is also usually based on fluid film theories applicable to bearings with large clearances.

### 9.6.1 | Mechanical Face Seals–Laminar Flow Analysis

A mechanical face seal consists of a flexibly mounted seal ring which usually has five degrees of freedom. It is basically a dynamic system in which the flexibly mounted primary seal ring is separated by a thin fluid film from the rotating seat as shown in Fig. 9.35.

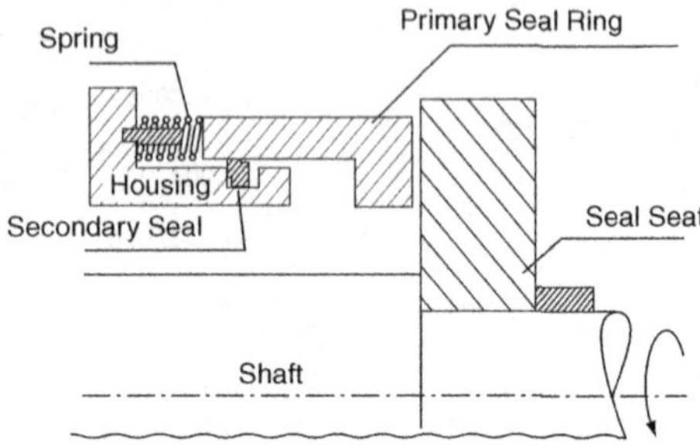


**Figure 9.35** | Radial Face Seal [Etsion and Dan, ASME JOLT, 1983]

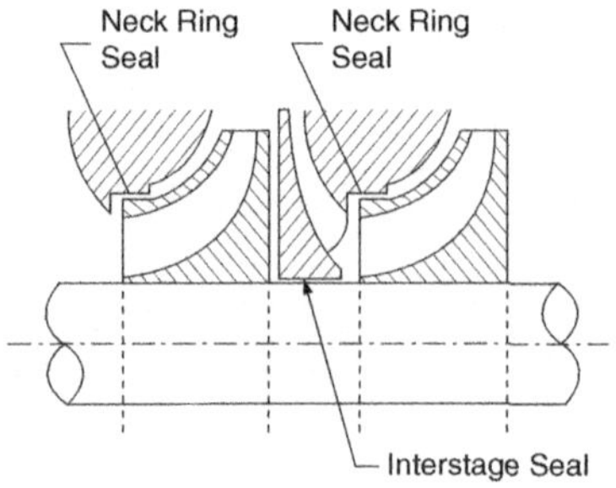


**Figure 9.36** | Annular Seal [Childs, ASME JOLT, 1983]

The presence of fluid film was established in the early 1960s by Denny ( 1961). Since then, various mechanism which provide sources of fluid film have been proposed. These mechanisms include surface waviness, angular misalignment, thermal and mechanical distortions, and phase change as summarized in the state-of-the-art review by Ludwig and Greiner (1978), Nau (1980), Lebeck (1980), and Hughes and Cho (1980). Sharoni and Etsion (1981) investigated the effects of diametric tilt and coning. Prior to these studies, Sneck (1968) presented a general analysis to determine seal leakage, load capacity, and torque in both laminar and turbulent flow including the effects of geometry and fluid inertia.

The analysis presented here follows the work of Sneck (1968) for laminar flow. The analysis is based on short bearing lubrication theory.

Geometry of the seal is shown in Fig. 9.37. $X$-axis is chosen so that it passes always through the point of minimum film thickness.

Expression for film thickness is written as

$$h = h_0 - r\cos\theta \tan\lambda \tag{9.133a}$$

$\lambda$ is angle of misalignment and $\theta$ the angular coordinate, $r$-radial coordinate, $z$-axial coordinate. However, for small angle of misalignment the expression of film can be written as:

$$h = h_0 - \lambda r \cos\theta \tag{9.133b}$$

Radial momentum equation including centrifugal inertia but neglecting gyroscopic effect in rotating coordinate system reduces to

$$-\rho\left(\frac{v_\theta^{\,2}}{r}\right)=-\frac{\partial p}{\partial r}+\eta\frac{\partial^2 v_r}{\partial z^2} \tag{9.134}$$

The tangential momentum for small leakage rates reduces to

$$\frac{\partial^2 v_\theta}{\partial z^2}=0 \tag{9.135}$$

$v_r$ and $v_\theta$ are radial and tangential velocity components of the fluid.

Referring to Fig. 9.37(b) the surface velocity of a point on upper disc is given by

$$V=\vec{\omega}x\vec{p}=(r\omega)\vec{e}_\theta+(r\omega\lambda\sin\theta)\vec{e}_z \tag{9.136}$$

$\vec{e}_\theta$ and $\vec{e}_z$ are unit vectors.

Solution of Equation (9.134) is of the form

$$v_\theta=f(r,\theta)z+g(r,\theta) \tag{9.137}$$

Application of the boundary conditions given by equation (9.136) gives

$$v_\theta=(r\omega)\frac{z}{h} \tag{9.138}$$

Thus, equation (9.134) with the substitution of (9.138) becomes

$$\eta\frac{\partial^2 v_r}{\partial z^2}=\frac{\partial p}{\partial r}-\rho r\omega^2\left(\frac{z}{h}\right)^2 \tag{9.139}$$

Integrating (9.139) twice with respect to $z$, one obtains

$$v_r=\frac{z^2}{2\eta}\frac{\partial p}{\partial r}-\frac{\rho r z^4}{12\eta h^2}\omega^2+f(r,\theta)z+g(r,\theta) \tag{9.140}$$

Applying the boundary conditions of (9.134) and (9.139), Equation (9.135) reduces to

$$v_r=\left(\frac{z^2-zh}{2\eta}\right)\frac{\partial p}{\partial r}-\frac{\rho r^4}{12\eta h^2}\omega^2\left(z^4-zh^3\right) \tag{9.141}$$

Reynolds equation for pressure distribution is obtained by integrating the continuity equation across the film thickness $h$ as follows:

$$\int_0^h\frac{1}{r}\frac{\partial}{\partial r}(rv_r)+\int_0^h\frac{1}{r}\frac{\partial v_\theta}{\partial\theta}dz+\int_0^h\frac{\partial v_z}{\partial z}dz=0 \tag{9.142}$$

The Reynolds equation is obtained for short bearing and steady state condition as

$$\frac{\partial}{\partial r}\left(rh^3\frac{\partial p}{\partial r}\right)=0.3\rho\omega^2\frac{\partial}{\partial r}\left(r^2h^3\right)+6\eta r\omega^2\frac{\partial h}{\partial\theta} \tag{9.143}$$

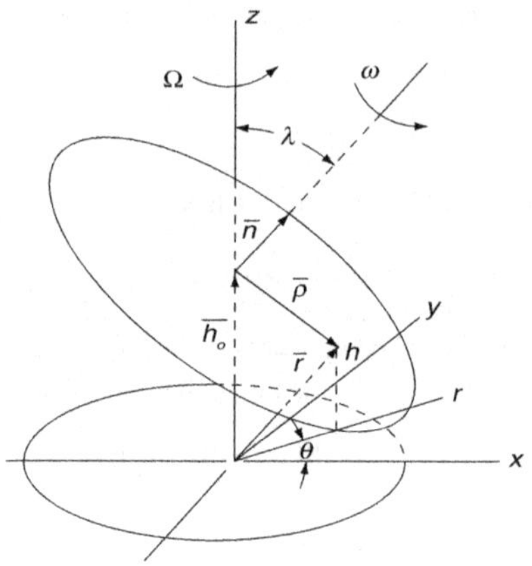


**Figure 9.37(a)** | Face Seal, Inertial Coordinate System [H. J. Sneck, ASME, JOLT, 1968]

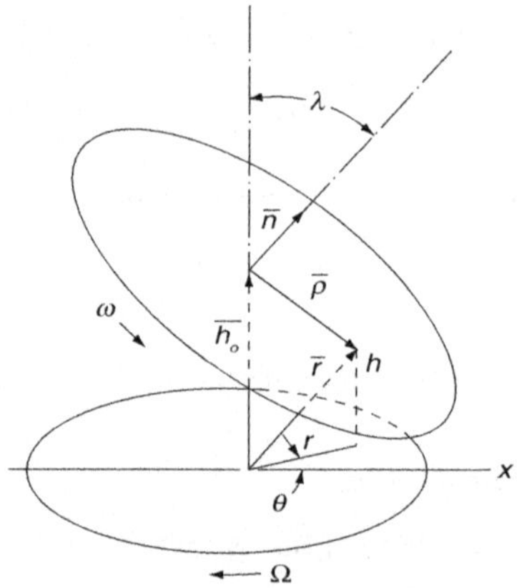


**Figure 9.37(b)** | Face Seal, Inertial Coordinate System [H. J. Sneck, ASME, JOLT, 1968]

Pressure distribution is obtained by integrating above equation twice with respect to $r$ as

$$p = 0.15\rho\omega^2 r^2 + 2\eta\lambda\omega\sin\theta\int\frac{r^2 dr}{h^3} + \phi(\theta)\int\frac{dr}{rh^3} + \psi(\theta) \tag{9.144}$$

The integrals are defined as

$$I_1(r,\theta) = \int\frac{r^2 dr}{h^3};\; I_2(r,\theta) = \int\frac{dr}{rh^3} \tag{9.145}$$

The boundary conditions are given as

$$\begin{aligned} p &= p_1 \quad at \quad r = r_1 \\ p &= p_2 \quad at \quad r = r_2 \end{aligned} \tag{9.146}$$

$\phi(\theta)$ and $\psi(\theta)$ are obtained as

$$\phi(\theta) = (p_2 - p_1)\frac{\begin{Bmatrix} 0.15\rho\omega^2(r_2^2 - r_1^2) + 2\eta\omega\lambda\sin\theta \\ [I_1(r_2,\theta) - I_1(r_1,\theta)] \end{Bmatrix}}{\{I_2(r_2,\theta) - I_2(r_1,\theta)\}} \tag{9.147}$$

$$\psi(\theta) = p_1 - 0.15\rho\omega^2 r_1^2 \tag{9.148}$$

## Seal Leakage Flow (Q)

Volumetric leakage through the seal is given by

$$Q = \int_0^{2\pi}\int_0^{h} v_r r dz d\theta = -\frac{1}{12\eta}\int_0^{2\pi}\phi(\theta)\, d\theta \tag{9.149}$$

Substituting for $\phi(\theta)$ and integrating the seal leakage flow rate is obtained as

$$\begin{aligned} Q &= \frac{\pi h_0^3}{6\eta\ln(r_2/r_1)}\left[p_1 - p_2 + 0.15\rho\omega^2(r_2^2 - r_1^2)\right] \\ &\times\left\{1 + \frac{1}{2}\left\{\left[\frac{3(r_2 - r_1)}{\ln(r_2/r_1)}\right]^2 - \left[\frac{3(r_2^2 - r_1^2)}{\ln(r_2/r_1)}\right]\right\}\left(\frac{\lambda}{h_0}\right)^2\right\} \\ &+ \ldots\ldots\ldots\ldots \end{aligned} \tag{9.150a}$$

$$\begin{aligned} \bar{Q} &= Q/\frac{\pi h_0^3}{6\eta\ln(r_2/r_1)}\left[p_1 - p_2 + 0.15\rho\omega^2(r_2^2 - r_1^2)\right] \\ &= \left\{1 + \frac{1}{2}\left\{\left[\frac{3(r_2 - r_1)}{\ln(r_2/r_1)}\right]^2 - \left[\frac{3(r_2^2 - r_1^2)}{\ln(r_2/r_1)}\right]\right\}\left(\frac{\lambda}{h_0}\right)^2\right\} \\ &+ \ldots\ldots\ldots\ldots \end{aligned} \tag{9.150b}$$

Above expression retains terms up to $\left(\frac{\lambda}{h_0}\right)^2$ only. Higher order terms have been neglected.

Plot of dimensionless leakage flow rate is shown in Fig. 9.38 against misalignment parameter. However, $\left(\frac{r_2\lambda}{h_0}\right)$ must be less than one to avoid physical contact between two surfaces. Increase in leakage rate is seen to be proportional to $\left(\frac{r_2\lambda}{h_0}\right)^2$, the misalignment parameter.

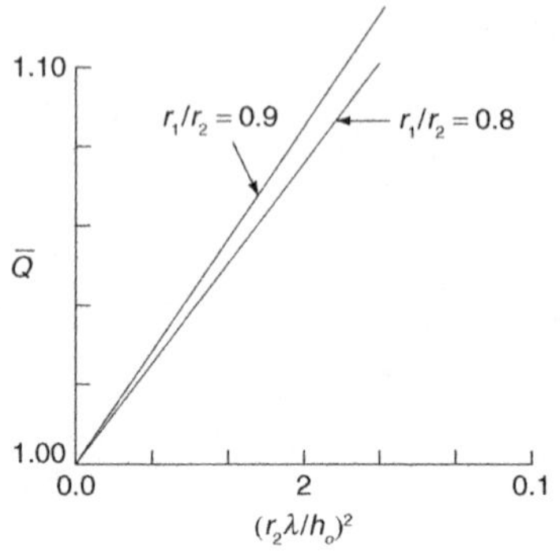


**Figure 9.38** | Dimensionless Leakage $\bar{Q}$ Rate Versus Misalignment Parameter

For zero misalignment leakage flow rate is:

$$Q = \frac{\pi h_0^3}{6\eta \ln\left(r_2/r_1\right)}\left[p_1 - p_2 + 0.15\rho\omega^2\left(r_2^{\,2} - r_1^{\,2}\right)\right] \tag{9.151}$$

## Seal Force (*F*)

Seal force due to fluid film pressure tending to separate the two surfaces is given by

$$\begin{aligned}
F = \int_{r_1}^{r_2}\int_0^{2\pi} pr\,dr\,d\theta &= \pi p_1\left(r_2^2 - r_1^2\right) + \frac{\pi}{2}\left(0.15\rho\omega^2\right)\left(r_2^2 - r_1^2\right)^2 \\
&+2\pi\left[\frac{p_2 - p_1 - \frac{3}{20}\rho\omega^2\left(r_2^2 - r_1^2\right)}{\ln\left(r_2/r_1\right)}\right]\times\left[\frac{r_2^2\ln\left(r_2/r_1\right)}{2} - \frac{r_2^2 - r_1^2}{4}\right] \\
&\times\left\{1+\left[\frac{1}{r_2^2\ln\left(r_2/r_1\right) - \left(\frac{r_2^2 - r_1^2}{2}\right)}\right]\times\left\{\frac{3}{4}\left(r_2^2 - r_1^2\right)^2 - \frac{3\left(r_2 - r_1\right)}{\ln\left(r_2/r_1\right)}\right.\right. \\
&\left.\left.\times\left[\left(r_2^3 - r_1^3\right) - \frac{3}{2}r_1\left(r_2^2 - r_1^2\right)\right]\right\}\right\} \\
&+\left[\left(\frac{3\left(r_2 - r_1\right)}{\ln\left(r_2/r_1\right)}\right)^2 - 3\left(\frac{\left(r_2^2 - r_1^2\right)}{\ln\left(r_2/r_1\right)}\right)\left[\frac{r_2^2\ln\left(r_2/r_1\right)}{2} - \left(\frac{\left(r_2^2 - r_1^2\right)}{4}\right)\right]\left(\frac{\lambda}{h_0}\right)^2\right] + \ldots
\end{aligned} \tag{9.152}$$

For an aligned seal the seal force $F$ is given as

$$F = \pi p_1 \left(r_2^2 - r_1^2\right) + \frac{\pi}{2}\left(0.15\rho\omega^2\right)\left(r_2^2 - r_1^2\right)^2$$

$$+ 2\pi \left[ \frac{p_2 - p_1 - \dfrac{3\rho}{20}}{\ln\left(\dfrac{r_2}{r_1}\right)} \right] \times \left[ \frac{r_2^2 \ln\left(r_2 / r_1\right)}{2} - \left(\frac{r_2^2 - r_1^2}{2}\right) \right] \tag{9.153}$$

## Seal Torque (*T*)

Seal torque $T$ is determined as

$$T = \int_0^{2\pi} \int_{r_1}^{r_2} \tau_{r\theta} r^2 dr d\theta$$

$$T = \int_0^{2\pi} \int_{r_1}^{r_2} \eta \left[ \frac{r}{h}\omega + \omega\lambda\cos\theta \right] r^2 dr d\theta \tag{9.154}$$

$$T = \frac{\pi\eta\omega\left(r_2^4 - r_1^4\right)}{2h_0} \times \left[1 + \frac{1}{3}\left(\frac{r_2^6 - r_1^6}{r_2^4 - r_1^4}\right)\left(\frac{\lambda}{h_0}\right)^2 + \frac{3}{16}\left(\frac{r_2^6 - r_1^6}{r_2^4 - r_1^4}\right)\left(\frac{\lambda}{h_0}\right)^4 + \ldots\right] \tag{9.155a}$$

$$\bar{T} = \frac{T}{\left(\dfrac{\pi\eta\omega\left(r_2^4 - r_1^4\right)}{2h_0}\right)} = \left[1 + \frac{1}{3}\left(\frac{r_2^6 - r_1^6}{r_2^4 - r_1^4}\right)\left(\frac{\lambda}{h_0}\right)^2 + \frac{3}{16}\left(\frac{r_2^6 - r_1^6}{r_2^4 - r_1^4}\right)\left(\frac{\lambda}{h_0}\right)^4 + \ldots\right] \tag{9.155b}$$

For zero misalignment seal torque reduces to:

$$T = \frac{\pi\eta\omega\left(r_2^4 - r_1^4\right)}{2h_0} \tag{9.156}$$

Dimensionless seal torque is plotted against misalignment parameter in Fig. 9.39.

Seal torque is proportional to misalignment parameter $\left(\dfrac{r_2\lambda}{h_0}\right)^2$ but is less sensitive to misalignment parameter. A small amount of misalignment is not avoidable in face seals. Film thickness in face seals varies radically due to thermal or elastohydrodynamic distortion which influence leakage flow rate. Besides, there is experimental evidence that face seals do not remain flat after long use and tangential waviness is observed in seals. Cavitation or regions of subambient pressures occur in face seals. In the regions of subambient pressure,

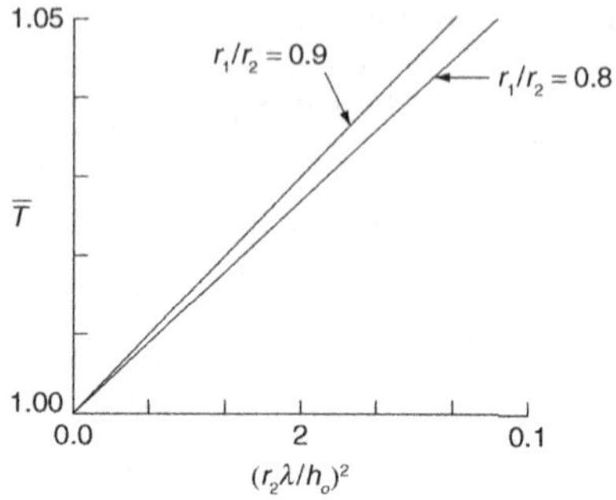


**Figure 9.39** | Dimensionless Seal Torque Versus Misalignment Parameter

it is assumed that dissolved gases in the fluid, i.e., air mostly come out and rupture the film. Cavitations region where the pressure is equal to the saturation pressure of the liquid with the air is observed to occur. Cavitations phenomena and two phase flow in seals have been investigated by Nau (1980) and Hughes *et al.*, Lebeck (1980). Dynamics of primary seal ring in face seals have been investigated Etsion and Dan (1981), and Etsion (1982).

### 9.6.2 | Analysis of Liquid Annular Seals

Liquid annular seals are similar to journal bearings with external pressurization or journal bearings with pressure induced axial flow. Fig. 9.40 shows the neck ring or wearing ring seal which prevent leakage flow from impeller discharge to inlet and interstage seal which prevent leakage between the stages of a multistage centrifugal pump. The full head of the pump is dropped across the balance-piston seal with resultant leakage bypassed to the pump inlet. The major difference between bearings and seals is in the radial clearance to journal radius ratio, i.e., *C/R* ratio. In bearings, it is of the order of 0.001, whereas in seals it is of the order of 0.003.

Due to large clearance and high pressure at the seal entrance, the flow in the seal clearance is often highly turbulent. It is generally accepted that Reynolds equation for turbulent lubrication is not adequate to analyze annular pump seals, since, it does not consider the fluid inertia effects. Due to high pressure, annular seals develop large direct stiffness. The early analysis of pump seals by Lomakin (1958) proved this. Bulk flow models which include fluid inertia effect were adopted by Black (1969), and Black and Jenssen (1970).

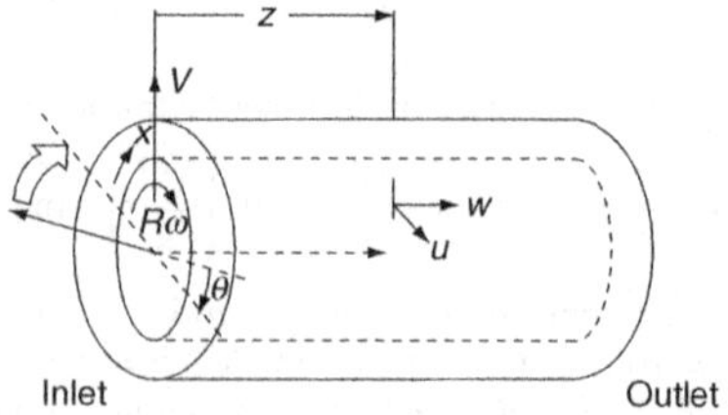


**Figure 9.40** | Annular Seal

Childs (1973) used Hirs' (1973) bulk flow theory to analyze turbulent flow in annular seals. In bulk flow theory, the velocity distribution across the seal clearance is neglected and an average velocity component is used to do the calculations. Shear stress as a function of local Reynolds number is used as a turbulence model.

The annular seal geometry is shown in Fig. 9.40. In bulk flow analysis usually adopted for plain liquid seals, the velocity components $u$, $w$ are the bulk flow velocities averaged over the clearance $h$ in the circumferential and axial directions, respectively. The shaft and seal axes are parallel; therefore, film thickness in the seal is not a function of axial coordinate $z$.

The continuity equation for an eccentric seal is written as

$$\frac{\partial}{\partial z}(hw)+\frac{1}{R}\frac{\partial}{\partial \theta}(hu)+\frac{\partial h}{\partial t}=0 \tag{9.157}$$

In steady state condition, $\frac{\partial h}{\partial t}=0$.

In the axial direction equation of motion is written as:

$$-h\frac{\partial p}{\partial z}=\left(\tau_{rz}+\tau_{sz}\right)+\rho h\left[w\frac{\partial w}{\partial z}+\frac{u}{R}\frac{\partial w}{\partial \theta}+\frac{\partial w}{\partial t}\right] \tag{9.158}$$

where $\tau_{rz}$ and $\tau_{sz}$ are turbulent shear stresses at the rotational and stationary surfaces, respectively.

Similarly, circumferential direction equation of motion is given as

$$-\frac{h}{R}\frac{\partial p}{\partial \theta}=\left(\tau_{r\theta}+\tau_{s\theta}\right)+\rho h\left[\frac{\partial u}{\partial t}+\frac{u}{R}\frac{\partial u}{\partial \theta}+w\frac{\partial u}{\partial z}\right] \tag{9.159}$$

where $\tau_{r\theta}$ and $\tau_{s\theta}$ are turbulent shear stresses at the rotational and stationary surfaces, respectively.

Hirs in his bulk flow model used a Blasius-type pipe friction model to describe the shear stresses in the annulus with inner cylinder rotating. Thus, shear stresses are expressed as

$$\tau_s=\rho f_s u_s^2/2 \text{ and } \tau_r=\rho f_r u_r^2/2 \tag{9.160}$$

where $f_s$ and $f_r$ are friction factors and $u_s$ and $u_r$ are bulk flow velocities at the stationary and rotating surfaces, respectively. Friction factors are defined as:

$$f_s=n_s\left(\mathrm{Re}\,hu_s\right)^{m_s} \text{ and } f_s=n_s\left(\mathrm{Re}\,hu_s\right)^{m_s} \tag{9.161}$$

where $n_s$ and $m_s$ characterize the flow resistance at the stationary surface, $n_r$ and $m_r$ characterize the flow resistance at the rotating surface. Re is the flow Reynolds number.

$$n_s=m_s=0.079 \text{ and } m_s=m_r=m=-0.25$$

are the values adopted for smooth pipes in Blasius-type friction model.

In case of centered annular seal under steady condition, the equations reduce to

$$-h\frac{\partial p}{\partial z}=\left(\tau_{rz}+\tau_{sz}\right)+\rho h\left[w\frac{\partial w}{\partial z}+\frac{u}{R}\frac{\partial w}{\partial \theta}\right] \tag{9.162}$$

$$-\frac{h}{R}\frac{\partial p}{\partial \theta}=\left(\tau_{r\theta}+\tau_{s\theta}\right)+\rho h\left[\frac{u}{R}\frac{\partial u}{\partial \theta}+w\frac{\partial u}{\partial z}\right] \tag{9.163}$$

Inlet pressure boundary condition is expressed in terms of kinetic head loss at the inlet as

$$p_i(0,\theta)=p_s-\frac{\rho}{2}(1+\xi)w^2(0,\theta) \tag{9.164}$$

$\xi$ is inlet loss coefficient which usually varies between 0 and 0.5.

Seal leakage for a known seal pressure $p_s$ can be determined from the solution of Equations (9.162) and (9.163) subject to boundary conditions given by Equation (9.164) simultaneously satisfying flow continuity equation given by Equation (9.157). For further details, readers can refer to the book by Childs (1993).

## Examples

***E.9.1*** A circular step externally pressurized capillary compensated thrust bearing has the following specifications:

Pad outer diameter, $d_0$ = 40 cm, diameter of the recess, $d_i$ = 20 cm, Journal speed, $N$ = 500 rpm, oil supply pressure, $p_s$ = 400 kN/m², viscosity of oil $\eta$ = 0.05 Pa-s, film thickness, $h$ = 0.08 mm. If the capillary length, $l_c$ = 20 mm and length to diameter ratio $l_c / d_c$ = 20, determine the recess pressure, load carrying capacity, stiffness, and oil flow rate of the bearing. What would be the value of film thickness for minimum power consumption?

***Solution:***

The given data is as follows: $d_0$ = 40 cm, $d_i$ = 20 cm, $N$ = 500 rpm, $p_s = 400 kN/m^2$, $\eta$ = 0.05 Pa-s, $h$ = 0.08 mm, $l_c = 20 mm$, $l_c / d_c = 20$ or $d_c = 1 mm$.

Using recess flow continuity equation for capillary compensation, the following relationship can be written

$$\frac{\pi d_c^4}{128 l_c}\left(p_s-p_r\right)=\frac{\pi h^3}{6\eta \ln\left(r_0/r_i\right)}$$

where

$$r_0=d_0/2=20\,\text{cm and } r_i=d_i/2=10\,\text{cm}$$

and $p_r$ is recess pressure.

Substituting the values of parameters in the above relationship, one obtains

$$\frac{p_s}{p_r}=1+\frac{128 l_c h^3}{6\ln\left(r_0/r_i\right)d_c^4}=1+\frac{128\times 20\times 10^{-3}\times\left(0.08\times 10^{-3}\right)^3}{6\ln(2)\times\left(0.8\times 10^{-3}\right)^4}=2.315$$

$$p_r=\frac{p_s}{2.315}=\frac{400}{2.315}=216\text{ kN/m}^2$$

### Load Capacity

$$W = \frac{\pi p_r \left(r_0^2 - r_i^2\right)}{2\ln\left(\frac{r_0}{r_i}\right)} = \frac{\pi \times 216 \times 10^3 \times \left(0.15^2 - 0.1^2\right)}{2\ln\left(\frac{0.15}{0.1}\right)} = 10,457.5 \text{ N}$$

$$W = 10.4575 \text{ kN}$$

### Stiffness of Bearing Pad

$$k = \frac{3W}{h} = \frac{3 \times 10457.5}{0.08 \times 10^{-3}} = 3.92 \times 10^8 \text{ N/m}$$

### Oil Flow Rate

$$Q = \frac{\pi h^3 p_r}{6\eta \ln\left(r_0 / r_i\right)} = \frac{\pi\left(0.08 \times 10^{-3}\right)^3 \times 216 \times 10^3}{6 \times 0.05 \times \ln\left(0.15 / 0.1\right)} = 2.855 \times 10^{-6} \text{m}^3/\text{sec}$$

$$Q = 2.855 \text{m}^3/\text{s}$$

**E.9.2** A three recess capillary compensated journal bearing has a length to diameter ratio 1. Axial length of recess is 0.75L where $L$ is axial length of the bearing. Total circumferential length of each recess is $0.25\pi D$, where $D$ is diameter of the journal. The supply pressure of the oil is 500 KN/m$^2$ and journal diameter is 60 mm. Determine for concentric position of the journal, the optimum speed for minimum power consumption when the radial clearance is 0.02 mm and the optimum radial clearance for minimum power consumption when the journal speed is 500 rpm. Determine the power consumption and oil flow rate for either case. The viscosity of the oil is given as 0.05 Pas.

**Solution:**

Number of recesses $N = 3$, axial length of the recess $l_r = 0.75L$, circumferential length of the recess $l_c = 0.25\pi D, D = 60$ mm, supply pressure $p_s = 500$ kN/m$^2$

Since $L_b/D$ is 1, the axial length of the bearing is 60 mm and radial clearance $C = 0.02$ mm. Thus,

$$l_r = 0.75L_b = 0.75 \times 60 = 45 \text{ mm}$$

axial land width

$$l = 0.125L_b = 0.125 \times 60 = 7.5 \text{ mm}$$

where $L_b$ is the axial width of the bearing.

The expression for power consumption is written as

$$\bar{P} = 1 + \frac{1}{6\pi\bar{Q}}\left(\frac{\omega}{mL}\right)^2 \left\{\frac{2\pi}{\sqrt{1-\varepsilon^2}} + mL^2 \sum_{n=1}^{N} \frac{1}{1 - \varepsilon\cos\left(\frac{2\pi}{N}n - \alpha\right)}\right\}$$

$\bar{Q} = \beta$ when the eccentricity ratio is zero, i.e., in concentric position of the journal center. Thus, optimum speed for minimum power consumption is obtained by substituting $\bar{P} = 4$.

$$\omega_0 = 3mL\left\{\frac{2\pi\beta}{2\pi + NmL^2}\right\}^{1/2} \text{ where } L = \frac{l_c}{l} \text{ and } m = \frac{ll_r}{Dl_c}$$

$$l_c = \frac{0.25\pi D}{3} = \frac{0.25\times\pi\times 60}{3} = 15.7\,\text{mm},$$

is width of circumferential land

$$m = \frac{7.5\times10^{-3}\times45\times10^{-3}}{60\times10^{-3}\times15.7\times10^{-3}} = 0.358,\ L = \frac{15.7}{7.5} = 2.1$$

Assuming the optimum value of concentric pressure ratio $\beta$ for capillary compensated bearing as 0.5, optimum value of dimensionless speed is obtained as:

$$\omega_0 = 3\times0.358\times2.18\times\left\{\frac{2\pi\times0.5}{2\pi+3\times0.358\times2.1\times2.1}\right\}^{1/2} = 1.25$$

Thus, optimum shaft speed $N_{s0}$ is determined as:

$$\omega_0 = \frac{6\pi\eta N_{s0} ll_r}{C^2 p_s}$$

$$N_{s0} = \frac{C^2 p_s \omega_0}{6\pi\eta ll_r}$$

Substituting appropriate values in the above expressions, optimum shaft speed is obtained as:

$$N_{s0} = \frac{(0.02\times10^{-3})^2\times500\times10^3\times1.25}{6\pi\times0.05\times7.5\times10^{-3}\times45\times10^{-3}} = 0.786\text{ rps}$$

For a known speed of the shaft, optimum clearance $C_0$ can be determined from the following expression:

$$\omega_0 = \frac{6\pi\eta N_s ll_r}{C_0^2 p_s} \text{ or } C_0^2 = \frac{6\pi\eta N_s ll_r}{\omega_0 p_s}$$

For shaft speed $N_s = 500$ rpm, optimum clearance can be determined by substituting

$$C_0^2 = \frac{6\pi\times0.05\times(500/60)\times7.5\times10^{-3}\times45\times10^{-3}}{1.25\times500\times10^3} = 42.4\times10^{-10}$$

$$C_0 = 0.065\times10^{-3}\,\text{m}$$

## Problems

**P.9.1** Derive the expressions for load capacity, oil flow rate, and power consumption for an annular recess externally pressurized circular thrust bearing. Determine the expressions for optimum film thickness and shaft speed for minimum power consumption. An annular recess hydrostatic thrust bearing has the following dimensions:

Outer radius of the bearing = 40 cm, inner radius of the bearing = 20 cm, recess inner and outer radii are 25 cm and 35 cm, respectively. Recess to supply pressure ratio is 0.65. Oil supply pressure is 400.0 KN/m$^2$. Assume a shaft speed of 200 rpm and film thickness = 0.05 mm determine load capacity, oil flow rate, and power consumption. Also determine the optimum shaft speed and film thickness for minimum power consumption.

**P.9.2** A centrifuge weighing 50 $N$ and rotating at $10^5$ rpm is supported by an uncompensated circular step hydrostatic thrust bearing. The recess pressure is 0.05 MPa and the ratio of outer to inner radii 1.67, while the film thickness is 50 micro meter. Calculate the bearing dimensions, lubricant flow rate, and friction torque. Lubricant viscosity is 0.05 Pas.

**P.9.3** A shaft is supported on two 3-recess hydrostatic journal bearings which are symmetrically placed with respect to a radial load $W$. The shaft rotates at an angular speed of $N_s$ rps. For a capillary compensated bearing, the supply pressure is $2.0 \times 10^6$ N/m$^2$. The shaft diameter is 6.0 cm. The bearing length $L$ is 6.0 cm. The width of the axial land is 0.1L and angular extent of circumferential land is 20$^0$. The operating eccentricity ratio is 0.4. Calculate the optimum radial clearance for minimum power consumption for a shaft speed of 500 rpm. Determine the recess pressures, load capacity, and oil flow rate. The viscosity of the oil is 0.05 Pas and oil density is 900 Kg/m$^3$.

**P.9.4** A single recess orifice compensated bearing has $L/D = 1$. The axial recess length is 0.8L and circumferential recess length is $0.8\pi D$. Journal diameter $D = 10$ cm and radial clearance is 0.02 mm. Find the load capacity, oil flow rate, and power loss if the oil supply pressure is 450 kN/m$^2$, operating eccentricity ratio = 0.3 and oil viscosity is 0.02 Pas. Assume concentric pressure ratio as 0.65 and that the journal is rotating at its optimum speed for minimum power loss. The load is acting through the center of the recess.

## REFERENCES


*Artiles, A., Walowit, J. and Shapiro, W.* (1982), Analysis of hybrid fluid film journal bearings with turbulence and inertia effects, *Advances in Computer Aided Bearing Design, ASME*, New York.

*Black, H.* (1969), Effect of hydraulic forces on annular pressure seals on the vibration of centrifugal pump rotors, *Journal of Mech. Engineering Science*, **11**, 206–213.

*Black, H. and Jenssen, D.* (1970), Dynamic hybrid properties of annular pressure seals, *Journal of Mechanical Engineering*, **184**, 92–100.

*Braun, M.J., Adams, M.L. and Mullen, R.L.* (1985), Analysis of a two row hydrostatic journal bearing with variable properties, inertia effects and surface roughness, *Israel Journal of Technology, 22*, 155–164.

*Brown, G.M.* (1961), Dynamic characteristics of hydrostatic thrust bearings, *Machine Tool Design & Research*, **1–2**, 157.

*Childs, D.* (1983), Dynamic analysis of turbulent annular seals based on Hirs' lubrication equation, *ASME Journal of Lubrication Technology*, **105**, 437–444.

*Childs, D.* (1993), *Turbo Machinery Rotor Dynamics* Phenomena, Modeling and Analysis, John Wiley & Sons Inc, New York, USA.

*Cusano, C. and Conry, T.F.* (1974), Design of multi recess hydrostatic journal bearings for minimum total power loss, *ASME, Journal for Industry*, 226–232.

*Davies, P.B. and Leonard, R.* (1969–70), The dynamic behavior of multirecess hydrostatic journal bearings, *Proc. I. Mech. Engineers*, London, U.K., **184,** 139–147.

*Davies, P.B.* (1969–70), A general analysis of multi-recess hydrostatic journal bearings, *Proc. Institution of Mech. Engineers*, London, UK, **184**, part I.

*Denny, D.F.* (1961), Some measurements of fluid pressures between plane parallel thrust surfaces with special reference to the behavior of radial face seals, *Wear*, **4**, 64–83.

*Dowson, D.* (1961), Inertia effects in hydrostatic thrust bearings, *ASME, Journal of Basic Engineering*, **83**, 227.

*Etsion, I.* (1982), Dynamic analysis of non contacting face seals, *ASME Journal of Lubrication Technology*, **104**, 460–468.

*Etsion, I. and Dan, Y.* (1981), An analysis of mechanical face seals vibration, *ASME Journal of Lubrication Technology*, **103**, 428–435.

*Ghigliazza, R. and Michelini, R.C.* (1968), Comparative investigation of friction in externally pressurized bearings, *International J. of Wear*, **12**.

*Ghosh, M.K. and Majumdar, B.C.* (1978), Dynamic characteristics of multi recess externally pressurized oil journal bearings, *Trans. ASME, Journal of Lubrication Technology*, **100**, 467–471.

*Ghosh, M.K. and Majumdar, B.C.* (1978), Stiffness and damping characteristics of hydrostatic multirecess oil journal bearings, *Machine Tool Design & Research*, **18**, 139–151.

*Ghosh, M.K. and Majumdar, B.C.* (1982), Dynamic stiffness and damping characteristics of compensated hydrostatic thrust bearings, *Trans. ASME, Journal of Lubrication Technology*, **104**, 491–496.

*Ghosh, M.K. and Vishwanath, N.S.* (1987), Recess volume fluid compressibility effect on the dynamic characteristics of multirecess hydrostatic journal bearings with journal rotation, *Trans ASME Journal of Tribology*, **109**, 417–426.

*Ghosh, M.K., Guha, S.K. and Majumdar, B.C.* (1989), Rotor dynamic coefficients of multirecess hybrid journal bearings Part I, *International Journal of Wear*, **129**, 245–259.

*Ghosh, M.K., Majumdar, B.C. and Rao, J.S.* (1979), Steady state and dynamic behavior of hybrid oil journal bearings, *Journal of Mech. Engineering Science, Inst. of Mech. Engineers*, London, U.K., **21**, 345–351.

*Heller, S.* (1974), Static and dynamic performance of externally pressurized fluid film journal bearings in the turbulent regime, *Trans. ASME, Journal of Lubrication Technology*, **96**, 381–390.

*Hirs, G.G.* (1973), A bulk flow theory for turbulence in lubricant film, *ASME Journal of Lubrication Technology*, **95**, 137–146.

*Hughes, W.F. and Cho, N.H.* (1980), Phase change in liquid seals II—Isothermal and adiabatic bounds with real fluids, *ASME, Journal of Lubrication Technology*, **102**, 350–367.

*Hunt, J.B.* (1964), Dynamic characteristic of hydrostatic thrust bearings, *Proc., I. Mech. E, London*, U.K. **178**, 246.

*Leabeck., A.O.* (1980), A mixed friction hydrostatic face seal model with phase change, *ASME, Journal of Lubrication Technology*, **102**, 133–138.

*Leonard, R. and Rowe, W.B.* (1973), Dynamic force coefficients and the mechanism of instability in hydrostatic journal bearings, *Wear*, **23**, 277–282.

*Licht, L. and Cooley, J.W.* (1964), Dynamics of externally pressurized sliders with incompressible and compressible films, *Trans ASME, Journal of Basic Engineering*, **86**, 396–404.

*Ling, M.T.S.* (1962), On the optimization of stiffness of externally pressurized bearings, *ASME, Journal of Basic Engineering*, 119–122.

*Ludwig, L.P. and Grainer, H.P.* (1978), Designing mechanical face seal for improved performance, Part 2: Lubrication, *Mech. Engineering*, **100**, 18–23.

*Nau, B.S.* (1980), Observation and analysis of mechanical seal film characteristic, *ASME, Journal of Lubrication Technology*, **102**, 341–349.

*O'Donoghue, J.P., Rowe, W.B. and Cameron, A.* (1970), Optimization of externally pressurized bearings for minimum power and low temperature rise, *Proc. I. Mech. Engineers, Tribology*, **3**, 153.

*Redcliffe, J.M. and Vohr, J.H.* (1969), Hydrostatic bearings for cryogenic rocket engine turbo pumps, *Trans. ASME, Journal of Lubrication Technology*, **91**, 557–575.

*Rhode, S.M. and Ezzat, H.A.* (1976), On the dynamic behavior of hybrid journal bearings, *Trans. ASME, Journal of Lubrication Technology*, **98**, 90–94.

*San Andres, L.* (1990), Turbulent hybrid bearings with fluid inertia effects, *ASME Journal of Tribology*, **112**, 699–707.

*Schwazenback, J. and Gill, K.F.* (1966), *Dynamic Characteristics of a Hydrostatic Thrust Bearing*, Engineer, 3009.

*Sharoni, A. and Etsion, I.* (1981), Performance of end face seals with diametric tilt and coning, *ASLE Trans*, **24**, 61–70.

*Shinkle, J.N. and Hornung, K.G.* (1965), Frictional characteristics of liquid hydrostatic journal bearings, *Trans. ASME, Journal of Engineering for Industry*, **87**, 163–169.

*Singh, D.V. et al.* (1979), Static and dynamic characteristics of an orifice compensated hydrostatic journal bearing, *Trans. ASLE*, **22**, 162–170.

*Sneck, H.J.* (1968), The effects of geometry and inertia on face seal performance-laminar flow, *Trans. ASME, Journal of Lubrication Technology*, **89**, 333–341.

Chapter 10

# Fluid Inertia Effects and Turbulence in Fluid Film Lubrication

Generally, the flow of lubricant in the clearance space of bearings and other mechanical components are categorized as slow viscous flow. The Reynolds numbers are low and the flow is in the laminar regime. Therefore, fluid inertia effects are neglected because the viscous forces are few orders of magnitude higher than the inertia forces. It has been shown earlier in Chapter 3 that this assumption is generally valid in majority of lubrication problems in mechanical components of various machines. However, at high speeds of operation, the Reynolds number increase and inertia effects become significant as well as comparable to viscous effect. Neglecting fluid inertia effect in such situations can lead to inaccurate results and inadequacies in the design of bearings. Both journal and thrust bearings of high speed turbomachinery, e.g., bearings of machines in power plants generating 500 megawatt or more often operate in flow regimes where fluid inertia effects cannot be neglected. The flow is often in the turbulent regime also.

Power losses in these bearings increase and the temperatures in the bearings also increase. Thrust bearings usually operate at higher Reynolds numbers than journal bearings. In either case, it is desired to achieve an efficient design that will reduce power loss and temperature rise. An efficient design should achieve higher load carried to power loss ratio and minimum film thickness of the lubricant should be decided on the basis of safe and reliable operation. It thus becomes imperative to account for fluid inertia effects and turbulence in fluid film bearings operating at high Reynolds numbers. There are other application domains also where Reynolds numbers can range from 5000–100 000, e.g., in cryogenic turbopumps, nuclear systems, etc. where low viscosity lubricants are generally used. Invariably a transition flow regime exists between laminar and turbulent flow regimes of lubrication which is characterized by formation of Taylor vortices. This transition regime is called vortex flow regime. In hydrodynamic journal bearings it occurs at Reynolds numbers from 1500–2000.

In the following section, theoretical approach to incorporate fluid inertia effects in the analysis of hydrodynamic bearings, squeeze film lubrication, and externally pressurized lubrication will be dealt with. Turbulent lubrication theory and its application to fluid film bearings shall be subsequently considered.

## 10.1 | Fluid Inertia Effects In Lubrication

At high Reynolds numbers, all the assumptions made in the derivation of Reynolds equation also remain valid except that fluid inertia forces are no more negligible in comparison to viscous forces. Therefore, the Governing equations for lubricant flow in the clearance space including fluid inertia terms can now be written for an incompressible fluid of constant viscosity as:

$$\rho\left\{\frac{\partial u}{\partial t}+u\frac{\partial u}{\partial x}+v\frac{\partial u}{\partial y}+w\frac{\partial u}{\partial z}\right\}=-\frac{\partial p}{\partial x}+\eta\frac{\partial^2 u}{\partial z^2} \tag{10.1}$$

$$\rho\left\{\frac{\partial v}{\partial t}+u\frac{\partial v}{\partial x}+v\frac{\partial v}{\partial y}+w\frac{\partial v}{\partial z}\right\}=-\frac{\partial p}{\partial y}+\eta\frac{\partial^2 u}{\partial z^2} \tag{10.2}$$

$$0=-\frac{\partial p}{\partial z} \tag{10.3}$$

Continuity equation for an incompressible fluid is written as:

$$\frac{\partial u}{\partial x}+\frac{\partial v}{\partial y}+\frac{\partial w}{\partial z}=0 \tag{10.4}$$

Following dimensionless variables are used to get dimensionless equations.

$$\bar{x}=\frac{x}{l},\bar{y}=\frac{y}{l},\bar{z}=\frac{z}{h},\bar{u}=\frac{u}{u_0},\bar{v}=\frac{v}{u_0}$$
$$\bar{w}=\frac{\bar{w}}{u_0},\bar{p}=\frac{ph^2}{\eta l u_0};\bar{\eta}=\frac{\eta}{\eta_0},\bar{\rho}=\rho/\rho_0$$

$\tau = \upsilon t$, where $\upsilon$ is frequency of vibration.

Equations (10.1), (10.2), and (10.3) can be further expressed as

$$\sigma\frac{\partial \bar{u}}{\partial \tau}+\mathrm{Re}\left(\bar{u}\frac{\partial \bar{u}}{\partial \bar{x}}+\bar{v}\frac{\partial \bar{u}}{\partial \bar{y}}+\bar{w}\frac{\partial \bar{u}}{\partial \bar{z}}\right)=-\frac{1}{\bar{\rho}}\frac{\partial \bar{p}}{\partial \bar{x}}+\frac{\bar{\eta}}{\bar{\rho}}\frac{\partial^2 \bar{u}}{\partial \bar{z}^2} \tag{10.5}$$

$$\sigma\frac{\partial \bar{v}}{\partial \tau}+\mathrm{Re}\left(\bar{u}\frac{\partial \bar{v}}{\partial \bar{x}}+\bar{v}\frac{\partial \bar{v}}{\partial \bar{y}}+\bar{w}\frac{\partial \bar{v}}{\partial \bar{z}}\right)=-\frac{1}{\bar{\rho}}\frac{\partial \bar{p}}{\partial \bar{y}}+\frac{\bar{\eta}}{\bar{\rho}}\frac{\partial^2 \bar{v}}{\partial \bar{z}^2} \tag{10.6}$$

$$-\frac{\partial \bar{p}}{\partial \bar{z}}=0 \tag{10.7}$$

The continuity equation reduces to

$$\frac{\partial \bar{v}}{\partial \bar{x}}+\frac{\partial \bar{v}}{\partial \bar{y}}+\frac{\partial \bar{w}}{\partial \bar{z}}=0 \tag{10.8}$$

Re is film Reynolds number and $\sigma$ is squeeze number as defined in Chapter 3.

$$\bar{\text{Re}} = \text{Re}\left(\frac{h}{l}\right)^2; \sigma = \rho_0 \upsilon h^2 / \eta_0$$

Re is flow Reynolds number which is usually less than 1000 for slow viscous flows in thin film lubrication. In slow viscous flows typical of film lubrication $\bar{\text{Re}} << 1$ and $\sigma$ is also of the same order. Therefore, all the inertia terms are neglected in comparison to pressure and viscous terms. In Equations (10.5) and (10.6), the terms on the left hand side consists of fluid acceleration terms which are referred as temporal inertia terms characterized by multiplication of $\sigma$ and convective inertia terms characterized by multiplication of $\bar{\text{Re}}$, respectively. In case of small amplitude high frequency oscillations, $\sigma >> \bar{\text{Re}}$ or $\sigma >> 1, \bar{\text{Re}} << 1$, the convective inertia terms become negligible and can be omitted. The Equations (10.1) to (10.3) reduce to:

$$\rho\frac{\partial u}{\partial t} = -\frac{\partial p}{\partial x} + \eta\frac{\partial^2 u}{\partial z^2} \tag{10.9}$$

$$\rho\frac{\partial v}{\partial t} = -\frac{\partial p}{\partial y} + \eta\frac{\partial^2 v}{\partial z^2} \tag{10.10}$$

$$0 = -\frac{\partial p}{\partial z} \tag{10.11}$$

Investigation of temporal inertia effects was carried out by several researchers. Kuhn and Yates (1964), Kuzma (1967), Tichy and Winer (1970), Fritz (1972), Modest and Tichy (1978), Mulachi (1980), Szeri *et al.* (1983), Ghosh *et al.* (1989), and San Andres (1990) have investigated problems related to circular plates in normal oscillation, squeeze film dampers, externally pressurized, and hybrid bearings. In these investigations, it has been shown that temporal inertia plays a dominant influence on the damping coefficients and also contributes toward added mass coefficient for bodies moving in fluid media. Added mass coefficient is regarded as the ratio of total mass of the fluid which has to be accelerated to the mass of fluid displaced by the body to accelerate in the media.

Contrary to the above situation, when $\bar{\text{Re}} >> 1$ and $\sigma << 1$, temporal inertia terms can be omitted and convective inertia terms are only retained. This situation occurs in high Reynolds number flows.

Investigation of convective inertia effect on the pressure generation and performance of journal thrust bearings have been dealt with adequately for the past several decades.

## 10.2 | Fluid Inertia Effect in Thrust Bearings

Navier–Stokes equations for an incompressible fluid of constant viscosity including both convective and temporal inertia terms can be expressed in cylindrical (r,θ,z) coordinates with usual assumptions of thin film lubrication as

$$\frac{\partial v_r}{\partial t} + v_r\frac{\partial v_r}{\partial r} + \frac{v_\theta}{r}\frac{\partial v_r}{\partial \theta} - \frac{v_\theta^{\,2}}{r} = -\frac{1}{\rho}\frac{\partial p}{\partial r} + \frac{\eta}{\rho}\frac{\partial^2 v_r}{\partial z^2} \tag{10.12}$$

$$\frac{\partial v_\theta}{\partial t} + v_r\frac{\partial v_\theta}{\partial r} + \frac{v_\theta}{r}\frac{\partial v_\theta}{\partial \theta} + \frac{v_r v_\theta}{r} = -\frac{1}{\rho r}\left(\frac{\partial p}{\partial \theta}\right) + \frac{\eta}{\rho}\frac{\partial^2 v_\theta}{\partial z^2} \tag{10.13}$$

$$0 = -\frac{\partial p}{\partial z} \tag{10.14}$$

Flow continuity equation is expressed as

$$\frac{\partial v_\theta}{\partial \theta} + \frac{\partial}{\partial r}(rv_r) + r\left(\frac{\partial v_z}{\partial z}\right) = 0 \tag{10.15}$$

where $v_r$, $v_\theta$ and $v_z$ are radial, circumferential, and axial components of fluid velocities.

It can be shown that besides centrifugal inertia effect, i.e., $\rho\frac{v_\theta^2}{r}$ and the term $\frac{\rho v_r v_\theta}{r}$, rest of the inertia terms are at least two orders of magnitude smaller and therefore can be neglected without any tangible error. In large thrust bearings, circumferential velocity is large whereas in mechanical face seals and rotating discs also centrifugal effect is significantly high. Therefore, influence of inertia forces in large thrust bearings, face seals, and rotating discs can result in modification of both velocity and pressure distributions. Usually, centrifugal inertia is the major contributor amongst the inertia effects. It has been investigated by Chen and Dareing (1976) and Pinkus and Lund (1981) in thrust bearings and seal-like configurations. It has also been shown that large areas of bearings and seals may also witness cavitation or otherwise get starved of lubricant. This may also lead to significant reduction in load carrying capacity. Existence of incomplete films was also observed by Etsion (1981) in sector shape thrust pads.

Thus, retaining only the centrifugal inertia terms, the momentum Equations (10.12) and (10.13) and continuity Equation (10.15) reduce to

### *Momentum Equations*

$$\eta\frac{\partial^2 v_\theta}{\partial z^2} = \frac{1}{r}\left(\frac{\partial p}{\partial \theta}\right) \tag{10.16}$$

$$\eta\frac{\partial^2 v_r}{\partial z^2} = \frac{\partial p}{\partial r} - \rho\frac{v_\theta^2}{r} \tag{10.17}$$

### *Continuity Equation*

$$\frac{\partial v_\theta}{\partial \theta} + \frac{\partial}{\partial r}(rv_r) + r\frac{\partial v_z}{\partial z} = 0 \tag{10.18}$$

Boundary conditions for Equations (10.16) and (10.17) are given as

$$v_r = 0, v_\theta = r\omega \text{ at } z = 0 \text{, at the runner surface}$$

$$v_r = 0, v_\theta = 0 \text{ at } z = h \text{, at the bearing surface} \tag{10.19}$$

$$v_z = \frac{\partial h}{\partial t} \text{ at } z = 0 \text{ and } v_z = 0 \text{ at } z = h$$

Integration of Equations (10.16) and (10.17) twice with respect to z and using the above boundary conditions yields:

$$v_\theta = \omega(h-z)\frac{r}{h} - \frac{z(h-z)}{2\eta}\left(\frac{\partial p}{\partial \theta}\right) \tag{10.20}$$

$$v_r = \left\{\frac{\rho r\omega^2 - \dfrac{\partial p}{\partial r}}{2\eta}\right\} z(h-z) + \frac{\rho z}{\eta r}\left[A_1(h^2-z^2) + A_2(h^3-z^3)A_3(h^4-z^4)A_4(h^5-z^5)\right] \tag{10.21}$$

where

$$A_1 = -\frac{\omega}{3h}\left[\left(\frac{\partial p}{\partial \theta}\right)\frac{h^2}{2\eta} + r^2\omega\right]$$

$$A_2 = \frac{1}{12}\left\{\frac{r^2\omega^2}{h^2} + \frac{1}{2\eta}\left(\frac{\partial p}{\partial \theta}\right)\cdot\left[\frac{1}{2\eta}\left(\frac{\partial p}{\partial \theta}\right)\frac{h^2}{r^2} + 4\omega\right]\right\}$$

$$A_3 = -\frac{(\partial p/\partial \theta)}{20\eta}\left[\frac{\left(\dfrac{\partial p}{\partial \theta}\right)h}{2\eta r^2} + \frac{\omega}{h}\right]$$

$$A_4 = \frac{1}{120}\frac{(\partial p/\partial \theta)^2}{\eta^2 r^2}$$

Flow continuity, Equation (10.18) can be written in the integral form by integrating across the film thicknesses as:

$$\int_o^h \frac{\partial}{\partial \theta}(v_\theta dz) + \int_o^h \frac{\partial}{\partial r}(rv_r dz) + \int_o^h r\frac{\partial v_z}{\partial z}dz = 0 \tag{10.22}$$

using Liebnitz's rule for integration it becomes,

$$\frac{\partial}{\partial \theta}\int_o^h v_\theta dz + \frac{\partial}{\partial r}\int_o^h \frac{\partial}{\partial r} rv_r dz - v_\theta(h,\theta)\frac{\partial h}{\partial \theta} - v_r(h,\theta)\frac{\partial h}{\partial r} - rv_z = 0 \tag{10.23}$$

where
$$v_z = \frac{\partial h}{\partial t}$$

Combining Equations (10.20), (10.21), and (10.23), modified Reynolds equation including centrifugal inertia of fluid is obtained as:

$$\frac{\partial}{\partial r}\left[\frac{rh^3}{\eta}\left(\frac{\partial p}{\partial r}\right)\right] + \frac{1}{r}\frac{\partial}{\partial \theta}\left[\frac{h^3}{\eta}\left(\frac{\partial p}{\partial \theta}\right)\right] = 6r\omega\frac{\partial h}{\partial \theta} + \left(\frac{3}{10}\rho\omega^2\right)\frac{\partial}{\partial r}\left(\frac{r^2h^3}{\eta}\right) - \frac{\rho}{10}\frac{\partial}{\partial r}\left\{\frac{h^5}{\eta^3}\left(\frac{\partial p}{\partial \theta}\right)\left[\eta\omega - \frac{3}{28}\left(\frac{h}{r}\right)^2\left(\frac{\partial p}{\partial \theta}\right)\right]\right\} \tag{10.24}$$

This is a general equation for incompressible lubrication which account for variable viscosity field. In other words, viscosity pressure and temperature effects can be incorporated in the fluid film lubrication problems. However, to incorporate thermal effect, temperature field will have to be determined by solving energy equation satisfying appropriate boundary conditions.

## 10.3 | Performance of Circular Step Hydrostatic Thrust Bearing Including Centrifugal Inertia and Using Bubbly Lubricant

Influence of presence of air bubbles in the lubricant has been investigated considering centrifugal inertia effect on the performance of hydrostatic thrust bearing. Bearing configuration is shown in Fig. 10.1. A mechanical face seal configuration is also similar to this.

The important aspect of hydrostatic thrust bearing and face seals at high rotational speed is that cavitations occur which reduce the film domain and become severe when centrifugal effect is taken into consideration. It occurs in the region of low pressure and high temperature near the outer periphery. Usually two types of cavitation are observed, viz., separation cavitation or true cavitation. Separation cavitation occurs when the bearing/seal is immersed in the sealing fluid. When the outer perimeter is exposed to air, the film may separate from the stationary surface and remain attached to the rotating surface. Otherwise, in the regions of subambient pressure dissolved air come out and the film raptures, which is known as true cavitation. This produces a cavity which extends completely across the clearance space. Cavity pressure is close to saturation pressure of the liquid. Therefore, theory dealing with hydrostatic step thrust bearings and mechanical face seals including centrifugal inertia must also involve thermal analysis. Thus, Reynolds equation and energy equation have to be solved simultaneously. Reynolds equation is rewritten as

$$\begin{aligned}\frac{\partial}{\partial r}\left\{\frac{rh^3}{\eta}\left(\frac{\partial p}{\partial r}\right)\right\}+\frac{1}{r}\frac{\partial}{\partial\theta}\left\{\frac{h^3}{\eta}\left(\frac{\partial p}{\partial\theta}\right)\right\}&=6r\omega\frac{\partial h}{\partial\theta}\\&+0.3\rho\omega^2\frac{\partial}{\partial r}\left(\frac{r^2h^3}{\eta}\right)-\frac{\rho\omega}{10}\frac{\partial}{\partial r}\left\{\frac{h^5}{\eta^2}\left(\frac{\partial p}{\partial\theta}\right)\right\}+\frac{3\rho}{280}\frac{\partial}{\partial r}\left\{\frac{h^7}{\eta^3r^2}\left(\frac{\partial p}{\partial\theta}\right)\right\}\end{aligned}\tag{10.25}$$

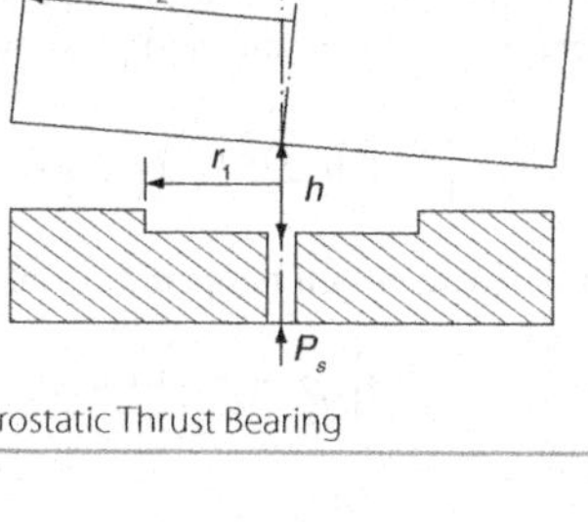


**Figure 10.1** | Tilted Circular Step Hydrostatic Thrust Bearing

and the corresponding energy equation is written as

$$\left(q_\theta \frac{\partial T}{\partial \theta} + q_r \frac{\partial T}{\partial r}\right) = \frac{\eta}{\rho c_p}\left\{\left(\frac{\partial v_r}{\partial z}\right)^2 + \left(\frac{\partial v_\theta}{\partial z}\right)^2\right\} \tag{10.26}$$

where $q_\theta$ and $q_r$ are lubricant volume flow rates in $\theta$ and $r$ directions which can be given by

$$q_\theta = \int_o^h v_\theta dz$$

$$q_r = \int_o^h v_r dz$$

Expressions for $v_\theta$ and $v_r$ are given in Equations (10.20) and (10.21), respectively. Substituting and integrating $q_\theta$ and $q_r$ are obtained as

$$q_\theta = \frac{rh\omega}{2} - \frac{h^3}{12\eta}\left(\frac{\partial p}{\partial \theta}\right) \tag{10.27}$$

$$q_r = \frac{h^3}{12\eta}\left[\left\{-\left(\frac{\partial p}{\partial r}\right) + 0.3\rho r\omega^2\right\} + \left\{\frac{3}{280}\frac{\rho h^4}{r^3\eta}\left(\frac{\partial p}{\partial \theta}\right) - \frac{1}{10}\frac{\rho h^2\omega}{r\eta}\left(\frac{\partial p}{\partial \theta}\right)\right\}\right] \tag{10.28}$$

Neglecting pressure and temperature effects the viscosity and density of a bubbly lubricant are given by following expressions given in Chapter 2 as:

$$\frac{1}{\eta} = \frac{x}{\eta_a} + \frac{1-x}{\eta_l} \tag{10.29}$$

$$\frac{1}{\rho} = \frac{x}{(\rho_a)} + \frac{1-x}{(\rho_l)} \tag{10.30}$$

where $x$ is volume fraction of air and symbols $a$ and $l$ stand for air and liquid, respectively. To investigate the influence of centrifugal inertia effects, Equation (10.25) and (10.26) have to be solved using available numerical procedures, e.g., finite difference method satisfying proper boundary conditions.

Lubricant film thickness expression of a misaligned or tilted seal configuration shown in Fig. 10.2 is written as

$$h = h_0 - r\cos\theta \tan\lambda \tag{10.31}$$

where $\lambda$ is angle of misalignment or tilt and $h_0$ is film thickness in parallel position.

Air viscosity and density are functions of temperature and pressure which can be written as

$$\rho_a = \frac{p}{RT}, \eta_a = \eta_{ia}\left[1 + 0.00275T - (4.75e-7)T^2\right] \tag{10.32a}$$

Lubricant density and viscosity are dependent on temperature only in following manner as given below:

$$\eta_l = \eta_{il}\exp\{\beta(T_{in} - T)\}, \rho_l = \rho_{il}\{1 + \alpha(T_{in} - T)\} \tag{10.32b}$$

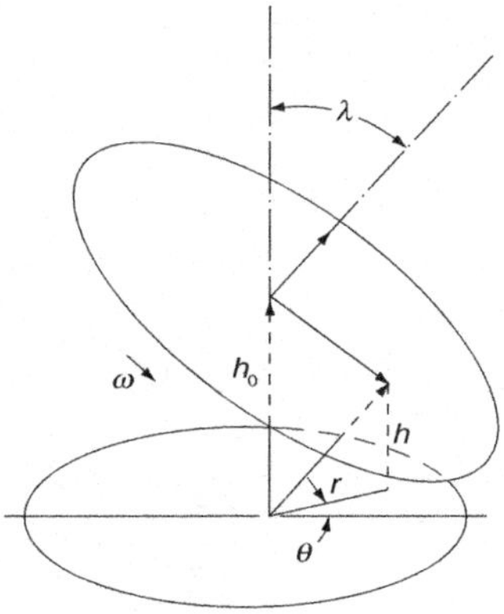


**Figure 10.2** | Coordinate System of Tilted Circular Step Hydrostatic Thrust Bearing

$T_{in}$ is the inlet temperature of the lubricant. β, α are viscosity and density temperature coefficients, respectively. Subscripts *a* and *l* refer to air and lubricant and *i* for inlet values. Boundary conditions used are expressed as:

$$p = p_s \left(\text{supply pressure}\right) \; at \; \mathrm{r} = \mathrm{r}_1$$

$$p = 0 \text{ at } r = r_2$$

$$p = \frac{dp}{d\theta} = \frac{dp}{dr} = 0 \text{, at the cavitation boundary} \tag{10.33}$$

$$T = T_{in} \text{ at } r = r_1$$

The algorithm and flowchart for the solution of the problem is given in Fig. 10.3. The input data used for the case study are given in Table 10.1.

Load capacity, frictional power loss, and mass flow rate are determined for known pressure distribution. Thus,

Load capacity,
$$W = \int_{r_1}^{r_2} \int_{0}^{2\pi} p r dr d\theta \tag{10.34}$$

Frictional power loss factor is determined from frictional torque as:

$$P_f = \int_{r_1}^{r_2} \int_{0}^{2\pi} \omega \tau_{r\theta} r^2 dr d\theta$$

$$P_f = \int_0^{2\pi} \int_{r_1}^{r^2} \eta \frac{r\omega^2}{h} r^2 dr d\theta \tag{10.35}$$

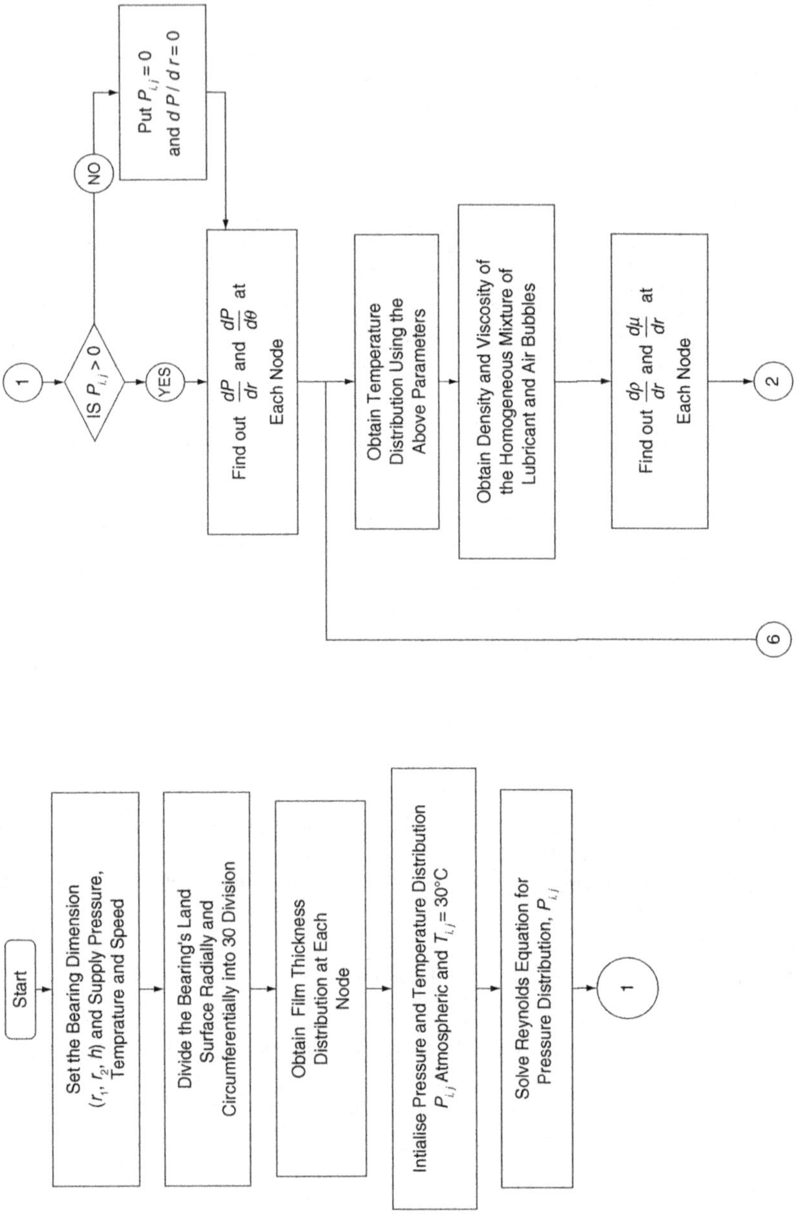
Start
Set the Bearing Dimension ($r_1$, $r_2$, $h$) and Supply Pressure, Temprature and Speed
Divide the Bearing's Land Surface Radially and Circumferentially into 30 Division
Obtain Film Thickness Distribution at Each Node
Intialise Pressure and Temperature Distribution $P_{i,j}$ Atmospheric and $T_{i,j}$ = 30°C
Solve Reynolds Equation for Pressure Distribution, $P_{i,j}$
1
1
IS $P_{i,j} > 0$
NO
Put $P_{i,j} = 0$ and $dP/dr = 0$
YES
Find out $\frac{dP}{dr}$ and $\frac{dP}{d\theta}$ at Each Node
Obtain Temperature Distribution Using the Above Parameters
Obtain Density and Viscosity of the Homogeneous Mixture of Lubricant and Air Bubbles
Find out $\frac{d\rho}{dr}$ and $\frac{d\mu}{dr}$ at Each Node
2
6

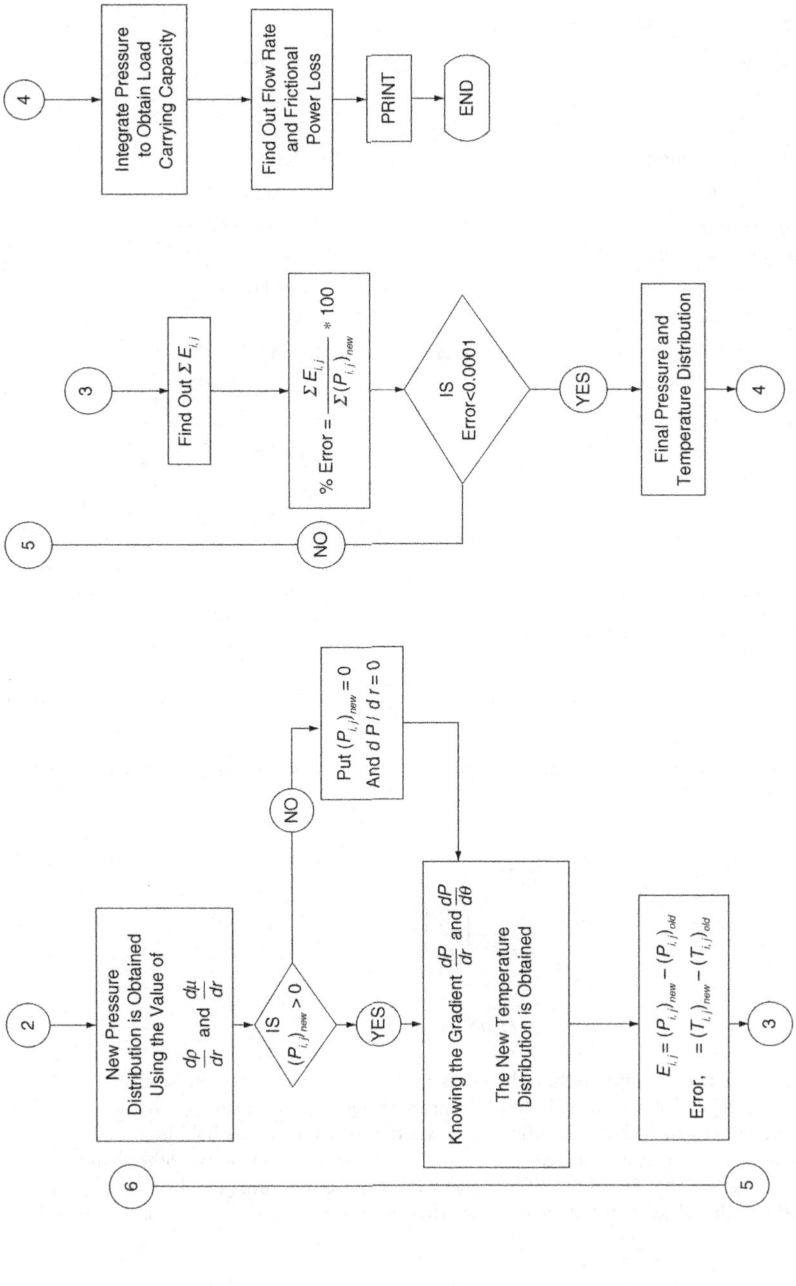


**Figure 10.3** | Flowchart

**Table 10.1** | Input Data

| | |
|---|---|
| Recess radius | 0.16 m |
| Outer radius | 0.237 m |
| Film clearance | $10^{-4}$ m |
| Radial extent of bearing | 0.077 m |
| Angular speed, $\omega$ | 0-500 rad/s |
| Supply pressure, $p_s$ | $1.5 \times 10^5$ N/m$^2$ |
| Lubricant inlet temperature, $T_{in}$ | 30$^0$C |
| Lubricant inlet density, $\rho_{il}$ | 855 Kg/m$^3$ |
| Lubricant inlet viscosity, $\eta_{il}$ | 0.05 Ns/m$^2$ |
| Coefficient of thermal expansion of lubricant, $\alpha$ | $7.34 \times 10^{-4}$/$^0$C |
| Temp. viscosity coefficient of lubricant, $\beta$ | $4.91 \times 10^{-2}$/$^0$C |
| Angle of misalignment, $\lambda$ | 0.001265 radian |
| Specific heat at constant pressure, $c_p$ | $2.02 \times 10^3$ Joule/Kg $^0$C |
| Gas constant of air, $R$ | 287 Joule/Kg mole/$^0$K |
| Air viscosity, $\eta_a$ | $1.9 \times 10^{-5}$Ns/m$^2$ |

Lubricant mass flow rate is determined as

$$M = \int_0^{2\pi} \int_0^{b} \rho v_r r d\theta dz \tag{10.36}$$

Dimensionless load capacity $\bar{W}$, frictional power loss $\bar{P}_f$ and mass flow rate $\bar{M}$ are determined from the following expressions.

$$\bar{W} = \frac{W}{p_s \pi r_2^2}; \tag{10.37a}$$

$$\bar{P}_f = \frac{P_f}{\eta_i r_2^3 \omega^2}\left(\frac{h_2}{r_2}\right); \tag{10.37b}$$

$$\bar{M} = \frac{M}{\rho_i L^2 \omega h_2} \tag{10.37c}$$

where $L = (r_2 - r_1)$ and $h_2$ is the minimum film thickness at the outer periphery of the bearing.

Performance of misaligned step bearing in terms of pressure, temperature, load capacity, power loss, and mass flow rate are presented which show the influence of speed of rotation and air bubble content.

Load capacity reduces with speed, whereas power loss and mass flow rate, on the other hand, increase. It is seen that air bubble content significantly influences the bearing performance. Improvement in load capacity is observed with increase in air bubble content, whereas power loss factor and mass flow rate reduce with increase in air bubble content. Thus, inclusion of air bubble in small amount may be beneficial toward improving the performance of the bearing. Typical pressure and temperature variation at different speeds of rotation are shown in Figs 10.4 and 10.5, respectively.

Figure 10.6 shows the influence of air bubble content on the pressure distribution. Pressure distribution for both isothermal and thermal analysis is shown in Fig. 10.7.

Similarly, influence of air bubble content on mass flow rate, load capacity and power loss are shown in Figs 10.8, 10.9, and 10.10, respectively. Influence of air bubble content on the cavitation zone can be seen in Fig. 10.11 in the bearing for both parallel and misaligned configurations. One can clearly see that cavitation zone is reduced with inclusion of air bubble in the lubricant.

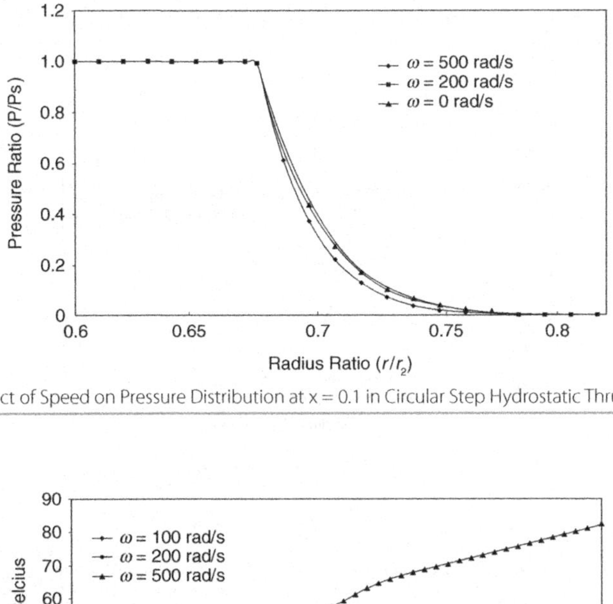


**Figure 10.4** | Effect of Speed on Pressure Distribution at x = 0.1 in Circular Step Hydrostatic Thrust Bearing

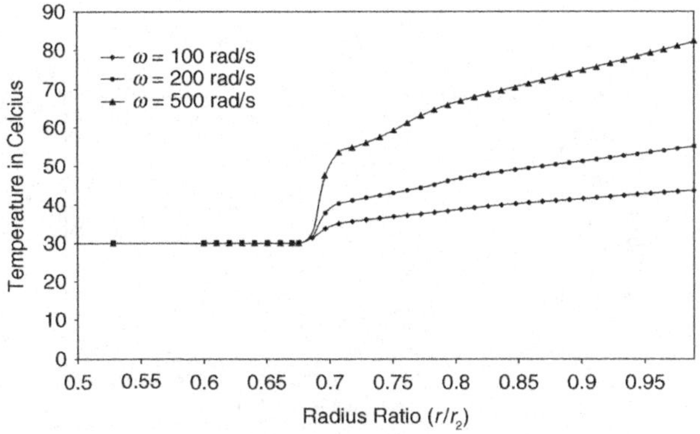


**Figure 10.5** | Effect of Speed on Temperature Distribution at $x = 0.1$ in Circular Step Hydrostatic Thrust Bearing

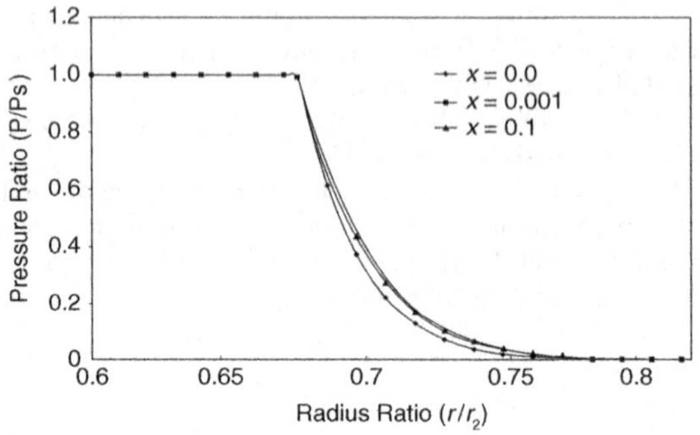


**Figure 10.6** | Effect of Air Bubble on the Pressure Distribution in Circular Step Hydrostatic Thrust Bearing at a Speed of 200 rad/sec

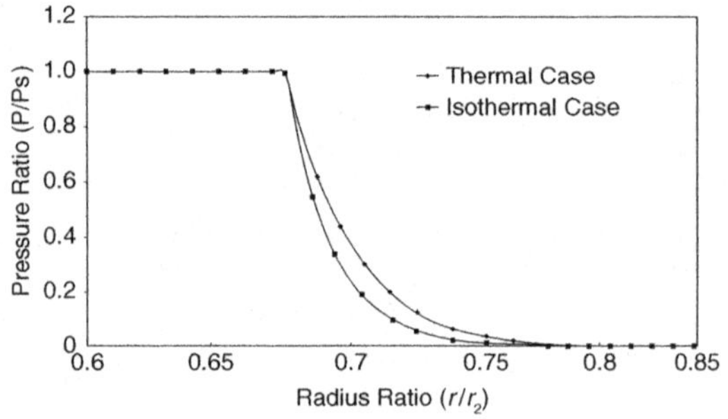


**Figure 10.7** | Pressure Distributions Along Radial Direction at a Speed of 200 rad/sec

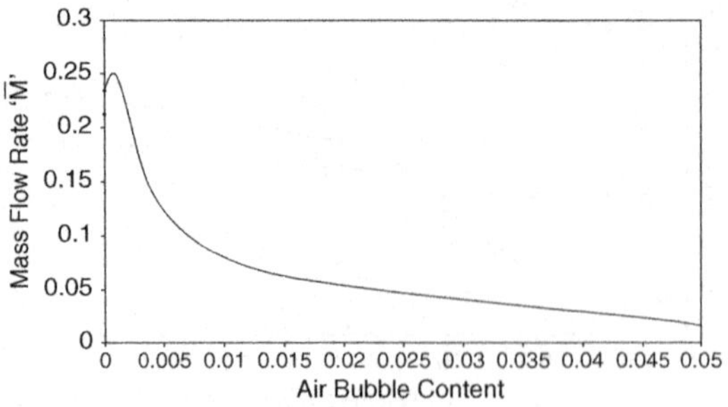


**Figure 10.8** | Variation in Mass Flow Rate due to Variation in Air Bubble Content

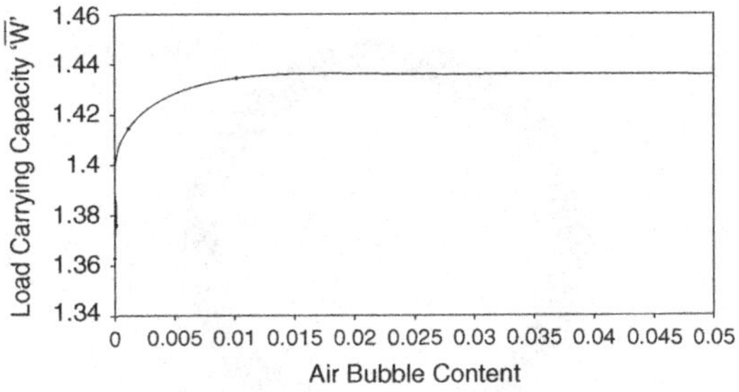


**Figure 10.9** | Variation of Load Carrying Capacity due to Variation in Air Bubble Content

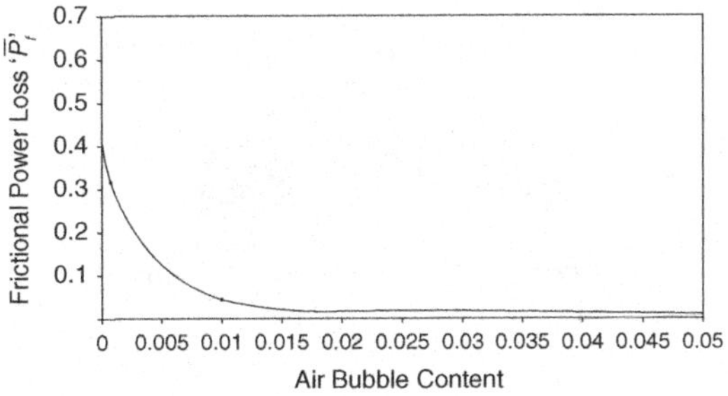


**Figure 10.10** | Variation in Frictional Power Loss due to Variation in Air Bubble Content

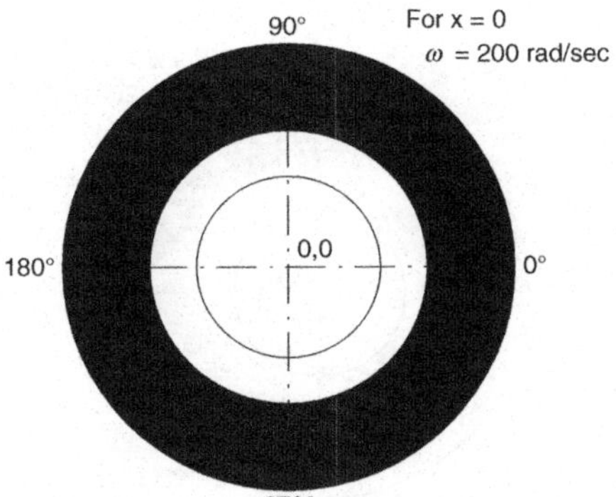


**Figure 10.11(a)** | Cavitations Zone on Bearing Land Surface in Circular Step Hydrostatic Thrust Bearing

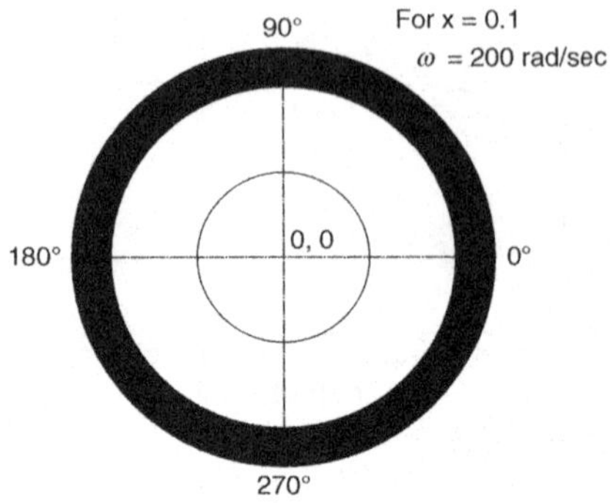


**Figure 10.11(b)** | Cavitations Zone on Bearing Land Surface for Circular Step Hydrostatic Thrust Bearing

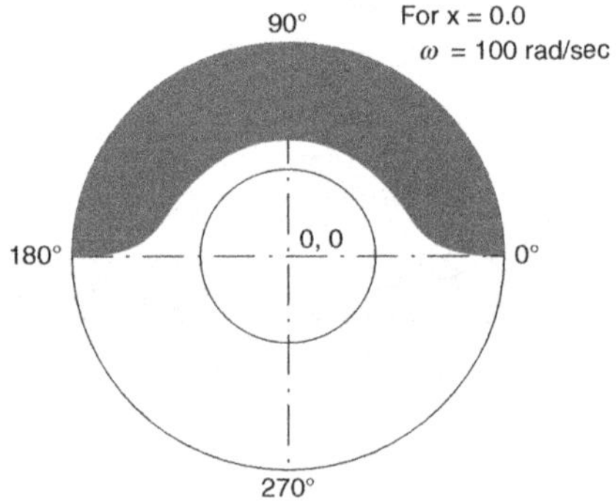


**Figure 10.11(c)** | Cavitations Zone on Bearing Land Surface for Tilted Hydrostatic Thrust Bearing ($\lambda = 0.001265$ rad )

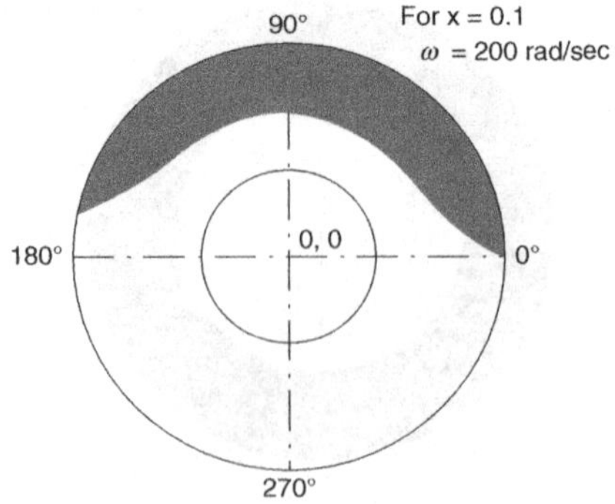


**Figure 10.11(d)** | Cavitations Zone on Bearing Land Surface for Tilted Hydrostatic Thrust Bearing ($\lambda = 0.001265$ rad)

## 10.4 | Reynolds Equation for Journal Bearings Including Fluid Inertia Effects

There are several applications which are of practical importance where Reynolds numbers are quite high, e.g., liquid metal bearings used in nuclear installation, after Re = 5000 in cryogenic application using liquid gases as lubricant where Re = 5000 and in some air bearings/air cushions where Re ≅ 200,000, etc. In other words, in applications where low viscosity fluid is used as lubricant, Reynolds number is likely to be high and convective inertia begins to influence the performance of the bearings. This problem was studied by several researchers, viz., Milne (1959), Osterle *et al.* (1957), Constantinescu (1970), etc. Effect of fluid inertia was reported to be generally small. However, theoretical procedures to incorporate fluid inertia effects in journal bearings have been developed by Reinhardt and Lund (1965) and Kakoty and Majumdar (2000) using perturbation method. The procedure is briefly presented below.

Kakoty and Majumdar (2000) followed a procedure discussed below to derive modified Reynolds equation that includes convective inertia effects. Referring to Fig. 10.12, dimensionless form of momentum equations and continuity equation are written as

$$\overline{\text{Re}}\left(\frac{\partial\bar{u}}{\partial\tau}+\bar{u}\frac{\partial\bar{u}}{\partial\theta}+\frac{D}{L}\bar{v}\frac{\partial\bar{u}}{\partial\bar{y}}+\bar{w}\frac{\partial\bar{u}}{\partial\bar{z}}=-\frac{\partial\bar{p}}{\partial\theta}+\frac{\partial^2\bar{u}}{\partial\bar{z}^2}\right) \tag{10.38}$$

$$\overline{\text{Re}}\left(\frac{\partial\bar{v}}{\partial\tau}+\bar{u}\frac{\partial\bar{v}}{\partial\theta}+\bar{v}\left(\frac{D}{L}\right)\frac{\partial\bar{v}}{\partial\bar{y}}+\bar{w}\frac{\partial\bar{v}}{\partial\bar{z}}=-\left(\frac{D}{L}\right)\frac{\partial\bar{p}}{\partial\bar{y}}+\frac{\partial^2\bar{v}}{\partial\bar{z}^2}\right) \tag{10.39}$$

$$0=-\frac{\partial\bar{p}}{\partial\bar{z}} \tag{10.40}$$

$$\frac{\partial\bar{u}}{\partial\theta}+\left(\frac{D}{L}\right)\frac{\partial\bar{v}}{\partial\bar{y}}+\frac{\partial\bar{w}}{\partial\bar{z}}=0 \tag{10.41}$$

where, $\theta=\dfrac{x}{R},\ \bar{y}=\dfrac{y}{(L/2)},\ \bar{z}=\dfrac{z}{C},\ \bar{h}=\dfrac{h}{C},$

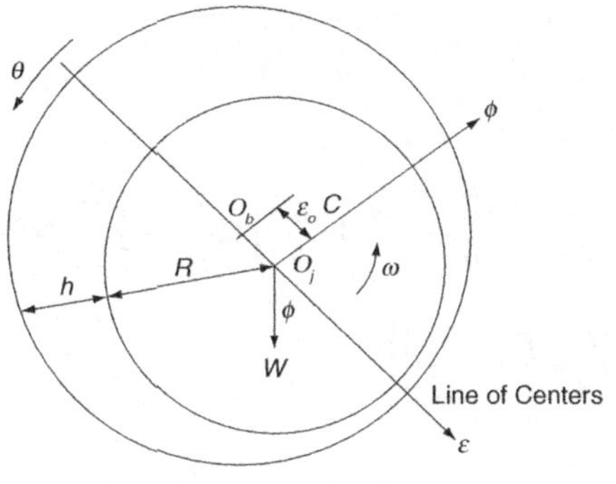


**Figure 10.12** | Schematic Diagram of a Journal Bearing

$$\bar{u}=\frac{u}{R\omega},\bar{v}=\frac{v}{R\omega},\bar{w}=\frac{w}{C\omega},\ \tau=\omega_p t,\ \text{and}\ \bar{p}=\frac{p}{\eta\omega\left(\frac{R}{C}\right)^2}$$

$\omega_p$ – angular velocity of whirl

Since the velocity profiles remain unaltered, i.e., parabolic, when inertia effect is included, velocity components are expressed as

$$\bar{u}=\frac{\bar{z}}{\bar{h}}+q_\theta\left(\frac{\bar{z}^2}{\bar{h}^2}-\frac{\bar{z}}{\bar{h}}\right) \tag{10.42}$$

$$\bar{v}=q_{\bar{y}}\left(\frac{\bar{z}^2}{\bar{h}^2}-\frac{\bar{z}}{\bar{h}}\right) \tag{10.43}$$

Substituting these into momentum equations and integrating yields

$$q_\theta=\frac{\bar{h}^2}{2}\frac{\partial p}{\partial\theta}+\overline{\text{Re}}\,I_x \tag{10.44}$$

$$q_{\bar{y}}=\frac{\bar{h}^2}{2}\left(\frac{D}{L}\right)\frac{\partial\bar{p}}{\partial\bar{y}}+\overline{\text{Re}}\,I_y \tag{10.45}$$

Using the continuity equation modified Reynolds equation including convective inertia can be obtained as

$$\frac{\partial}{\partial\theta}\left(\bar{h}^3\frac{\partial\bar{p}}{\partial\theta}\right)+\left(\frac{D}{L}\right)^2\frac{\partial}{\partial\bar{y}}\left(\bar{h}^3\frac{\partial\bar{p}}{\partial\bar{y}}\right)=12\bar{V}-6\frac{\partial\bar{h}}{\partial\theta}-2\overline{\text{Re}}\left[\frac{\partial\left(\bar{h}I_x\right)}{\partial\theta}+\left(\frac{D}{L}\right)\frac{\partial}{\partial\bar{y}}\left(I_y\bar{h}\right)\right] \tag{10.46}$$

where

$$I_x=\frac{\bar{h}}{2}\left[-\frac{1}{2}\left(\Omega\frac{\partial\bar{h}}{\partial\tau}+\frac{4}{3}\frac{\partial\bar{h}}{\partial\theta}-2\bar{V}+\frac{D}{L}\frac{\bar{h}}{6}\frac{\partial q_{\bar{y}}}{\partial\bar{y}}\right)-\frac{q_\theta}{6}\left(\Omega\frac{\partial\bar{h}}{\partial\tau}+\frac{\partial\bar{h}}{\partial\theta}-\frac{D}{L}\frac{\bar{h}}{5}\frac{\partial q_{\bar{y}}}{\partial\bar{y}}\right)\right.$$

$$\left.-\frac{\bar{h}}{6}\times\left(\Omega\frac{\partial q_\theta}{\partial\tau}+\frac{\partial q_\theta}{\partial\theta}-\frac{D}{L}\frac{q_{\bar{y}}}{5}\frac{\partial q_\theta}{\partial\bar{y}}\right)+\frac{q_\theta}{15}\left(\bar{h}\frac{\partial q_\theta}{\partial\theta}+\frac{q_\theta}{2}\frac{\partial\bar{h}}{\partial\theta}\right)\right] \tag{10.47}$$

and

$$I_y=\frac{\bar{h}}{2}\left[-\frac{q_{\bar{y}}}{6}\left(\Omega\frac{\partial\bar{h}}{\partial\tau}+\frac{1}{2}\frac{\partial\bar{h}}{\partial\theta}-\frac{q_\theta}{5}\frac{\partial\bar{h}}{\partial\theta}\right)-\frac{\bar{h}}{6}\times\left(\Omega\frac{\partial q_{\bar{y}}}{\partial\tau}+\frac{1}{2}\frac{\partial q_{\bar{y}}}{\partial\theta}-\frac{q_{\bar{y}}}{5}\frac{\partial q_\theta}{\partial\theta}\right)\right.$$

$$\left.+\frac{\bar{h}}{15}\left(\frac{q_\theta}{2}\cdot\frac{\partial q_{\bar{y}}}{\partial\theta}+\frac{D}{L}q_{\bar{y}}\frac{\partial q_{\bar{y}}}{\partial\bar{y}}\right)\right] \tag{10.48}$$

where $\Omega=\frac{\omega_p}{\omega}$, whirl ratio.

It is assumed that the journal whirls about its mean steady state position given by $\varepsilon_0$ and $\varphi_0$. Therefore, for first order perturbation pressure $\bar{p}$, flow parameters $q_\theta$ and $q_{\bar{y}}$, and film thickness $\bar{h}$ can be written as:

$$\bar{p} = \bar{p}_0 + \bar{p}_x \varepsilon_1 e^{i\tau} + \bar{p}_z \varepsilon_0 \varphi_1 e^{i\tau} + i\bar{p}_{\dot{x}} \varepsilon_1 e^{i\tau} + i\bar{p}_{\dot{z}} \varepsilon_0 \varphi_1 e^{i\tau} \tag{10.49}$$

$$q_\theta = q_{\theta 0} + q_{\theta x} \varepsilon_1 e^{i\tau} + q_{\theta z} \varepsilon_0 \varphi_1 e^{i\tau} + iq_{\theta \dot{x}} \varepsilon_1 e^{i\tau} + iq_{\theta \dot{z}} \varepsilon_0 \varphi_1 e^{i\tau} \tag{10.50}$$

$$q_{\bar{y}} = q_{\bar{y}0} + q_{\bar{y}x} \varepsilon_1 e^{i\tau} + q_{\bar{y}z} \varepsilon_0 \varphi_1 e^{i\tau} + iq_{\bar{y}\dot{x}} \varepsilon_1 e^{i\tau} + iq_{\bar{y}\dot{z}} \varepsilon_0 \varphi_1 e^{i\tau} \tag{10.51}$$

$$\bar{h} = \bar{h}_0 + \varepsilon_1 e^{i\tau} \cos\theta + \varepsilon_0 \varphi_1 e^{i\tau} \sin\theta \tag{10.52}$$

where

$$\varepsilon = \varepsilon_0 + \varepsilon_1 e^{i\tau}, \; \varepsilon_1 << \varepsilon_0 \tag{10.53}$$

$$\varphi = \varphi_0 + \varphi_1 e^{i\tau}, \; \varphi_1 << \varphi_0 \tag{10.54}$$

Substituting these equations into Equation (10.44) to (10.46), we can obtain perturbed equations retaining only up to first order terms. Thus, five sets of equations are obtained as:

***First Set***

$$q_{\theta 0} = \frac{\bar{h}_0^2}{2} \frac{\partial \bar{p}_0}{\partial \theta} + \overline{\mathrm{Re}}\, I_{x0} \tag{10.51}$$

$$q_{\bar{y}0} = \frac{\bar{h}_0^2}{2} \frac{D}{L} \frac{\partial \bar{p}_0}{\partial \bar{y}} + \overline{\mathrm{Re}}\, I_{y0} \tag{10.52}$$

$$\frac{\partial}{\partial\theta}\left(\bar{h}_0^3 \frac{\partial \bar{p}_0}{\partial \theta}\right) + \left(\frac{D}{L}\right)^2 \frac{\partial}{\partial \bar{y}}\left(\bar{h}_0^3 \frac{\partial \bar{p}_0}{\partial \bar{y}}\right) = 6\frac{\partial \bar{h}}{\partial \theta} - 2\overline{\mathrm{Re}}\left[\frac{\partial\left(\bar{h}_0 I_{x0}\right)}{\partial \theta} + \left(\frac{D}{L}\right)\frac{\partial}{\partial \bar{y}}\left(I_{y0}\bar{h}_0\right)\right] \tag{10.53}$$

where

$$I_{x0} = \bar{h}_0\left(\frac{1}{6} - \frac{q_{\theta 0}}{12} + \frac{q_{\theta 0}^2}{60}\right)\frac{\partial \bar{h}_0}{\partial \theta} - \bar{h}_0^2\left(\frac{1}{24} - \frac{q_{\theta 0}}{60}\right)\frac{D}{L}\frac{\partial q_{\bar{y}0}}{\partial \bar{y}} + \bar{h}_0^2\left(\frac{q_{\theta 0}}{30} - \frac{1}{12}\right)\frac{\partial q_{\theta 0}}{\partial \theta} + \frac{\bar{h}_0^2 q_{\bar{y}0}}{60}\frac{D}{L}\frac{\partial q_{\bar{y}0}}{\partial \bar{y}} \tag{10.54}$$

$$I_{y0} = \bar{h}_0 q_{\bar{y}0}\left(-\frac{1}{24} + \frac{q_{\theta 0}}{60}\right)\frac{\partial \bar{h}_0}{\partial \theta} - \bar{h}_0^2\left(\frac{1}{24} - \frac{q_{\theta 0}}{60}\right)\frac{\partial q_{\bar{y}0}}{\partial \theta} + \frac{\bar{h}_0^2 q_{\bar{y}0}}{60}\left(\frac{\partial q_{\theta 0}}{\partial \theta} + 2\frac{D}{L}\frac{\partial q_{\bar{y}0}}{\partial \bar{y}}\right) \tag{10.55}$$

Boundary conditions for steady state pressure distribution are given below

$$\bar{p}_0(\theta, \bar{y}) = 0 \text{ for } \theta = \theta_1 \text{ and } \theta_2$$

$$\bar{p}_0(\theta, \pm 1) = 0$$

$$\frac{\partial \bar{p}_0}{\partial \bar{y}}(\theta, 0) = 0 \tag{10.56}$$

$$\frac{\partial \bar{p}_0}{\partial \theta} = 0 \text{ at } \theta = \theta_2$$

where $\theta_1$ and $\theta_2$ are the coordinates at which film begin and cavitates, respectively.

***Second Set***

$$q_{\theta x} = \bar{h}_0 \frac{\partial \bar{p}_0}{\partial \theta} \cos\theta + \frac{\bar{h}_0^2}{2} \frac{\partial \bar{p}_x}{\partial \theta} + \overline{\mathrm{Re}}\, I_{xx} \tag{10.57}$$

$$q_{\bar{y}x} = \frac{D}{L} \bar{h}_0 \frac{\partial \bar{p}_0}{\partial \bar{y}} \cos\theta + \frac{\bar{h}_0^2}{2} \frac{D}{L} \frac{\partial \bar{p}_x}{\partial \bar{y}} + \overline{\mathrm{Re}}\, I_{yx} \tag{10.58}$$

$$\frac{\partial}{\partial\theta}\left(3\bar{h}_0^2 \cos\theta \frac{\partial \bar{p}_0}{\partial\theta}\right) + \frac{\partial}{\partial\theta}\left(\bar{h}_0^3 \frac{\partial \bar{p}_x}{\partial\theta}\right) + \left(\frac{D}{L}\right)^2 \frac{\partial}{\partial\bar{y}}\left(\bar{h}_0^3 \frac{\partial \bar{p}_x}{\partial\bar{y}}\right) + \left(\frac{D}{L}\right)^2 \frac{\partial}{\partial\bar{y}}\left(3\bar{h}_0^2 \cos\theta \frac{\partial \bar{p}_0}{\partial\bar{y}}\right)$$

$$= -6\sin\theta - 2\overline{\mathrm{Re}}\times\left[\frac{\partial\left(\bar{h}_0 I_{xx}\right)}{\partial\theta} + \left(\frac{D}{L}\right)\frac{\partial}{\partial\bar{y}}\left(I_{yx}\bar{h}_0\right) + \frac{\partial}{\partial\theta}\left(I_{x0}\cos\theta\right) + \frac{D}{L}\frac{\partial}{\partial\bar{y}}\left(I_{y0}\cos\theta\right)\right] \tag{10.59}$$

where

$$I_{xx} = C_1 \sin\theta + C_2 \cos\theta + C_3 q_{\theta x} + C_4 q_{\bar{y}x} + C_5 \frac{D}{L}\frac{\partial q_{\bar{y}x}}{\partial\bar{y}} + C_6 \frac{D}{L}\frac{\partial q_{\theta x}}{\partial\bar{y}} + C_7 \frac{\partial q_{\theta x}}{\partial\theta} \tag{10.60}$$

$$I_{yx} = K_1 \sin\theta + K_2 \cos\theta + K_3 q_{\theta x} + K_4 q_{\bar{y}x} + K_5 \frac{\partial q_{\bar{y}x}}{\partial\theta} + K_6 \frac{\partial q_{\theta x}}{\partial\theta} + K_7 \frac{D}{L}\frac{\partial q_{\bar{y}x}}{\partial\bar{y}} \tag{10.61}$$

***Third Set***

$$q_{\theta z} = \bar{h}_0 \frac{\partial \bar{p}_0}{\partial \theta} \sin\theta + \frac{\bar{h}_0^2}{2} \frac{\partial \bar{p}_z}{\partial \theta} + \overline{\mathrm{Re}}\, I_{xz} \tag{10.62}$$

$$q_{\bar{y}z} = \frac{D}{L} \bar{h}_0 \frac{\partial \bar{p}_0}{\partial \bar{y}} \sin\theta + \frac{\bar{h}_0^2}{2} \frac{D}{L} \frac{\partial \bar{p}_z}{\partial \bar{y}} + \overline{\mathrm{Re}}\, I_{yz} \tag{10.63}$$

$$\frac{\partial}{\partial\theta}\left(3\bar{h}_0^2 \sin\theta \frac{\partial \bar{p}_0}{\partial\theta}\right) + \frac{\partial}{\partial\theta}\left(\bar{h}_0^3 \frac{\partial \bar{p}_y}{\partial\theta}\right) + \left(\frac{D}{L}\right)^2 \frac{\partial}{\partial\bar{y}}\left(\bar{h}_0^3 \frac{\partial \bar{p}_y}{\partial\bar{y}}\right) + \left(\frac{D}{L}\right)^2 \frac{\partial}{\partial\bar{y}}\left(3\bar{h}_0^2 \sin\theta \frac{\partial \bar{p}_0}{\partial\bar{y}}\right)$$

$$= 6\cos\theta - 2\overline{\mathrm{Re}}\times\left[\frac{\partial\left(\bar{h}_0 I_{xz}\right)}{\partial\theta} + \left(\frac{D}{L}\right)\frac{\partial}{\partial\bar{y}}\left(I_{yz}\bar{h}_0\right) + \frac{\partial}{\partial\theta}\left(I_{x0}\sin\theta\right) + \frac{D}{L}\frac{\partial}{\partial\bar{y}}\left(I_{y0}\sin\theta\right)\right] \tag{10.64}$$

where

$$I_{xz} = -C_1 \cos\theta + C_2 \sin\theta + C_3 q_{\theta z} + C_4 q_{\bar{y}z} + C_5 \frac{D}{L}\frac{\partial q_{\bar{y}z}}{\partial\bar{y}} + C_6 \frac{D}{L}\frac{\partial q_{\theta z}}{\partial\bar{y}} + C_7 \frac{\partial q_{\theta z}}{\partial\theta} \tag{10.65}$$

$$I_{yz} = -K_1 \cos\theta + K_2 \sin\theta + K_3 q_{\theta z} + K_4 q_{\bar{y}z} + K_5 \frac{\partial q_{\bar{y}z}}{\partial\theta} + K_6 \frac{\partial q_{\theta z}}{\partial\theta} + K_7 \frac{D}{L}\frac{\partial q_{\bar{y}z}}{\partial\bar{y}} \tag{10.66}$$

### Fourth Set

$$q_{\theta\dot{x}} = -\frac{\bar{h}_0^2}{2}\frac{\partial \bar{p}_{\dot{x}}}{\partial \theta} + \bar{\mathrm{Re}}\, I_{x\dot{x}} \tag{10.67}$$

$$q_{\bar{y}\dot{x}} = -\frac{\bar{h}_0^2}{2}\frac{D}{L}\frac{\partial \bar{p}_{\dot{x}}}{\partial \bar{y}} + \bar{\mathrm{Re}}\, I_{y\dot{x}} \tag{10.68}$$

$$\frac{\partial}{\partial\theta}\left(\bar{h}_0^3\frac{\partial \bar{p}_{\dot{x}}}{\partial \theta}\right) + \left(\frac{D}{L}\right)^2\frac{\partial}{\partial \bar{y}}\left(\bar{h}_0^3\frac{\partial \bar{p}_{\dot{x}}}{\partial \bar{y}}\right) = 12\Omega\cos\theta - 2\bar{\mathrm{Re}}\left[\frac{\partial\left(\bar{h}_0 I_{x\dot{x}}\right)}{\partial\theta} + \left(\frac{D}{L}\right)\frac{\partial}{\partial \bar{y}}\left(I_{y\dot{x}}\bar{h}_0\right)\right] \tag{10.69}$$

where

$$I_{x\dot{x}} = A_1\cos\theta + A_2 q_{\theta x} + A_3 q_{\theta\dot{x}} + A_4 q_{\bar{y}\dot{x}} + A_5\frac{D}{L}\frac{\partial q_{\bar{y}\dot{x}}}{\partial \bar{y}} + A_6\frac{D}{L}\frac{\partial q_{\theta\dot{x}}}{\partial \bar{y}} + A_7\frac{\partial q_{\theta\dot{x}}}{\partial\theta} \tag{10.70}$$

$$I_{y\dot{x}} = B_1\cos\theta + B_2 q_{\bar{y}x} + B_3 q_{\theta\dot{x}} + B_4 q_{\bar{y}\dot{x}} + B_5\frac{\partial q_{\bar{y}\dot{x}}}{\partial\theta} + B_6\frac{D}{L}\frac{\partial q_{\bar{y}\dot{x}}}{\partial \bar{y}} + B_7\frac{\partial q_{\theta\dot{x}}}{\partial\theta} \tag{10.71}$$

### Fifth Set

$$q_{\theta\dot{z}} = -\frac{\bar{h}_0^2}{2}\frac{\partial \bar{p}_{\dot{z}}}{\partial \theta} + \bar{\mathrm{Re}}\, I_{x\dot{z}} \tag{10.72}$$

$$q_{\bar{y}\dot{z}} = -\frac{\bar{h}_0^2}{2}\frac{D}{L}\frac{\partial \bar{p}_{\dot{z}}}{\partial \bar{y}} + \bar{\mathrm{Re}}\, I_{y\dot{z}} \tag{10.73}$$

$$\frac{\partial}{\partial\theta}\left(\bar{h}_0^3\frac{\partial \bar{p}_{\dot{z}}}{\partial \theta}\right) + \left(\frac{D}{L}\right)^2\frac{\partial}{\partial \bar{y}}\left(\bar{h}_0^3\frac{\partial \bar{p}_{\dot{z}}}{\partial \bar{y}}\right) = 24\Omega\sin\theta - 2\bar{\mathrm{Re}}\left[\frac{\partial\left(\bar{h}_0 I_{x\dot{z}}\right)}{\partial\theta} + \left(\frac{D}{L}\right)\frac{\partial}{\partial \bar{y}}\left(I_{y\dot{z}}\bar{h}_0\right)\right] \tag{10.74}$$

where

$$I_{x\dot{z}} = \left(A_1 + \Omega\frac{\bar{h}_0}{2}\right)\sin\theta + A_2 q_{\theta z} + A_3 q_{\theta\dot{z}} + A_4 q_{\bar{y}\dot{z}} + A_5\frac{D}{L}\frac{\partial q_{\bar{y}\dot{z}}}{\partial \bar{y}} + A_6\frac{D}{L}\frac{\partial q_{\theta\dot{z}}}{\partial \bar{y}} + A_7\frac{\partial q_{\theta\dot{z}}}{\partial\theta} \tag{10.75}$$

$$I_{y\dot{z}} = B_1\sin\theta + B_2 q_{\bar{y}z} + B_3 q_{\theta\dot{z}} + B_4 q_{\bar{y}\dot{z}} + B_5\frac{\partial q_{\bar{y}\bar{y}}}{\partial\theta} + B_6\frac{D}{L}\frac{\partial q_{\bar{y}\dot{z}}}{\partial \bar{y}} + B_7\frac{\partial q_{\theta\dot{z}}}{\partial\theta} \tag{10.76}$$

The coefficients $A_1, A_2 \ldots B_1, B_2 \ldots\ldots, C_1, C_2 \ldots K_1, K_2 \ldots$ are given in the Appendix 10.1 at the end of this chapter.

The boundary conditions for the above sets of equations are as follows

- $\bar{p}_i(\theta,\bar{y}) = 0$ for $\theta = \theta_1$ and $\theta_2$,
- $\bar{p}_i(\theta,\pm 1) = 0$,
- $\partial\bar{p}_i/\partial\bar{y}(\theta,0) = 0$, $i = x,\dot{x},z,\dot{z}$

The dimensionless dynamic forces along $\varepsilon$ and $\varphi$ directions (Fig.10.10) are given as follows

$$\left(\bar{F}_{\varepsilon}\right)_{d} = \left(-k_{\varepsilon\varepsilon} - i\Omega b_{\varepsilon\varepsilon} + \Omega^{2} a_{\varepsilon\varepsilon}\right)\varepsilon_{1} e^{i\tau} + \left(-k_{\varepsilon\varphi} - i\Omega b_{\varepsilon\varphi} + \Omega^{2} a_{\varepsilon\varphi}\right)\varepsilon_{0}\varphi_{1} e^{i\tau} \tag{10.77}$$

$$\left(\bar{F}_{\varphi}\right)_{d} = \left(-k_{\varphi\varepsilon} - i\Omega b_{\varphi\varepsilon} + \Omega^{2} a_{\varphi\varepsilon}\right)\varepsilon_{1} e^{i\tau} + \left(-k_{\varphi\varphi} - i\Omega b_{\varphi\varphi} + \Omega^{2} a_{\varphi\varphi}\right)\varepsilon_{0}\varphi_{1} e^{i\tau} \tag{10.78}$$

Under steady state condition

$$\left(\bar{F}_{\varepsilon}\right)_{0} = -\bar{W}_{0}\cos\varphi_{0} \tag{10.79}$$

$$\left(\bar{F}_{\varphi}\right)_{0} = \bar{W}_{0}\sin\varphi_{0} \tag{10.80}$$

where

$$\left(\bar{F}_{r}\right)_{0} = \left(\frac{F_{\varepsilon}C^{2}}{\eta\omega R^{3}L}\right) = \int_{0}^{1}\int_{\theta_{1}}^{\theta_{2}} \bar{p}\cos\theta\, d\theta\, d\bar{y} \tag{10.81}$$

$$\left(\bar{F}_{\varphi}\right)_{0} = \left(\frac{F_{\varphi}C^{2}}{\eta\omega R^{3}L}\right) = \int_{0}^{1}\int_{\theta_{1}}^{\theta_{2}} \bar{p}\sin\theta\, d\theta\, d\bar{y} \tag{10.82}$$

$$\bar{W}_{0} = \left[\left(\bar{F}_{\varepsilon}\right)_{0}^{2} + \left(\bar{F}_{\varphi}\right)_{0}^{2}\right]^{1/2} \quad \text{and} \quad \varphi_{0} = \tan^{-1}\left[-\frac{\left(F_{\varphi}\right)_{0}}{\left(F_{\varepsilon}\right)_{0}}\right] \tag{10.83}$$

where $\theta_1$ and $\theta_2$ are the coordinates at which the film starts and cavitates, respectively. The steady load can be expressed in terms of Sommerfeld number $S$ where $S = 1/\left(\pi \bar{W}_0\right)$. The dynamic coefficients $k_{..}$, $b_{..}$, $a_{..}$ ....which are dimensionless stiffness, damping, and inertia coefficients, respectively, can be calculated from $\bar{p}_x, \bar{p}_y, \bar{p}_{\dot{x}}, \bar{p}_{\dot{z}}$ and are given in Appendix 10.2 at the end of this chapter.

Using the dynamic coefficients stability of a rigid rotor can be analyzed in the following manner.

The equations of motions of a rigid rotor supported on two identical bearings under steady load are expressed for small amplitude oscillations about the equilibrium position as

$$\Omega^{2}\bar{M}\bar{W}_{0}\left(\ddot{\varepsilon} - \varepsilon\dot{\varphi}^{2}\right) = \bar{F}_{\varepsilon} + \bar{W}_{0}\cos\varphi \tag{10.84}$$

$$\Omega^{2}\bar{M}\bar{W}_{0}\left(\varepsilon\ddot{\varphi} + 2\dot{\varepsilon}\dot{\varphi}\right) = \bar{F}_{\varphi} - \bar{W}_{0}\sin\varphi \tag{10.85}$$

Since

$$\bar{F}_{\varepsilon} = \left(\bar{F}_{\varepsilon}\right)_{0} + \left(\bar{F}_{\varepsilon}\right)_{d} \quad \text{and} \quad \bar{F}_{\varphi} = \left(\bar{F}_{\varphi}\right)_{0} + \left(\bar{F}_{\varphi}\right)_{d} \tag{10.86}$$

Substituting Equations (10.53) and (10.54) into Equations (10.77) to (10.80), the equations of motion can be expressed in the following form:

$$\left(k_{\varepsilon\varepsilon} + i\Omega b_{\varepsilon\varepsilon} - \Omega^{2} a_{\varepsilon\varepsilon} - \Omega^{2}\bar{M}\bar{W}_{0}\right)\varepsilon_{1} e^{i\tau} + \left(k_{\varepsilon\phi} + i\Omega b_{\varepsilon\phi} - \Omega^{2} a_{\varepsilon\phi} + \frac{\bar{W}_{0}}{\varepsilon_{0}}\sin\phi_{0}\right)\varepsilon_{0}\phi_{1} e^{i\tau} = 0 \tag{10.87}$$

$$\left(k_{\phi\varepsilon} + i\Omega b_{\phi\varepsilon} - \Omega^2 a_{\phi\varepsilon}\right)\varepsilon_1 e^{i\tau} + \left(k_{\phi\phi} + i\Omega b_{\phi\phi} - \Omega^2 a_{\phi\phi} - \Omega^2 \bar{M}\bar{W}_0 + \frac{\bar{W}_0}{\varepsilon_0}\cos\phi_0\right)\varepsilon_0\phi_1 e^{i\tau} = 0 \quad (10.88)$$

For a nontrivial solution the determinant of the two equations should be zero. Equating the imaginary and real parts of the determinant separately to zero following two equations are obtained.

$$\begin{aligned} &k_{\varepsilon\varepsilon} b_{\varphi\varphi} + k_{\varphi\varphi} b_{\varepsilon\varepsilon} + b_{\varepsilon\varepsilon}\frac{\bar{W}_0}{\varepsilon_0}\cos\varphi_0 - k_{\varphi\varepsilon} b_{\varepsilon\varphi} - k_{\varepsilon\varphi} b_{\varphi\varepsilon} \\ &-\frac{\bar{W}_0}{\varepsilon_0} b_{\varphi\varepsilon}\sin\varphi_0 + \Omega^2\left(-b_{\varepsilon\varepsilon} a_{\varphi\varphi} - b_{\varepsilon\varepsilon}\bar{M}\bar{W}_0 - b_{\varphi\varphi} a_{\varepsilon\varepsilon}\right. \\ &\left.-\bar{W}_0\bar{M} b_{\varphi\varphi} + b_{\varphi\varepsilon} a_{\varepsilon\varphi} + b_{\varepsilon\varphi} a_{\varphi\varepsilon}\right) = 0 \end{aligned} \quad (10.89)$$

$$\begin{aligned} &\left(k_{\varepsilon\varepsilon} k_{\varphi\varphi} + k_{\varepsilon\varepsilon}\frac{\bar{W}_0}{\varepsilon_0}\cos\varphi_0 - k_{\varphi\varepsilon} k_{\varepsilon\varphi} - \frac{\bar{W}_0}{\varepsilon_0}\sin\varphi_0\right) \\ &+\Omega^2\left(-k_{\varepsilon\varepsilon} a_{\varphi\varphi} - \bar{M}\bar{W}_0 k_{\varepsilon\varepsilon} b_{\varphi\varphi} b_{\varepsilon\varepsilon} - k_{\varphi\varphi} a_{\varepsilon\varepsilon}\right. \\ &-a_{\varepsilon\varepsilon}\frac{\bar{W}_0}{\varepsilon_0}\cos\varphi_0 - \bar{M}\bar{W}_0 k_{\varphi\varphi} - \frac{\bar{W}_0^2}{\varepsilon_0}\bar{M}\cos\varphi_0 \\ &\left.+k_{\varphi\varepsilon} a_{\varepsilon\varphi} + b_{\varphi\varepsilon} b_{\varepsilon\varphi} + k_{\varepsilon\varphi} a_{\varphi\varepsilon} - \frac{\bar{W}_0}{\varepsilon_0} a_{\varphi\varepsilon}\sin\varphi_0\right) \\ &+\Omega^4\left(a_{\varepsilon\varepsilon} a_{\varphi\varphi} + \bar{M}\bar{W}_0 a_{\varepsilon\varepsilon} + \bar{M}\bar{W}_0 a_{\varphi\varphi} + \bar{M}^2\bar{W}_0^2 + a_{\varphi\varepsilon} a_{\varepsilon\varphi}\right) = 0 \end{aligned} \quad (10.90)$$

Solution of first set of equations satisfying the boundary conditions determines the steady state pressure and flow parameters. Newton–Raphson method and finite difference scheme are used to obtain converged solution of set of equations. Convergence was achieved up to a value of $\bar{\text{Re}} = 1.5$. Load capacity and attitude angle are calculated using the pressure distribution. After determining steady state pressure distribution, the remaining four sets of equations are solved satisfying the boundary conditions to determine perturbed pressure distributions and flow parameters. Finite difference scheme with successive over-relaxation is used to obtain converged solutions of the set of equations. These pressure distributions are then used to evaluate dynamic coefficients. Dynamic coefficients are initially evaluated for an assumed value of whirl ratio. These dynamic coefficients are then used to determine mass parameter using Equation (10.89). The evaluated mass parameter and assumed whirl must satisfy Equation (10.90). If it does not then another value of whirl ratio is assumed and the process is repeated till Equation (10.90) is satisfied. Results of this analysis are given in Table 10.2 for steady state performance in terms of load capacity, attitude angle for various eccentricity ratios, and Reynolds numbers.

In general, it is observed that fluid inertia effect is not very significant. Table 10.3 gives the results of mass parameters and whirl ratios for various lengths to diameter ratios, eccentricity ratios, and Reynolds numbers including fluid inertia effects and neglecting fluid inertia effects. It is seen that fluid inertia considerably influences the magnitude of critical mass. Therefore, to ascertain the stability of the rotor fluid inertia effect need to be taken into account when the Reynolds number is high.

**Table 10.2** | Comparison of Steady State Characteristics of Journal Bearings for $L/D = 1$

| Re* | $\varepsilon_0$ | $\overline{W}_0$ (Present) | $\overline{W}_0$ [Chen *et al.*] | $\phi^0$ (Present) | $\phi^0$ [Chen *et al.*] |
|---|---|---|---|---|---|
| 0 | 0.2 | 0.5042 | 0.5013 | 73.71 | 73.9 |
| | 0.5 | 1.7903 | 1.779 | 56.64 | 56.8 |
| | 0.8 | 7.4597 | 7.146 | 34.66 | 36.2 |
| | 0.9 | 17.7139 | 16.982 | 23.90 | 26.4 |
| 0.28 | 0.2 | 0.5055 | 0.5041 | 73.75 | 74.2 |
| | 0.5 | 1.8058 | 0.5041 | 73.75 | 57.0 |
| | 0.8 | 7.4837 | 7.151 | 34.72 | 36.3 |
| | 0.9 | 17.7615 | 16.993 | 23.93 | 26.4 |
| 0.56 | 0.2 | 0.5070 | 0.5051 | 73.79 | 74.5 |
| | 0.5 | 1.8058 | 1.790 | 56.79 | 57.2 |
| | 0.8 | 7.5081 | 7.159 | 34.78 | 36.4 |
| | 0.9 | 17.809 | 17.002 | 23.97 | 26.4 |
| 1.4 | 0.2 | 0.5112 | 0.5086 | 73.95 | 75.3 |
| | 0.5 | 1.837 | 1.587 | 57.05 | 58.0 |
| | 0.8 | 7.5852 | 7.187 | 35.02 | 36.7 |
| | 0.9 | — | 17.030 | — | 26.6 |

## 10.5 | Influence of Temporal Inertia on the Performance of Journal Bearings

Temporal inertia effects are due to fluid accelerations and are generally seen under unsteady conditions when flow velocities become time dependent. It has been seen that convective inertia effects are usually small and therefore, to investigate the influence of temporal inertia, convective inertia terms may be omitted. Navier–Stokes equations with usual assumptions made to derive Reynolds equation can thus be written for incompressible flow neglecting convective inertia effects as (San Andres and Vance, 1987).

$$\mathrm{Re}\frac{\partial \overline{u}}{\partial \tau} = -\frac{\partial \overline{p}}{\partial \theta} + \frac{\partial^2 \overline{u}}{\partial \overline{z}^2} \tag{10.91}$$

$$\mathrm{Re}\frac{\partial \overline{v}}{\partial \tau} = -\frac{\partial \overline{p}}{\partial \overline{y}} + \frac{\partial^2 \overline{v}}{\partial \overline{z}^2} \tag{10.92}$$

Continuity equation becomes

$$\frac{\partial \overline{u}}{\partial \theta} + \frac{\partial \overline{v}}{\partial \overline{y}} + \frac{\partial \overline{w}}{\partial \overline{z}} = 0 \tag{10.93}$$

**Table 10.3** | Comparison of Critical Mass Parameter and Whirl Ratio Calculated by Incorporating and Neglecting Acceleration Coefficients

| | | | Considering Acceleration Coefficients | | Neglecting Acceleration Coefficients | |
|---|---|---|---|---|---|---|
| *L/D* | Re* | $\varepsilon_0$ | $\overline{M}$ | $\Omega$ | $\overline{M}$ | $\Omega$ |
| 0.5 | 0.5 | 0.1 | 8.20 | 0.530 | 9.54 | 0.530 |
| 0.5 | 0.5 | 0.3 | 7.78 | 0.550 | 8.17 | 0.550 |
| 0.5 | 0.5 | 0.6 | 7.52 | 0.570 | 7.64 | 0.570 |
| 1.0 | 0.5 | 0.1 | 7.44 | 0.536 | 8.53 | 0.536 |
| 1.0 | 0.5 | 0.3 | 7.45 | 0.532 | 7.87 | 0.530 |
| 1.0 | 0.5 | 0.6 | 7.39 | 0.549 | 7.58 | 0.547 |
| 2.0 | 0.5 | 0.1 | 6.10 | 0.558 | 7.17 | 0.556 |
| 2.0 | 0.5 | 0.3 | 5.78 | 0.553 | 6.23 | 0.530 |
| 2.0 | 0.5 | 0.6 | 7.27 | 0.506 | 7.55 | 0.504 |
| 0.5 | 1.0 | 0.1 | 7.69 | 0.530 | 9.88 | 0.530 |
| 0.5 | 1.0 | 0.3 | 7.76 | 0.543 | 8.49 | 0.540 |
| 0.5 | 1.0 | 0.6 | 7.57 | 0.570 | 7.72 | 0.570 |
| 1.0 | 1.0 | 0.1 | 6.98 | 0.538 | 9.19 | 0.536 |
| 1.0 | 1.0 | 0.3 | 7.09 | 0.536 | 7.89 | 0.533 |
| 1.0 | 1.0 | 0.6 | 7.14 | 0.554 | 7.48 | 0.557 |
| 2.0 | 1.0 | 0.1 | 5.89 | 0.563 | 8.16 | 0.560 |
| 2.0 | 1.0 | 0.3 | 5.36 | 0.542 | 6.27 | 0.537 |
| 2.0 | 1.0 | 0.6 | 6.95 | 0.513 | 7.50 | 0.509 |

where

$$\mathrm{Re} = \frac{\rho C^2 \omega}{\eta},\ \bar{u} = \frac{u}{\omega R},\ \bar{v} = \frac{v}{\omega R},\ \bar{w} = \frac{w}{\omega C},\ \bar{p} = \frac{p}{\eta\omega\left(R/C\right)^2},$$

$$\theta = \frac{x}{R}, \bar{y} = \frac{y}{R}, \bar{z} = \frac{z}{C},$$

$C$ is radial clearance of the bearing and $\omega$ is the characteristic frequency.

Boundary conditions for velocity components are in the dimensionless form as

$$\bar{u} = \bar{v} = \bar{w} = 0 \ \ at \ \ \bar{z} = 0$$

$$\bar{u} = \bar{v} = 0; \bar{w} = \frac{\partial H}{\partial \tau} \ \ at \ \ \bar{z} = H$$

where, $H = \dfrac{h}{C}$ and $\tau = \omega t$ , $h$ is film thickness.

It is assumed that for small amplitude motions, the flow within the clearance space of the bearing is governed by Equations (10.91) to (10.93). It is considered that flow in the clearance space is governed by following mean flow equations

Continuity equation:
$$\frac{\partial q_\theta}{\partial \theta}+\frac{\partial q_{\bar{y}}}{\partial \bar{y}}=-\frac{\partial H}{\partial \tau} \tag{10.94}$$

Momentum equation:
$$\mathrm{Re}\frac{\partial q_\theta}{\partial \tau}=-H\frac{\partial \bar{p}}{\partial \theta}+\Delta\tau_{\theta \bar{z}} \tag{10.95}$$

$$\mathrm{Re}\frac{\partial q_{\bar{y}}}{\partial \tau}=-H\frac{\partial \bar{p}}{\partial \bar{y}}+\Delta\tau_{\bar{y}\bar{z}} \tag{10.96}$$

where $q_\theta=\int_0^H \bar{u}d\bar{z}$ and $q_{\bar{y}}=\int_0^H \bar{v}d\bar{z}$, $\alpha$ is inertial wall shear stress coefficient for small amplitude eccentric motions.

$q_\theta, q_{\bar{y}}$ are local dimensionless flow rates and $\Delta\tau_{\theta\bar{z}}, \Delta\tau_{\bar{y}\bar{z}}$ dimensionless wall shear stress difference in circumferential and axial direction, respectively. It is generally assumed that velocity fields are altered by fluid inertia at low or moderate Reynolds numbers.

Therefore, wall shear stress differences can be expressed as

$$\Delta\tau_{\theta\bar{z}}=-\frac{12q_\theta}{H^2}+\mathrm{Re}(1-\alpha)\frac{\partial q_\theta}{\partial \tau} \tag{10.97}$$

$$\Delta\tau_{\bar{y}\bar{z}}=-\frac{12q_{\bar{y}}}{H^2}+\mathrm{Re}(1-\alpha)\frac{\partial q_{\bar{y}}}{\partial \tau} \tag{10.98}$$

$\alpha$ is a function of Re. For long and short bearings, $\alpha$ takes the asymptotic values as $\alpha \to 1.2$ as $\mathrm{Re} \to 0$ and $\alpha \to 1.0$ as $\mathrm{Re} \to \infty$. However, these values are generally found to give good result up to $\mathrm{Re} = 25$, which cover the range where the flow remains laminar.

Substituting Equations (10.97) and (10.98) into (10.95) and (10.96) give

$$\alpha\mathrm{Re}\frac{\partial q_\theta}{\partial \tau}=-H\frac{\partial \bar{p}}{\partial \theta}-12q_\theta/H^2 \tag{10.99}$$

$$\alpha\mathrm{Re}\frac{\partial q_{\bar{y}}}{\partial \tau}=-H\frac{\partial \bar{p}}{\partial \bar{y}}-12q_{\bar{y}}/H^2 \tag{10.100}$$

Equations (10.99) and (10.100) combined with Equation (10.94) can be reduced to the following modified form of Reynolds equation which includes temporal inertia effect

$$\frac{\partial}{\partial \theta}\left(H^3\frac{\partial \bar{p}}{\partial \theta}\right)+\frac{\partial}{\partial \bar{y}}\left(H^3\frac{\partial \bar{p}}{\partial \bar{y}}\right)=12\frac{\partial H}{\partial \tau}+\alpha\mathrm{Re}\,H^2\left[\frac{\partial^2 H}{\partial \tau^2}+\frac{2}{H}\left(\frac{\partial H}{\partial \tau}\right)^2\right] \tag{10.101}$$

## 10.6 | Temporal Inertia Effect on the Dynamic Performance of Multi-recess Hydrostatic Journal Bearing

Ghosh *et al.* (1988) investigated on the influence of temporal inertia on the dynamic performance of multi-recess hydrostatic journal bearing. Reynolds Equation (10.66) can be expressed with some modification for $\alpha = 1$ as

$$\frac{\partial}{\partial\theta}\left(H^3\frac{\partial\bar{p}}{\partial\theta}\right)+\left(\frac{D}{L}\right)^2\frac{\partial}{\partial\bar{y}}\left(H^3\frac{\partial\bar{p}}{\partial\bar{y}}\right)=\sigma\frac{\partial H}{\partial\tau}+\frac{\sigma\,\mathrm{Re}\,H^2}{12}\left[\frac{\partial^2 H}{\partial\tau^2}+\frac{2}{H}\left(\frac{\partial H}{\partial\tau}\right)^2\right] \tag{10.102}$$

where $\sigma = 12\eta\omega / p_s (C/R)^2$

$$\bar{p}=p/p_s,\ \bar{y}=\frac{2y}{L},\ H=\frac{h}{C},\ \theta=\frac{x}{R},\quad \mathrm{Re}=\rho C^2\omega/\eta$$

$p_s$ – supply pressure of oil

Perturbation method is adapted to express pressure and local film thickness assuming that the journal center executes harmonic motion with $\mathrm{Re}\{\varepsilon_1 e^{i\tau}\}$ around its steady state position given by eccentricity ratio, $\varepsilon_0$ as shown in Fig. 10.12.

Thus, pressure and film thickness are written neglecting higher order terms as:

$$\bar{p}=\bar{p}_0+\varepsilon_1 e^{i\tau}\bar{p}_1;\quad \bar{H}=\bar{H}_0+\varepsilon_1 e^{i\tau}\cos\theta \tag{10.103}$$

where $\bar{H}_0 = 1+\varepsilon_0\cos\theta$

Substituting Equation (10.103) into Reynolds Equation (10.102) and collecting zeroth and first order terms we get:

$$\frac{\partial}{\partial\theta}\left(H_0^3\frac{\partial\bar{p}_0}{\partial\theta}\right)+\left(\frac{D}{L}\right)^2\frac{\partial}{\partial\bar{y}}\left(H_0^3\frac{\partial\bar{p}}{\partial\bar{y}}\right)=0 \tag{10.104}$$

$$\frac{\partial}{\partial\theta}\left(H_0^3\frac{\partial\bar{p}_1}{\partial\theta}\right)+\left(\frac{D}{L}\right)^2\frac{\partial}{\partial\bar{y}}\left(H_0^3\frac{\partial\bar{p}_1}{\partial\bar{y}}\right)=i\sigma\cos\theta-\frac{\mathrm{Re}}{12}\sigma H_0^2\cos\theta \tag{10.105}$$

The above equations are solved to determine pressure distribution $\bar{p}_0(\theta,\bar{y})$ and $\bar{p}_1(\theta,\bar{y})$ over the bearing land area following finite difference method using Gauss–Seidel iterative procedure with an over-relaxation factor satisfying proper boundary conditions.

The boundary conditions for Equations (10.107) and (10.108) are given as

For $\bar{p}_0$

- $\bar{p}_0(\theta,-1)=\bar{p}_0(\theta,+1)=0$
- $\dfrac{\partial\bar{p}_0}{\partial\bar{y}}(\theta,1)=0$
- $\bar{p}_0=\bar{p}_{r0}$ at the r$^{\text{th}}$ recess
- $\bar{p}_0(\theta,\bar{y})=\bar{p}_0(\theta+2\pi,\bar{y})$ (10.106)

For $\bar{p}_1$

- $\bar{p}_1(\theta,-1) = \bar{p}_1(\theta,+1) = 0$
- $\dfrac{\partial \bar{p}_1}{\partial \bar{y}}(\theta,0) = 0$
- $\bar{p}_1 = \bar{p}_{r1}$ at the r$^{th}$ recess
- $\bar{p}_1(\theta,\bar{y}) = \bar{p}_1(\theta+2\pi,\bar{y})$ (10.107)

Steady state recess pressure ($\bar{p}_{r0}$) and dynamic recess pressure ($\bar{p}_{r1}$) are given by the following equations

$$\bar{p}_{r1} = -\left\{\frac{\bar{Q}_{r1}\bar{P}_{r0}\lambda + \psi^2\gamma\cos\theta_r}{\lambda^2+\psi^2\gamma^2}\right\} - i\left\{\frac{\psi\lambda\cos\theta_r - \bar{Q}_{r1}\bar{P}_{r0}\psi\gamma}{\lambda^2+\psi^2\gamma^2}\right\}$$

***For Orifice Compensation***

$$\bar{p}_{r0} = -0.5\left(\frac{\delta o}{\bar{Q}_{r0}}\right) + 0.5\left[\left(\frac{\delta_0}{\bar{Q}_{r0}}\right)^4 + 4\left(\frac{\delta_0}{\bar{Q}_{r0}}\right)^2\right]^{1/2} \tag{10.108}$$

$$\lambda = 0.5\delta_0\left(1-\bar{p}_{r0}\right)^{0.5} + \bar{Q}_{r0}$$

***For Capillary Compensation***

$$\bar{p}_{r0} = \frac{\delta_c}{\lambda} \text{ and } \lambda = \left(\bar{Q}_{r0} + \delta_c\right) \tag{10.109}$$

Also, $\gamma = \bar{V}_0\beta p_s$, recess volume compressibility parameter and $\psi = \sigma A / R^2$, recess frequency parameter. Derivation of Equations (10.108) and (10.109) has been described in Chapter 9.

### 10.6.1 | Dynamic Load Capacity

Dynamic load capacity of the bearing is evaluated using the known dynamic pressure distribution as:

$$W_d\varepsilon_1 e^{i\tau} = -2\int_0^{L/2}\int_0^{2\pi} \varepsilon_1 e^{i\tau} p_1 \cos\theta R d\theta dy \tag{10.110}$$

Since the journal executes harmonic motion with amplitude $y = C\varepsilon_1 e^{i\omega t}$ the dynamic load capacity ($W_d$) can be expressed as:

$$W_d e^{i\omega t} = -Ky - B\dot{y} - E\ddot{y} \tag{10.111}$$

$$W_d = -KC - iBC\omega + EC\omega^2 \tag{10.112}$$

In the dimensionless form it is expressed as

$$\bar{W}_d = \frac{W_d}{LDp_s\varepsilon_1} = -\frac{KC}{LDp_s} - i\frac{BC\omega}{LDp_s} + \frac{EC\omega^2}{LDp_s} \tag{10.113}$$

where

Stiffness coefficient, $$K_s = -\frac{KC}{LDp_s},$$

Damping coefficient, $$B_s = -\frac{\mathrm{Im}(\bar{W}_d)}{\sigma} = \frac{B}{24\eta L\left(\frac{R}{C}\right)^3}$$

Inertia coefficient, $$E_s = \frac{\mathrm{Re}\,\eta\omega E}{CLDp_s}$$

Temporal inertia effect is seen in the form of added mass coefficient or inertia coefficient ($E_s$) which is dependent on Reynolds number. It can be combined with stiffness coefficient $K_s$ and together can be expressed as dynamic stiffness coefficient $K_d$ which would be dependent on Reynolds number. Thus,

$$K_d = -\mathrm{Re}(\bar{W}_d) = (K_s - E_s)$$

Result of dynamic stiffness ($K_d$) of capillary and orifice compensated bearings against Reynolds number (Re) are shown in Figs 10.13 to 10.17 for various values of $\sigma$ and $\gamma$. Influence of Reynolds number for various values of recess volume parameter ($\gamma$) from 0.01 to 0.05 are shown in Fig. 10.13. Role of $\gamma$ appears to be very

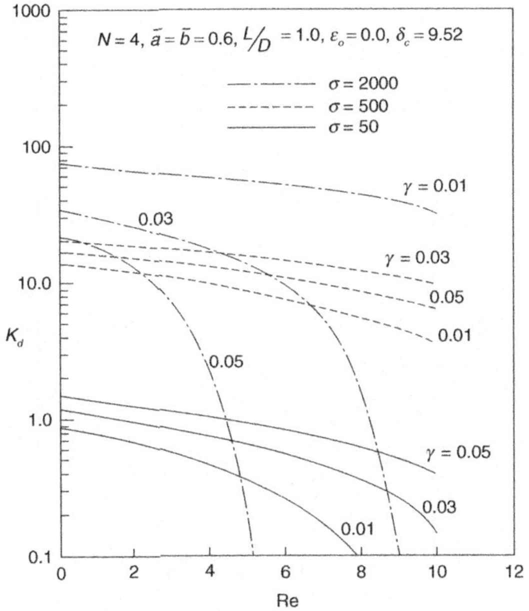


**Figure 10.13** | $K_d$ versus Re for Various Values of $\gamma$ a Capillary Compensated Bearing

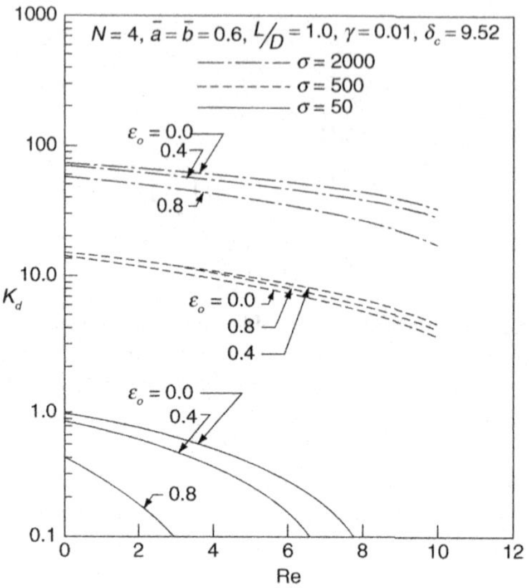


**Figure 10.14** | $K_d$ versus Re for Various Values of $\varepsilon_0$ a Capillary Compensated Bearing

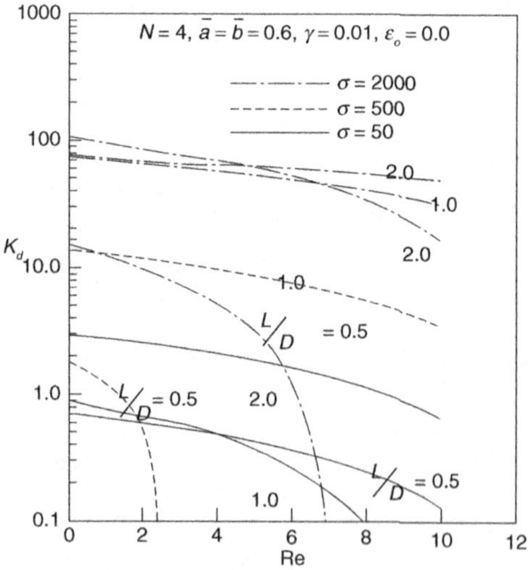


**Figure 10.15** | $K_d$ versus Re for Various Values of $L/D$ a Capillary Compensated Bearing

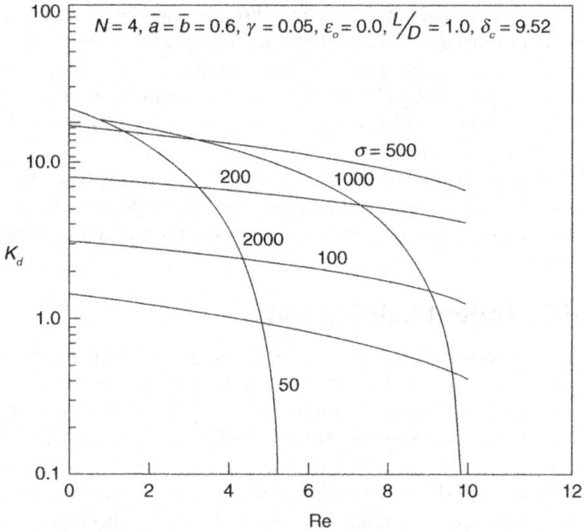


**Figure 10.16** | $K_d$ versus Re for Various Values of $\sigma$ a Capillary Compensated Bearing

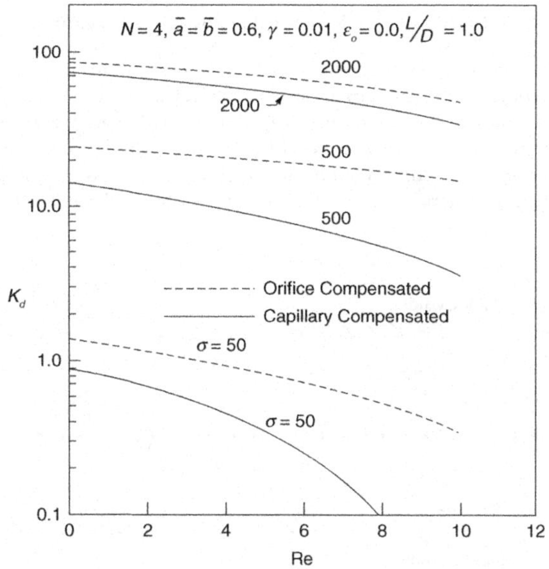


**Figure 10.17** | $K_d$ versus Re for Various Values of $\sigma$ a Capillary and an Orifice Compensated Bearing

significant. Figure 10.17 shows comparison between orifice and capillary compensated bearings. It can be seen that orifice compensated bearing is better and give higher values of dynamic stiffness.

A suitable bearing design must account for fluid inertia effect and choose a suitable value of $\gamma$. There is a short range of $Re$ and $\sigma$ for a particular value of $\gamma$ when there is a sharp decline in the dynamic stiffness of the bearing. Effect of eccentricity ratio and $L/D$ ratio can be seen in Figs 10.14 and 10.15. A short $L/D$ ratio appears to witness larger effect of fluid inertia.

Figure 10.17 shows relative performance of capillary and orifice compensated bearings. Influence of fluid inertia is more pronounced in capillary compensated bearings. When Re $\rightarrow$ 0 and recess volume compressibility effect is neglected, i.e., $\gamma = 0$, dynamic stiffness is equivalent to the static stiffness of the bearing.

## 10.7 | Theory of Turbulent Lubrication

Transition from laminar flow to turbulence usually occurs when Reynolds number is >2100 in pipe flow. In journal bearings, this transition occurs through a regime known as vortex flow which is characterized by appearance of Taylor vortices and is observed at a Reynolds number >1700. Taylor (1923) showed that when laminar flow of a viscous fluid between two concentric cylinders becomes unstable, toroidal vortices equally spaced along the axis of the cylinder is witnessed as shown in Fig. 10.18. Taylor showed that this occurs due to centrifugal inertia and that when the radial clearance between the cylinders is small as is the case in journal bearings, this transition occurs at a Reynolds number=1700. This is also known as Taylor number ($T_a$) or critical Reynolds number (Re) critical.

Taylor number is expressed as

$$T_a = 41.2\left(\frac{R}{C}\right)^{\frac{1}{2}}, \text{ R– journal radius, C– radial clearance}$$

Instability of laminar flow between eccentric cylinders due to rotation of inner cylinder with outer cylinder stationary have been investigated by Diprima (1963), Kamal (1966) and Dai *et al.* (1992). Castle and Mobbs (1961, 1968) and Mobbs and Younes (1973) conducted experiments on vortex flow. However, when Reynolds number exceeds 2000, flow in case of concentric cylinders becomes turbulent. Turbulence may be visualized as an irregular condition of flow in which velocity, pressure, etc. show random variations in time and space. It occurs in nature more often and also in flows through closed space, i.e., confined

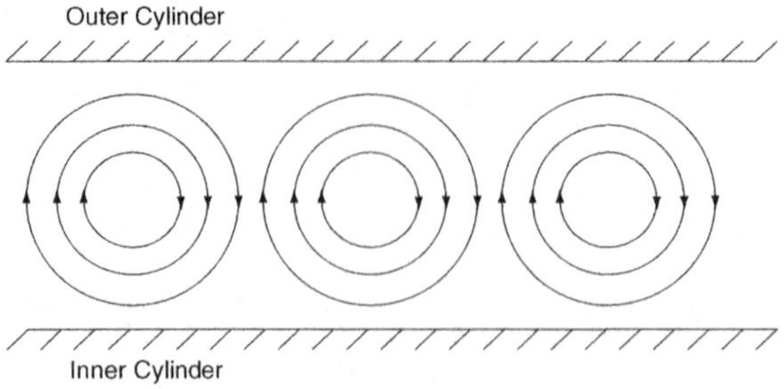


**Figure 10.18** | Taylor Vortices

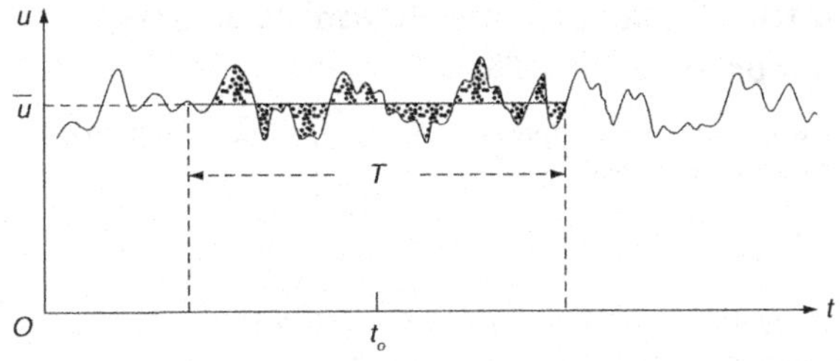


**Figure 10.19** | Velocity Profile

between boundaries such as ducts, pipes, concentric cylinders, rotating discs, etc. It is usually seen that in turbulent flows velocities, pressures, etc. show random variations superposed over a distinct statistical average value as depicted in Fig. 10.19. It is generally known that turbulence in lubrication of bearings and seals occur due to use of low dynamic viscosity process fluids, e.g., in bearings of cryogenic turbomachinery using liquid hydrogen and oxygen, in nuclear applications where liquid metals (sodium) is employed, water lubricated bearings, bearings and seals in advanced gas turbines, etc. These applications drew the attention of engineers toward the need to develop adequate theoretical models to address the issue of turbulence in lubrication.

## 10.8 | Fluctuations and Average Values in Turbulent Flow

Random fluctuations in turbulent flow are at a scale with order of magnitude comparable to bulk motion of the fluid. Fluctuating velocities can be measured with precision instruments such as hot wire anemometer, Laser doppler anemometer, etc. Taylor and Von Karman suggested that turbulence can be generated by the flow of layers of fluids at various velocities past over another or by fluid flow over solid surfaces. Based on this, it can be said that there are two types of turbulence, i.e., turbulence generated due to viscous effect in flow past or over solid walls known as wall turbulence and free turbulence generated by flow of layers of fluids in free stream at different velocities. In lubrication of bearings and seals, turbulence originates from solid surfaces and is designated as wall turbulence.

Let $u$, $v$, and $w$ be the fluid velocity components in Cartesian coordinates and are functions of time. Then mean or average velocities at a time '$t_0$' are defined as:

$$\bar{u} = \frac{1}{T}\int_{t_0-T/2}^{t_0+T/2} u\,dt; \quad \bar{v} = \frac{1}{T}\int_{t_0-T/2}^{t_0+T/2} v\,dt;$$
$$\bar{w} = \frac{1}{T}\int_{t_0-T/2}^{t_0+T/2} w\,dt \tag{10.114}$$

where $T$ is the sampling period.

Thus, fluctuating components of the velocities are $u' = \{u(t)-\bar{u}\}, v'\{v(t)-\bar{v}\}$ and $w' = \{w(t)-\bar{w}\}$, and it can be said that mean or average of the fluctuations are zero over the sampling period, T.

## 10.9 | Momentum Equations and Reynolds Stresses for an Incompressible Flow

Multiplying continuity equation for incompressible flow by $\rho u$ and adding to the momentum equation in $x$-direction neglecting body force, we get

$$\rho\frac{\partial u}{\partial t} = \frac{\partial}{\partial x}\left(\sigma_{xx} - \rho u^2\right) + \frac{\partial}{\partial y}\left(\tau_{yx} - \rho uv\right) + \frac{\partial}{\partial z}\left(\tau_{zx} - \rho uw\right) \tag{10.115}$$

In a similar manner, multiplying continuity equation for incompressible by $\rho v$ and $\rho w$ and adding to momentum equations in $y$ and $z$ directions, respectively, one would obtain

$$\rho\frac{\partial v}{\partial t} = \frac{\partial}{\partial x}\left(\tau_{xy} - \rho vu\right) + \frac{\partial}{\partial y}\left(\sigma_{yy} - \rho v^2\right) + \frac{\partial}{\partial z}\left(\tau_{zy} - \rho vw\right) \tag{10.116}$$

$$\rho\frac{\partial w}{\partial t} = \frac{\partial}{\partial x}\left(\tau_{xz} - \rho wu\right) + \frac{\partial}{\partial y}\left(\tau_{yz} - \rho wv\right) + \frac{\partial}{\partial z}\left(\sigma_{zz} - \rho w\right) \tag{10.117}$$

These are momentum equations written in another manner. If now all terms in these equations and the continuity equation are averaged, one would get for quasi steady flows the momentum equations when

$$\frac{\partial \bar{u}}{\partial t} = \frac{\partial \bar{v}}{\partial t} = \frac{\partial \bar{w}}{\partial t} = 0 \text{ as}$$

$$\rho\left(\bar{u}\frac{\partial \bar{u}}{\partial x} + \bar{v}\frac{\partial \bar{u}}{\partial y} + \bar{w}\frac{\partial \bar{u}}{\partial z}\right) = \frac{\partial}{\partial x}\left(\sigma_{xx} - \rho\overline{u'^2}\right) + \frac{\partial}{\partial y}\left(\bar{\tau}_{xy} - \rho\overline{u'v'}\right) + \frac{\partial}{\partial z}\left(\bar{\tau}_{zx} - \rho\overline{u'w'}\right) \tag{10.118}$$

$$\rho\left(\bar{u}\frac{\partial \bar{v}}{\partial x} + \bar{v}\frac{\partial \bar{v}}{\partial y} + \bar{w}\frac{\partial \bar{v}}{\partial z}\right) = \frac{\partial}{\partial x}\left(\bar{\tau}_{xy} - \underline{\rho\overline{u'v'}}\right) + \frac{\partial}{\partial y}\left(\bar{\sigma}_{yy} - \rho\overline{v'^2}\right) + \frac{\partial}{\partial z}\left(\bar{\tau}_{yx} - \rho\overline{v'w'}\right) \tag{10.119}$$

$$\rho\left(\bar{u}\frac{\partial \bar{w}}{\partial x} + \bar{v}\frac{\partial \bar{w}}{\partial y} + \bar{w}\frac{\partial \bar{w}}{\partial z}\right) = \frac{\partial}{\partial x}\left(\bar{\tau}_{zy} - \underline{\rho\overline{u'w'}}\right) + \frac{\partial}{\partial y}\left(\bar{\tau}_{yz} - \rho\overline{v'w'}\right) + \frac{\partial}{\partial z}\left(\bar{\sigma}_{zz} - \rho\overline{w'^2}\right) \tag{10.120}$$

and continuity equation as

$$\frac{\partial \bar{u}}{\partial x} + \frac{\partial \bar{v}}{\partial y} + \frac{\partial \bar{w}}{\partial z} = 0 \tag{10.121}$$

Substituting for stresses in terms of pressure and velocity gradients using average velocities, the momentum equations for Newtonian fluids can be written as

$$\rho\left(\bar{u}\frac{\partial \bar{u}}{\partial x} + \bar{v}\frac{\partial \bar{u}}{\partial y} + \bar{w}\frac{\partial \bar{u}}{\partial z}\right) = -\frac{\partial p}{\partial x} + \eta\left(\frac{\partial^2 \bar{u}}{\partial x^2} + \frac{\partial^2 \bar{u}}{\partial y^2} + \frac{\partial^2 \bar{u}}{\partial z^2}\right)$$
$$-\rho\left[\frac{\partial}{\partial x}\left(\underline{\rho\overline{u'^2}}\right) + \frac{\partial}{\partial y}\left(\underline{\rho\overline{u'v'}}\right) + \frac{\partial}{\partial z}\left(\underline{\rho\overline{u'w'}}\right)\right] \tag{10.122}$$

The terms underlined represent Reynolds stresses or stresses due to exchange of energy in the turbulent mixing process. Equations in $y$ and $z$ directions can be written in a similar manner.

It is almost impossible to solve the above equations. Therefore, models have been developed which provide solutions to engineering problems but may not be physically correct. In general, it is considered that very close to a wall or a solid boundary there is a layer known as viscous sub layer in which viscous stresses are much larger than turbulent stresses followed by an intermediate layer where viscous and turbulent stresses are of the same order. Beyond this, there is a free turbulence region where turbulent stresses are much larger than viscous stresses and the mean velocity distribution is described by a logarithmic type of law of the wall. This region is larger than the other two regions. This type of turbulent flow is generally seen over a flat plate. Fully developed boundary layer flow in pipes and ducts are similar to flat plate flow. However, shear stresses vary linearly with distance from the wall. Flow through bearings and seals are similar to flow through an annulus. A detailed survey of turbulent lubrication theory and its application to design was presented by Taylor and Dowson (1974), and Wilcock (1981) discussed aspects of efficient design of turbulent thrust bearings.

## 10.10 | Turbulent Lubrication Theories

Generally, three models have been developed which relate Reynolds stresses to mean or average flow velocities. However, these models can be applied to two-dimensional flows only. These models are as follows:

### 10.10.1 | Prandtl Mixing Length Theory

Prandtl proposed a momentum mixing length concept for modeling wall turbulence which originates from a solid boundary. He assumed that there is a length between layers where transfer of momentum or mixing occurs which is given by,

$$-\bar{\rho}\ \overline{u'w'} = \rho l^2 \left|\frac{\partial \bar{u}}{\partial z}\right| \frac{\partial \bar{u}}{\partial z} \tag{10.123}$$

where $l$ is the mixing length and is related to independent variable by a constant $k$.

Constantinescu adopted this approach in the analysis of bearings operating in turbulent regime. Thus, he assumed strong Couette flow and mixing length $l$ is given by $l = kz$ where $k$ is usually taken as 0.4.

### 10.10.2 | Eddy Viscosity or Turbulent Viscosity Model

The most widely accepted theory for turbulent flow was given by Boussinesq. According to this theory, in two-dimensional turbulent flows, effective shear stress is given by

$$\tau_{ij} = (\eta + \rho \in)\frac{\partial \bar{u}_i}{\partial x_j} = \eta\left(1 + \frac{\in}{\nu}\right)\frac{\partial \bar{u}_i}{\partial x_j} \tag{10.124}$$

where $\in$ is known as eddy viscosity and is expressed in terms of Reynolds stresses as:

$$\in = -\frac{1}{\partial \bar{u}_i / \partial x_j}\overline{u_i' u_j'} \tag{10.125}$$

Thus, the effective shear stress can be expressed in terms of mean or average velocities. However, $\in$ is different for different flow conditions and not constant for a given flow. It is dependent on local conditions or in other words varies spatially.

A turbulent flow which experiences a shear stress $\tau_w$ at the impermeable stationary wall at $z = 0$, law of wall correlates mean velocity profile near the wall by

$$u^+ = g(z^+) \tag{10.126}$$

where $$u^+ = u/u_\tau,\ z^+ = \frac{u_\tau z}{\nu},\ u_\tau = \left(\left|\tau_w\right|/\rho\right)^{1/2}$$

This correlation can be defined as eddy viscosity, thus

$$-u'w' = \in \frac{\partial u}{\partial z} \tag{10.127}$$

Reichardt in 1951 proposed following empirical relation for eddy-diffusivity:

$$\frac{\in}{v} = k\left(z^+ - \delta_l^+ \tanh\frac{z^+}{\delta_l^+}\right) \tag{10.128}$$

where $k$ and $\delta_l^+$ were chosen as 0.4 and 10.7 by Ng (1964) which correlate with known data. Ng and Pan (1965) derived Reynolds equation for turbulent lubrication using Reichardt's formula for eddy diffusivity and generalized law of wall given above.

Hirs (1973) proposed a theory for turbulent lubrication based on bulk flow. This is an extension of basic work of Blasius which uses relationship between wall shear stress and mean velocity of flow relative to wall. According to this, the following simple equation is assumed for all types of flow, viz., pressure flow, drag flow, or combination of both ,i.e.,

$$\frac{\tau}{\frac{1}{2}\rho\bar{u}^2} = n\left(\frac{\rho\bar{u}h}{\eta}\right)^m \tag{10.129}$$

where $\tau$ = wall shear stress, $\rho$ = density of fluid, $\eta$ = viscosity of fluid, $\bar{u}$ = mean velocity of flow relative to wall or surface at which shear stress acts, $h$ = film thickness, $n$ and $m$ are empirical constants, $\tau/\frac{1}{2}\rho\bar{u}^2$ = friction factor. $n$ and $m$ values were obtained by curve fitting data from experimental results for different flows. For details, refer to the papers by Hirs (1973, 1974).

## 10.11 | Derivation of Reynolds Equation for Turbulent Lubrication

This section deals with the derivation of Reynolds equation for turbulent lubrication neglecting convective and temporal inertia effects. The Navier–Stokes equations for turbulent flow is thus written as:

$$0 = -\frac{\partial p}{\partial x} + \eta\frac{\partial^2 \bar{u}}{\partial z^2} + \frac{\partial}{\partial z}\left(-\bar{\rho}\overline{u'w'}\right) \tag{10.130}$$

$$0 = -\frac{\partial p}{\partial y} + \eta\frac{\partial^2 \bar{v}}{\partial z^2} + \frac{\partial}{\partial z}\left(-\bar{\rho}\overline{v'w'}\right) \tag{10.131}$$

General assumptions made in the derivation of Reynolds equation for laminar lubrication in Chapter 3 are also valid for the above equations.

Using the concept of eddy-diffusivity, the above equations can be rewritten as

$$0 = -\frac{\partial \bar{p}}{\partial x} + \frac{\partial}{\partial z}\left\{\eta\left(1+\frac{\in}{\nu}\right)\frac{\partial \bar{u}}{\partial z}\right\} \tag{10.132}$$

$$0 = -\frac{\partial \bar{p}}{\partial y} + \frac{\partial}{\partial z}\left\{\eta\left(1+\frac{\in}{\nu}\right)\frac{\partial \bar{v}}{\partial z}\right\} \tag{10.133}$$

For details, refer to the papers by Ng and Pan (1965) and Elrod and Ng (1967). It is necessary to mention that while applying Reichardt's formula, it is considered that each half channel has its own wall and therefore,

$$z^+ = \begin{cases} \xi\dfrac{h}{\nu}\sqrt{\dfrac{|\tau_w|}{\rho}} & \text{for } 0 \le \xi = \dfrac{z}{h} \le 1/2 \\ (1-\xi)\dfrac{h}{\nu}\sqrt{\dfrac{|\tau_w|}{\rho}} & \dfrac{1}{2} \le \xi \le 1 \end{cases} \tag{10.134}$$

An assumption is made that terms corresponding to mean inertia and variation of shear stresses along the surface coordinates, thus

$$\frac{\partial \tau_{xz}}{\partial z} = \frac{\partial p}{\partial x} \tag{10.135}$$

$$\frac{\partial \tau_{yz}}{\partial z} = \frac{\partial p}{\partial y} \tag{10.136}$$

Integrating Equations (10.135) and (10.136) with respect to z, we get

$$\tau_{xz}(z) = \tau_{xz}\left(\frac{h}{2}\right) + \left(z - \frac{h}{2}\right)\frac{\partial p}{\partial x} = \eta\left(1+\frac{\in}{\nu}\right)\frac{\partial u}{\partial z} \tag{10.137}$$

$$\tau_{yz}(z) = \tau_{yz}\left(\frac{h}{2}\right) + \left(z - \frac{h}{2}\right)\frac{\partial p}{\partial y} = \eta\left(1+\frac{\in}{\nu}\right)\frac{\partial v}{\partial z} \tag{10.138}$$

$\tau_{xz}\left(\frac{h}{2}\right)$ and $\tau_{yz}\left(\frac{h}{2}\right)$ are constants of integration which are to be determined using velocity boundary condition given below

$$\begin{aligned} u(h) &= U \\ v(h) &= 0 \end{aligned} \tag{10.139}$$

In the turbulent flow, it is also assumed that flow field is a small perturbation of Couette flow and therefore shear stresses can be expressed as

$$\tau_{xz} = \tau_c + \delta\tau_x \text{ and } \tau_{yz} = \delta\tau_y \tag{10.140}$$

since

$$|\tau_w| = \sqrt{(\tau_{xz})^2 + (\tau_{yz})^2} = \tau_c + \delta\tau_x + 0(\delta^2) \tag{10.141}$$

where $\tau_c$ is shear stress in Couette flow. Eddy viscosity can also be approximated using small perturbation as

$$\frac{\in}{v}(\xi,|\tau_w|) = \frac{\in}{v}(\xi,\tau_c) + \delta\tau_x \left.\frac{\partial\left(\frac{\in}{v}\right)}{\partial|\tau_w|}\right|_{\tau=\tau_c} + 0(\delta^2)$$

$$= f_c(\xi) - 1 + \frac{\delta\tau_x}{\tau_c} g_c(\xi) + 0(\delta^2) \tag{10.142}$$

where

$$f_c(\xi) - 1 = k\left[\xi h_c^+ - \delta_l^+ \tanh\left(\xi \frac{h_c^+}{\delta_l^+}\right)\right] \tag{10.143}$$

$$g_c(\xi) = \frac{k}{2}\xi h_c^+ \tanh^2\left(\eta \frac{h_c^+}{\delta_l^+}\right) \tag{10.144}$$

with

$$h_c^+ = \frac{h}{v}\sqrt{\frac{\tau_c}{\rho}}$$

ξ should be replaced by (1 – ξ) in the range of 1/2 ≤ ξ ≥ 1 which makes $f_c(\xi)$ and $g_c(\xi)$ symmetrical with respect to ξ = 1/2. Integrating Equations (10.137) and (10.138) and substituting Equations (10.142) to (10.144) neglecting 0(δ) terms one obtains

$$u(\xi) = \frac{h\tau_c}{\eta}\int_0^\xi \frac{d\xi'}{f_c(\xi')} + \left(\frac{h}{\eta}\right)\left[\frac{1}{2}\tau_{xz} - \tau_c\right]\int_0^\xi \frac{1}{f_c(\xi')}\left[1 - \frac{g_c(\xi')}{f_c(\xi')}\right]d\xi'$$

$$-\frac{h^2}{\eta}\frac{\partial p}{\partial x}\int_0^\xi \frac{\left(\frac{1}{2} - \xi'\right)}{f_c(\xi')}\left[1 - \frac{g_c(\xi')}{f_c(\xi')}\right]d\xi' \tag{10.145}$$

$$v(\xi) = \frac{\left(h/2\right)\tau_{yz}}{\eta}\int_0^\xi \frac{d\xi'}{f_c(\xi')} - \frac{h^2}{\eta}\frac{\partial p}{\partial y}\int_0^\xi \frac{\left(\frac{1}{2} - \xi'\right)}{f_c(\xi')}d\xi' \tag{10.146}$$

It can be shown that, since, $f_c(\xi)$ and $g_c(\xi)$ are symmetrical while $\frac{1}{2} - \xi$ is antisymmetrical about $\xi = 1/2$ it is found that

$$\tau_{xz}(1/2) = \tau_c \text{ and } \tau_{yz}(1/2) = 0 \tag{10.147}$$

For Couette-dominated turbulent flow Equation (10.145) reduces to

$$u_c|\xi| = \frac{h\tau_c}{\eta}\int_0^\xi \frac{d\xi'}{f_c(\xi)} \tag{10.148}$$

and

$$u_c(1) = U = \frac{h\tau_c}{\eta}\int_0^1 \frac{d\xi}{f_c(\xi)} \tag{10.149}$$

or, in the dimensionless form it is,

$$R_h = \frac{Uh}{v} = \left(h_c^+\right)^2 \int_0^1 \frac{d\xi}{f_c(\xi)} \tag{10.150}$$

$R_h$ is Couettte Reynolds number in term of $h_c^+$

Equation (10.145) and (10.146) are simplified using Equation (10.148) to (10.150)

$$\frac{u}{U} = \frac{h\tau_c}{\eta U}\int_0^{\xi} \frac{d\xi'}{f_c(\xi')} - \frac{h^2}{\eta U}\frac{\partial p}{\partial x}\int_0^{\xi}\left(\frac{\frac{1}{2}-\xi'}{f_c(\xi')}\right)\left[1-\frac{g_c(\xi')}{f_c(\xi')}\right]d\xi'$$

or

$$\frac{u}{U} = \frac{1}{2} + \frac{\left(h_c^+\right)^2}{R_h}\int_{\frac{1}{2}}^{\xi} \frac{d\xi'}{f_c(\xi')} - \frac{h^2}{\eta U}\frac{\partial p}{\partial x}\int_0^{\xi}\frac{\left(\frac{1}{2}-\xi'\right)}{f_c(\xi')}\left[1-\frac{g_c(\xi')}{f_c(\xi')}\right]d\xi' \tag{10.151}$$

$$\frac{v}{U} = -\frac{h^2}{\eta U}\frac{\partial p}{\partial y}\int_0^{\xi}\frac{\left(\frac{1}{2}-\xi'\right)}{f_c(\xi')}d\xi' \tag{10.152}$$

Equations (10.151) and (10.152) when integrated across the channel width or film thickness yields

$$\int_0^h u\,dz = Uh\left(\frac{1}{2} - G_x\frac{h^2}{\eta U}\frac{\partial p}{\partial x}\right) \tag{10.153}$$

and

$$\int_0^h v\,dz = -Uh\left(G_y\frac{h^2}{\eta U}\frac{\partial p}{\partial y}\right) \tag{10.154}$$

where $G_x$ and $G_y$ are defined as

$$G_x = \int_0^1 d\xi \int_0^{\xi} \frac{(1/2)-\xi'}{f_c(\xi')}\left[1-\frac{g_c(\xi')}{f_c(\xi')}\right]d\xi'$$

$$G_y = \int_0^1 d\xi \int_0^{\xi} \frac{(1/2)-\xi'}{f_c(\xi')}d\xi'$$

Substituting Equations (10.153) and (10.154) into flow continuity equation for an incompressible fluid film, i.e.,

$$\frac{\partial}{\partial x}\left(\int_0^h u\,dz\right) + \frac{\partial}{\partial y}\left(\int_0^h v\,dz\right) + \frac{\partial h}{\partial t} = 0 \tag{10.155}$$

Reynolds equation for turbulent lubrication is obtained as:

$$\frac{\partial}{\partial x}\left(\frac{h^3}{\eta}G_x\frac{\partial p}{\partial x}\right) + \frac{\partial}{\partial y}\left(\frac{h^3}{\eta}G_y\frac{\partial p}{\partial y}\right) = \left(\frac{U}{2}\frac{\partial h}{\partial x} + \frac{\partial h}{\partial t}\right) \tag{10.156}$$

$G_x$ and $G_y$ depend on Reynolds number $R_h$ and are thus functions of local film thickness $h$. The above equation is similar to Reynolds equation for laminar flow. Variation of $G_x$ and $G_y$ depend on Reynolds number Re and are thus functions of local film thickness $h$. The above equation is in a form similar to Reynolds equation for laminar flow. Variation of $G_x$ and $G_y$ with Reynolds number is shown in Fig. 10.20 for $\text{Re} \le 2 \times 10^5$.

For dominant Couette flows empirical expressions for $K_x = \frac{1}{G_x}$ and $K_y = \frac{1}{G_y}$ were developed based on the work of Ng and Pan by Constantinescu (1970) as given below

$$K_x = 12 + 0.0136(\text{Re})^{0.9}$$
$$K_y = 12 + 0.0043(\text{Re})^{0.96}$$

Elrod and $N_g$ (1967) extended the work of Ng and Pan to turbulent hydrostatic and hybrid bearings, where $K_x$ and $K_y$ depend on the pressure gradient, $\frac{\rho h^3}{\eta^2}\nabla p$ and

$$\nabla p = \left[\left(\frac{\partial p}{\partial x}\right)^2 + \left(\frac{\partial p}{\partial y}\right)^2\right]^{1/2}$$

For dominant Poiseuille flow the values of $K_x$ and $K_y$ are given as below (Taylor and Dowson 1974)

$$K_x = K_y = \frac{\left(\text{Re}_p\right)^{0.681}}{0.68}$$

where $\text{Re}_p$ is Poiseuille Reynolds number determined using mean flow velocity.

Hirs (1974) also gave following relationships for $K_x$ and $K_y$ for dominant Couette flows as

$$K_x = \frac{1}{G_x} = 0.0687(\text{Re})^{0.75}$$

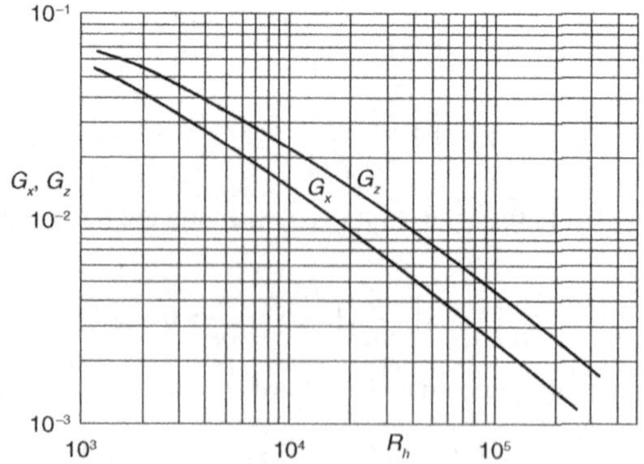


**Figure 10.20** | $G_x$ and $G_z$ versus $R_h$ [NG and Pan, ASME, *J* of Basic Engineering, 1965]

$$K_y = \frac{1}{G_y} = 0.0392(\mathrm{Re})^{0.75}$$

The values of $K_x$ and $K_y$ must not be less than 12. $K_x = 12$ when Re = 977 and $K_y = 12$ when Re = 2060. Hirs (1974) used $n = 0.066$ and $m = -0.25$ for smooth surfaces and Re $\leq 10^5$.

However, it has been observed that at high Reynolds numbers the values obtained by Ng, Pan, and Elrod and by Hirs are almost identical. Discrepancies occur in the transition regime when the flow changes from turbulent to laminar, i.e., at low Reynolds numbers.

Steady state analysis of plain journal bearings in turbulent flow was done by Kumar and Rao (1995). Reynolds Equation (10.156) with $\frac{\partial h}{\partial t} = 0$ was solved using finite difference method satisfying proper boundary conditions and performance characteristics of journal bearings in terms of Sommerfeld number (S) and friction parameter $\left(\frac{R}{C} f\right)$ versus eccentricity ratio ($\varepsilon$) were presented for various Reynolds numbers and $L/D$ ratios as shown in Figs 10.21 to 10.22. Turbulence is equivalent to operating with a lubricant of increased viscosity, other factors remaining same. Therefore, it manifests in increased load capacity and frictional drag for the bearing. Whereas increased load capacity is beneficial, on the other hand, increased frictional drag means more power consumption which is a major drawback.

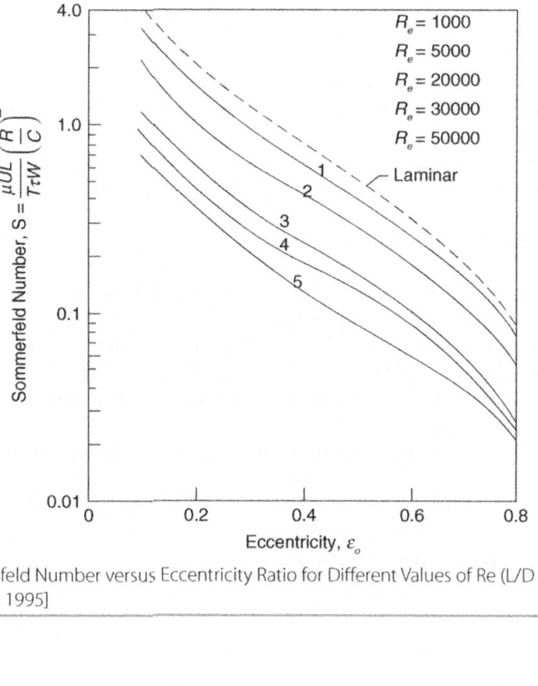


**Figure 10.21** | Sommerfeld Number versus Eccentricity Ratio for Different Values of Re (L/D = 0.5) [ Kumar and Rao, IJE & MS, 1995]

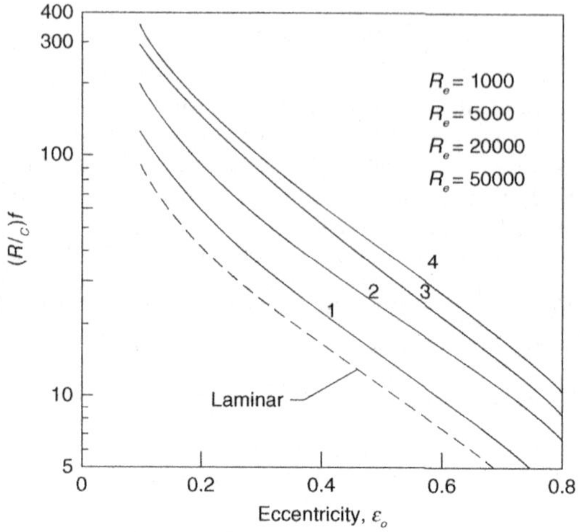


**Figure 10.22** | Coefficient of Friction versus Eccentricity Ratio for Different Values of Re L/D = 0.5 [Kumar and Rao, IJE & MS, 1995]

## REFERENCES


*Chen, C.H. and Chen, C.K.* (1989), The influence of fluid inertia on the dynamic characteristics of finite journal bearings, *Wear*, 131, 229–240.

*Chen, Chung-Moon and Dareing, D.W.* (1976), The contribution of fluid film inertia to the thermohydrodynamic lubrication of sector-pad thrust bearings, *ASME Journal of Lubrication Technology*, 98, 125–132.

*Constantinescu, V.N.* (1959), On turbulent lubrication, *Proceedings of Institution of Mechanical Engineers*, U.K., 173, 881–900.

*Constantinescu, V.N.* (1970), On the influence of inertia forces in turbulent and laminar self-acting films, *ASME Journal of Lubrication Technology*, 92, 473–481.

*Dai, R.X., Dong, Q.M. and Szeri, A.Z.* (1992), Flow between eccentric rotating cylinders: bifurcation and stability, *International Journal of Engineering Science*, 30, 1323–1340.

*DiPrima, R.C.* (1960), The stability of a viscous flow between rotating cylinders with axial flow, *Journal of Fluid Mechanics*, 9, 621–631.

*DiPrima, R.C.* (1963), A note on the stability of flow in loaded journal bearings, *ASLE Transactions*, 6, 249–253.

*Elrod, H.G. Jr. and Ng, C.W.* (1967), A theory for turbulent fluid films and its application to bearings, *ASME Journal of Lubrication Technology*, 89, 346–367.

*Etsion, I. and Barkan, I.* (1981), Analysis of a hydrodynamic thrust bearing with incomplete film, *ASME Journal of Lubrication Technology*, 103,355–360.

*Fritz, R.J.* (1972), The effect of liquids on the dynamic motion of immersed solids, *ASME Journal of Engineering for Industry*, 94, 167–173.

*Ghosh, M.K., Guha, S.K. and Majumdar,* B.C. (1989), Fluid inertia effect on the dynamic stiffness of hydrostatic journal bearings, Paper No. 89-Trib-7, STLE/ASME *Joint Tribology Conference*, Fort Lauderdale, Florida, Oct. 16–19.

*Hirs, G.G.* (1973), A bulk flow theory for turbulence in lubricating films, *ASME Journal of Lubrication Technology*, 95,137–146.

*Hirs, G.G.* (1974), A systematic study of turbulent film flow, *ASME Journal of Lubrication Technology*, 96,118–126.

*Kakoty, S.K. and Majumdar, B.C.* (2000), Effect of fluid inertia on the dynamic coefficients and stability of journal bearings, *Proc. I. Mechanical Engineers*, U.K., 214 Part J, 229–242.

*Kamal, M.M.* (1966), Separation in the flow between eccentric rotating cylinders, *ASME Journal of Basic Engineering*, 88, 717-724.

*Kuhn, E.C. and Yates, C.C.* (1964), Fluid inertia effect on the film pressure between axially oscillating circular plates, *ASLE Transactions*, 7, 299–303.

*Kumar, A. and Rao, N.S.* (1995), Steady state analysis of plain journal bearings in turbulent regime, *Indian Journal of Engineering & Materials Sciences*, 2, 163–166.

*Kuzma, D.C.* (1967), Fluid inertia effects in squeeze films, *Applied Scientific Research*, 18, 15–20.

*Milne, A.A.* (1959), On the effect of lubricant inertia in the theory of hydrodynamic lubrication, *ASME Journal of Basic Engineering*, 81, 239–244.

*Mobbs, F.R. and Younes, M.S.* (1974), The Taylor vortex regime in the flow between eccentric rotating cylinders, *ASME Journal of Lubrication Technology*, 96, 127–134.

*Modest, M.F. and Tichy, J.A.* (1978), Squeeze film flow in arbitrary shaped journal bearings subject to oscillations, *ASME Journal of Lubrication Technology*, 100, 323–329.

*Mulachy, T.M.* (1980), Fluid forces on rods vibrating in finite length annular regions, *Journal of Applied Mechanics*, 47, 234–240.

*Ng, C.W.* (1964), Fluid dynamic foundation of turbulent lubrication theory, *ASLE Transactions*, 7, 311–321.

*Ng, C.W. and Pan C.H.T.* (1965), A linearized turbulent lubrication theory, *ASME Journal of Basic Engineering*, 87, 675–688.

*Osterle, J.F., Chou, Y.T. and Saibel, E.A.* (1957), Effect of lubricant inertia in journal bearing lubrication, *Journal of Applied Mechanics*, 24, 494–496.

*Pinkus, O. and Lund, J.W.* (1981), Centrifugal effects in thrust bearings and seals under laminar conditions, *ASME Journal of Lubrication Technology*, 103,126–135.

*Reinhardt, E. and Lund, J.W.* (1975), The influence of fluid inertia on the dynamic properties of journal bearings, *ASME Journal of Lubrication Technology*, 97, 159–167.

*San Andres, L.A.* (1990), Turbulent hybrid bearings with fluid inertia effects, *ASME Journal of Tribology*, 112, 699–707.

*San Andres, L.A. and Vance, J.M.* (1987), Force coefficients for open ended squeeze film dampers executing small amplitude motions about an off-center equilibrium position, *ASLE Transactions*, 30, 69–76.

*Szeri, A.Z., Raimondi, A.A. and Giron-Duarte, A.* (1983), Linear force coefficients for squeeze film dampers, *ASME Journal of Lubrication Technology*, 105, 326–334.

*Taylor, C.M. and Dowson, D.* (1974), Turbulent lubrication theory-application to design, *ASME Journal of Lubrication Technology*, 96, 36–47.

*Taylor, G.I.* (1923), Stability of a viscous liquid contained between two rotating cylinders, *Phil. Trans. Roy. Society*, A, 223, 289–343.

*Tichy, J.A. and Winer, W.* (1970), Inertial considerations in parallel circular squeeze film bearings, *ASME Journal of Lubrication Technology*, 92, 588–592.

*Wilcock, D.F.* (1981), Design of efficient turbulent thrust bearings iii. self-pressurized design, *ASME Journal of Lubrication Technology,* 103,459–466.

## APPENDIX 10.1

$$A_1 = \frac{\Omega h_0}{4}\left(1 - \frac{q_{\theta 0}}{3}\right)$$

$$A_2 = \frac{\Omega h_0^2}{12}$$

$$A_3 = h_0\left(-i\Omega\frac{h_0}{1`2} - \frac{1}{12}\frac{\partial h_0}{\partial \theta} + \frac{h_0}{60}\frac{L}{D}\frac{\partial q_{\bar{z}0}}{\partial \bar{z}} + \frac{h_0}{30}\frac{\partial q_{\theta 0}}{\partial \theta} + \frac{q_{\theta 0}}{30}\frac{\partial h_0}{\partial \theta}\right)$$

$$A_4 = h_0\frac{h_0^2}{60}\frac{L}{D}\frac{\partial q_{\theta 0}}{\partial \theta}$$

$$A_5 = \frac{h_0^2}{24}\left(-1 + \frac{\partial q_{\theta 0}}{5}\right)$$

$$A_6 = \frac{h_0^2 q_{\bar{z}0}}{60}$$

$$B_1 = \frac{\Omega h_0 q_{\bar{z}0}}{12}$$

$$B_1 = A_2$$

$$B_3 = \frac{h_0}{60}\left(q_{\bar{z}0}\frac{\partial h_0}{\partial \theta} + h_0\frac{q_{\bar{z}0}}{\partial \theta}\right)$$

$$B_3 = h_0\left(-i\Omega\frac{h_0}{1`2} - \frac{1}{24}\frac{\partial h_0}{\partial \theta} + \frac{h_0}{30}\frac{L}{D}\frac{\partial q_{\bar{z}0}}{\partial \bar{z}} + \frac{h_0}{60}\frac{\partial q_{\theta 0}}{\partial \theta} + \frac{q_{\theta 0}}{30}\frac{\partial h_0}{\partial \theta} + \frac{q_{\theta 0}}{60}\frac{\partial h_0}{\partial \theta}\right)$$

$$B_5 = A_5$$

$$B_6 = 2A_6$$

$$B_7 = A_6$$

$$C_1 = \frac{h_0}{6}\left(-1 + \frac{q_{\theta 0}}{2} - \frac{q^2_{\theta 0}}{10}\right)$$

$$C_2 = \frac{1}{6}\left(\frac{\partial \bar{h}}{\partial \theta} - \frac{h_0}{2}\frac{\partial q_{\bar{z}0}}{\partial \bar{z}} - h_0\frac{\partial q_{\theta 0}}{\partial \theta}\right) - \frac{q_{\theta 0}}{12}\frac{\partial h_0}{\partial \theta} + \frac{q_{\theta 0}h_0}{30}\frac{L}{D}\frac{\partial q_{\bar{z}0}}{\partial \bar{z}} + \frac{q^2_{\theta 0}}{60}\frac{\partial h_0}{\partial \theta} + \frac{q_{\theta 0}h_0}{15}\frac{\partial q_{\theta 0}}{\partial \theta} + \frac{q^2_{\theta 0}}{60}\frac{\partial h_0}{\partial \theta}$$

$$C_3 = -\frac{h_0}{12}\frac{\partial h_0}{\partial \theta} + \frac{h_0^2}{60}\frac{L}{D}\frac{\partial q_{\bar{z}0}}{\partial \bar{z}} + \frac{h_0 \partial q_{\theta 0}}{30}\frac{\partial h_0}{\partial \theta}\frac{h_0^2}{30}\frac{\partial q_{\theta 0}}{\partial \theta}$$

$$C_4 = A_4$$

$$C_5 = A_5$$

$$B_6 = A_6$$

$$C_7 = 2C_5$$

$$K_1 = \frac{h_0 q_{\bar{z}0}}{12}\left(\frac{1}{2} - \frac{q_{\theta 0}}{5}\right)$$

$$K_2 = \frac{q_{\bar{z}0}}{12}\left(-\frac{1}{2} + 2\frac{L}{D}\frac{\partial q_{\theta 0}}{\partial \theta}\right)\left(1 - \frac{\partial q_{\theta 0}}{5}\right) + \frac{h_0 q_{\bar{z}0}}{30}\left(\frac{\partial q_{\theta 0}}{\partial \theta} + \frac{D}{L}\frac{\partial q_{\theta 0}}{\partial \bar{z}}\right)$$

$$K_3 = \frac{h_0}{60}\left(q_{\bar{z}0}\frac{\partial h_0}{\partial \theta} + h_0\frac{\partial q_{\bar{z}0}}{\partial \theta}\right)$$

$$K_4 = \frac{h_0}{12}\frac{\partial h_0}{\partial \theta}\left(-\frac{1}{2} + \frac{q_{\theta 0}}{5}\right) + \frac{h_0^2}{60}\left(\frac{\partial q_{\theta 0}}{\partial \theta} + 2\frac{D}{L}\frac{\partial q_{\bar{z}0}}{\partial \bar{z}}\right)$$

$$K_5 = A_5$$

$$K_6 = A_6$$

$$K_7 = 2K_6$$

## APPENDIX 10.2

The dynamic coefficients are

$$k_{\varepsilon\varepsilon} = -\left(\int_0^1\int_{\theta 1}^{\theta 2} \bar{p}_x \cos\theta \; d\theta \, d\bar{z}\right)$$

$$k_{\varphi\varepsilon} = -\left(\int_0^1\int_{\theta 1}^{\theta 2} \bar{p}_x \sin\theta \; d\theta \, d\bar{z}\right)$$

$$k_{\varepsilon\varphi} = -\left(\int_0^1 \int_{\theta 1}^{\theta 2} \bar{p}_5 \cos\theta \ \ d\theta \, d\bar{z}\right)$$

$$k_{\varphi\varphi} = -\left(\int_0^1 \int_{\theta 1}^{\theta 2} \bar{p}_5 \sin\theta \ \ d\theta \, d\bar{z}\right)$$

$$b_{\varepsilon\varepsilon} = -\frac{1}{\Omega} \mathrm{Re}\left(\int_0^1 \int_{\theta_1}^{\theta 2} \bar{p}_{\dot{x}} \cos\theta \, d\theta \, d\bar{z}\right)$$

$$b_{\varphi\varepsilon} = -\frac{1}{\Omega} \mathrm{Re}\left(\int_0^1 \int_{\theta_1}^{\theta 2} \bar{p}_{\dot{x}} \sin\theta \, d\theta \, d\bar{z}\right)$$

$$b_{\varepsilon\varphi} = -\frac{1}{\Omega} \mathrm{Re}\left(\int_0^1 \int_{\theta_1}^{\theta 2} \bar{p}_{\dot{y}} \cos\theta \, d\theta \, d\bar{z}\right)$$

$$b_{\varphi\varphi} = -\frac{1}{\Omega} \mathrm{Re}\left(\int_0^1 \int_{\theta_1}^{\theta 2} \bar{p}_{\dot{y}} \sin\theta \, d\theta \, d\bar{z}\right)$$

$$a_{\varepsilon\varepsilon} = -\frac{1}{\Omega^2} \mathrm{Im}\left(\int_0^1 \int_{\theta_1}^{\theta 2} \bar{p}_{x} \cos\theta \, d\theta \, d\bar{z}\right)$$

$$a_{\varphi\varepsilon} = -\frac{1}{\Omega^2} \mathrm{Im}\left(\int_0^1 \int_{\theta_1}^{\theta 2} \bar{p}_{\dot{x}} \sin\theta \, d\theta \, d\bar{z}\right)$$

$$a_{\varepsilon\varphi} = -\frac{1}{\Omega^2} \mathrm{Im}\left(\int_0^1 \int_{\theta_1}^{\theta 2} \bar{p}_{\dot{y}} \cos\theta \, d\theta \, d\bar{z}\right)$$

$$a_{\varphi\varphi} = -\frac{1}{\Omega^2} \mathrm{Im}\left(\int_0^1 \int_{\theta_1}^{\theta 2} \bar{p}_{\dot{y}} \sin\theta \, d\theta \, d\bar{z}\right)$$

Chapter 11

# Gas-lubricated Bearings

## 11.1 | Introduction

The operating principle of gas-lubricated bearings is essentially same as that of liquid-lubricated bearings. In hydrodynamic gas bearings (sometimes called aerodynamic bearings), the pressure is generated by the action of shearing and squeezing the environmental gas between the surfaces in relative motion, whereas in externally pressurized gas bearings (aerostatic bearings) pressurized gas is fed from a compressor for development of pressure. Hence, both axial and radial loads can be supported avoiding metal-to-metal contact. Like liquid-lubricated bearings, gas bearings can operate entirely aerodynamically or aerostatically or by combination of both. In some applications, it becomes necessary to supply externally pressurized gas to aerodynamic bearing to prevent instabilities.

Some of the exclusive advantages of gas bearings are that gases are chemically stable for a wide range of temperature and have extremely low viscosities. Gas-bearing electric motors with ceramic windings can be operated at temperature up to 500° C for long periods. It has also been reported that aerostatic ceramic bearings were operated at temperatures up to 800° C at a speed of 65,000 rpm. Various types of expansion turbines, gas liquefiers, and refrigeration plants use gas bearings at low temperatures. A small high-speed expansion turbine developed by the British Oxygen Company operates at 350 000 rpm using helium gas at a temperature between 50° and 13° K. The low viscosity of gases as compared to liquids are exploited in specialized applications such as dynamometers, wind-tunnel balances, and mechanical instruments which operate at extreme low-static frictional torque. When air is used as a lubricant, a special advantage can be gained since the exhaust gas can be released to the surroundings giving rise to large flow rates. Low-friction characteristics of aerostatic bearings can also be exploited in the slideways of machine tools.

Like oil-lubricated bearings, the total power consumption (power loss due to viscous friction and power supply to the pressure sources) is not necessarily low. However, there are several disadvantages too. These bearings are more prone to instability. In gas bearings, the film thickness is much thinner than bearings using oil as lubricant. Therefore, the surfaces are to be finished with close tolerances so that the minimum film thickness is not of the same order as the roughness of the surfaces.

For successful operation of gas bearings, a few precautions are to be taken such as the surfaces are to be finished with high accuracy, there should be no misalignment between the journal and bearing, operating speed must be high, and applied load is kept at a relatively low level.

While dealing with oil-lubricated bearings as discussed in the previous chapters, the density of the lubricant has been treated as constant. However, in gas-lubricated bearings the density variation with pressure should be taken into account. At low speeds, the density may be kept as constant, but at high speed the density variation with pressure must be considered. The expansion of gas, in most cases, can be taken as isothermal, although at moderate speeds there will be some amount of friction loss, thereby the slight increase in temperature. This is advantageous since the viscosity of gas increases with increase in the temperature.

The Reynolds equation has been derived under no slip condition, that is, fluid particles adhering to the bearing surfaces will have the velocities of the surfaces. This condition is valid for continuum flow theory. When the film thickness is comparable to mean free molecular path of the gas, the continuum flow theory is not valid and the slip flow will occur.

When Knudsen number, $K_n < 0.01$, flow may be treated as continuum. Knudsen number $K_n$ is defined as $K_n = \frac{\lambda}{h}$, where $\lambda$ is the mean free molecular path and $h$ is the film thickness. If $0.01 < K_n < 15$, slip flow will occur. The load capacity decreases with Knudsen number.

The effect of compressibility on liquid and gas lubricated bearings can be analyzed in the following ways. It is known that fluid velocity in the direction of motion of a plane slider is given by:

$$u = \frac{1}{2\eta}\frac{dp}{dx}\left(z^2 - hz\right) + U\left(1 - \frac{z}{h}\right) \tag{11.1}$$

The mass flow rate per width can be written as

$$m = \rho\int_0^h u\,dz$$

$$= \rho\int_0^h \left(\frac{1}{2\eta}\frac{dp}{dx}\left(z^2 - hz\right) + U\left(1 - \frac{z}{h}\right)\right)dz$$

$$= \rho\left(\frac{Uh}{2} - \frac{h^3}{12\eta}\frac{dp}{dx}\right) \tag{11.2}$$

or

$$\frac{dp}{dx} = \frac{12\eta}{h^3}\left(\frac{Uh}{2} - \frac{m}{\rho}\right) \tag{11.3}$$

For liquid lubricated bearing $\rho$ is constant, hence the volume rate of flow per unit width is constant.

For gases, the density increases as the pressure increases. Thus, $\frac{m}{\rho}$ term in Equation (11.3) decreases and $\frac{dp}{dx}$ increases. Hence is the variation of pressure (Fig. 11.1). The rise in pressure curve is less abrupt in the region $\frac{dp}{dx} > 0$, whereas the drop is significant when $\frac{dp}{dx} < 0$. One can observe that the shape of pressure

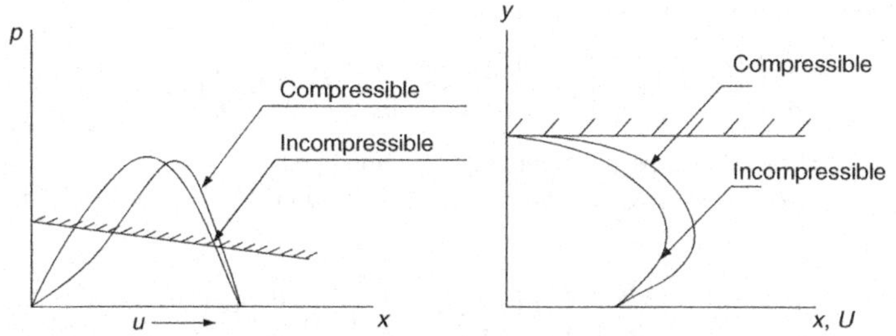


**Figure 11.1** | Pressure and Velocity Profile of Incompressible and Compressible Lubricant

variation in the direction of motion is not the same for the two bearings. In case of gas bearings, the effect of ambient pressure is noticeable if the pressure variation in the bearing is large; the effect of compressibility is more when compared with the ambient pressure. On the other hand, when the pressure variation is small compared to the ambient pressure, the compressibility is quite less.

For a given operating condition, $U$ increases due to compressibility. The increase inside the lubricant film is significant.

## 11.2 | Governing Equations

The generalized Reynolds equation which is already derived earlier can be written as

$$\frac{\partial}{\partial x}\left(\frac{\rho h^3}{\eta}\frac{\partial p}{\partial x}\right)+\frac{\partial}{\partial y}\left(\frac{\rho h^3}{\eta}\frac{\partial p}{\partial y}\right)=6U\frac{\partial(\rho h)}{\partial x}+12\frac{\partial(\rho h)}{\partial t} \tag{11.4}$$

Equation (11.4) is the governing equation for the gas-lubricated bearings. Here viscosity varies insignificantly with pressure. Viscosity can be assumed to be constant. To consider the variation of density with pressure, it is assumed that gas bearings operate isothermally and for perfect gas, the relationship of density with pressure can be given by $p = \rho RT$, where $R$ is the gas constant and $T$ is the absolute temperature.

If it is assumed that a gas obeys a polytropic relation,

then

$$p\rho^{-n} = \text{constant} \tag{11.5}$$

$n$ = polytropic gas-expansion exponent.

The polytropic index, $n = c_p/c_v$.

When the flow is isothermal, $n = 1$.

Substituting Equation (11.5) into Equation (11.4),

$$\frac{\partial}{\partial x}\left(p^{\frac{1}{n}}h^3\frac{\partial p}{\partial x}\right)+\frac{\partial}{\partial y}\left(p^{\frac{1}{n}}h^3\frac{\partial p}{\partial y}\right)=6\eta U\frac{\partial}{\partial x}\left(p^{\frac{1}{n}}h\right)+12\eta\frac{\partial}{\partial t}\left(p^{\frac{1}{n}}h\right) \tag{11.6}$$

For an isothermal film, Equation (11.6) will reduce to

$$\frac{\partial}{\partial x}\left(ph^3\frac{\partial p}{\partial x}\right)+\frac{\partial}{\partial y}\left(ph^3\frac{\partial p}{\partial y}\right)=6\eta U\frac{\partial}{\partial x}(ph)+12\eta\frac{\partial}{\partial t}(ph) \tag{11.7}$$

Equation (11.7) is a nonlinear partial differential equation in $p$.

Substituting $\bar{p}=p/p_a$, $\bar{h}=h/h_2$, $\bar{x}=x/B$, $\bar{y}=y/L$ in equation (11.7) under steady-state condition, we get

$$\frac{\partial}{\partial \bar{x}}\left(\bar{p}\bar{h}^3\frac{\partial \bar{p}}{\partial \bar{x}}\right)+\left(\frac{B}{L}\right)^2\frac{\partial}{\partial \bar{y}}\left(\bar{p}\bar{h}^3\frac{\partial \bar{p}}{\partial \bar{y}}\right)=\Lambda\frac{\partial}{\partial \bar{x}}(\bar{p}\bar{h}) \tag{11.8}$$

where $\Lambda=\dfrac{6\eta UB}{p_a h_2^2}$ is called bearing number, which is mainly a function of speed.

## 11.3 | Limiting Solution

Before obtaining the general solution two limiting solutions, for small $\Lambda$ (i.e., for small speed) and large $\Lambda$ (a very large speed) are tried here. Equation (11.8) can be written as

$$\frac{\partial}{\partial \bar{x}}\left(\bar{h}^3\frac{\partial \bar{p}}{\partial \bar{x}}\right)+\left(\frac{B}{L}\right)^2\frac{\partial}{\partial \bar{y}}\left(\bar{h}^3\frac{\partial \bar{p}}{\partial \bar{y}}\right)+\frac{\bar{h}^3}{\bar{p}}\left[\left(\frac{\partial \bar{p}}{\partial \bar{x}}\right)^2+\left(\frac{\partial \bar{p}}{\partial \bar{z}}\right)^2\right]=\Lambda\left[\frac{\partial \bar{h}}{\partial \bar{x}}+\frac{\bar{h}}{\bar{p}}\frac{\partial \bar{p}}{\partial \bar{x}}\right] \tag{11.9}$$

For small bearing numbers, Equation (11.9) can be approximated and it can be shown that the differential equation for gas and oil-lubricated bearings will be identical.

When $\Lambda\to 0, \bar{p}\to 1$ and the pressure rise, $\Delta\bar{p}\to 0$.

The terms $\dfrac{\partial \bar{p}}{\partial \bar{x}}$ and $\dfrac{\partial \bar{p}}{\partial \bar{y}}$ are small, and $\left(\dfrac{\partial \bar{p}}{\partial \bar{x}}\right)^2$ and $\left(\dfrac{\partial \bar{p}}{\partial \bar{y}}\right)^2$ are still smaller.

Again $\dfrac{\bar{h}}{\bar{p}}\dfrac{\partial \bar{p}}{\partial \bar{x}} << \dfrac{\partial \bar{h}}{\partial \bar{x}}$

Therefore, Equation (11.9) for small bearing numbers reduces to

$$\frac{\partial}{\partial \bar{x}}\left(\bar{h}^3\frac{\partial \bar{p}}{\partial \bar{x}}\right)+\left(\frac{B}{L}\right)^2\frac{\partial}{\partial \bar{y}}\left(\bar{h}^3\frac{\partial \bar{p}}{\partial \bar{y}}\right)=\Lambda\frac{\partial \bar{h}}{\partial \bar{x}} \tag{11.10}$$

Equation (11.10) corresponds to the Reynolds equation using incompressible lubricant.

For large bearing numbers, i.e., for $\Lambda\to\infty$, pressure can remain finite only if $\dfrac{\partial}{\partial \bar{x}}(\bar{p}\bar{h})\to 0$ i.e. $\bar{p}\bar{h}$ = constant. Therefore, at values of high bearing number, pressures generated are independent of bearing number.

There exists a fundamental difference between oil lubrication and gas lubrication. In oil lubrication, the pressure is proportional to speed and viscosity of lubricant and is independent of ambient pressure, whereas in self-acting gas-lubricated bearings at high speeds, the pressure is independent of speed, and viscosity is dependent on ambient pressure.

It follows from the above discussion that for small values of $\Lambda$ the effect of compressibility can be ignored. The equations of gas-lubricated bearings become identical with those of oil-lubricated bearings. Despite this fact, the two phenomena are not identical, because some characteristic properties of gases continue to manifest themselves even though the effect of compressibility is small. This is mainly due to expansibility of gas.

For an example, if we consider the case of an infinitely long journal bearing the pressure distribution using boundary condition, $p(\theta=0)=p(2\pi)$, can be given by

$$p = p_a\left(1+\frac{\Lambda\varepsilon}{2+\varepsilon^2}\frac{(2+\varepsilon\cos\theta)\sin\theta}{(1+\varepsilon\cos\theta)^2}\right)$$

where

$$\Lambda = \frac{6\eta\omega}{p_a\left(\frac{C}{R}\right)^2}$$

We can see $\frac{(p-p_a)}{p_a}$ increases linearly with $\Lambda$. The pressure variation is antisymmetric with respect to the line $\theta=\pi$ (Fig.11.2). In case of liquid lubricant, the region $\pi\le\theta\le 2\pi$ is inactive as the cavitation sets in and film breaks down. On the other hand, for gases there can be depressions, but the expression for pressure distribution given above is valid for the entire circumference of the bearing $(\pi\le\theta\le 2\pi)$. Hence the film is full.

For oil lubricated bearings, we can find the load capacity considering $0-\pi$ film. Similarly, the load capacity can be found for gas-lubricated bearing for $0-\pi$ film. From these, the ratio of load capacity of gas-lubricant to that of oil lubricant will be given by $2\sqrt{\frac{(1-\varepsilon^2)}{1-\left(1-\left(4/\pi^2\right)\varepsilon^2\right)}}$ for small values of $\Lambda$.

This ratio is larger than unity for $\varepsilon\le 0.94$. Hence, the load capacity of a gas-lubricated bearing at $\Lambda\to 0$ is relatively larger than the load corresponding to an oil lubricant of same viscosity $\eta$. However, in practice one can hardly find a gas lubricant having same viscosity of that of oil.

Because of nonlinearity of differential equation, it is difficult to obtain a closed form solution. Therefore, several methods of linearizing the differential equations are adopted.

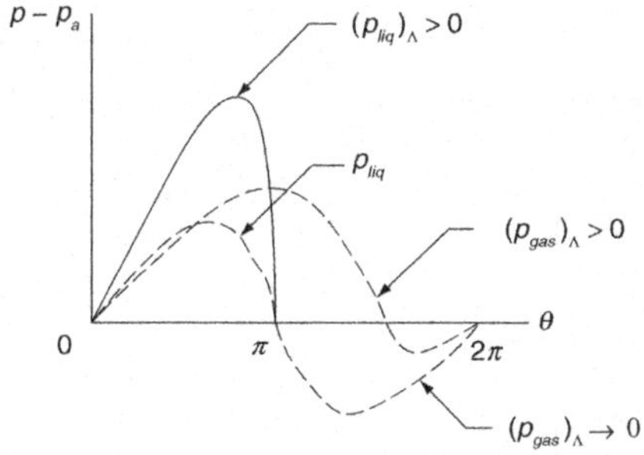


**Figure 11.2** | Variation of Pressure at Low Speeds for Liquid and Gas Lubricants

## 11.4 | Infinitely Long Plane Slider

For an infinitely long plane slider bearing shown in Fig.11.3, the fundamental equation for pressure distribution under steady-state condition can be written as

$$\frac{\partial}{\partial x}\left(\frac{\rho h^3}{\eta}\frac{\partial p}{\partial x}\right)=6U\frac{\partial(\rho h)}{\partial x} \tag{11.11}$$

For an isothermal flow and using constant viscosity, Equation (11.11) reduces to

$$\frac{\partial}{\partial x}\left(\rho h^3\frac{\partial p}{\partial x}\right)=6\eta U\frac{\partial(\rho h)}{\partial x} \tag{11.12}$$

With the following substitutions, $\bar{x}=x/B$ $\bar{p}=p/p_a$, and $\bar{h}=h/h_2$, nondimensional form of Equation (11.12) is given by

$$\frac{\partial}{\partial \bar{x}}\left(\bar{p}\bar{h}^3\frac{\partial \bar{p}}{\partial \bar{x}}\right)=\Lambda\frac{\partial(\bar{p}\bar{h})}{\partial \bar{x}} \tag{11.13}$$

where

$$\Lambda=\frac{6\eta UB}{p_a h_2^2}$$

The dimensionless film thickness $\bar{h}$ is given by

$$\bar{h}=n-n(n-1)\bar{x} \tag{11.14}$$

where

$$n=\frac{h_1}{h_2} \quad \text{and} \quad \bar{x}=\frac{x}{B}$$

Integrating Equation (11.13),

$$\bar{p}\bar{h}^3\frac{\partial \bar{p}}{\partial \bar{x}}=\Lambda\left(\bar{p}\bar{h}-C_1\right) \tag{11.15}$$

where $C_1$ = constant.

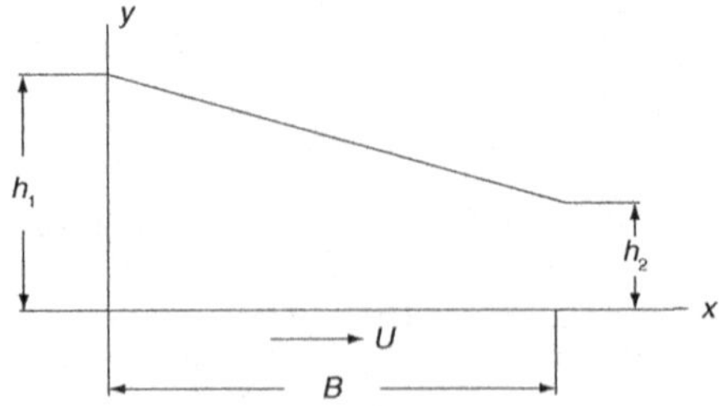


**Figure 11.3** | An Infinitely Long Plane Slider

Substituting $\Phi = \bar{p}\bar{h}$, Equation (11.15) then becomes

$$\frac{\Phi d\Phi}{\Phi^2 - \bar{\Lambda}\Phi + \bar{\Lambda}C_1} = \frac{d\bar{h}}{\bar{h}} \tag{11.16}$$

where $$\bar{\Lambda} = \frac{\Lambda}{n-1}.$$

The boundary conditions are:

$$\Phi = n \text{ at } \bar{x} = 0 \qquad \text{and} \qquad \Phi = 1 \text{ at } \bar{x} = 1 \tag{11.17}$$

Integrating Equation (11.16), we obtain

$$\Phi^2 - \bar{\Lambda}\Phi + \bar{\Lambda}C_1 = C_2\bar{h}^2\Psi(\Phi) \tag{11.18}$$

where $C_2$ is a constant of integration and $\Psi(\Phi)$ is given by the following form:

$$\Psi_1(\Phi) = exp - \left\{ \frac{2\bar{\Lambda}}{\left[\bar{\Lambda}(4C_1 - \bar{\Lambda})\right]^{\frac{1}{2}}} \tan^{-1} \frac{2\Phi - \bar{\Lambda}}{\left[\bar{\Lambda}(4C_1 - \bar{\Lambda})\right]^{\frac{1}{2}}} \right\} \text{for } \bar{\Lambda} < 4C_1$$

$$\Psi_c(\Phi) = exp \frac{8C_1}{2\Phi - \bar{\Lambda}} \quad \text{for } \bar{\Lambda}_c = 4C_1$$

and $$\Psi_2(\Phi) = \left\{ \frac{2\Phi - \bar{\Lambda} + \left[\bar{\Lambda}(4C_1 - \bar{\Lambda})\right]^{\frac{1}{2}}}{2\Phi - \bar{\Lambda} - \left[\bar{\Lambda}(4C_1 - \bar{\Lambda})\right]^{\frac{1}{2}}} \right\}^{\bar{\Lambda}/\left[\bar{\Lambda}(4C_1 - \bar{\Lambda})\right]^{1/2}} \qquad \text{for } \bar{\Lambda} > 4C_2$$

$C_1$ and $C_2$ can be evaluated using the boundary conditions given in Equation (11.17).

The load-carrying capacity is

$$W = L\int_0^B (p - p_a)dx \tag{11.19}$$

$$= LBp_a\int_0^1 (\bar{p} - 1)d\bar{x}$$

or $$\bar{W} = \left(\frac{W}{LBp_a}\right) = \int_0^1 (\bar{p} - 1)d\bar{x}$$

or $$\bar{W} = \int_0^1 \left[\frac{\Phi}{n - (n-1)\bar{x}} - 1\right] d\bar{x} \tag{11.20}$$

Table 11.1 | Load Capacity and Centre of Pressure for Infinitely Long Plane Sliders Under Isothermal Condition

| $n$ | $\Lambda$ | $\bar{W}$ | $\bar{x}_c$ |
|---|---|---|---|
| 1.5 | 0.5 | 0.01091 | 0.5456 |
| | 1.0 | 0.02172 | 0.5520 |
| | 5.0 | 0.09570 | 0.5861 |
| | 10.0 | 0.14860 | 0.6168 |
| | 25.0 | 0.19420 | 0.6585 |
| 2.0 | 0.5 | 0.01323 | 0.5724 |
| | 1.0 | 0.02640 | 0.5761 |
| | 5.0 | 0.12340 | 0.6008 |
| | 10.0 | 0.21240 | 0.6235 |
| | 50.0 | 0.36390 | 0.6870 |
| 3.0 | 0.5 | 0.01232 | 0.6095 |
| | 1.0 | 0.02063 | 0.6116 |
| | 5.0 | 0.12010 | 0.6264 |
| | 10.0 | 0.22520 | 0.6402 |
| | 50.0 | 0.56180 | 0.6950 |

Substituting $(\Phi)$ in Equation (11.19) and integrating numerically gives load capacity. For bearing number approaching 0 and $\infty$ the load capacities are given by

$$\left.\begin{aligned} \bar{W}_{\Lambda\to 0} &= \Lambda\left(n^2-1\right)^{-1}\left[\ln n - 2(n-1)(n+1)^{-1}\right]-1 \\ \text{and}\quad \bar{W}_{\Lambda\to\infty} &= \frac{n}{n-1}-1 \end{aligned}\right\} \tag{11.21}$$

The isothermal load-carrying capacity and centre of pressure for infinitely long slider bearings for various attitudes and speeds are given in Table 11.1

## 11.5 | Finite Journal Bearings

The Reynolds equation under steady-state condition for isothermal film is

$$\frac{\partial}{\partial x}\left(\rho h^3\frac{\partial p}{\partial x}\right)+\frac{\partial}{\partial y}\left(\rho h^3\frac{\partial p}{\partial y}\right)=6\eta U\frac{\partial}{\partial x}(\rho h) \tag{11.22}$$

Equation (11.22) can be nondimensionalized with the following substitutions:

$$\theta = x/R,\ \bar{y} = y/R,\ \bar{h} = h/C,\ \bar{p} = p/p_a, \quad \text{and} \quad U = \omega R.$$

The dimensionless form of Equation (11.22) will be

$$\frac{\partial}{\partial\theta}\left(\bar{p}\bar{h}^3\frac{\partial p}{\partial\theta}\right)+\frac{\partial}{\partial\bar{y}}\left(\bar{p}\bar{h}^3\frac{\partial p}{\partial\bar{y}}\right)=\Lambda\frac{\partial}{\partial\theta}\left(\bar{p}\bar{h}\right) \tag{11.23}$$

where

$$\Lambda = \frac{6\eta\omega}{p_a\left(C/R\right)^2}.$$

Equation (11.23) is a nonlinear partial differential equation in $\bar{p}$ and its solution satisfying the boundary conditions is extremely difficult. For this reason, some approximate methods of solution are tried.

### 11.5.1 | The Perturbation Method [Ausman (1957)]

The perturbation method is one of the techniques employed to linearize Equation (11.23) in order to obtain an approximate solution. The pressure is expressed in power series of increasing eccentricity ratio $\varepsilon$.

$$\bar{p} = 1+\varepsilon\bar{p}_1+\varepsilon^2\bar{p}_2+\varepsilon^3\bar{p}_3+\ldots.. \tag{11.24}$$

where $\bar{P}_1 = P_1/P_{a,}\ \bar{P}_2 = P_2/P_{a,}\ \bar{P}_3 = P_3/P_a$ etc.

The film thickness is expressed by

$$\bar{h} = 1+\varepsilon\cos\theta \tag{11.25}$$

Substituting Equations (11.24) and (11.25) in Equation (11.23), the following set of linear differential equations is obtained:

$$\frac{\partial^2\bar{p}_1}{\partial\theta^2}+\frac{\partial^2\bar{p}_1}{\partial\bar{y}^2}=\Lambda\left(\frac{\partial\bar{p}_1}{\partial\theta}-\sin\theta\right) \tag{11.26}$$

$$\frac{\partial^2\bar{p}_2}{\partial\theta^2}+\frac{\partial^2\bar{p}_2}{\partial\bar{y}^2}+\left[\left(\frac{\partial\bar{p}_1}{\partial\theta}\right)^2+\left(\frac{\partial\bar{p}_1}{\partial\bar{y}}\right)^2\right]-3\frac{\partial\bar{p}_1}{\partial\theta}\sin\theta=\Lambda\left[\frac{\partial^2\bar{p}_2}{\partial\theta}+\frac{3}{2}\sin2\theta-2\frac{\partial\bar{p}_1}{\partial\theta}\cos\theta-\bar{h}\frac{\partial\bar{p}_1}{\partial\theta}\right] \tag{11.27}$$

Similarly, a differential equation in $\bar{P}_3$ can be found. Equation (11.26) can be solved for $\bar{P}_1$ which can be substituted in Equation (11.27) as a known function. Equation (11.27) can be solved for $\bar{P}_2$, which in turn can be used for the equation in $\bar{P}_3$ and so on. It will be shown later that $\bar{P}_1$ comes out to be quite lengthy and its substitution in the right hand side of Equation (11.27) makes the equation complex. This does not permit to get $\bar{P}_2, \bar{P}_3$,etc. Moreover, the $\bar{P}_2$ does not contribute to the load carrying capacity because of orthogonality. Thus, load capacity can be calculated using the first-order perturbation solution with fair accuracy for small eccentricity ratios $(\varepsilon \le 0.3)$.

The first-order solution.

Let $\bar{p}_1$ be taken as

$$\bar{p}_1 = \bar{p}_1(\infty) + Y(\bar{y})\Theta(\theta) \tag{11.28}$$

where $Y$ is a function of $\bar{y}$ alone, $\Theta$ is a function of $\theta$ alone, and $\bar{p}_1(\infty)$ is the first-order perturbation solution for an infinitely long journal bearing.

The $\bar{p}_1(\infty)$ is given by

$$\bar{p}_1(\infty) = \frac{\Lambda}{1+\Lambda^2}(\sin\theta - \Lambda\cos\theta) \tag{11.29}$$

On substitution of Equation (11.28) into Equation (11.29), a homogeneous equation in $Y$ and $\Theta$ is obtained. It can be then solved by separation of variables method.

The boundary conditions are:

- $\bar{p}_1(\theta, \bar{y}) = \bar{p}_1(\theta + 2\pi, \bar{y})$
- $\dfrac{\partial \bar{p}_1}{\partial \bar{y}} = 0$ at $\bar{y} = 0$ (11.30)
- $\bar{p}_1 = 0$ at $\bar{y} = \pm\dfrac{L}{D}$

Solution of Equation (11.26) with the help of Equations (11.28), (11.29) and the first two boundary condition of Equation (11.30) gives pressure distribution.

The load components parallel to and perpendicular to the line of centres are calculated from

$$W_r = -R^2 \int_{-L/D}^{L/D}\int_0^{2\pi} p\cos\theta \, d\theta \, d\bar{y} \tag{11.31}$$

$$W_\theta = R^2 \int_{-L/D}^{L/D}\int_0^{2\pi} p\sin\theta \, d\theta \, d\bar{y}$$

The load components in terms of perturbed pressure $\bar{p}_1$ are

$$W_r = -R^2 p_a \int_{-L/D}^{L/D}\int_0^{2\pi} \varepsilon\bar{p}_1 \cos\theta \, d\theta \, d\bar{y} \tag{11.32}$$

$$W_\theta = R^2 p_a \int_{-L/D}^{L/D}\int_0^{2\pi} \varepsilon\bar{p}_1 \sin\theta \, d\theta \, d\bar{y}$$

Substitution of perturbed pressures and integration yields

$$\frac{W_r}{\pi\varepsilon p_a RL} = \frac{\Lambda}{1+\Lambda^2}\left[\Lambda + \frac{(\alpha - \beta\Lambda)\sin(2\beta L/D) - (\alpha\Lambda + \beta)\sinh(2\alpha L/D)}{\left(\dfrac{L}{D}\right)\sqrt{1+\Lambda^2}\,[\cosh(2\alpha L/D) + \cos(2\beta L/D)]}\right] \tag{11.33a}$$

and $$\frac{W_\theta}{\pi\varepsilon p_a RL} = \frac{\Lambda}{1+\Lambda^2}\left[1 - \frac{(\alpha-\beta\Lambda)\sinh\left(\frac{2\alpha L}{D}\right)+(\alpha\Lambda+\beta)\sin\left(\frac{2\alpha L}{D}\right)}{\left(\frac{L}{D}\right)\sqrt{1+\Lambda^2}\,[\cosh(2\alpha L/D)+\cos(2\beta L/D)]}\right] \tag{11.33b}$$

where $$\alpha^2 = \frac{\sqrt{1+\Lambda^2}+1}{2} \quad \text{and} \quad \beta^2 = \frac{\sqrt{1+\Lambda^2}-1}{2}$$

The total load $$W = \sqrt{W_r^2 + W_\theta^2} \tag{11.34}$$

and the attitude angle $$\phi = \tan^{-1}\left(\frac{W_\theta}{W_r}\right) \tag{11.35}$$

For an infinitely long journal bearing, $\frac{L}{D} \to \infty$, the load components for this case will be

$$\frac{W_r(\infty)}{\pi\varepsilon p_a RL} = \frac{\Lambda^2}{1+\Lambda^2} \tag{11.36a}$$

and $$\frac{W_\theta(\infty)}{\pi\varepsilon p_a RL} = \frac{\Lambda}{1+\Lambda^2} \tag{11.36b}$$

The above results are true for small $\varepsilon(\varepsilon \le 0.3)$. The dimensionless load and attitude angle against bearing number are shown in Fig. 11.4

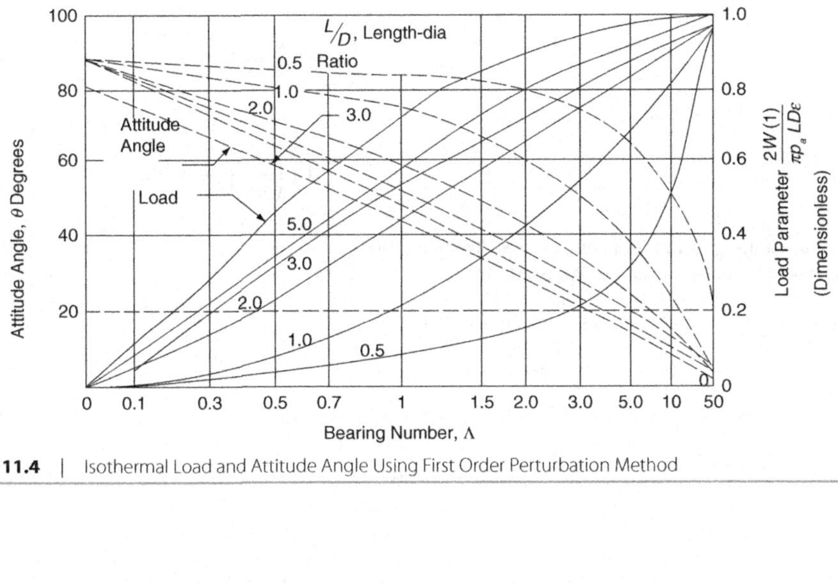


**Figure 11.4** | Isothermal Load and Attitude Angle Using First Order Perturbation Method

### 11.5.2 | The Linearized 'ph' Method [Ausman (1960)]

The linearized 'ph' method corrects the deficiency of the first-order perturbation method at high eccentricities. The general method of linearization is essentially the same as the perturbation method except that the 'ph' is considered to be the dependent variable.

Letting $\Phi = \bar{p}\bar{h}$, $\theta = \frac{x}{R}$, $\bar{y} = \frac{y}{R}$, $U = \omega R$ and

$\bar{h} = 1 + \varepsilon \cos\theta$, the dimensionless form of the Reynolds equation is

$$\frac{\partial^2 \Phi}{\partial \theta^2} + \frac{\partial^2 \Phi}{\partial y^2} = \Lambda \frac{\partial \Phi}{\partial \theta} - \varepsilon \cos\theta \tag{11.37}$$

where

$$\Lambda = \frac{6\eta\omega}{p_a \left(\frac{C}{R}\right)^2}$$

The boundary conditions are:

- $\Phi = \bar{h}$ at $\bar{y} = \pm \frac{L}{D}$
- $\Phi(\theta, \bar{y}) = \Phi(\theta + 2\pi, \bar{y})$ (11.38)
- $\frac{\partial \Phi}{\partial y} = 0$ at $\bar{y} = 0$

Solution of Equation (11.37) with the above boundary conditions $\bar{\Phi}$ is obtained.

The load components $W_r$ and $W_\theta$, parallel and perpendicular to the line of centres, are given by Equation (11.29), which upon substitution of $\bar{p}\left(=\frac{\Phi}{\bar{h}}\right)$ and subsequent integration yields

$$W_r = \frac{2\pi R^2 p_a}{\varepsilon} \frac{1-\sqrt{1-\varepsilon^2}}{\sqrt{1-\varepsilon^2}} \int_{-\frac{L}{D}}^{\frac{L}{D}} \left(1 - \frac{\Lambda}{1+\Lambda^2} g_2 \bar{y}\right) d\bar{y}$$

and

$$W_\theta = \frac{2\pi R^2 p_a}{\varepsilon} \left(1-\sqrt{1-\varepsilon^2}\right) \int_{-\frac{L}{D}}^{\frac{L}{D}} \left(\frac{\Lambda}{1+\Lambda^2} g_1 \bar{y}\right) d\bar{y}$$

For functions $g_1\bar{y}$ and $g_2\bar{y}$ one can see Ausman (1960).

The load components can be expressed as

$$W_r = W_r^{(1)} \left(\frac{2}{\varepsilon^2}\right) \frac{1-\sqrt{1-\varepsilon^2}}{\sqrt{1-\varepsilon^2}} \tag{11.39a}$$

$$W_\theta = W_\theta^{(1)} \left(\frac{2}{\varepsilon^2}\right) \left(1-\sqrt{1-\varepsilon^2}\right) \tag{11.39b}$$

where $W_r^{(1)}$ and $W_\theta^1$ are the first-order perturbation solutions.

The total load and attitude angle are

$$W = W^{(1)}\left(\frac{2}{\varepsilon^2}\right)\frac{\left(1-\sqrt{1-\varepsilon^2}\right)}{\sqrt{1-\varepsilon^2}}\sqrt{1-\varepsilon^2 \sin^2 \phi^{(1)}} \tag{11.40}$$

and

$$\tan\phi = \sqrt{1-\varepsilon^2}\tan\phi^{(1)} \tag{11.41}$$

where $W^{(1)}$ and $\phi^{(1)}$ are load and attitude angle, $\Phi$ obtained from the first-order perturbation solution. The $W^{(1)}$ and $\Phi^{(1)}$ can be found from Fig. 11.4.

## 11.6 | Externally Pressurized Gas Bearings

The operating principle of an externally pressurized (hydrostatic) gas bearing is basically same as that of a hydrostatic oil bearing. Here the compressed gas at a constant supply pressure via a restrictor (usually an orifice) is fed to the bearing clearance which, in turn, takes the load avoiding metal-to-metal contact and finally it exhausts at ambient pressure from the bearing ends.

### 11.6.1 | Circular Step Thrust Bearings

For a circular thrust bearing shown in Fig.11.5, the volume rate of flow of gas

$$Q = -\frac{h^3}{12\eta}\frac{\partial p}{\partial r}.2\pi r \tag{11.42}$$

$$pQ = p_r Q_r = \text{constant} \tag{11.43}$$

From continuity consideration

$$Q_r \frac{p_r}{p} = -\frac{2\pi r h^3}{12\eta}\frac{\partial p}{\partial r} \tag{11.44}$$

Since

$$pQ = p_r Q_r = \text{constant}$$

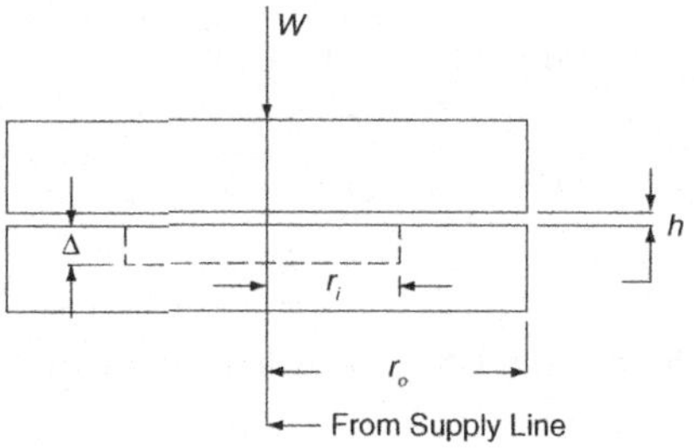


**Figure 11.5** | A Circular Step Thrust Gas Bearing

Separating the variables and integrating

$$p = p_r\left(1 - \frac{12\eta Q_r}{\pi h^3 p_r}\ln\frac{r}{r_i}\right)^{1/2} \tag{11.45}$$

At

$$r = r_0,\ p = p_a,$$

$$\frac{p_a}{p_r} = \left(1 - \frac{12\eta Q_r}{\pi h^3 p_r}\ln\frac{r_0}{r_i}\right)^{1/2} \tag{11.46}$$

From which $Q_r$ can be found and when it is substituted in Equation (11.45), one obtains

$$p = p_r\left[1 - \frac{1-\left(\frac{p_a}{p_r}\right)^2}{\ln\left(\frac{r_0}{r_i}\right)}\ln\frac{r}{r_i}\right]^{\frac{1}{2}} \tag{11.47}$$

or

$$p^2 = -\left(p_r^2 - p_a^2\right)\frac{\ln\left(\frac{r}{r_i}\right)}{\ln\left(\frac{r_0}{r_i}\right)} + p_r^2 \tag{11.48}$$

The load capacity $W$ is

$$W = \pi p_r r_1^2 - \pi p_a r_0^2 + \int_{r_1}^{r_0} 2\pi rp\,dr \tag{11.49}$$

where $p_r$ is recess pressure.

As it is difficult to perform integration, it is assumed that the pressure drops linearly from $r_i$ to $r_0$,

$$p = p_r - p_a \quad 0 \le r \le r_i$$

$$p = \left(p_r - p_a\right)\left(\frac{r_0 - r}{r_0 - r_i}\right) \quad r_i \le r \le r_0$$

and

$$W = \pi\left(p_r - p_a\right)r_i^2 + 2\pi\int_{r_i}^{r_0} pr\,dr$$

or

$$W = \frac{\pi}{3}\left(p_r - p_a\right)\frac{r_0^3 - r_i^3}{r_0 - r_i} \tag{11.50}$$

### 11.6.2 | Stiffness

Using an orifice as a restrictor, the mass flow rate, $G$, into the bearing through a single orifice when unchoked flow occurs, i.e.,

for

$$\frac{p_r}{p_s} \ge \left(\frac{2}{k+1}\right)^{k/k-1} \tag{11.51a}$$

$$G = C_d A_0 p_s \left[\frac{2k}{RT(k-1)}\right]^{1/2} \left(\frac{p_r}{p_s}\right)^{(2/k)} - \left[\left(\frac{p_r}{p_s}\right)^{(k+1)/k}\right]^{1/2}$$

for choked flow condition,

i.e., for $$\frac{p_r}{p_s} < \left(\frac{2}{k+1}\right)^{k/k-1}$$

$$G = C_d A_0 p_s \left(\frac{k}{RT}\right)^{1/2} \left(\frac{2}{k+1}\right)^{\frac{k+1}{2(k-1)}} \tag{11.51b}$$

$C_d$ = coefficient of discharge of orifice restrictor and
$A_0$ = area of orifice restrictor,

For an adiabatic flow of air through the orifice, $k = 1.4$. When $\frac{p_r}{p_s} > 0.528$ the flow is unchoked, and for $\frac{p_r}{p_s} < 0.528$, the flow is constant and depends on supply pressure.

The mass rate of flow through the bearing is

$$G = \frac{\pi h^3 \left(p_r^2 - p_a^2\right)}{12\eta RT \ln\left(\frac{r_0}{r_i}\right)} \tag{11.52}$$

This equation can be written as

$$G = \frac{h^3 \left(p_r^2 - p_a^2\right)\bar{G}}{24\eta RT} \tag{11.53}$$

where $$\bar{G} = \frac{2\pi}{\ln\left(\frac{r_0}{r_i}\right)}$$

From continuity consideration flow through the orifice should be equal to the bearing outflow. For $\frac{p_r}{p_s} \geq 0.528$

$$\delta\left(\frac{2}{k-1}\right)^{\frac{k+1}{2(k-1)}} \left[\left(\frac{p_r}{p_s}\right)^{2/k} - \left(\frac{p_r}{p_s}\right)^{k+1/k}\right]^{1/2} - \bar{G}\left[\left(\frac{p_r}{p_s}\right)^2 - \left(\frac{p_a}{p_s}\right)^2\right] = 0 \tag{11.54a}$$

For $$\frac{p_r}{p_s} < 0.528,\ \delta\left(\frac{2}{k+1}\right)^{\frac{k+1}{2(k-1)}} - \bar{G}\left[\left(\frac{p_r}{p_s}\right)^2 - \left(\frac{p_a}{p_s}\right)^2\right] = 0 \tag{11.54b}$$

where $$\delta = \frac{24\eta C_d A_0 (kRT)^{1/2}}{h^3 p_s}$$

Equation (11.54) can be solved numerically using Newton–Raphson method in order to obtain the pressure at the downstream of the orifice for various values of restrictor parameter $\delta$ and pad shape ratio.

Thus, knowing the pressure in the recess $p_r$ the load capacity and flow rate of compensated bearing can be found.

The static stiffness $S$ can be calculated from

$$S = -\frac{dW}{dh} \tag{11.55}$$

## 11.7 | Journal Bearings

An approximate method of solution is adopted here by treating the feed sources as a 'line source'. At small eccentricity ratio, this analysis considers only axial flow. Heinrich (1959) used this approach for evaluating load, flow, and stiffness characteristics of this type of bearing.

For a stationary journal (Fig. 11.6) the governing differential equation under isothermal condition is

$$\frac{\partial}{\partial x}\left(h^3 \frac{\partial p^2}{\partial x}\right) + \frac{\partial}{\partial y}\left(h^3 \frac{\partial p^2}{\partial y}\right) = 0 \tag{11.56}$$

If flow through the bearing is purely axial which is a reasonable assumption for small eccentricity ratios, Equation (11.56) reduces to

$$\frac{\partial^2 p^2}{\partial y^2} = 0 \tag{11.57}$$

Here $h$ is replaced by radial clearance $C$.

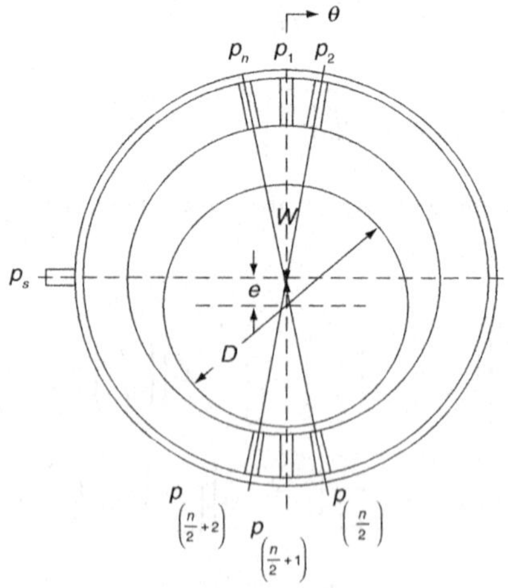


**Figure 11.6** | An Externally Pressurized Gas Journal Bearing

Integrating Equation (11.57) with the boundary conditions

$$(i)\ p = p_r \text{ at } y = 0 \text{ and } (ii)\ p = p_a \text{ at } y = \frac{L}{2},$$

we get,

$$p^2 = \left(p_a^2 - p_r^2\right)\frac{y}{L/2} + p_r^2$$

or

$$\bar{p}^2 = \left(1 - \bar{p}_r^2\right)\bar{y} + \bar{p}_r^2 \tag{11.58}$$

where

$$\bar{p} = \frac{p}{p_a}, \bar{p}_r = \frac{p_r}{p_a} \text{ and } \bar{y} = \frac{y}{L/2}.$$

Equating the mass rate of flow per unit circumferential length at the mid plane,

$$\frac{-2C^3 p_a^2}{24\eta RT(L/2)}\frac{\partial \bar{p}^2}{\partial \bar{y}}\bigg|_{\bar{y}=0} = \frac{N}{\pi D}G \tag{11.59}$$

where $N$ is the number of orifices and $G$ is the mass rate of flow through a single orifice.

Using flow through orifice for unchoked and choked conditions, one obtains

for

$$\frac{p_r}{p_s} \geq \left(\frac{2}{k+1}\right)^{k/k-1}$$

$$N\left(\frac{L}{D}\right)\delta\left[\frac{2}{k-1}\right]^{1/2}\left[\left(\frac{p_r}{p_s}\right)^{2/k} - \left(\frac{p_r}{p_s}\right)^{\frac{k+1}{k}}\right]^{1/2} = 4\pi\left[\left(\frac{p_r}{p_s}\right)^2 - \left(\frac{p_a}{p_s}\right)^2\right] \tag{11.60a}$$

and for

$$\frac{p_r}{p_s} < \left(\frac{2}{k+1}\right)^{k/k-1}$$

$$N\left(\frac{L}{D}\right)\delta\left(\frac{k}{k+1}\right)^{\frac{k+1}{2(k-1)}} = 4\pi\left[\left(\frac{p_r}{p_s}\right)^2 - \left(\frac{p_a}{p_s}\right)^2\right] \tag{11.60b}$$

where

$$\delta = \frac{24\eta C_d A_0 (kRT)^{1/2}}{C^3 p_s}$$

Equations (11.60a & b) are nonlinear algebraic equations in $\left(\frac{p_r}{p_s}\right)$ and can be solved using Newton–Raphson iteration method for constant values of $N, \frac{L}{D}, \delta$ and $\left(\frac{p_a}{p_s}\right)$. The pressure distribution can be calculated from Equation (11.58).

The load capacity $W$ is

$$W = -L\int_0^{2\pi} pR\cos\theta d\theta \tag{11.61}$$

or

$$\bar{W}\left(=\frac{W}{LDp_a}\right) = -\frac{1}{2}\int_0^{2\pi} \bar{p}\cos\theta d\theta \tag{11.62}$$

Introducing $\bar{p}$ and integrating Equation (11.62), dimensionless load can be calculated.

When the journal rotates with constant angular velocity, $\omega$, the differential equation under isothermal condition is

$$\frac{\partial}{\partial x}\left(h^3\frac{\partial p^2}{\partial x}\right) + \frac{\partial}{\partial y}\left(h^3\frac{\partial p^2}{\partial y}\right) = 6\eta\omega R\frac{\partial(ph)}{\partial x} \tag{11.63}$$

For small eccentricity ratios, Equation (11.63) can be linearized by using first-order perturbation method with respect to $\varepsilon$. Representing discrete feeding holes by a continuous line feed, Lund (1964) obtained a solution using pressure perturbation theory. When Equation (11.63) is normalized, it will reduce to

$$\frac{\partial}{\partial\theta}\left(\bar{h}^3\frac{\partial\bar{p}^2}{\partial\theta}\right) + \left(\frac{D}{L}\right)^2\bar{h}^3\frac{\partial\bar{p}^2}{\partial\bar{y}^2} = \Lambda\frac{\partial(\bar{p}\bar{h})}{\partial\theta} \tag{11.64}$$

where

$$\Lambda = \frac{6\eta\omega}{p_a(C/R)^2}$$

The pressure and film thickness can be written as

$$\bar{p}(\theta,\bar{y}) = \bar{p}_0(\bar{y}) + \varepsilon\,\bar{p}_1(\theta,\bar{y}) \tag{11.65}$$

and

$$\bar{h} = 1 + \varepsilon\cos\theta \tag{11.66}$$

Substituting Equations (11.65) and (11.66) into Equation (11.64) and retaining up to first linear terms,

$$\frac{\partial^2 p_0^2}{\partial y^2} = 0 \tag{11.67}$$

and

$$\frac{\partial^2(\bar{p}_0\bar{p}_1)}{\partial\theta^2} + \left(\frac{D}{L}\right)^2\frac{\partial^2(\bar{p}_0\bar{p}_1)}{\partial\bar{y}^2} = \Lambda\left[\frac{1}{\bar{p}_0}\frac{\partial(\bar{p}_0\bar{p}_1)}{\partial\theta} - \bar{p}_0\sin\theta\right] \tag{11.68}$$

Following an approach used by Heinrich (1959) for nonrotating journal, Lund (1964) solved $\bar{p}_0$ and $\bar{p}_1$ and calculated load capacity and attitude angle for various operating conditions. A more accurate solution considering discreteness of feed holes is available in (Majumdar 1972).

## 11.8 | Porous Gas Bearings

### 11.8.1 | Journal Bearings

Gas-lubricated bearings using porous surfaces are being employed in recent years in many applications. A few analyses on the subject are available in studies conducted in 1962 and 1964. Constantinescu (1962) obtained a closed form solution of pressure distribution and load capacity using short bearing theory. Sneck and Yen (1964) provided a solution taking $\varepsilon$ as a perturbation parameter. In 1975, the static characteristics were found numerically for any eccentricity ratio and porous bushing thickness considering three- dimensional flow of gas in porous medium.

An analytical solution (Majumdar 1977) considering one-dimensional (radial) gas flow through the porous bushing is obtained here. The flow through the porous medium is governed by Darcy's law. The porous bushing has both the ends sealed for economy of gas consumption.

Using conventional assumption of gas-lubricated porous bearings under static condition for the bearing shown in Fig. 11.7, the governing equations in porous medium and bearing clearance are:

$$\frac{\partial^2 \bar{p}^2}{\partial \bar{z}^2} = 0 \tag{11.69}$$

$$\frac{\partial}{\partial \theta}\left(\bar{h}^3 \frac{\partial \bar{p}^2}{\partial \theta}\right) + \left(\frac{D}{L}\right)^2 \bar{h}^3 \left(\frac{\partial^2 \bar{p}^2}{\partial \bar{y}^2}\right) = \Lambda_p \left(\frac{\partial \bar{p}'^2}{\partial \bar{z}}\right)_{\bar{z}=1} \tag{11.70}$$

where $\bar{p}'$ and $\bar{p}$ are dimensionless pressures in the porous medium and bearing clearance, respectively, and $\Lambda_p$ is feeding parameter which is given by $\Lambda_p = \dfrac{12k_z R^2}{C^3 H}$, $k_z$ is the permeability coefficient of porous material in the radial direction, H is the thickness of porous bush.

The boundary conditions for Equations (11.69) and (11.70) are:

- at $\bar{z} = 0, \bar{p}' = \bar{p}_s$
- at $\bar{z} = 1, \bar{p}' = \bar{p}$
- $\bar{p}(\theta, \bar{y}) = \bar{p}(\theta + 2\pi, \bar{y})$ (11.71)
- $\bar{p}(\theta, \pm 1) = 1$

Equations (11.69) and (11.70) are linearized by a first-order perturbation method with respect to $\varepsilon$.

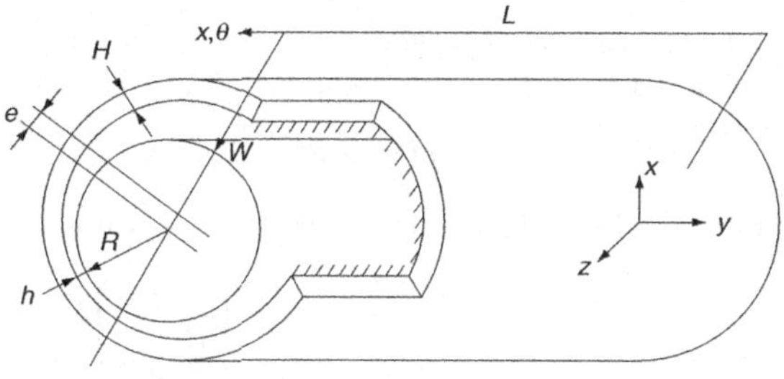


**Figure 11.7** | A Porous Gas Bearing

$$\bar{p}' = \bar{p}_0'(\bar{z}) + \varepsilon \bar{p}_1'(\bar{z})$$

$$\bar{p}' = \bar{p}_0'(\bar{y}) + \varepsilon \bar{p}_1'(\theta, \bar{y}) \tag{11.72}$$

Substitution of Equations (11.72) into Equations (11.69) and (11.70) results in the following equations:

$$\frac{\partial^2 \bar{p}_0'^2}{\partial \bar{z}^2} = 0 \tag{11.73}$$

$$\frac{\partial^2 (\bar{p}_0', \bar{p}_1')}{\partial \bar{z}^2} = 0 \tag{11.74}$$

$$\left(\frac{D}{L}\right)^2 \frac{\partial^2 \bar{p}_0'}{\partial \bar{y}^2} - \Lambda_p \left(\frac{\partial \bar{p}_0'}{\partial \bar{z}}\right)_{\bar{z}=1} = 0 \tag{11.75}$$

and

$$\frac{\partial^2 (\bar{p}_0 \bar{p}_1^2)}{\partial \theta^2} + \left(\frac{D}{L}\right)^2 \frac{\partial^2 (\bar{p}_0 \bar{p}_1^2)}{\partial \bar{y}^2} + \frac{3}{2}\left(\frac{D}{L}\right)^2 \frac{\partial^2 \bar{p}_0^2}{\partial \bar{y}^2} \cos\theta = \Lambda_p \left[\frac{\partial (\bar{p}_0 \bar{p}_1')}{\partial \bar{z}}\right]_{\bar{z}=1} \tag{11.76}$$

The boundary conditions of Equation (11.72) are accordingly modified. After evaluating $\frac{\partial \bar{p}_0^2}{\partial \bar{z}}$ and $\frac{\partial (\bar{p}_0 \bar{p}_1')}{\partial \bar{z}}$ at $\bar{z} = 1$, Equation (11.75) and (11.76) reduce to

$$\left(\frac{D}{L}\right)^2 \frac{d^2 \bar{p}_0^2}{d\bar{z}^2} - \Lambda_p (\bar{p}_0^2 - \bar{p}_s^2) = 0 \tag{11.77}$$

and

$$\frac{\partial^2 (\bar{p}_0 \bar{p}_1)}{\partial \theta^2} + \left(\frac{D}{L}\right)^2 \frac{\partial^2 (\bar{p}_0 \bar{p}_1)}{\partial \bar{y}^2} + \frac{3}{2}\left(\frac{D}{L}\right)^2 \frac{d^2 \bar{p}_0^2}{d\bar{y}^2} \cos\theta - \Lambda_p (\bar{p}_0 \bar{p}_1) = 0 \tag{11.78}$$

Equation (11.77) when solved using the conditions that $\bar{p}_0 = 1$ at $\bar{y} = 1$ and $\frac{d\bar{p}_0}{d\bar{y}} = 0$ at $\bar{y} = 0$ results

$$\bar{p}_0^2 = \frac{(1 - \bar{p}_s^2)\cosh\left(\frac{L}{D}\sqrt{\Lambda_p}\,\bar{y}\right)}{\cosh\left(\frac{L}{D}\sqrt{\Lambda_p}\right)} + \bar{p}_s^2 \tag{11.79}$$

Equation (11.78) is solved in the following way:
Assuming a solution of the type

$$\bar{p}_0 \bar{p}_1 = g(\bar{y}) \cos\theta \tag{11.80}$$

Using Equations (11.79) and (11.80), Equation (11.78) will be obtained as

$$\left(\frac{D}{L}\right)^2 \frac{d^2 g}{d\bar{y}^2} - (1 - \Lambda_p) g + \frac{1.5\Lambda_p (1 - \bar{p}_s^2)\cosh\left(\frac{L}{D}\sqrt{\Lambda_p}\right)}{\cosh\left(\frac{L}{D}\sqrt{\Lambda_p}\right)} = 0 \tag{11.81}$$

Equation (11.81) when solved yields

$$g = \frac{1}{\frac{L}{D}\sqrt{1+\Lambda_p}} \int_0^{\bar{y}} \sinh\left\{\left(\frac{L}{D}\sqrt{1+\Lambda_p}\right)(\bar{y}-u)\right\} f(u)\,du$$
$$+A\cosh\left(\frac{L}{D}\sqrt{1+\Lambda_p \bar{y}}\right) + B\sinh\left(\frac{L}{D}\sqrt{1+\Lambda_p \bar{y}}\right) \tag{11.82}$$

where

$$f(u) = \frac{1.5\left(\frac{L}{D}\right)^2 \Lambda_p \left(1-\bar{p}_s^2\right)\cosh\left(\frac{L}{D}\sqrt{\Lambda_p u}\right)}{\cosh\left(\frac{L}{D}\sqrt{\Lambda_p}\right)} \tag{11.83}$$

and A and B constants, are evaluated from the conditions

- at $\bar{y} = 1, g = 0$
- at $\bar{y} = 0, \frac{dg}{d\bar{y}} = 0.$ (11.84)

The $g$ will come out to be

$$g = \frac{1.5\,\Lambda_p\left(1-\bar{p}_s^2\right)}{\cosh\left(\frac{L}{D}\sqrt{\Lambda p}\right)}\left[\frac{\cosh\left(\frac{L}{D}\sqrt{1+\Lambda_p}\ \bar{y}\right)}{\cosh\left(\frac{L}{D}\sqrt{1+\Lambda_p}\right)}\left\{\cosh\left(\frac{L}{D}\sqrt{1+\Lambda_p}\right)\right\}\right.$$
$$\left. -\cosh\left(\frac{L}{D}\sqrt{\Lambda_p}\right) - \left\{\cosh\left(\frac{L}{D}\sqrt{1+\Lambda_p}\ \bar{y}\right) - \cosh\left(\frac{L}{D}\sqrt{\Lambda_p}\right)\right\}\right] \tag{11.85}$$

The load capacity $W$ can be found from

$$W = -2\int_0^{L/2}\int_0^{2\pi} \varepsilon p_1 R\cos\theta\, d\theta\, dy \tag{11.86}$$

Equation (11.86) can be written as

$$\bar{W}\left(=\frac{W}{LD\,p_a\varepsilon}\right) = -\frac{\pi}{2}\int_0^1 \frac{g}{\bar{p}_0}\, d\bar{y} \tag{11.87}$$

Introducing $g$ and $\bar{p}_0$ in Equation (11.87) and subsequent integration gives the dimensionless load capacity. The load capacity versus feeding parameter for various $\frac{L}{D}$ ratios is plotted in Fig. 11.8.

The mass rate of flow from bearing ends can be calculated from

$$G = -\frac{2C^3}{12\eta RT}\int_0^{2\pi} p\left.\frac{\partial p}{\partial y}\right|_{y=L/2} (1+\varepsilon\cos\theta)^3\, r\,d\theta \tag{11.88}$$

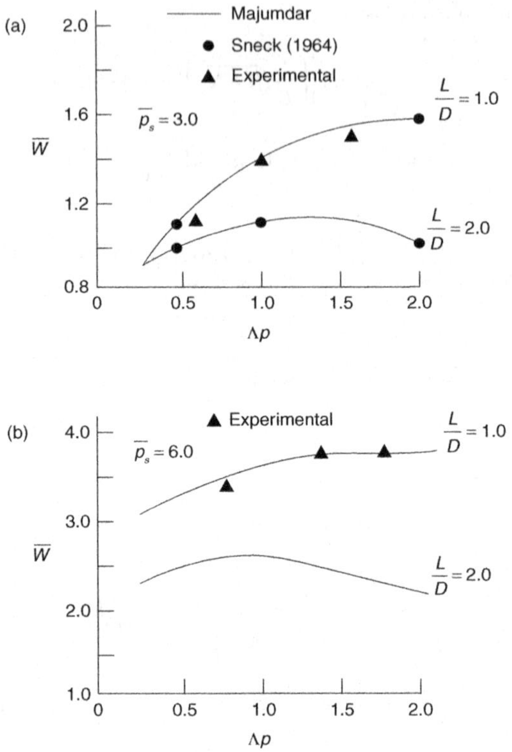


**Figure 11.8** | Load Capacity of Journal Bearing

The variation of flow is insignificant with $\varepsilon$. The flow of concentric position will be nearly equal to that of any small eccentricity ratio. Thus, $G$ can be written as

$$G = -\frac{2C^3 p_a^2}{24\eta RT\left(\frac{L}{D}\right)} \int_0^{2\pi} \frac{\partial \bar{p}_0^2}{\partial \bar{y}}\Bigg|_{\bar{y}=1} \tag{11.89}$$

After substituting the value of $\bar{P}_0$ in Equation (11.89) and performing integration

$$\bar{G}\left(= \frac{24\eta RTG}{C^3 p_a^2 \left(\bar{p}_s^2 - 1\right)}\right) = 4\pi\sqrt{\Lambda_p}\,\tanh\left(\frac{L}{D}\sqrt{\Lambda_p}\right) \tag{11.90}$$

For a rotating journal bearing the dimensionless Reynolds equation in the bearing clearance is

$$\frac{\partial}{\partial\theta}\left(\bar{h}^3\frac{\partial\bar{p}^2}{\partial\theta}\right)+\left(\frac{D}{L}\right)^2\bar{h}^3\frac{\partial^2\bar{p}^2}{\partial\bar{y}^2}=2\Lambda\frac{\partial(\bar{p}\bar{h})}{\partial\theta}+\Lambda_p\left(\frac{\partial\bar{p}'^2}{\partial\bar{z}}\right)_{\bar{z}=1} \tag{11.91}$$

The pressure distribution in the porous medium will be Equation (11.64). The boundary conditions for these equations are still given by Equation (11.71). The method follows, in general, from Majumdar (1978).

The solution can be obtained using a first-order perturbation theory with respect to $\varepsilon$. The pressures $\bar{P}'$ and $\bar{P}$ are written in the same form as given by Equation (11.72). With the introduction of Equation (11.72) in Equations (11.69) and (11.91), one can obtain Equation (11.73), (11.74), and (11.75) and

$$\frac{\partial^2(\bar{p}_0\bar{p}_1)}{\partial\theta^2}+\left(\frac{D}{L}\right)^2\frac{\partial^2(\bar{p}_0\bar{p}_1)}{\partial\bar{y}^2}+1.5\left(\frac{D}{L}\right)^2\frac{\partial^2(\bar{p}_0{}^2)}{\partial\bar{y}^2}\cos\theta$$
$$=\Lambda\left[\frac{1}{\bar{p}_0}\frac{\partial(\bar{p}_0\bar{p}_1)}{\partial\theta}-\bar{p}_0\sin\theta\right]+\Lambda_p\left[\frac{\partial(\bar{p}_0{}'\bar{p}_1{}')}{\partial\bar{z}}\right]_{\bar{z}=1} \tag{11.92}$$

The steady-state pressure in the bearing clearance will be same as that of Equation (11.79). Now for the solution of Equation (11.92), assume

$$\bar{p}_0\bar{p}_1=\mathrm{Re}\{g(\bar{y})e^{-i\theta}\} \tag{11.93}$$

Substituting Equation (11.93) and replacing $\frac{\partial^2\bar{p}_0^2}{\partial\bar{y}^2}$ by $\left(\frac{L}{D}\right)^2\Lambda_p(\bar{p}_0^2-\bar{p}_s^2)$, Equation (11.92) boils down to

$$\left(\frac{D}{L}\right)^2\frac{d^2g}{d\bar{y}^2}-\left[1+\Lambda_p-\frac{i\Lambda}{\bar{p}_0}\right]g+1.5\Lambda_p(\bar{p}_0^2-\bar{p}_s^2)+i\Lambda\bar{p}_0 \tag{11.94}$$

Equation (11.94) is solved numerically by finite difference scheme with the boundary conditions $g=0$ at $\bar{y}=1$ and $\frac{dg}{d\bar{y}}=0$ at $\bar{y}=0$.

The two load components are calculated from:

$$W_r=(W\cos\phi)=-2\int_0^{L/2}\int_0^{2\pi}\varepsilon p_1 R\cos\theta d\theta dy$$

and

$$W_\theta=(W\sin\phi)=2\int_0^{L/2}\int_0^{2\pi}\varepsilon p_1 R\sin\theta d\theta dy \tag{11.95}$$

Equation (11.95) can be written as

$$\overline{W}_r\left(=\frac{W_r}{LDp_a\in}\right)=-\frac{\pi}{2}\int_0^1\frac{1}{\overline{p}_0}\mathrm{Re}(g)d\overline{y} \tag{11.96}$$

$$\overline{W}_\theta\left(=\frac{W_\theta}{LDp_a\in}\right)=\frac{\pi}{2}\int_0^1\frac{1}{\overline{p}_0}\mathrm{Im}(g)d\overline{y}$$

On performing intergrations of Equation (11.96), the load capacity and attitude angle can be calculated from:

$$\overline{W}=\sqrt{\overline{W}_r^2+\overline{W}_\theta^2} \tag{11.97}$$

and

$$\phi=\tan^{-1}\left(\overline{W}_\theta/\overline{W}_r\right) \tag{11.98}$$

The load capacity and attitude angle are shown in Fig. 11.9 for various $\frac{L}{D}$ ratios and journal speeds.

In the preceding section, the static characteristics of externally pressurized gas journal bearings have been given assuming flow of gas occurring only in the radial direction of the porous matrix. If such an assumption is not made and gas is considered to flow in circumferential, axial, and radial directions in the porous medium, the performance characteristics are given in Majumdar (1976).

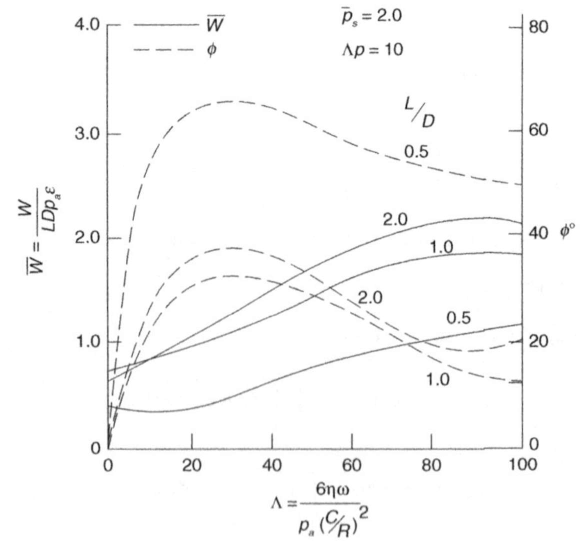


**Figure 11.9** | Load Capacity and Attitude Angle for Various Journal Speeds and $\frac{L}{D}$ Ratios

## 11.9 | Circular Porous Thrust Bearing

For a porous bearing having circular shape as shown in Fig. 11.10, the governing equation at the porous matrix is

$$\frac{1}{r}\frac{\partial}{\partial r}\left(r\frac{\partial p'^2}{\partial r}\right)+\frac{\partial p'^2}{\partial z^2}=0 \tag{11.99}$$

The Reynolds equation in the clearance gap is:

$$\frac{h^3}{12\eta r}\frac{d}{dr}\left(r\frac{dp^2}{dr}\right)=\left.\frac{k_z}{\eta}\frac{dp'^2}{dz}\right|_{z=H} \tag{11.100}$$

The associated boundary conditions of bearing are:

- $p' = p_s$ at $z = 0$ and $0 \le r \le r_0$
- $p' = p$ at $z = H$ and $0 \le r \le r_0$
- $p = p_a$ at $r = r_0$
- $\frac{\partial p}{\partial r} = 0$ at $r = 0$ (11.101)

Integrating Equation (11.95) and using the first two boundary conditions of Equation (11.101),

$$\left.\frac{\partial p'^2}{\partial z}\right|_{z=H}=\frac{\left(p^2-p_s^2\right)}{H}-\frac{H}{2}\frac{1}{r}\frac{d}{dr}\left(r\frac{dp^2}{dr}\right) \tag{11.102}$$

Using Equation (11.102), Equation (11.100) can be written as

$$\frac{1}{r}\frac{d}{dr}\left(\bar{r}\frac{d\bar{p}^2}{\partial\bar{r}}\right)=\Lambda_p\left(\bar{p}^2-\bar{p}_s^2\right) \tag{11.103}$$

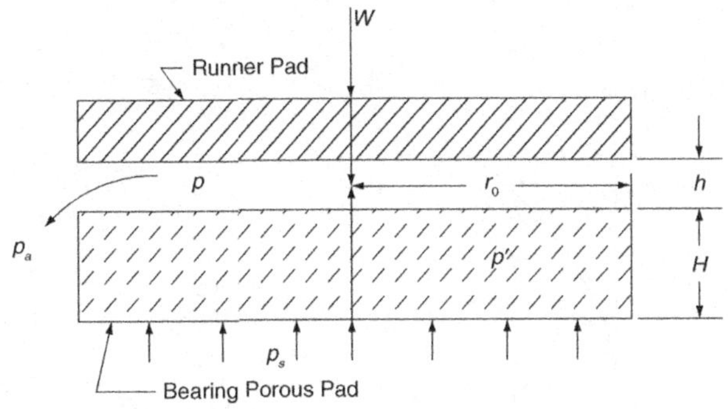


**Figure 11.10** | A Circular Porous Thrust Bearing

where $\bar{r} = r/r_0, \Lambda_p = \dfrac{12k_z r_o^2}{Hh^3}, \bar{p} = p/p_a, \bar{p}_s = p_s/p_a$

The solution of Equation (11.103) and using the last two boundary condition of Equation (11.101) is given by

$$\bar{p}^2 = \bar{p}_s^2 + (1-\bar{p}_s^2)\frac{I_0\left(\sqrt{\Lambda_p}\bar{r}\right)}{I_0\sqrt{\Lambda_p}} \tag{11.104}$$

where $I_0$ is the modified Bessel function of first kind. The load capacity and mass rate of flow are

$$\bar{W} = \frac{W}{\pi r_0^2(p_s - p_a)} = \frac{2}{(\bar{p}_s - 1)}\int_0^1(\bar{p}-1)\bar{r}d\bar{r} \tag{11.105}$$

and

$$\bar{G} = \frac{12\eta RTG}{\pi h^3(p_s^2 - p_a^2)} = \frac{\sqrt{\Lambda_p}I_1\sqrt{\Lambda_p}}{I_0\sqrt{\Lambda_p}} \tag{11.106}$$

The stiffness $S$ can be calculated from

$$\begin{aligned} S &= -\frac{dW}{dh} = -\frac{dW}{d\Lambda_p}\frac{d\Lambda_p}{dh} \\ &= 3\frac{dW}{d\Lambda_p}\frac{\Lambda_p}{h} \end{aligned} \tag{11.107}$$

or

$$\bar{S} = \frac{Sh}{\pi r_0^2(p_s - p_a)} = 3\Lambda_p\frac{d\bar{W}}{d\Lambda_p} \tag{11.108}$$

Thus, the stiffness can be calculated from the slope of $\bar{W}$ vs. $\Lambda_p$ curve.

The load capacity, flow rate, and stiffness for different operation conditions are given in Figs 11.11 and 11.12.

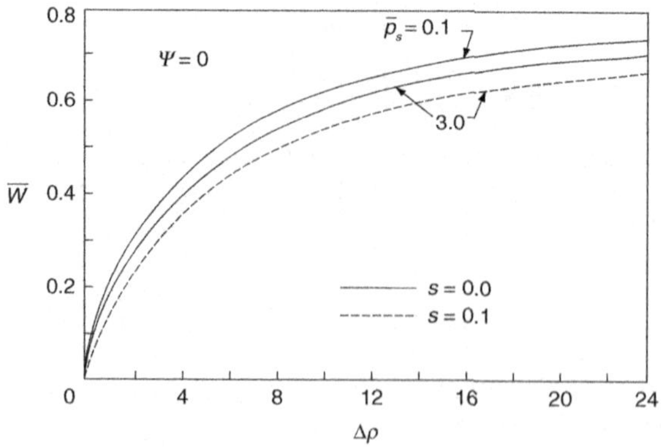


**Figure 11.11** | Variation of Load Capacity With Feeding Parameter and Supply Pressure

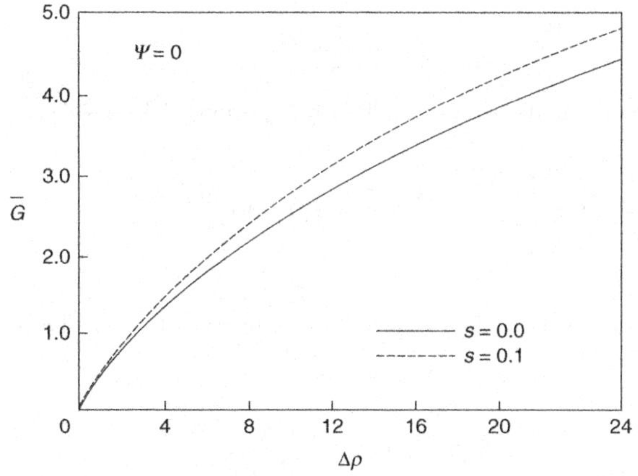


**Figure 11.12** | Variation of Flow Rate With Feed Parameter and Supply Pressure

### 11.9.1 | Effect of Slip Flow on Porous Bearings

In all previous solutions concerning porous gas bearings, the no-slip condition was assumed. In the following analysis, modified Beavers–Joseph (1972) slip velocity condition is used to find the static characteristics of a circular thrust bearing. The effect of slip on the performance characteristics will be shown (Rao 1979).

Here no-slip condition is assumed at the impermeable runner surface and the modified Beavers–Joseph boundary condition is applied at the porous surface of the bearing pad. This condition is

$$\frac{\partial u}{\partial y} = \frac{\alpha}{\sqrt{k_y}} \text{ at } z = H \tag{11.109}$$

where $\alpha$ is a dimensionless constant which depends on the characteristics of porous material and does not depend on the lubricant properties and film thickness.

The flow through the porous medium is governed by the steady-state mass continuity equation

$$\frac{1}{r}\frac{\partial}{\partial r}(\rho r u') + \frac{1}{r}\frac{\partial}{\partial \theta}(\rho v') + \frac{\partial}{\partial z}(\rho w') = 0 \tag{11.110}$$

Using Darcy's law equation of state and taking $\frac{\partial p'}{\partial \theta} = 0$, Equation (11.110) reduces to

$$\frac{1}{r}\frac{\partial}{\partial r}\left(r\frac{\partial p'^2}{\partial r}\right) + \frac{\partial p'^2}{\partial y^2} = 0 \tag{11.111}$$

The modified Reynolds equation considering slip velocity of Equation (11.109) and Darcy's law is

$$\frac{h^3}{12\eta}\frac{(4s+1)}{s+1}\frac{1}{r}\frac{d}{dr}\left(r\frac{dp^2}{dr}\right) = \frac{k_y}{\eta}\frac{\partial p^{2'}}{\partial z}\bigg|_{z=H} \tag{11.112}$$

where $s$ a slip parameters, is $\frac{\sqrt{k_y}}{\alpha h}$.

Assuming that pressure gradient is varying linearly, Equation (11.112) will be

$$\frac{1}{\bar{r}}\frac{d}{d\bar{r}}\left(\bar{r}\frac{d\bar{p}^2}{d\bar{r}}\right)=\frac{\Lambda_p}{\left[(4s+1)/(s+1)+\psi\right]}\left(\bar{p}^2-\bar{p}_s^2\right) \tag{11.113}$$

where $\psi = 6k_z H/h^3$

When $\psi = 0$ the flow will be purely axial and $s = 0$ corresponds to no-slip boundary. The solution of Equation (11.113) is

$$\bar{p}^2 = \bar{p}_s^2 + \left(1-\bar{p}_s^2\right)\frac{I_0\left(\sqrt{\Lambda_p^*}\,\bar{r}\right)}{I_0\sqrt{\Lambda_p^*}} \tag{11.114}$$

where

$$\Lambda_p^* = \frac{\Lambda_p}{(4s+1)(s+1)+\psi} \tag{11.115}$$

The load, mass rate of flow, and stiffness can be calculated with the usual method, and $\bar{W}$, $G$, and $\bar{S}$ will be given by Equation (12.105) to (12.108) where $\Lambda_p$ is to be replaced by $\Lambda_p^*$. The effect of slip on the load and flow is shown in Figs 11.11 and 11.12, respectively.

In the recent year, Saha (2004) has obtained a solution of two-layered porous gas bearings (Fig. 11.13). Table 11.2 provides the steady-state characteristics of these bearings. These bearings also provide better stability (Fig. 11.14).

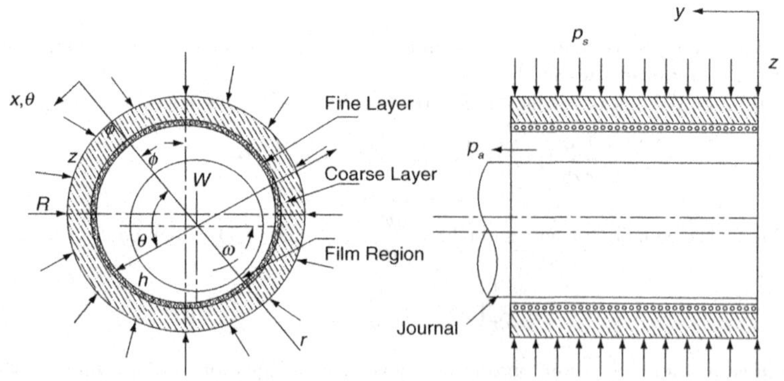


**Figure 11.13** | An Externally Pressurized Two-layered Porous Gas Journal Bearing

**Table 11.2** | Comparison of Steady State Performance of Single-Layered With Those of Two-Layered Porous Journal Bearing

| $L/D = 1.0$, $H/R = 0.1$, $\bar{P}_s = 2.0$, $\beta = 1.0$ (wrt 2-layered) $\bar{K}_{x1} = \bar{K}_{x2} = \bar{K}_{z1} = \bar{K}_{z2} = 1.0$ | | | | | |
|---|---|---|---|---|---|
| $\varepsilon_0$ | $\Lambda$ | $\bar{W}_0$ | $\phi_0$ | $\mu_f(R/C)$ | $\bar{G}$ |
| 0.2 | 0.0 | 0.1144 (0.0803) | 0.0000 (0.0000) | 0.0000 (0.0000) | 50.7117 (72.5467) |
| | 1.0 | 0.1162 (0.0812) | 8.0518 (7.0878) | 4.6062 (6.5908) | 50.7058 (72.5439) |
| | 2.0 | 0.1214 (0.0836) | 15.5513 (13.8454) | 8.8175 (12.7956) | 50.6936 (72.5370) |
| | 3.0 | 0.1249 (0.0875) | 20.4272 (20.0088) | 12.8566 (18.3386) | 50.6028 (72.5246) |
| | 4.0 | 0.1400 (0.0927) | 27.1733 (25.4347) | 15.2937 (23.0852) | 50.6259 (72.5081) |
| | 5.0 | 0.1520 (0.0989) | 31.2167 (30.0772) | 17.5999 (27.0469) | 50.5762 (72.4865) |
| 0.4 | 0.0 | 0.2241 (0.1582) | 0.0000 (0.0000) | 0.0000 (0.0000) | 51.6123 (74.4645) |
| | 1.0 | 0.2277 (0.598) | 8.2823 (7.1539) | 2.5233 (3.5864) | 51.5904 (74.4550) |
| | 2.0 | 0.2336 (0.1647) | 14.4051 (13.9753) | 4.9156 (6.9628) | 51.4727 (74.4268) |
| | 3.0 | 0.2542 (0.1724) | 22.3972 (20.2020) | 6.7816 (9.9782) | 51.4087 (74.3796) |
| | 4.0 | 0.2751 (0.1825) | 27.8400 (25.6904) | 8.3544 (12.5618) | 51.2652 (74.3140) |
| | 5.0 | 0.2991 (0.1948) | 32.0113 (30.3962) | 9.6036 (14.7154) | 51.0763 (74.2302) |
| 0.6 | 0.0 | 0.3246 (0.2306) | 0.0000 (0.0000) | 0.0000 (0.0000) | 53.0321 (77.7918) |
| | 1.0 | 0.3299 (0.2331) | 8.5983 (7.2846) | 2.0062 (2.8274) | 52.9864 (77.7719) |
| | 2.0 | 0.3454 (0.2402) | 16.5895 (14.2330) | 3.8326 (5.4876) | 52.8578 (77.7116) |
| | 3.0 | 0.3694 (0.2515) | 23.3761 (20.5785) | 5.3747 (7.8612) | 52.6252 (77.6111) |
| | 4.0 | 0.4001 (0.2664) | 28.9481 (26.1794) | 6.6153 (9.8917) | 52.3136 (77.4714) |
| | 5.0 | 0.4359 (0.2844) | 33.3600 (30.9900) | 7.5894 (11.5821) | 51.9231 (77.2928) |
| 0.8 | 0.0 | 0.4110 (0.2945) | 0.0000 (0.0000) | 0.0000 (0.0000) | 55.0223 (82.8624) |
| | 1.0 | 0.4173 (0.2977) | 9.1633 (7.4969) | 2.1230 (2.9578) | 55.0537 (82.8291) |
| | 2.0 | 0.4295 (0.3069) | 15.5978 (14.6444) | 4.1175 (5.7374) | 54.8085 (82.7294) |
| | 3.0 | 0.4701 (0.3217) | 24.8833 (21.1770) | 5.6526 (8.2102) | 54.4652 (82.5637) |
| | 4.0 | 0.5101 (0.3412) | 30.6473 (26.9388) | 6.9446 (10.3208) | 53.9465 (82.3327) |
| | 5.0 | 0.5621 (0.3647) | 35.2747 (31.8951) | 7.8775 (12.0712) | 53.0489 (82.0374) |

**Note:** Single-layered results are given within brackets.

For a single-layered porous bearings working at $5\mu m$ to $15\mu m$ clearance, the permeability coefficient ranges from $3.1\times10^{-15}$ to $8.4\times10^{-14}$m$^2$. For two-layered bearing, the fine layer has a typical permeability coefficient ranging from $2.6\times10^{-14}$ to $7.0\times10^{-15}$ m$^2$. The coarse layer should have permeability coefficient at least 200 times that of the fine layer, so that most of the pressure drop occurs across the thin restricting layer. If $\bar{k}_{z2}$ is taken as 200, $\rho$ for the two-layered decreases by 20 times as compared to conventional (single-layered) porous bearing for $\frac{H_1}{H} = 1$ . Some typical results are shown in Fig. 11.14 in which one can see a better load carrying capacity for the two-layered bearing. The friction variable has been reduced considerably for two-layered bearings. Thus, by introducing a fine layer, the load bearing capacity improves and friction variable reduces considerably.

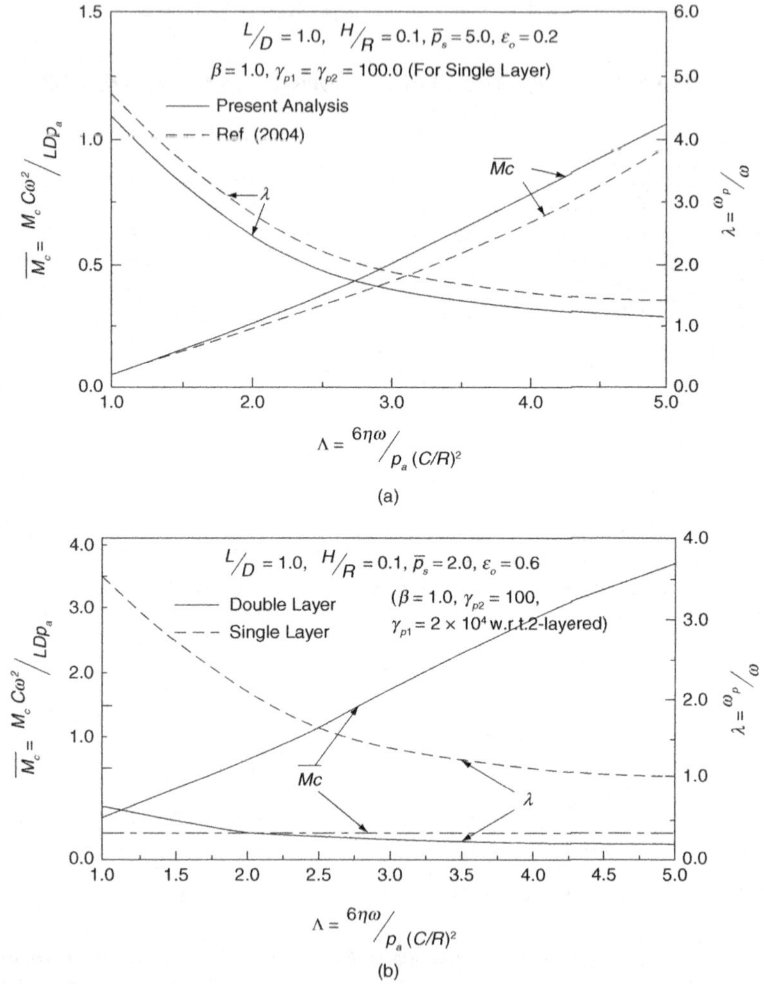


**Figure 11.14** | Comparison of Critical Mass Parameters and Whirl Ratio of a Single Layer With Those of a Double Layer

The two-layered structure would allow the designer the freedom to optimize the geometry of the coarse layer with low permeability to suit the flow restriction performance, provided the pressure drops across the coarse substrate is insignificant compared with that across the restricting layer. By choosing a sufficiently large particle size ratio between the coarse substrate and the thin surface layer, it is possible to control the fluid flow through the two-layered structure such that over 90% – 95% of the pressure occurs in the fine layer, even when its thickness is only 5%–10% of that of the coarse layer.

## 11.10 | Dynamic Characteristics of Gas-lubricated Bearings

As externally pressurized gas bearings give zero friction at zero speed and its use in low-speed instruments has enabled stick-slip to be completely eliminated. At high speeds, these bearings offer friction torque but its relative magnitude is extremely small compared to bearings using oil as lubricant.

Due to the above advantages, gas bearings are being used in high speed applications. But these bearings are prone to instability. Over a wide range of speeds there are a number of critical speeds in which large amplitude of whirl develops. Whirl may be synchronous or self-excited. Synchronous whirl is excited by unbalance of rotor and self-excited (so-called half-speed) whirl is self-excited at small eccentricities. In aerodynamic bearings, occurrence of self-excited whirl limits the operating speeds. Thus, an understanding of whirl instabilities is of great importance in the design of a bearing for a particular application. However, accurate balancing of a rotor can avoid synchronous whirl.

In externally pressurized gas bearings, another type of instability known as 'pneumatic' or 'air hammer' may occur even when there is no relative motion at zero speed. In thrust bearings, this instability produces axial vibration of the rotor; while in journal bearings, it causes radial vibration. Usually the pneumatic instability is a kind of self-excited instability at very low speed or zero speed. When such an instability occurs at finite speeds, it is termed as 'hybrid instability'.

In the following sections, a theoretical estimate of identifying the stable operation limit of externally pressurized thrust and journal bearings at zero speed is given. The method of analysis involves a lot of mathematical complexity, and in most cases one has to go for numerical solution.

### 11.10.1 | A Circular Step Thrust Bearing

The steady-state characteristics of a circular step thrust bearing have been found earlier. It is assumed that the pressure drops linearly from radius $r_i$ to radius $r_o$ (Fig. 11.5); and for small deviation from equilibrium point, this type of pressure distribution is likely to be preserved. There is no external damping and the motion of the runner is assumed to be vertical.

Let $m$ be the mass of the runner, the equation of motion using linear pressure drop can be written as

$$m\ddot{h} = 2\pi\left[\int_0^{r_i} p_r r\,dr + \int_{r_i}^{r_0} pr\,dr\right]$$

or

$$m\ddot{h} = A_e p_r \tag{11.116}$$

where

$$A_e = \frac{\pi}{3}\left(\frac{r_0^3 - r_i^3}{r_0 - r_i}\right)$$

or

$$p_r = \frac{m}{A_e}\ddot{h} \tag{11.117}$$

From mass continuity consideration

$$G_{in} - G_{out} = \frac{\partial G}{\partial t} \tag{11.118}$$

When an orifice is used as a restrictor, $G_{in}$ is given by Equation (11.51) for unchoked and choked flow conditions. The outflow $G_{out}$ can be calculated from Equation (11.52). Here $G_{in}$ is a function of $p_r$ only. $G_{out}$ depends on both $p_r$ and $h$. These can be expressed in the following form:

$$G_{in} = -\alpha p_r \tag{11.119}$$

$$G_{out} = \beta p_r + \theta h \tag{11.120}$$

where $\alpha$, $\beta$ and $\theta$ are all positive

and $$\alpha = \left(\frac{dG_{in}}{dp_r}\right)_0, \beta = \left(\frac{dG_{out}}{dp_r}\right)_0, \theta = \left(\frac{dG_{out}}{dh}\right)_0$$

The values of $\alpha$, $\beta$ and $\theta$ can be easily calculated from Equations (11.51) and (11.52). The suffix 0 denotes equilibrium condition.

Now $$\frac{\partial G}{\partial t} = \frac{\partial}{\partial t}\left[(\Delta + h)\pi r_i^2 \rho\right] + \frac{\partial}{\partial t}\left[2\pi \int_{r_i}^{r_0} \rho p_r dr\right] \tag{11.121}$$

or $$\frac{\partial G}{\partial t} = \left(\frac{\partial G}{\partial p_r}\right)_0 \frac{\partial p_r}{\partial t} + \left(\frac{\partial G}{\partial h}\right)_0 \frac{\partial h}{\partial t}$$

where $$\left(\frac{\partial G}{\partial p_r}\right)_0 = \frac{A_e h_0 + \Delta \pi r_i^2}{RT} \ (=q \text{ say})$$

$$\left(\frac{\partial G}{\partial h}\right)_0 = \frac{A_e (p_{ro} - p_a) + \pi r_i^2 p_a}{RT} \ (=s \text{ say})$$

Hence $$\frac{\partial G}{\partial t} = q\frac{\partial p_r}{\partial t} + s\frac{\partial h}{\partial t} \tag{11.122}$$

Substituting Equations (11.119), (11.120), and (11.122) in Equation (11.118), one obtains

$$q\dot{p}_r + s\dot{h} + (\alpha + \beta) p_r + \theta h = 0 \tag{11.123}$$

From Equation (11.117), $$\dot{p}_r = \frac{m}{A_e}\dddot{h} \tag{11.124}$$

Equation (11.123) can now be written as

$$\frac{m}{A_e} q\dddot{h} + (\alpha + \beta)\frac{m}{A_e}\ddot{h} + s\dot{h} + \theta h = 0 \tag{11.125}$$

or $$\dddot{h} + C_2\ddot{h} + C_1\dot{h} + C_0 h = 0 \tag{11.126}$$

where $$C_0 = \frac{\theta A_e}{mq}, \quad C_1 = \frac{sA_e}{mq} \tag{11.127}$$

and $$C_2 = \frac{(\alpha+\beta)}{q}$$

The coefficients $C_{0,}$ $C_{1,}$ and $C_2$ are all positive.

Applying Routh's stability criteria for stability

$$C_1 C_2 > C_0 \tag{11.128}$$

or $$\frac{\alpha+\beta}{\theta} > \frac{q}{s} \tag{11.129}$$

From Equation (11.129) it is seen that the values of $p_{r0}$ and $h_0$ have an opposite effect on the magnitude of the ratios from the two sides of inequality. If $p_{r0}$ and $h_0$ are held constant, it is clear that for better stability $\Delta$ should be as small as possible. From the physical point of view, this can also be explained. If the load is increased, the bearing surfaces move closer, the flow through the feed hole is reduced, and the recess pressure is increased. In a recess with a relatively small volume, the recess pressure will be in a position to adjust quickly to changes of clearance. The change in pocket pressure can lag to such an extent that following a disturbance from equilibrium position, a vibration of higher amplitude can be generated. Thus, it causes pneumatic instability.

### 11.10.2 | Orifice-compensated Externally Pressurized Journal Bearings

The study of pneumatic instability made by Lund (1967) for orifice-compensated externally pressurized journal bearings assumed a 'line source' supply and a suitable correction factor between source feeding and line feeding was used. The present analysis does not make such an assumption (Nidhi *et al.* 1979). The configuration to be analyzed is that of a journal bearing that has two circumferential rows of orifices at quarter station. It is assumed that journal executes a small harmonic plane oscillation with frequency $\omega_p$ in the absence of shaft rotation.

The Reynolds equation for isothermal case can be written as

$$\frac{\partial}{\partial\theta}\left(\bar{h}^3\frac{\partial\bar{p}^2}{\partial\theta}\right)+\left(\frac{D}{L}\right)^2\bar{h}^3\frac{\partial^2\bar{p}^2}{\partial\bar{y}^2} = 2\sigma\frac{\partial(\bar{p}\bar{h})}{\partial\tau} \tag{11.130}$$

where $$\sigma \text{ (squeeze number)} = \frac{12\eta\omega_p}{p_a(C/R)^2}$$

As the journal executes harmonic oscillations with $\mathrm{Re}\left[\varepsilon e^{i\tau}\right]$ and since for small vibrations first order perturbation is valid, pressure and local film thickness can be expressed as

$$\bar{p} = \bar{p}_0 + \varepsilon e^{i\tau}\bar{p}_1$$

and $$\bar{h} = 1 + \varepsilon e^{i\tau}\cos\theta \tag{11.131}$$

where $$\bar{p} = \bar{p}(\theta,\bar{y},\tau),\ \bar{p}_0 = \bar{p}_0(\theta,\bar{y}) \text{ and } \bar{p}_1 = \bar{p}_1(\theta,\bar{y})$$

Substituting Equation (11.131) in Equation (11.130) and neglecting higher order terms

$$\frac{\partial^2\bar{p}^2}{\partial\theta^2}+\left(\frac{D}{L}\right)^2\frac{\partial^2\bar{p}^2}{\partial\bar{y}^2} = 0 \tag{11.132}$$

and
$$\frac{\partial^2(\bar{p}_0\bar{p}_1)}{\partial\theta^2}+\left(\frac{D}{L}\right)^2\frac{\partial^2(\bar{p}_0\bar{p}_1)}{\partial\bar{y}^2}-\frac{3}{2}\frac{\partial\bar{p}_0^2}{\partial\theta}\sin\theta=i\sigma\left[\frac{\bar{p}_0\bar{p}_1}{\bar{p}}+\bar{p}_0\cos\theta\right] \quad (12.133)$$

Solution of Equation (11.132) satisfying the appropriate boundary conditions yields static pressure $\bar{P}_0$ in the clearance space from which the lubricant flow can be calculated. The solution of Equation (11.133) gives dynamic pressure $\bar{P}_1$ from which the stiffness and damping coefficients can be found. These coefficients will then be used for identifying pneumatic instability.

The boundary conditions of Equations (11.132) and (11.133) are:

- $\bar{p}_0(\theta,\pm 1)=1$
- $\frac{\partial\bar{p}_0}{\partial\bar{y}}(\theta,0)=0$
- $\bar{p}_{q0}=p_{q0}/p_a$ at $q$th station
- $\bar{p}_0(\theta,\bar{y})=\bar{p}_0(\theta+2\pi,\bar{y})$ (11.134)

and

- $\bar{p}_0(\theta,\pm 1)=1$
- $\frac{\partial\bar{p}_1}{\partial\bar{y}}(\theta,0)=0$
- $\bar{p}_{q1}=p_{q1}/p_a$ at $q$th station
- $\bar{p}_1(\theta,\bar{y})=\bar{p}_1(\theta+2\pi,\bar{y})$ (11.135)

The pressures $P_{q0}$ and $P_{q1}$ of Equations (11.134) and (11.135) are unknown. These are first estimated and then solution of Equations (11.132) and (11.133) will be obtained.

Equation (11.132) is solved numerically using finite difference methods using $\bar{P}_0=1$ at $(\theta,\pm 1)$ and with an arbitrary value of $\bar{P}_{q0}$ at each orifice station, when the mass rate of flow from the bearing ends is calculated from

$$G_{q0}=-\frac{2C^3}{24\eta RT}\int_0^{2\pi}\left.\frac{\partial p_0^{\,2}}{\partial y}\right|_{y=\frac{L}{2}} r\,d\theta \quad (11.136)$$

from which the dimensionless flow is

$$\bar{G}_{q0}=\frac{24\eta RTG_{q0}}{C^3\left(p_{q0}^{\,2}-p_a^{\,2}\right)} \quad (11.137)$$

From continuity consideration

$$G=G_q+\frac{\partial}{\partial t}\left[2\rho Na(\Delta+h)\right] \quad (11.138)$$

at $q$th station, where $G$ is the mass rate of flow through orifices and $G_q$ at $q$th station a = area of recess and $\Delta$ is the depth of recess.

Using the mass rate of flow through $2N$ number of orifices (such equations are given earlier) and perturbing $p_q$, $G_q$, and $h_q$ with following relations:

$$\bar{p}_q = \bar{p}_{q0} + \varepsilon e^{i\tau}\bar{p}_{q1}$$

$$\bar{G}_q = \bar{G}_{q0} + \varepsilon e^{i\tau}\bar{G}_{q1} \tag{11.139}$$

and

$$\bar{h}_q = 1 + \varepsilon e^{i\tau}\cos\theta_q$$

where

$$\bar{G}_{q1} = \frac{\partial \bar{G}_q}{\partial \varepsilon}\bigg|_{\varepsilon=0} = 3\bar{G}_{q0}\cos\theta_q$$

The continuity equations will come out as

for

$$\left(\frac{p_{q0}}{p_s}\right) \geq \left(\frac{2}{k+1}\right)^{\frac{k}{k-1}}$$

$$2N\delta\left(\frac{k}{k-1}\right)^{1/2}\left[\left(\frac{p_{q0}}{p_s}\right)^{2/k} - \left(\frac{p_{q0}}{p_s}\right)^{\frac{k+1}{k}}\right]^{1/2} - \bar{G}_{q0}\left[\left(\frac{p_{q0}}{p_s}\right)^2 - \left(\frac{p_a}{p_s}\right)^2\right] = 0 \tag{11.140}$$

$$N\delta\left(\frac{2}{k-1}\right)^{1/2}\left[\left(\frac{p_{q0}}{p_s}\right)^{2/k} - \left(\frac{p_{q0}}{p_s}\right)^{k+1/k}\right]^{-\frac{1}{2}}\left[\frac{2}{k}\left(\frac{p_{q0}}{p_s}\right)^{\frac{2-k}{k}} - \left(\frac{k+1}{k}\right)\left(\frac{p_{q0}}{p_s}\right)^{\frac{1}{k}}\right]\left(\frac{p_{q1}}{p_s}\right)$$

$$-2\bar{G}_{q0}\left(p_{q1}/p_s\right)\left(p_{q0}/p_s\right) - \bar{G}_{q1}\left[\left(\frac{p_{q0}}{p_s}\right)^2 - \left(\frac{p_a}{p_s}\right)^2\right]$$

$$-i2N\psi\left[\left(\frac{p_{q1}}{p_s}\right) + \frac{C}{\Delta}\left\{\left(\frac{p_{q1}}{p_s}\right) + \left(\frac{p_{q0}}{p_s}\right)\cos\theta_a\right\}\right] = 0 \tag{11.141}$$

where

$$\psi = \frac{2\eta a\omega_p\Delta}{C^3 p_s}$$

Similar equations for $\left(p_{q0}/p_s\right) < \left(\dfrac{2}{k+1}\right)^{\frac{k}{k-1}}$ can be written.

The above equations are solved by Newton–Raphson iteration method for various values of $\dfrac{L}{D}$, $\delta$, $\psi$, $\dfrac{C}{\Delta}$ and $p_a/p_s$. Equations (11.132) and (11.133) are now solved with real-boundary values.

The dynamic load capacity $W_1$ can be calculated from

$$W_1 = 2\int_0^{L/2}\int_0^{2\pi} p_1 \cos\theta r\, d\theta\, dy \tag{11.142}$$

Since $p_1$ is complex, the dynamic load $W_1$ can be expressed in terms of real and imaginary part as

$$W_1 = \mathrm{Re}(W_1) + i\,\mathrm{Im}(W_1) \tag{11.143}$$

Again as the journal executes small harmonic motion, the dynamic load can also be written as

$$W_1 \varepsilon e^{i\tau} = -SZ - B\frac{dZ}{dt} \tag{11.144}$$

where, $S$ and $B$ are stiffness and damping coefficients of gas film.

The amplitude of journal is given by

$$Z = C\varepsilon e^{i\tau} \tag{11.145}$$

Substituting Equation (12.146) in Equation (12.144)

$$\overline{W}_1 = \frac{W_1}{LDp_a} = -\frac{SC}{LDp_a} - i\frac{BC\omega_p}{LDp_a} \tag{11.146}$$

Dimensionless stiffness and damping can be defined as

$$\overline{S} = -\mathrm{Re}(\overline{W}_1)$$

and

$$\overline{B} = -\frac{12\,\mathrm{Im}(W_1)}{\sigma} = \frac{B}{2\eta L\left(R/C\right)^3} \tag{11.147}$$

Equation of motion for rotor mass can be written as

$$m\frac{d^2Z}{dt^2} = W_1 \varepsilon e^{i\tau} \tag{11.148}$$

or

$$-\frac{mC\omega_p^{\,2}}{LDp_a} = \overline{W}_1$$

or

$$\overline{m} = -\mathrm{Re}(\overline{W}_1) - i\,\mathrm{Im}(\overline{W}_1) \tag{11.149}$$

where

$$\overline{m} = \frac{mC\omega_p^{\,2}}{LDp_a}$$

At the threshold of instability,

$$\overline{m} = -\mathrm{Re}(\overline{W}_1) \tag{11.150}$$

and

$$\mathrm{Im}(\overline{W}_1) = 0 \tag{11.151}$$

Equation (11.150) can be written as

$$\overline{m} = \overline{S}$$

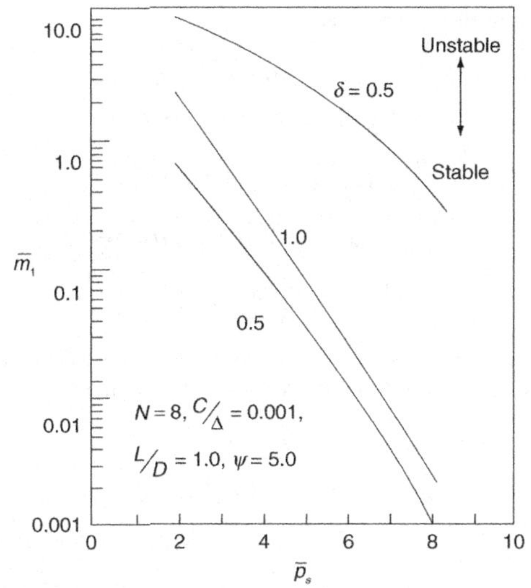


**Figure 11.15** | Stability Characteristics

As both $\bar{m}$ and $\bar{S}$ are dependent on $\omega_p$, another mass parameter $\bar{m}_1$ which is independent of $\omega_p$ can be defined as

$$\bar{m}_1 = \frac{mp_a(C/R)^5}{L\eta^2} = 288\frac{\bar{S}}{\sigma^2} \tag{11.152}$$

The variation of mass parameter $\bar{m}_1$ for various $\bar{P}_s$ and $\delta$ is shown in Fig. 11.15. For stability, one should operate the bearing with small supply pressures, small feeding parameter, and small design parameters.

### 11.10.3 | Porous Gas Journal Bearings

An externally pressurized porous gas journal bearing is shown in Fig. 11.7. The journal is stationary and its axis is parallel to the bearing axis. When no-slip condition is assumed, the equation of continuity of flow through porous medium and the modified Reynolds equation in the bearing clearance are

$$\frac{\partial^2 \bar{p}'^2}{\partial \bar{z}^2} - 2\sigma\gamma\frac{\partial \bar{p}'}{\partial \tau} = 0 \tag{11.153}$$

$$\frac{\partial}{\partial\theta}\left(\bar{h}^3\frac{\partial \bar{p}^2}{\partial\theta}\right) + \left(\frac{D}{L}\right)^2 \bar{h}^3\frac{\partial^2 \bar{p}^2}{\partial \bar{y}^2} = 2\sigma\frac{\partial(\bar{p}\bar{h})}{\partial\tau} + \Lambda p\left.\frac{\partial \bar{p}'^2}{\partial \bar{z}}\right|_{z=1} \tag{11.154}$$

where $\gamma = \dfrac{\mu C^2 H^2}{12k_z R^2}$, a porosity parameter

and $\mu$ = porosity of bushing material

The boundary conditions of the bearing are:

$$\bar{p}' = \bar{p}_s \text{ at } \bar{z} = 0 \text{ and } 0 \le \theta \le 2\pi$$

$$\bar{p}' = \bar{p} \text{ at } \bar{z} = 1 \text{ and } 0 \le \theta \le 2\pi$$

$$\frac{\partial \bar{p}}{\partial \bar{z}} = 0 \text{ at } \bar{y} = 0 \text{ and } 0 \le \theta \le 2\pi$$

$$\bar{p} = 1 \text{ at } \bar{y} = \pm 1 \text{ and } 0 \le \theta \le 2\pi$$

$$\bar{p}(\theta, \bar{y}) = p(\theta + 2\pi, \bar{y}) \tag{11.155}$$

A first-order perturbation method is used to linearize the differential Equations (11.153) and (11.154). The perturbation equation of pressure and film thickness can be written as

$$\bar{p} = \bar{p}_0 + \varepsilon_1 e^{ir} \bar{p}_1$$
$$\bar{p}' = \bar{p}_0' + \varepsilon_1 e^{i\tau} \bar{p}_1' \tag{11.156}$$
$$\bar{h} = \bar{h}_0 + \varepsilon_1 e^{i\tau} \cos\theta$$

where $\bar{h}_0 = 1 + \varepsilon_0 \cos\theta$, $\varepsilon_1$ is a perturbation parameter and $\varepsilon_0$ is steady-state eccentricity ratio.

Substituting Equation (11.156) in Equations (11.153) and (11.154) and neglecting higher powers of $\varepsilon_1$, the two sets of linear differential equations will be obtained

$$\frac{\partial^2 \bar{p}_0'^2}{\partial \bar{z}^2} = 0 \tag{11.157}$$

$$\frac{\partial^2 (\bar{p}_0 \bar{p}_1)}{\partial \bar{z}^2} - i\frac{\sigma\gamma}{\bar{p}_0'}(\bar{p}_0' \bar{p}_1') = 0 \tag{11.158}$$

$$\frac{\partial}{\partial \theta}\left(\bar{h}^3 \frac{\partial \bar{p}_0^2}{\partial \theta}\right) + \left(\frac{D}{L}\right)^2 \bar{h}_0^3 \frac{\partial^2 \bar{p}_0^2}{\partial \bar{y}^2} = \Lambda_p \left.\frac{\partial \bar{p}_0'}{\partial \bar{z}}\right|_{\bar{z}=1} \tag{11.159}$$

and

$$\frac{\partial^2 (\bar{p}_0 \bar{p}_1)}{\partial \bar{z}^2} + \frac{3}{\bar{h}_0}\frac{\partial \bar{h}_0}{\partial \theta}\frac{\partial (\bar{p}_0 \bar{p}_1)}{\partial \theta} + \left(\frac{D}{L}\right)^2 \frac{\partial^2 (\bar{p}_0 \bar{p}_1)}{\partial \bar{y}^2}$$

$$+ \frac{1.5}{\bar{h}_0}\frac{\partial^2 \bar{p}_0^2}{\partial \theta^2}\cos\theta + \left(\frac{3}{\bar{h}_0^2}\frac{\partial \bar{h}_0}{\partial \theta}\cos\theta - \frac{1.5}{\bar{h}_0 \sin\theta}\right)\frac{\partial \bar{p}_0^2}{\partial \theta}$$

$$+ \frac{1.5}{\bar{h}_0}\left(\frac{D}{L}\right)^2 \frac{\partial^2 \bar{p}_0^2}{\partial \bar{y}^2}\cos\theta = \frac{i\sigma}{\bar{h}_0^2}\frac{(\bar{p}_0 \bar{p}_1)}{\bar{p}_0}$$

$$+ \frac{i\sigma}{\bar{h}_0^3 \bar{p}_0}\cos\theta + \frac{\Lambda_p}{\bar{h}_0^3}\left.\frac{\partial (\bar{p}_0' \bar{p}_1')}{\partial \bar{z}}\right|_{\bar{z}=1} \tag{11.160}$$

Solution of Equation (11.160) with the boundary conditions $\bar{p}' = \bar{p}_s$ at $\bar{z} = 0$ and $\bar{p}' = \bar{p}$ at $\bar{z} = 1$,

we get
$$\bar{p}_0'^2 = \left(\bar{p}_0^2 - \bar{p}_s^2\right)\bar{z} + \bar{p}_s^{\,2} \tag{11.161}$$

and
$$\left.\frac{\partial \bar{p}_0'^2}{\partial \bar{z}}\right|_{\bar{z}=1} = \left(\bar{p}_0^2 - \bar{p}_s^2\right) \tag{11.162}$$

Using Equation (11.162) and noting $\frac{\partial \bar{h}_0}{\partial \theta} = -\varepsilon_0 \sin\theta$, Equation (11.159) can be written as

$$\frac{\partial^2 \bar{p}_0^2}{\partial \theta^2} + \left(\frac{D}{L}\right)^2 \frac{\partial^2 \bar{p}_0^2}{\partial \bar{y}^2} - \left(\frac{3\varepsilon_0 \sin\theta}{1+\varepsilon_0 \cos\theta}\right)\frac{\partial \bar{p}_0^2}{\partial \theta} = \Lambda_p \frac{\left(\bar{p}_0^2 - \bar{p}_s^2\right)}{\left(1+\varepsilon_0 \cos\theta\right)^3} \tag{11.163}$$

Equation (11.163) is solved numerically by finite difference method to obtain $\bar{P}_0$. Equation (11.158) is solved using a W.K.B.J. approximation.

Using the modified boundary conditions $\bar{p}_0' = 0$ at $\bar{z} = 0$ and $\bar{p}_1' = \bar{p}_1$ at $\bar{z} = 1$, one gets

$$\left.\frac{\partial\left(\bar{p}_0'\bar{p}_1'\right)}{\partial \bar{z}}\right|_{\bar{z}=1} = \left(\frac{i\sigma\gamma}{\bar{p}_0}\right)^{1/2} \bar{p}_0\bar{p}_1 \tag{11.164}$$

After substituting Equation (11.164) in Equation (11.160), this equation is solved numerically by finite difference method to obtain $\bar{p}_1$.

The dynamic force $W_1$ along the line of centers is

$$W_1 = 2\int_0^{L/2}\int_0^{2\pi} p_1 \cos\theta r d\theta dy \tag{11.165}$$

or
$$\bar{W}_1 \left(= \frac{W_1}{LDp_a}\right) = \frac{1}{2}\int_0^{1}\int_0^{2\pi} \bar{p}_1 \cos\theta r d\theta d\bar{y} \tag{11.166}$$

Following the approach used in the previous problem,

$$\bar{m} = -\mathrm{Re}\left(\bar{W}_1\right) = \bar{S} \tag{11.167}$$

and
$$\bar{m}_1 = \frac{288\bar{S}}{\sigma^2} \tag{11.168}$$

The effect of supply pressure, feeding parameter, and porosity parameter on the mass parameter $\bar{m}_1$ is shown in Figs 11.16 and 11.17. It may be noted that though the porosity $\mu$ decreases the threshold mass parameter $\bar{m}_1$, thereby decreasing the stability.

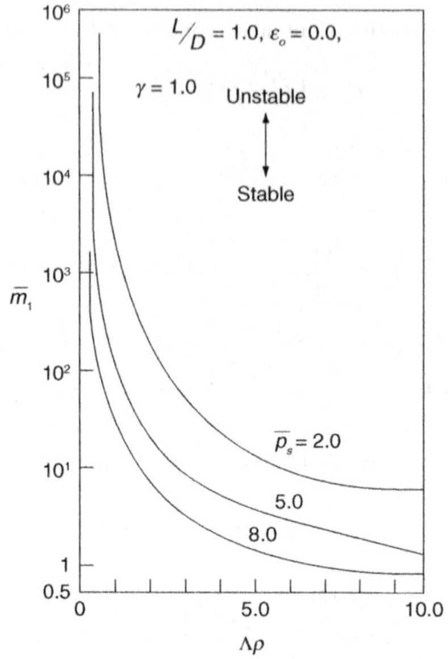


**Figure 11.16** | Stability Versus Feeding Parameter for Various Supply Pressures

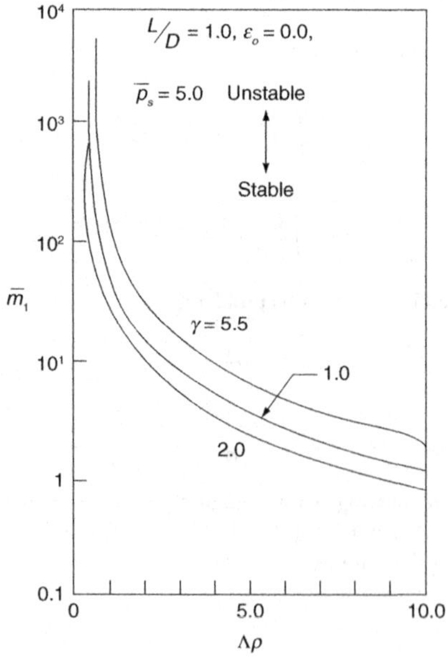


**Figure 11.17** | Stability Versus Feeding Parameter for Various Porosity Parameters

## 11.11 | Whirl Instability of Gas Bearings

Reynolds equation under dynamic condition is

$$\frac{\partial}{\partial\theta}\left(\bar{h}^3\frac{\partial\bar{p}^2}{\partial\theta}\right)+\left(\frac{D}{L}\right)^2\bar{h}^3\frac{\partial^2\bar{p}^2}{\partial\bar{y}}=2\Lambda\left(1-2\dot{\phi}\lambda\right)\frac{\partial(\bar{p}\bar{h})}{\partial\tau}\frac{\partial(\bar{p}\bar{h})}{\partial\theta}+4\Lambda\lambda\frac{\partial(\bar{p}\bar{h})}{\partial\tau} \tag{11.169}$$

where $\lambda=\dfrac{\omega_p}{\omega}.$

It is assumed that at the onset of whirl the position of the journal center can be defined as a steady-state value $(e_0,\phi_0)$ together with harmonic vibration of frequency $\omega_p$, thus

$$\varepsilon=\left|\varepsilon_0+\varepsilon_1 e^{i\tau}\right| \tag{11.170}$$

$$\phi=\phi_0+\phi_1 e^{i\tau}$$

where $\left|\varepsilon_1\right|\ll\varepsilon_0$

and $\left|\varepsilon_0\phi_1\right|\ll\varepsilon_0$

The pressure and local film thickness are

$$\bar{p}=\bar{p}_0+\varepsilon_1 e^{i\tau}\bar{p}_1+\varepsilon_0\phi_1 e^{i\tau}\bar{p}_2 \tag{11.171}$$

and

$$\bar{h}=\bar{h}_0+\varepsilon_0\cos\theta+\varepsilon_0\phi_1 e^{i\tau}\sin\theta \tag{11.172}$$

Substitution of Equations (11.171) and (11.172) into Equation (11.169) yields

$$\bar{h}_0^3\frac{\partial^2\bar{p}_0^2}{\partial\theta^2}+3\bar{h}_0^2\frac{\partial\bar{h}_0}{\partial\theta}\frac{\partial\bar{p}_0^2}{\partial\theta}+\bar{h}_0^3\left(\frac{D}{L}\right)^2\frac{\partial^2\bar{p}_0^2}{\partial\bar{y}^2}-2\Lambda p_0\frac{\partial\bar{h}_0}{\partial\theta}-2\Lambda\bar{h}_0\frac{\partial\bar{p}_0}{\partial\theta}=0 \tag{11.173}$$

$$3\bar{h}_0^2\cos\theta\frac{\partial^2\bar{p}_0^2}{\partial\theta^2}+2\bar{h}_0^3\frac{\partial^2\bar{Q}_1^2}{\partial\theta^2}+\left[6\bar{h}_0^3\cos\theta\frac{\partial\bar{h}_0}{\partial\theta}-3\bar{h}_0^3\sin\theta\right]\frac{\partial\bar{p}_0^2}{\partial\theta}+6\bar{h}_0^2\frac{\partial\bar{h}_0}{\partial\theta}\frac{\partial\bar{Q}_1}{\partial\theta}$$

$$+3\bar{h}_0^2\left(\frac{D}{L}\right)^2\cos\theta\frac{\partial^2\bar{p}_0^2}{\partial\bar{y}^2}+2\bar{h}_0^2\left(\frac{D}{L}\right)^2\frac{\partial^2 Q_1}{\partial\bar{y}^2}+2\Lambda\sin\theta\bar{p}_0-2\Lambda\frac{\partial\bar{h}_0}{\partial\theta}\frac{\bar{Q}_1}{\bar{p}_0}$$

$$-2\Lambda\cos\theta\frac{\partial\bar{p}_0}{\partial\theta}-2\Lambda\frac{\bar{h}_0}{\bar{p}_0}\frac{\partial\bar{Q}_1}{\partial\theta}-i4\Lambda\lambda\bar{h}_0\frac{\bar{Q}_1}{\bar{p}_0}-i4\Lambda\lambda\bar{p}_0\cos\theta=0 \tag{11.174}$$

and

$$3\bar{h}_0^2 \sin\theta \frac{\partial^2 \bar{p}_0^2}{\partial\theta^2} + 2\bar{h}_0^3 \frac{\partial^2 \bar{Q}_2}{\partial\theta^2} + \left[3\bar{h}_0^2 \cos\theta + 6\bar{h}_0 \sin\theta \frac{\partial \bar{h}_0}{\partial\theta}\right] \frac{\partial \bar{p}_0^2}{\partial\theta}$$

$$+6\bar{h}_0 \frac{\partial \bar{h}_0}{\partial\theta} \frac{\partial \bar{Q}_2}{\partial\theta} + 3\bar{h}_0^2 \left(\frac{D}{L}\right)^2 \sin\theta \frac{\partial^2 \bar{p}_0^2}{\partial \bar{y}^2} + 2\bar{h}_0^2 \left(\frac{D}{L}\right)^2 \frac{\partial^2 \bar{Q}_2}{\partial \bar{y}^2}$$

$$-2\Lambda\bar{p}_0 \cos\theta - 2\Lambda \frac{\partial \bar{h}_0}{\partial\theta} \frac{\bar{Q}_2}{\bar{p}_0} - \frac{2\Lambda\bar{h}_0}{\bar{p}_0} \frac{\partial \bar{Q}_2}{\partial\theta} \tag{11.175}$$

$$-2\Lambda \sin\theta \frac{\partial \bar{p}_0}{\partial\theta} - i4\lambda\Lambda\bar{h}_0 \frac{\bar{Q}_2}{\bar{p}_0} - i4\lambda\Lambda\bar{p}_0 \sin\theta$$

$$+i4\lambda\Lambda \frac{\bar{p}_0}{\varepsilon_0} \frac{\partial \bar{h}_0}{\partial\theta} + i4\lambda\Lambda \frac{\bar{h}_0}{\varepsilon_0} \frac{\partial \bar{p}_0}{\partial\theta} = 0$$

where $\bar{Q}_1 = \bar{p}_0 \bar{p}_1$ and $\bar{Q}_2 = \bar{p}_0 \bar{p}_2$

Solution of Equation (11.173) gives a steady-state pressure. These are used to find $\bar{Q}_1$ and $\bar{Q}_2$.

Nonlinearity of Equation (11.173) poses a great difficulty in finding analytical solution. Hence, Equation (11.173) is solved by Newton–Raphson iterative (NRI) method.

### 11.11.1 | NRI Method

Pressure in the clearance space can be written as

$$(\bar{p}_0)_{new} = (\bar{p}_0)_{old} + \Delta\bar{p}_1 \text{ where } \Delta\bar{p}_1 \ll \bar{p}_0 \tag{11.176}$$

Here $\bar{P}_1$ represents a small increment. Since increment is small equation (11.177) can be expressed as:

$$\bar{p}^2 = \bar{p}_0 + 2\bar{p}_0 \Delta\bar{p}_1 \tag{11.177}$$

Substituting Equation (11.177) in Equation (11.173), we get

$$\bar{h}_0^3 \left[\frac{\partial^2 \bar{p}_0^2}{\partial\theta^2} + 2\left(\bar{p}_0 \frac{\partial^2 (\Delta\bar{p}_1)}{\partial\theta^2} + \bar{p}_1 \frac{\partial^2 \bar{p}_0}{\partial\theta^2}\right)\right]$$

$$-3\bar{h}_0^2 \varepsilon_0 \sin\theta \left[\frac{\partial^2 \bar{p}_0^2}{\partial\theta^2} + 2\left(\bar{p}_0 \frac{\partial(\Delta\bar{p}_1)}{\partial\theta} + \bar{p}_1 \frac{\partial \bar{p}_0}{\partial\theta}\right)\right]$$

$$+\left(\frac{D}{L}\right)^2 \bar{h}_0^3 \left[\frac{\partial^2 \bar{p}_0^2}{\partial \bar{y}^2} + 2\left(\bar{p}_0 \frac{\partial^2 (\Delta\bar{p}_1)}{\partial \bar{y}^2} + \bar{p}_1 \frac{\partial^2 \bar{p}_0}{\partial \bar{y}^2}\right)\right]$$

$$-2\Lambda\left[-\varepsilon_0 \sin\theta\left(\bar{p}_0 + (\Delta\bar{p}_1)\right) + \bar{h}_0 \left(\frac{\partial \bar{p}_0}{\partial\theta} + \frac{\partial(\Delta\bar{p}_1)}{\partial\theta}\right)\right] = 0 \tag{11.178}$$

Equation (11.178) is a linear equation in $\bar{p}_1$ for assumed values of $\bar{p}_0$ and is solved by finite difference method with successive over-relaxation scheme. The new values of $\bar{p}_0$ for the next iteration are determined in the following manner: Let $\bar{p}_0^{N+1}$ be the guess value of the $N^{th}$ iteration. Hence,

$$\bar{p}_0^{N+1} = \bar{p}_0^{N} + \Delta\bar{p}_1^{N+1}$$

The convergence criterion in pressure used is

$$\left|\left(\Sigma \bar{p}_{old} / \Sigma \bar{p}_{new}\right)\right| \leq 0.001$$

The associated boundary conditions are:

**For $\bar{p}_0$**

$$\bar{p}_0(\theta, \bar{y}) = \bar{p}_0(\theta + 2\pi, \bar{y})$$

$$\bar{p}_0(\theta, \pm 1) = 1$$

$$\frac{\partial \bar{p}_0}{\partial \bar{y}}(\theta, 0) = 0 \tag{11.179}$$

**For $\bar{p}_1$**

$$\Delta\bar{p}_1(\theta, \bar{y}) = \bar{p}_1(\theta + 2\pi, \bar{y})$$

$$\Delta\bar{p}_1(\theta, \pm 1) = 0$$

$$\frac{\partial(\Delta\bar{p}_1)}{\partial \bar{y}}(\theta, 0) = 0 \tag{11.180}$$

The boundary conditions for $\bar{Q}_1$ and $\bar{Q}_2$ are:

$$\bar{Q}_1(\theta, \bar{y}) = \bar{Q}_1(\theta + 2\pi, \bar{y})$$

$$\bar{Q}_2(\theta, \bar{y}) = \bar{Q}_2(\theta + 2\pi, \bar{y})$$

$$\bar{Q}_1(\theta, +1) = 0$$

$$\bar{Q}_2(\theta, +1) = 0$$

$$\frac{\partial \bar{Q}_1}{\partial \bar{y}}(\theta, 0) = 0$$

$$\frac{\partial \bar{Q}_2}{\partial \bar{y}}(\theta, 0) = 0 \tag{11.181}$$

Following the similar approach as given in for an oil bearing, we can obtain the stiffness and damping coefficients.

Unlike oil bearings, the damping coefficients for gas bearings are frequency dependent $\left(\omega_p\right)$. The stiffness and damping coefficients thus obtained can be used to study the stability of a rigid rotor.

Following the similar line as given for oil bearings, we get the following two equations:

$$\bar{M} = \frac{1}{\lambda^2\left(\bar{D}_{\phi\phi} + \bar{D}_{rr}\right)}\begin{bmatrix}\left(\bar{D}_{rr}\bar{K}_{\phi\phi} + \bar{D}_{\phi\phi}\bar{K}_{rr}\right) - \left(\bar{D}_{r\phi}\bar{K}_{\phi r} + \bar{D}_{\phi r}\bar{K}_{r\phi}\right) \\ + \dfrac{\bar{W}_0}{\varepsilon_0}\left(\bar{D}_{rr}\cos\phi_0 + \bar{D}_{\phi r}\sin\phi_0\right)\end{bmatrix} \tag{11.182}$$

$$\bar{M}^2\lambda^4 - \lambda^2\left[\bar{M}\left(\frac{\bar{W}_0\cos\phi_0}{\varepsilon_0} + \bar{K}_{\phi\phi} + \bar{K}_{rr}\right) + \left(\bar{D}_{rr}\bar{D}_{\phi\phi} - \bar{D}_{\phi r}\bar{D}_{r\phi}\right)\right]$$

$$+\left(\bar{K}_{rr}\bar{K}_{\phi\phi} + \bar{K}_{\phi r}\bar{K}_{r\phi}\right) + \frac{\bar{W}_0}{\varepsilon_0}\left(\bar{K}_{rr}\cos\phi_0 + \bar{K}_{\phi r}\sin\phi_0\right) = 0 \tag{11.183}$$

The mass parameter $\bar{M}\left(= \dfrac{MC\omega^2}{W_0}\right)$ is calculated in the following way. The dynamic coefficients are initially evaluated for an assumed value of $\lambda$. These coefficients are used to find the mass parameter from Equation (11.182). The mass parameter so evaluated and the assumed whirl ratio should satisfy Equation (11.183). If not, choose a new value of $\lambda$ and continue till it is nearly equal to zero. Thus, $\bar{M}$ and $\lambda$ are critical values.

## REFERENCES

*Ausman, J.S.* (1957), The finite gas-lubricated journal bearing, *Conf. on Lubrication and Wear, I.Mech. E., London,* paper No. 22.

*Ausman, J.S.* (1960), An improved analytical solution for self-acting gas-lubricated journal bearings of finite length, *ASME, paper no. 60-LUB-9*

*Constantinescu, V.N.* (1962), Some characteristics regarding the design of bearings fed under pressure through a large number of holes thorough porous surface, *Studii si Crecetari de Mecanica, Academica Repuslicii Populare Rominie,* **13**, 175–191.

*Heinrich, G.* (1959), The theory of externally pressurized bearing with compressible lubricant, *Proc. First Intl. Symposium on Gas-lubricated bearings, ONR/ACR*-49, 261–265

*Lund, J.W.* (1964), The hydrostatic gas journal bearing with journal rotation and vibration, *J. Basic Engg., Trans, ASME, F,* **86**, 328–332.

*Lund, J.W. (1967)*, A theoretical analysis of whirl instability and pneumatic hammer for a rigid rotor in pressurized gas journal bearings, *J. Lub. Tech., Trans. ASME, F.* **89**, 154–166.

*Majumdar, B.C.* (1972), On the general solution of externally pressurized gas journal bearings, *J. Lub. Tech., Trans., ASME, F,* **94**, 328–332.

*Majumdar, B.C.* (1975), Analysis of externally pressurized porous gas journal bearings- *I Wear*, **33**, 25–36.

*Majumdar, B.C.* (1976), Design of externally pressurized gas-lubricated porous journal bearings, *Tribology International,* April, 71–74.

*Majumdar, B.C.* (1977), Porous journal bearings-a semi-analytical solution, *J. Lub. Tech. Trans, ASME, F*, **99**, 487–489.

*Majumdar, B.C.* (1978), On the analytical solution of externally pressurized porous gas journal bearings, *J. LUB. Tech., Trans, ASME, F,* **100**, 442–444.

*Nidhi, S.C., Basu, S.K.* and *Majumdar, B.C.* (1979), Pneumatic instability of externally pressurized hole-admission gas journal bearings, *Proc. National Conference on Industrial Tribology, I.I.P., Dehra Dun,* March, 27–36.

*Rao, N.S. (1979),* Effect of slip flow on static characteristics of externally pressurized gas porous thrust bearings, *Proc. National Conference on Industrial Tribology, I. I. P. Dehra Dun,* March, 37–46.

*Saha, N.* and *Majumdar, B.C.* (2004), Steady state and stability characteristics of hydrostatic two-layered porous oil journal bearings, *J. Engg. Tribology, IMECHE,* **218**, 99–108.

*Sneck, H.J. and Yen, K. T.* (1964), The externally pressurized porous wall gas-lubricated journal bearing I, *ASLE Trans.* **7**, 288–298.

Chapter 12

# Hydrodynamic Lubrication of Rolling Contacts

## 12.1 | Introduction

Hydrodynamic lubrication regime is generally encountered in lightly and moderately loaded nonconformal contacts, viz., in line and point contacts in rolling element bearings, gears, cams, etc. Pressure in the contact is not high enough to cause significant elastic deformation to alter the shape of lubricant film in the contact. These conjunctions are considered as rigid contacts for the purpose of analyzing fluid film lubrication problems. For very lightly loaded contacts, isoviscous lubrication regime prevails, whereas in moderately loaded contacts, piezoviscous effects become significant and lubrication of the contact is influenced by variation of viscosity due to pressure in the contact. Hydrodynamic flow condition prevails in the lubrication of roller-cage and ball-cage contacts. Viscous losses are also important in these contacts.

## 12.2 | Lubrication of Rolling Rigid Cylinders

Lubrication of rigid cylinders in rolling motion has been a vexing issue in the development of railroad and automobile industry. To maintain a minimum film thickness in the contact under load has been an utmost necessity for the safe operation of roller bearings, gears, etc. Martin (1916) developed a solution to determine the expressions for minimum film thickness and friction for lightly loaded cylinders operating in isoviscous lubrication regime. Rolling and sliding speeds were not high enough to cause significant thermal effects. The solution was developed for isothermal conditions. Geometry of the rolling contact is shown in Fig. 12.1.

1. Neglecting side leakage the Reynolds equation for isoviscous incompressible lubrication in pure rolling is written as

$$\frac{d}{dx}\left(h^3 \frac{dp}{dx}\right) = 6\eta U \frac{dh}{dx} \tag{12.1}$$

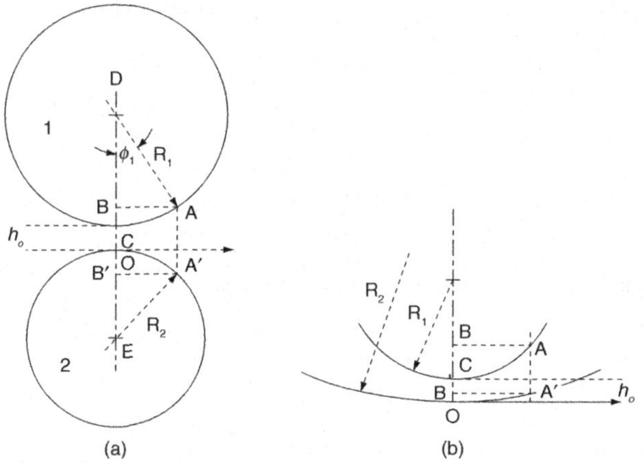


**Figure 12.1** | Cylindrical Rollers in Contact Separated by Oil Film

where $U = u_1 + u_2$, $u_1$ and $u_2$ are velocities of surfaces 1 and 2, respectively. In pure rolling $u_1 = u_2 = u$.

Referring to Fig. 12.1 expression for variation of film thickness in terms of $x$ can be determined. Thus,

$OB = OC + CB = h_0 + \left(DC - DA\cos\phi_1\right)$, $h_0$ is film thickness at the center of the contact

with
$$\cos\phi = -\frac{\phi^2}{2} + \frac{\phi^2}{4} - \frac{\phi^6}{6} + \ldots$$

If $\phi$ is small then only first two terms may be retained. With $\phi_1 = x/R_1$ and $\phi_2 = x/R_2$ where $x$ is measured from the center of the contact. $R_1$, $R_2$ are radii of rollers 1 and 2, respectively.

Therefore,

$$OB = h_0 + R_1\left\{1 - \left(1 - \frac{x^2}{2R_1}\right)\right\}$$

Similarly, we get

$$0B' = R_2\left\{1 - \left(1 - \frac{x^2}{2R_1}\right)\right\}$$

Film thickness

$$h = AA' = OB + OB'$$

and

$$h = h_0 + \frac{x^2}{2R_1} + \frac{x^2}{2R_2} = h_0 + \frac{x^2}{2R} \tag{12.2}$$

Where

$$\frac{1}{R} = \frac{1}{R_1} + \frac{1}{R_2}$$

$R$ is the equivalent radius of a roller on a plane.

It can be shown that in the case of rollers in internal contact (Fig. 12.1), the expression for film thickness is expressed as

$$h = h_0 + \frac{x^2}{2R} \tag{12.3}$$

where

$$\frac{1}{R} = \frac{1}{R_1} - \frac{1}{R_2}$$

Integrating Reynolds Equation (8.1) once with respect to $x$, we get

$$\frac{dp}{dx} = 6U\eta \frac{h - \bar{h}}{h^3} \tag{12.4}$$

where $\frac{dp}{dx} = 0;\ \text{at } h = \bar{h}$

To determine pressure distribution, Equation (12.4) is further integrated with respect to $x$. The following substitution is made to integrate Equation (12.4), i.e.,

$$\tan\gamma = \frac{x}{\sqrt{2Rh_0}} \tag{12.5}$$

With the above substitution, the expression for film thickness reduces to

$$h = h_0\left(1 + \tan^2\gamma\right) = h_0 \sec^2\gamma \tag{12.6}$$

and $$d\gamma = \frac{dx}{\sqrt{2Rh_0}} \frac{1}{\sec^2\gamma}$$

Integrating Equation (8.4) with the above substitution, we can obtain

$$dp = 6\eta U \frac{h_0 \sec^2\gamma - h_0 \sec^2\bar{\gamma}}{h_0^3 \sec^6\gamma} \sqrt{2Rh_0} \sec^2\gamma d\gamma \tag{12.7}$$

Let $$\bar{p} = \frac{h_0^2 p}{6U\eta\sqrt{2Rh_0}} \tag{12.8}$$

$$d\bar{p} = \cos^2\gamma d\gamma - \frac{\cos^4\gamma}{\cos^2\bar{\gamma}} d\gamma$$

Integrating we get

$$\bar{p} = \frac{\gamma}{2} + \frac{\sin 2\gamma}{4} - \frac{1}{\cos^2\bar{\gamma}}\left[\frac{3\gamma}{8} + \frac{\sin 2\gamma}{4} + \frac{\sin 4\gamma}{32}\right] + c \tag{12.9}$$

In the case of fully flooded lubrication where it is believed that copious amount of lubricant is available to lubricate the contact and therefore theoretically it can be assumed that pressure build up will start at a location far to the left of center of the contact, i.e., at $x = -\infty$ or $\gamma = -\pi/2$.

Therefore, inlet boundary condition is given as:

$$p = \bar{p} = 0 \text{ at } \gamma = -\pi/2$$

This gives constant of integration, $c = \frac{\pi}{4} - \frac{1}{\cos^2 \bar{\gamma}}\left(\frac{3\pi}{16}\right)$

Hence,

$$\bar{p} = \frac{\gamma}{2} + \frac{\pi}{4} + \frac{\sin 2\gamma}{4} - \frac{1}{\cos^2 \bar{\gamma}}\left[\frac{3}{8}\left(\gamma + \frac{\pi}{2}\right) + \frac{\sin 2\gamma}{4} + \frac{\sin 4\gamma}{32}\right] \tag{12.10}$$

Boundary condition at the exit or where the lubricant film breaks down is given by Reynolds boundary condition as

$$\bar{p} = \frac{d\bar{p}}{d\gamma} = 0 \text{ i.e. } p = \frac{dp}{dx} = 0 \text{ at } x = \bar{x} \text{ or } h = \bar{h},\ \gamma = \bar{\gamma}$$

Substituting

$$\bar{p} = 0 \text{ at } \gamma = \bar{\gamma}, \text{ we get}$$

$$\frac{1}{2}\left[\bar{\gamma} + \frac{\pi}{2} + \frac{\sin 2\bar{\gamma}}{2} - \frac{1}{\cos^2 \bar{\gamma}}\left\{\frac{3}{4}\left(\bar{\gamma} + \frac{\pi}{2}\right) + \frac{\sin 2\bar{\gamma}}{2} + \frac{\sin 4\bar{\gamma}}{16}\right\}\right] = 0$$

Above is a transcendental equation which can be solved by successive approximation to give

$$\sec^2 \bar{\gamma} = 1.22575, \text{ or } \bar{\gamma} = 25^{0}\,25'$$

Thus,

$$\bar{p} = \frac{1}{2}\left[\bar{\gamma} + \frac{\pi}{2} + \frac{\sin 2\gamma}{2} - 1.226\left\{\frac{3}{4}\left(\gamma + \frac{\pi}{2}\right) + \frac{\sin 2\gamma}{2} + \frac{\sin 4\gamma}{16}\right\}\right] \tag{12.11}$$

Pressure distribution obtained using Equation (8.11) is shown in Fig. 12.2. $\gamma$ and $\phi$ coincide only at $\phi = 0$ and $\pi/2$, i.e., when $\phi = 0$, respectively. At other angles both differ from each other.

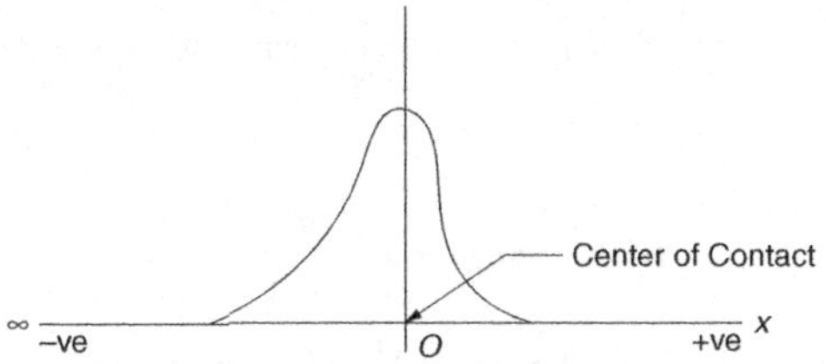


**Figure 12.2** | Pressure Distribution for Reynolds Boundary Condition

Load capacity is obtained by integrating pressure over the film domain on surface 2, therefore load capacity is given as

$$W = L\int_{-\infty}^{x_e} p dx \tag{12.12a}$$

$$W = L\int_{-\pi/2}^{\bar{\gamma}} \frac{12U\eta R}{h_0}\bar{p}\sec^2\gamma d\gamma \tag{12.12b}$$

Substituting for $\bar{p}$ using Equation (12.11) and $\bar{\gamma} = 0.44355$, $\sec^2\gamma = 1.226$, $\tan\gamma = 0.47513$, the load capacity per unit width of the roller is obtained as:

$$\frac{W}{L} = 2.44748\frac{UR\eta}{h_0} \tag{12.13}$$

where $U = u_1 + u_2$ and $R$ is composite radius as defined in Equations (12.2) and (12.3).

The viscous drag forces on the two surfaces of rollers are given as

$$f_1 = L\int_{-\infty}^{x_e}(\tau_{xz})dx \text{ and } f_2 = L\int_{-\infty}^{x_e}(\tau_{xz})_2 dx \tag{12.14}$$

The shear stresses $\tau_{xz}$ on surfaces 1 and 2 are expressed as

$$(\tau_{xz})_1 = -\left(\eta\frac{\partial u}{\partial z}\right)_{z=h} = -\frac{h}{2}\frac{\partial p}{\partial x} + \eta\frac{(u_2 - u_1)}{h} \tag{12.15a}$$

$$(\tau_{xz})_2 = -\left(\eta\frac{\partial u}{\partial z}\right)_{z=0} = -\frac{h}{2}\frac{\partial p}{\partial x} - \eta\frac{(u_2 - u_1)}{h} \tag{12.15b}$$

Substituting for shear stresses using Equation (12.15) into (12.14) it can be shown that friction forces per unit width of roller can be obtained as

$$\frac{f_1}{L} = -\frac{W_x}{2} + 2.84\eta(u_2 - u_1)\left(\frac{R}{h_0}\right)^{1/2}$$

$$\frac{f_2}{L} = -\frac{W_x}{2} - 2.84\eta(u_2 - u_1)\left(\frac{R}{h_0}\right)^{1/2} \tag{12.16}$$

where $W_x$ is the horizontal component of the force per unit length of roller due to fluid pressure which is given by

$$W_x = -L\int_{h_i}^{h_e} p dh = -R\int_{\phi_i}^{\phi_e} p\sin\phi d\phi \tag{12.17}$$

and is obtained as:

$$W_x = 4.58\eta(u_1 + u_2)\left(\frac{R}{h_0}\right)^{1/2} \tag{12.18}$$

The above results are very commonly known as Martin's solution and minimum film thickness in the contact is proportional to speed and inversely to load as can be seen in Equation (12.13).

2. For short cylinders, when L/D << 1 the short bearing approximation or Ocvirk's theory may be used and the Reynolds equation is thus written as

$$\frac{\partial}{\partial y}\left(h^3\frac{\partial p}{\partial y}\right)=6\eta\left(u_1+u_2\right)\frac{dh}{dx} \tag{12.19}$$

assuming $$\frac{\partial p}{\partial x}=0.$$

The pressure distribution is obtained after integrating with respect to $y$ with boundary conditions that $p = 0$ for $y = \pm$ L/2 as

$$p=\frac{3\eta\left(u_1+u_2\right)}{h^3}\frac{dh}{dx}\left(y^2-L^2/4\right) \tag{12.20}$$

From Equation (12.3) we obtain

$$\frac{dh}{dx}=x/R,$$

and the pressure distribution is given by

$$p=\frac{3\eta\left(u_1+u_2\right)x}{R\left(h_0+\dfrac{x^2}{2R}\right)^3}\left(y^2-L^2/4\right) \tag{12.21}$$

Pressure is zero at $x = 0$ or at the center of the contact.

The vertical load capacity is obtained by integrating Equation (12.21) as

$$W=\int_{-L/2}^{L/2}\int_{-\infty}^{0} p\,dx\,dy \tag{12.22}$$

$$W=\frac{\eta\left(u_1+u_2\right)L^3}{4h_0^2} \tag{12.23}$$

It has been assumed that film breaks down at the center of the contact, i.e., at $x = 0$ where it starts diverging. This assumption is similar to $\pi$ film assumption made for short bearing solution of journal bearings. The load capacity is independent of size or diameter of rollers. Dowson and Whomes (1967) presented solution of Reynolds equation for finite length rollers operating in isoviscous lubrication regime using finite difference method.

## 12.3 | Isoviscous Lubrication of Rigid Spherical Bodies in Rolling

Hydrodynamic lubrication of ellipsoidal solids in rolling is important in many engineering applications. These contacts are encountered in general in ball bearings of all types, spiral and hypoid gears, etc. Kapitza (1955) was the first to determine an expression for film thickness in case of isoviscous lubrication of spheres. Later Brewe *et al.* (1979) investigated the problem theoretically in detail and developed expressions for minimum film thickness in the contact including geometrical parameters for fully flooded and starved contacts as well in 1982.

A pair of ellipsoidal solids in contact is shown in Fig. 12.3. Contact geometry of the two rigid solids separated by a lubricant film is shown in Fig. 12.4.

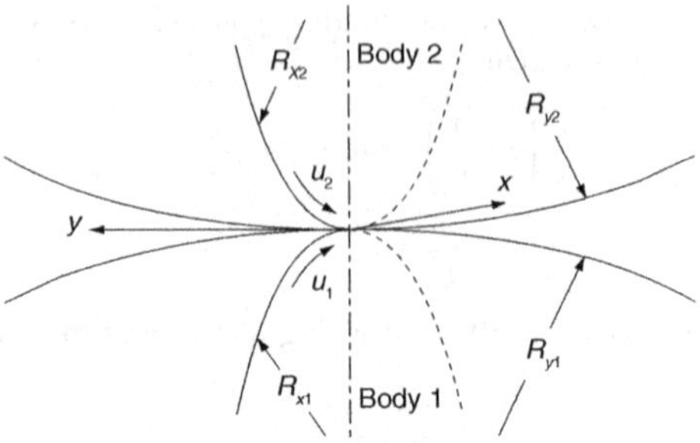


**Figure 12.3** | Elliptical Solids in Contact

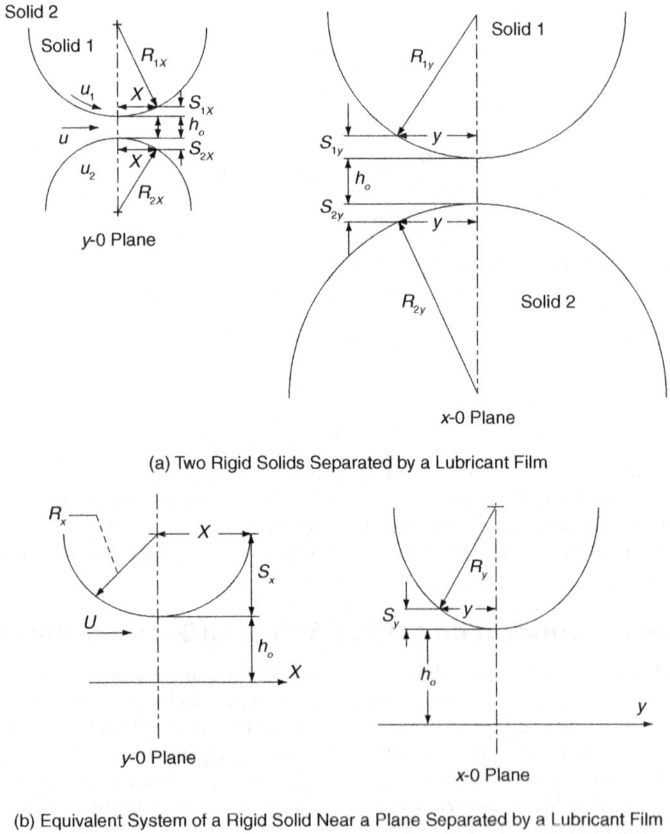


**Figure 12.4** | Contact Geometry

Expression for film thickness can be written as

$$h(x, y) = h_0 + S(x, y) \tag{12.24}$$

where $h_0$ is minimum film thickness at the center of the contact and $S(x,y)$ is the separation due to geometry of solids.

It can be shown in a manner similar to what has been done for rollers in contact looking into the curvatures of solids about two perpendicular axes $x$, $y$, that geometrical separation $S(x, y)$ can be written as

$$S(x, y) \cong \frac{x^2}{2R_x} + \frac{y^2}{2R_y} \tag{12.25a}$$

where

$$\frac{1}{R_x} = \frac{1}{R_{x1}} + \frac{1}{R_{x2}}$$
$$\frac{1}{R_y} = \frac{1}{R_{y1}} + \frac{1}{R_{y2}} \tag{12.25b}$$

and

$$\alpha = \frac{R_y}{R_x}$$

Therefore, film thickness $h(x, y)$ can be expressed as

$$h(x, y) = h_0 + \frac{x^2}{2R_x} + \frac{y^2}{2\alpha R_x} \tag{12.26}$$

Reynolds equation for isothermal, isoviscous, and incompressible lubrication is expressed as

$$\frac{\partial}{\partial x}\left(h^3 \frac{\partial p}{\partial x}\right) + \frac{\partial}{\partial y}\left(h^3 \frac{\partial p}{\partial y}\right) = 6\eta(u_1 + u_2)\frac{\partial h}{\partial x} \tag{12.27}$$

The equation can be written in the dimensionless form using the following substitution

$$\bar{h} = h / R_x, \bar{x} = \frac{x}{R_x}, \bar{y} = \frac{y}{R_x}, \alpha = \frac{R_y}{R_x},$$
$$u = \frac{(u_1 + u_2)}{2} \text{ and } \bar{p} = \frac{pR_x}{\eta u} \tag{12.28}$$

With the above substitutions, Equation (12.27) reduces to

$$\frac{\partial}{\partial \bar{x}}\left(\bar{h}^3 \frac{\partial \bar{p}}{\partial \bar{x}}\right) + \frac{\partial}{\partial \bar{y}}\left(\bar{h}^3 \frac{\partial \bar{p}}{\partial \bar{y}}\right) = 12\frac{\partial \bar{h}}{\partial \bar{x}} \tag{12.29}$$

The boundary conditions for Equation (12.29) are given by:

$$\bar{p} = 0 \text{ at } \bar{x} = -\bar{x}_{in}; \text{ or } \bar{h} = \bar{h}_{in} \text{ and } \bar{p} = \frac{\partial \bar{p}}{\partial x} = 0$$

Reynolds' film rupture boundary condition at exit, i.e., at $\bar{x} = \bar{x}_e$.

It is assumed that the contact is fully flooded, i.e., enough lubricant is available to the contact for film build up. More lubricant than this amount, if made available, would not increase the minimum film thickness in the contact.

Brewe *et al.* (1979) used finite difference method to solve the Reynolds equation (12.29) satisfying appropriate boundary conditions mentioned. Variable mesh was used to provide more number of meshes around the pressure peak. Coarse mesh size of 0.1 and fine mesh size of 0.002 was used for highly peaked pressure distribution. For the known pressure distribution due to a known value of minimum film thickness at the center of the contact, normal load component is determined by integrating pressure over the film domain as expressed below

$$w = \iint p\,dx\,dy$$

or

$$w = \eta u R_x \iint \bar{p}\,d\bar{x}\,d\bar{y} \tag{12.30}$$

Brewe *et al.* (1974) defined dimensionless load and speed parameters as

$$W = \frac{w}{E'R_x^2}, U = \frac{\eta u}{E'R_x} \tag{12.31}$$

Pressure distribution in the contact is shown in Fig. 12.5 for two different values of $\alpha = 1.0$ and 36.54.

The expression for minimum film thickness at the center of contact was determined by Brewe *et al.* (1979) after determining the load capacity for various rolling speeds and geometry parameters for known values of film thickness at the center of the contact. The expression for minimum film thickness as a function of dimensionless rolling speed ($U$), dimensionless load ($W$), and geometry parameter ($\alpha$) was determined following regression analysis of the data generated through theoretical analysis. Minimum film thickness at the center of the contact is given by

$$\bar{h}_{\min} = \bar{h}_0 = 128\alpha \left(\frac{\phi U L}{W}\right)^2 \tag{12.32}$$

where

$$L = 0.131 \tan^{-1}\frac{\alpha}{2} + 1.683$$

and $\phi = \left(1 + \frac{2}{3}\alpha\right)^{-1}$, is referred as Archard–Cowking side leakage factor.

The minimum film thickness equation developed by Kapitza (1955) using half-Sommerfeld boundary condition at the exit, i.e., $p = 0$ at $x = 0$, is given as

$$\bar{h}_{\min} = \bar{h}_0 = 128\alpha \left(\frac{\varphi U}{W}\frac{\pi}{2}\right)^2 \tag{12.33}$$

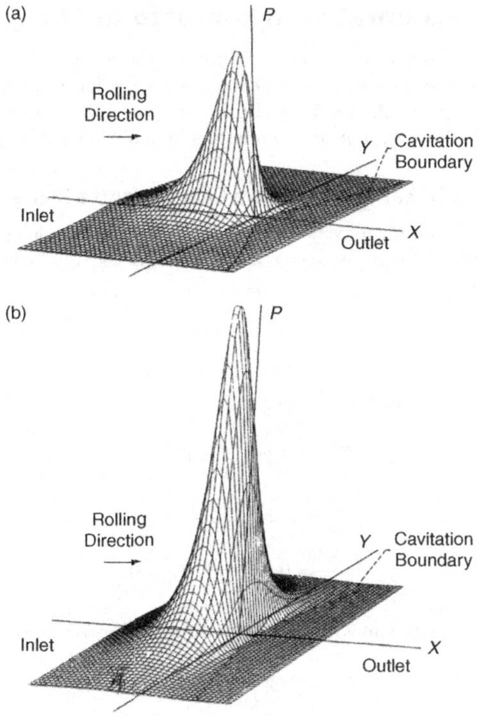


**Figure 12.5** | Three-dimensional Representations of Pressure Distributions as Viewed from Outlet Region for Two Radius $\alpha$. **(a)** $\alpha = 1.00$; **(b)** $\alpha = 36.54$ [From Brewe et al. (1979)]

Brewe and Hamrock (1982) modified the equation to include starvation effect into it. The minimum film formula was modified as Integrating Equation (12.35) with respect to $x$

$$\bar{h}_{min} = \bar{h}_0 = \left[ \frac{W/U}{\phi L (128\alpha)^{1/2}} + 3.02 \right]^{-2} \tag{12.34}$$

Experiments conducted by Boness (1970), Wedeven (1971), and theoretical analysis of Chiu (1974) revealed that often the contacts operate under starved condition of lubrication when adequate amount of lubricant is not available to the contact. Starvation is also caused by kinematic conditions due to high rolling speed and reverse flow in the inlet region which reduces the supply of lubricant to the contact. Starvation also significantly affects the load capacity and friction in the contact.

## 12.4 | Squeeze Film Lubrication in Nonconformal Contacts

Squeeze film lubrication is a mode of lubrication when the surfaces approach each other in the absence of relative sliding velocity. Positive pressures are generated when the lubricant held between the surfaces is squeezed out due to normal approach velocity and this is responsible to carry the load and prevent direct contact between the surfaces. This kind of situation occurs in many practical applications.

### 12.4.1 | Squeeze Film Lubrication Between a Cylinder and a Plane

The geometry of cylinder and coordinate system are shown in Fig. 12.6. The length of the cylinder is assumed to be large relative to the radius of the cylinder so that the side leakage can be neglected.

The governing equation is

$$\frac{d}{dx}\left(h^3\frac{dp}{dx}\right) = -12\eta V \tag{12.35}$$

Integrating Equation (12.35) with respect to $x$

$$\frac{dp}{dx} = -12\eta V\frac{x}{h^3} + C_1 \tag{12.36}$$

Since

$$\frac{dp}{dx} = 0 \text{ at } x = 0,\ C_1 = 0$$

Integrating Equation (12.36) again with respect to $x$, pressure distribution is obtained as

$$p = -12\eta V\int\frac{x}{h^3}dx + C_2 \tag{12.37}$$

The film thickness $h$ is a function of $x$ and an approximate equation of $h$ is given by Equation (12.3) and is reproduced here below

$$h = h_0 + \frac{1}{2}\frac{x^2}{R} \tag{12.38}$$

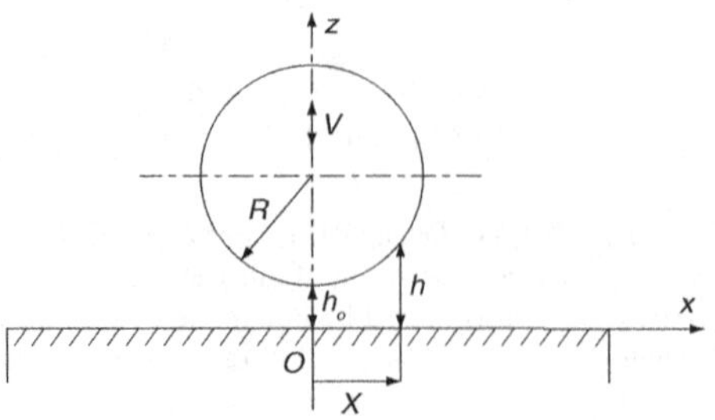


**Figure 12.6** | Squeeze Film Between a Cylinder and a Plane

Therefore,

$$\frac{dh}{dR} = x/R \text{ or } Rdh = xdx \tag{12.39}$$

Using Equation (12.39), Equation (12.37) can be written as

$$p = -12\eta VR\int\frac{dh}{h^3} + C_2$$

or

$$p = \frac{6\eta VR}{h^2} + C_2 \tag{12.40}$$

As $h \to \infty, p \to 0$, we find $C_2 = 0$.

Therefore,

$$p = \frac{6\eta VR}{h^2} = \frac{6\eta VR}{h_0^2}\frac{1}{\left(1+\dfrac{x^2}{2Rh_0}\right)^2} \tag{12.41}$$

The load capacity $W$ can be calculated from using Equation (12.41) for $p$ as

$$W = L\int_{-\infty}^{\infty} pdx = \frac{6\eta VRL}{h_0^2}\int_{-\infty}^{\infty}\frac{dx}{\left(1+\dfrac{x^2}{2Rh_0}\right)^2} \tag{12.42}$$

To solve

$$\int\frac{dx}{\left(1+\dfrac{x^2}{2Rh_0}\right)^2}, \text{ let } \frac{x^2}{2Rh_0} = \tan^2\psi$$

Then

$$dx = \sqrt{2Rh_0}\sec^2\psi d\psi$$

and

$$\left(1+\frac{x^2}{2Rh_0}\right)^2 = \left(1+\tan^2\psi\right)^2 = \sec^4\psi$$

Now

$$\int_{-x}^{x}\frac{dx}{\left(1+\dfrac{x^2}{2Rh_0}\right)^2} = \sqrt{2Rh_0}\int_{-\pi/2}^{\pi/2}\cos^2\psi d\psi$$

$$= \sqrt{2Rh_0}\left[\frac{\psi}{2}+\frac{\sin 2\psi}{4}\right]_{-\frac{\pi}{2}}^{\frac{\pi}{2}} = \frac{\pi}{2}\sqrt{2Rh_0}$$

$$= \sqrt{2Rh_0}\left[\frac{\psi}{2}+\frac{\sin 2\psi}{4}\right]_{\frac{\pi}{2}}^{\frac{\pi}{2}} = \frac{\pi}{2}\sqrt{2Rh_0}$$

Therefore, the squeeze load capacity $W$ is obtained as

$$W = \frac{3\pi\eta VRL}{h_0}\left(\frac{2R}{h_0}\right)^{1/2} \tag{12.43}$$

The time of approach can be found from

$$\frac{W}{3\pi\eta VRL\sqrt{2R}}\int dt = \int \frac{dh_0}{h_0^{3/2}} = \frac{2}{h_0^{1/2}}$$

and

$$\Delta t = \frac{6\sqrt{2}\pi\eta RL}{W}\left[\left(\frac{R}{h_{o2}}\right)^{1/2} - \left(\frac{R}{h_{o1}}\right)^{1/2}\right] \tag{12.44}$$

where $h_{01}$ and $h_{02}$ are values of film thickness $h_0$ at time $t_1$ and $t_2$, respectively. If $h_{02} << h_{01}$ then

$$\Delta t = \frac{6\sqrt{2}\pi\eta RL}{W}\left(\frac{R}{h_{o2}}\right)^{1/2} \tag{12.45}$$

### 12.4.2 | Squeeze Film Lubrication Between a Sphere and a Plane

It is assumed that a rigid sphere of radius $R$ is approaching a plane as shown in Fig. 12.7.

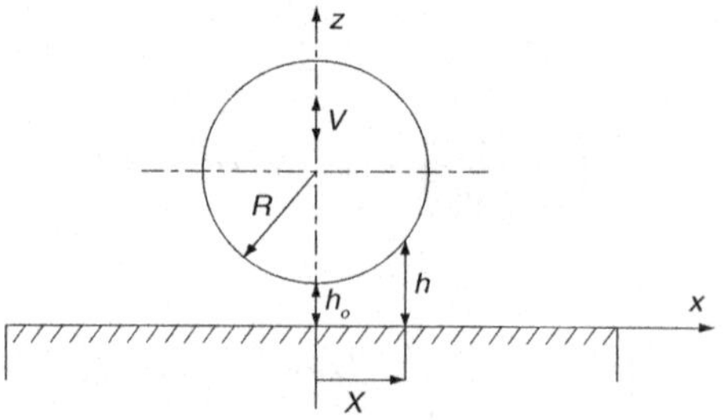


**Figure 12.7** | Squeeze Film Between a Sphere and a Plane

The fundamental equation in this case may be rewritten as

$$\frac{dp}{dr} = -\frac{6\eta rV}{h^3} \tag{12.46}$$

Integrating Equation (12.46) with respect to $r$, we get

$$p = -6\eta V \int \frac{rdr}{h^3} + C_2 \tag{12.47}$$

The film thickness $h$ can be expressed as

$$h = h_0 + R - \sqrt{R^2 - r^2} \tag{12.48}$$

and

$$\frac{dh}{dr} = \frac{r}{\sqrt{R^2 - r^2}} = \frac{r}{\left(h_0 + R - h\right)} \tag{12.49}$$

Making use of Equation (12.49), Equation (12.47) can be written as

$$p = -6\eta V \int \frac{\left(h_0 + R - h\right)}{h^3} dh + C_2$$

and

$$p = -3\eta V \left[\frac{2h - h_0 - R}{h^2}\right] + C_2 \tag{12.50}$$

Using the boundary condition that $p = 0$ at $h = h_0 + R$ in Equation (12.50), the constant $C_2$ becomes

$$C_2 = \frac{3\eta V}{h_0 + R}$$

Therefore,

$$p = \frac{3\eta V}{h_0 + R}\left[1 - \frac{h_0 + R}{h}\right]^2 \tag{12.51}$$

It is known that pressure acts normal to the surface of the sphere. Therefore, defining $\theta$ as the angle between line of pressure and the vertical, the load capacity $W$ is determined as:

$$W = \int_0^{R/2} \int_0^R 2\pi rp \cos\theta d\theta dr = \int_0^R 2\pi rpdr = \frac{6\pi\eta V}{\left(h_0 + R\right)} \int_0^R \left[1 - \frac{h_0 + R}{h}\right]^2 rdr$$

As $rdr = \left(h_0 + R - h\right) dh$, the above expression can be transformed to

$$W = \frac{6\pi\eta V}{\left(h_0 + R\right)} \int_{h_0}^{h_0+R} \left\{\left(h_0 + R - h\right)\left[1 - \frac{2\left(h_0 + R\right)}{h} + \frac{\left(h_0 + R\right)^2}{h^2}\right]\right\} dh \tag{12.52}$$

The integration of right hand side of Equation (12.52) yields

$$W = 6\pi\eta V h_0 \left\{\frac{3}{2}\left(\frac{R}{h_0}-1\right)-\left(1+\frac{R}{h_0}\right)\ln\left(1+\frac{R}{h_0}\right)+\frac{1}{2\left(1+\frac{R}{h_0}\right)}\left[1+2\left(1+\frac{R}{h_0}\right)^3\right]\right\} \tag{12.53}$$

In most cases

$$\frac{R}{h_0} >> 1,$$

Hence,

$$W = 6\pi\eta V h_0 \left(\frac{R}{h_0}\right)^2 \tag{12.54}$$

The time of approach may be found after substituting $V = -\frac{dh}{dt}$ in Equation (12.54).

$$\int dt = -\frac{6\pi\eta R^2}{W}\int\frac{dh_0}{h_0} = -\frac{6\pi\eta R^2}{W}\ln h_0$$

Therefore, the time of approach for film thickness to reduce from $h_{01}$ to $h_{02}$ is given by

$$\Delta t = \frac{6\pi\eta R^2}{W}\ln\left(\frac{h_{01}}{h_{02}}\right)$$

## 12.5 | Effect of Squeeze Motion on the Lubrication of Rigid Solids

Under dynamic loads, the solid surfaces either approach each other at a velocity or separate from each other. Rollers and balls in rolling elements move around the races in rolling and thus pass through loaded and unloaded regions. Load variations also induce squeeze motion in the rollers and balls which bring the surfaces either closer or separate them. Squeeze motion is superposed over the entrainment motion and the action thus combines the two motions. This influences the film thickness and results in variations in the load capacity of the bearing.

Sasaki *et al.* (1962) presented a solution for the isothermal lubrication of rigid cylinders subject to sinusoidal load for a non-Newtonian fluid by using superposition of pressure curves generated by normal approach and entraining velocities. Vichard (1971) analyzed theoretically the transient effect associated with squeeze film action under both hydrodynamic and elastohydrodynamic lubrication. It was observed that at low film thickness the damping phenomenon associated with normal approach was more important under elastohydrodynamic conditions than in hydrodynamic condition.

Dowson, Markho, and Jones (1976) presented a general theoretical analysis of the hydrodynamic lubrication of rigid cylindrical contacts by an isoviscous lubricant in combined rolling and normal. Experimental investigations were also reported by Markho and Dowson (1976). Results showed that normal motion significantly influences the load capacity and film rupture boundary. Ghosh *et al.* (1985) investigated in detail the effect of normal motion on the performance of rigid nonconformal contacts in combined rolling and normal motion for both fully flooded and starved contacts.

## 12.6 | Hydrodynamics of Rigid Point Contacts in Combined Rolling and Normal Motion

Figure 12.8 shows the lubricant flow in the contact under rolling and normal motion.

Reynolds equation for hydrodynamic lubrication of two rigid spherical solids separated by incompressible, isoviscous lubricant film is expressed as:

$$\frac{\partial}{\partial x}\left(h^3\frac{\partial p}{\partial x}\right)+\frac{\partial}{\partial y}\left(h^3\frac{\partial p}{\partial y}\right)=12\eta U\frac{\partial h}{\partial \overline{x}}+12\eta U_N \qquad (12.55)$$

where

$$h(x, y)=h_0+S(x, y)=h_0+\frac{x^2}{2R_x}+\frac{y^2}{2R_y} \qquad (12.56)$$

Expressions for $R_x$ and $R_y$ are given in Equation (12.25b). Substituting

$$\overline{x}=x/R_x, \overline{y}=y/R_x, H=h/R_x, \overline{p}=\frac{pR_x}{\eta U}$$
$$\alpha=R_y/R_x, U=\frac{1}{2}(u_a+u_b), \overline{U}_N=U_N/U$$

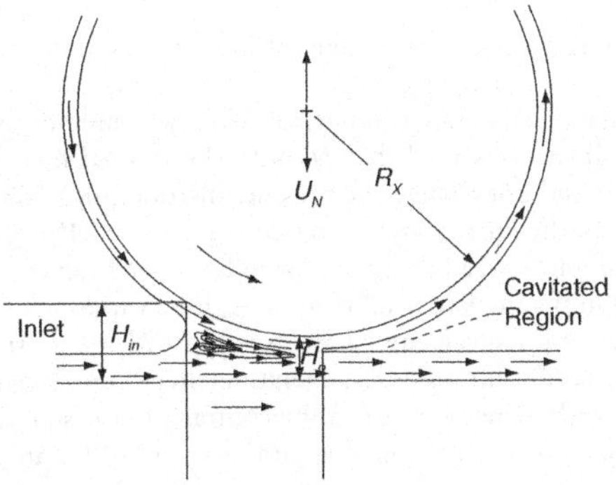


**Figure 12.8** | Lubricant Flow for a Rolling–Sliding Contact

Equation (12.55) can be expressed in dimensionless form as

$$\frac{\partial}{\partial \bar{x}}\left(H^3 \frac{\partial \bar{p}}{\partial \bar{x}}\right)+\frac{\partial}{\partial \bar{y}}\left(H^3 \frac{\partial \bar{p}}{\partial \bar{y}}\right)=12\left(\frac{\partial H}{\partial \bar{x}}+q\sqrt{2H_0}\right) \tag{12.57}$$

where $q = \bar{U}_N / \sqrt{2H_0}$ is the dimensionless normal velocity parameter which incorporates both normal velocity $\bar{U}_N$ and central film thickness $H_0$ in it.

Boundary conditions of Equation (12.57) are given as $\bar{p}=0$ at the inlet boundary, i.e., at $H = H_{in}$; $\bar{p}=\frac{\partial \bar{p}}{\partial N}=0$ at the cavitations boundary (i.e., Reynolds boundary condition), where $N$ is normal to the boundary. Film shape is described as

$$H = H_0 + \frac{\bar{x}^2}{2} + \frac{\bar{y}^2}{2\alpha} \tag{12.58}$$

A pressure distribution satisfying above boundary conditions is determined for a given speed, viscosity, central film thickness, fluid inlet level, and normal velocity parameter using the Gauss–Siedel iterative procedure with over-relaxation following a finite difference scheme of discretization. A fine meshing of 0.001 and a coarse mesh spacing of 0.01 was used for $\alpha = 1$, whereas for other values of $\alpha$, boundaries of computation zone were located according to $\bar{y} = \alpha^{0.5}\bar{x}$ and correspondingly coarse grid were used for $\alpha$ values higher than 1.0.

Hydrodynamic load capacity is determined as

$$w = \oint\!\!\oint_A p\,dx\,dy \tag{12.59}$$

where $A$ is the domain of integration which is dependent on both the fluid inlet level and cavitations boundary.

Instantaneous load carrying capacity is expressed as a ratio

$$\beta = \frac{w}{w_{q=0}} \tag{12.60}$$

$w_{q=0}$ represents the dimensionless steady-state value of load for identical conditions to that of instantaneous load.

Computations were done for a large range of normal velocity parameter $q$ varying between −1.0 and 0.75 for fully flooded lubrication. It was observed that $H_{in} = 0.035$ ensured that fully flooded condition existed in the contact, i.e., $H_{in} > 0.035$ did not change the pressure distribution in the contact and the performance parameters as well. Central film thickness film was varied from $10^{-3}$ to $10^{-5}$, geometry parameter $\alpha$ was varied from 0.2 to 35.0. Dimensionless speed parameter or rolling speed values ensured that film thickness and lubrication domain remained in the isoviscous lubrication regime. Variation of dynamic load ratio $\beta$ is shown in Fig. 12.9 with dimensionless normal velocity parameter $|q|$ for $H_0 = 1 \times 10^{-4}$, $\alpha = 1.0$ and $H_{in} = 0.035$.

Parameter $q$ clearly has a significant pressure generating effect in normal approach and thereby increasing the dynamic load ratio $\beta$ with increase in $q$. On the contrary, $\beta$ was significantly reduced during normal separation, i.e., for $|q| > 0$. Pressure distributions for various values of $|q|$ are shown in Fig. 12.10. It can be seen that during normal approach the film rupture boundary moves down stream into the exit region or away from the center of the contact.

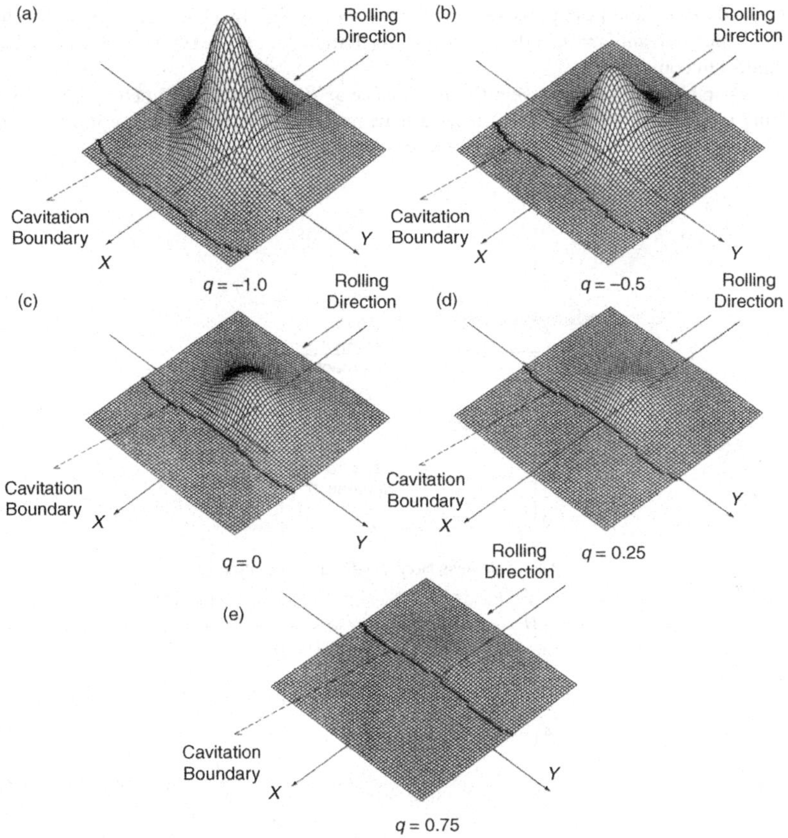


**Figure 12.9** | Pressure Distribution in Contact for Various Values of Dimensionless Normal Velocity Parameter $q$. Dimensionless Central Film Thickness $H_0 = 1 \times 10^{-4}$; Dimensionless Geometry Parameter $\alpha = 1.0$; and Dimensionless Inlet Starvation Parameter $H_{in} = 0.035$

During separation it moves upstream toward the inlet side. The location of pressure peak and the entire pressure distribution in the contact also shifts accordingly.

Peak pressure ratio, $\xi = \dfrac{p_{\max}}{p_{\max}\big|_{q=0}}$ variation with $|q|$ is shown in Fig. 12.10 against $|q|$. Pressures of the order of three or four times the peak pressure in steady-state situation are generated during normal approach. On the other hand, similar reductions are observed during separation. Effect of geometry parameter $\alpha$ on the dynamic load ratio $\beta$ and peak pressure ratio $\xi$ are shown in Figs 12.11 and 12.12 against $\alpha$ for $q = -1.0$ during normal approach and $q = 0.75$ during normal separation. In case of $\alpha >> 1.0$ resembles the usual case of rigid cylinders in contact.

Geometry parameter $\alpha$ has a very significant influence on $\beta$ and $\xi$ both. Parameters $\beta$ and $\xi$ both increase with $\alpha$ during normal approach until it reaches a maximum value. Further increase in $\alpha$ does not change $\beta$ and $\xi$. Reverse trend is seen in case of normal separation.

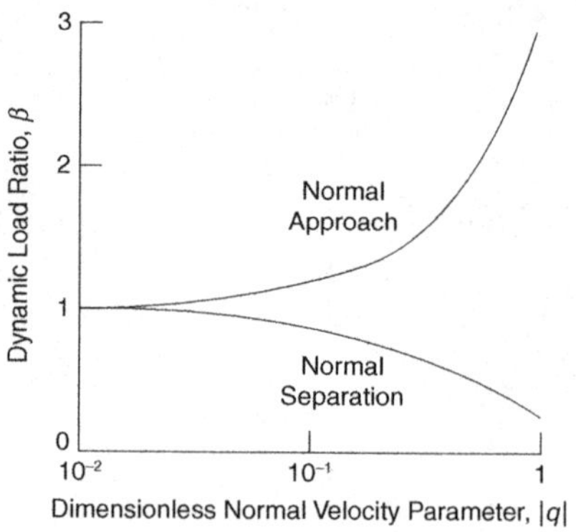


**Figure 12.10** | Variation of Dynamic Load Ratio with Dimensionless Normal Velocity Parameter. Dimensionless Central Film Thickness $H_0 = 1 \times 10^{-4}$; Dimensionless Geometry Parameter $\alpha = 1.0$; and Dimensionless Inlet Starvation Parameter $H_{in} = 0.035$ [Ghosh and Hamrock, ASME, 1985]

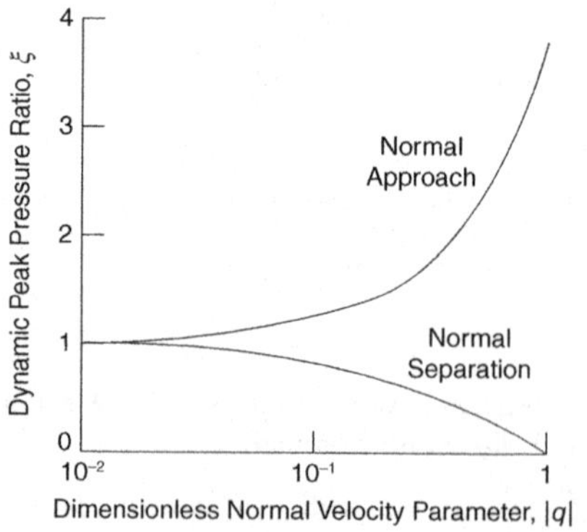


**Figure 12.11** | Variation of Dynamic Peak Pressure Ratio with Dimensionless Normal Velocity Parameter. Dimensionless Central Film Thickness $H_0 = 1 \times 10^{-4}$; Dimensionless Geometry Parameter $\alpha = 1.0$; and Dimensionless Inlet Starvation Parameter $H_{in} = 0.035$ [Ghosh and Hamrock, ASME, 1985]

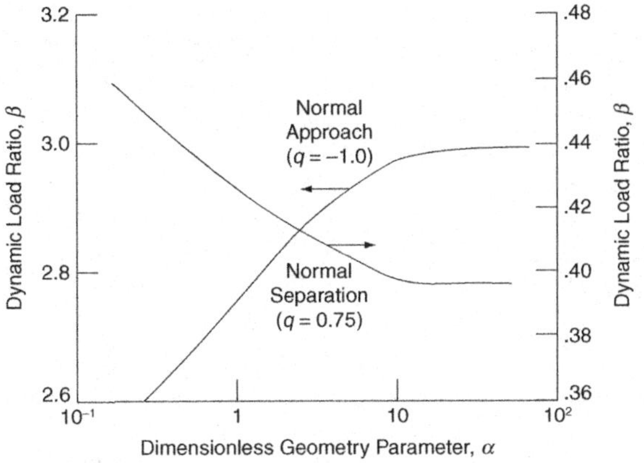


**Figure 12.12** | Variation of Dynamic Load Ratio with Dimensionless Geometry Parameter. Dimensionless Central Film Thickness $H_0 = 1 \times 10^{-4}$; and Dimensionless Inlet Starvation Parameter $H_{in} = 0.035$ [Ghosh and Hamrock, ASME, 1985]

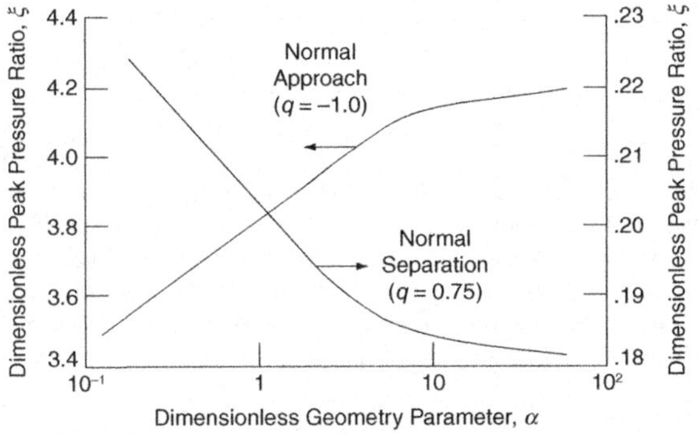


**Figure 12.13** | Variation of Dynamic Peak Pressure Ratio with Dimensionless Geometry Parameter. Dimensionless Central Film Thickness $H_0 = 1 \times 10^{-4}$; and Dimensionless Inlet Starvation Parameter $H_{in} = 0.035$ [Ghosh and Hamrock, ASME, 1985]

Ghosh *et al.* (1985) developed the following relationships for dynamic load ratio $\beta$ and peak pressure ratio $\xi$ as given below by curve fitting large number of data generated by computation.

$$\beta = \left\{\alpha^{-0.028} \sec h \; 1.68q\right\}^{1/q} \tag{12.61}$$

$$\xi = \left\{\alpha^{-0.032} \sec h(2q)\right\}^{1/q} \tag{12.62}$$

Influence of starvation on the dynamic load ration $\beta$ and peak pressure ratio $\xi$ are shown in Figs 12.13 and 12.14, respectively. It is observed that starvation significantly affects the dynamic load ratio and peak pressure ratio.

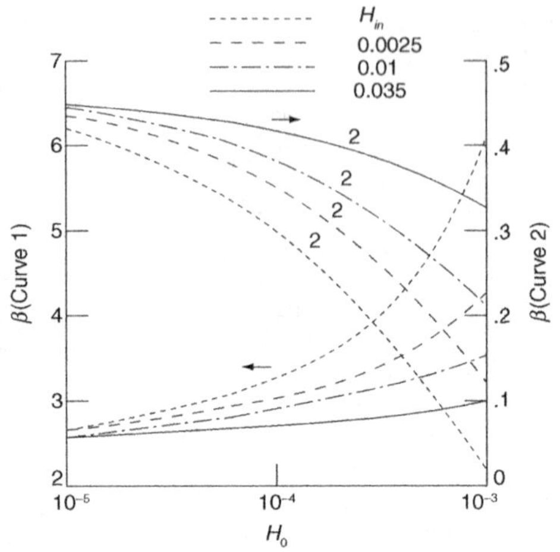


**Figure 12.14** | Variation of Dynamic Load Ratio, $\beta$, with Central Film Thickness, $H_0$ for Various Inlet Starvation Parameters, for Geometry Parameter, $\alpha = 1.0$. Curve 1 for Normal Approach, $q = -1.0$. Curve 2 for Normal Separation, $q = 0.75$. [Ghosh and Hamrock, ASLE, 1987]

## Problems

***P.12.1*** A 5 cm diameter steel roller 2.5 cm width is held between two parallel flat steel plates. The plates are considered infinitely wide and long enough to cover its ends. The top plate is held on the roller by a normal force 10 N and the oil viscosity at the contact surfaces is 0.35 Poise. Estimate the oil film thickness using Martin's theory between the roller and the top plate if it moves at 25 m/sec in positive $x$ direction while the lower plate is held stationary. Assume the surfaces to be rigid and viscosity of oil constant. Compare the magnitude of film thickness obtained using short roller approximation.

***P.12.2*** A wheel on a rail contact can be treated similar to a cylinder of radius 0.5 m rolling on a track with transverse radius of 0.3 m. The radial load on each wheel is 100 KN and the rolling speed is 50 m/sec. The contact between rail and wheel may be treated as rigid. If the rail track is covered with oil of viscosity 0.01 Pas, determine the minimum film thickness using rigid contact lubrication theory. However, if the track is covered with water which has a viscosity 0.001 Pas what will be the ratio of minimum film thickness compared to the oil on the track.

***P12.3*** Two rollers each of 50 mm length and 25 mm radius are rotating at an angular speed 30 rev/s. The viscosity of the lubricant used is 0.01 Ns/m$^2$. Calculate the minimum film thickness using long bearing theory when the load acting on the rollers is 3.0 KN. If the two rollers approach each other at a normal velocity, determine the time taken for the film thickness to reduce from 1 to 0.1 micro meter when the rollers are not rotating about their axes.

***P.12.4*** A cylindrical roller bearing has following specifications: Inner race diameter = 64 mm, outer race diameter = 96 mm, diameter of rollers = 16 mm, axial length of rollers = 16 mm, number of rollers = 9, inner race angular velocity = 524 rad/sec, viscosity of oil = 0.01 Ns/m$^2$, radial load on the bearing = 10.8 KN, outer race is fixed. Calculate the minimum film thickness at roller inner race and outer race using Martin's formula.

## REFERENCES

*Boness, R.J.* (1970), The effect of oil supply on cage and roller motion in a lubricated roller bearing, *ASME Journal of Lubrication Technology*, **92**, 237–246.

*Brewe, D.E. and Hamrock, B.J.* (1982), Analysis of starvation effects on hydrodynamic lubrication in nonconforming contacts, *ASME, Journal of Lubrication Technology*, **104**, 410–417.

*Brewe, D.E., Hamrock, B.J. and Taylor, C.M.* (1979), Effect of geometry on hydrodynamic film thickness, *ASME, Journal of Lubrication Technology*, **101**, 231–239.

*Chiu, Y.P.* (1974), A theory of hydrodynamic friction forces in starved point contacts considering cavitation, *ASME Journal of Lubrication Technology*, **96**, 237–246.

*Dowson, D. and Whomes, T.L.* (1967), Side leakage factors for a rigid cylinder lubricated by an isovisous fluid, *Lubrication and Wear, Fifth Convention, Proc. I. Mech. Eng*, U.K., **181**, 165–176.

*Dowson, D., Markho, P.H. and Jones, D.A.*, The lubrication of lightly loaded cylinders in combined rolling, sliding and normal motion—Part I—Theory, *ASME, Journal of Lubrication Technology*, **98**, 509–516.

*Ghosh, M.K., Hamrock, B.J. and Brewe, D.E.* (1987), Starvation effects on the hydrodynamic lubrication of rigid nonconformal contacts in combined rolling and normal motion, *ASLE Transactions*, **30**, 91–99.

*Ghosh, M.K., Hamrock, B.J. and Brewe, D.E.* (1985), Hydrodynamic lubrication of rigid nonconformal contacts in combined rolling and normal motion, *ASME, Journal of Tribology*, **107**, 97–103.

*Kapitza, P.L.* (1955), Hydrodynamic theory of lubrication during rolling, *Zh. Tekg. Fiz.*, **25**, 747–762.

*Markho, P.H. and Dowson, D.* (1976), The lubrication of lightly loaded cylinders in combined rolling, sliding and normal motion—Part II—Experiment, *ASME, Journal of Lubrication Tech.*, **98**, 517–523.

*Martin, H.M.* (1916), Lubrication of gear teeth, *Engineering* (London), **102**, 119–121.

*Sasaki, T., Mori, H. and Okino, N.* (1962), Fluid lubrication theory of roller bearing, Part I—Fluid lubrication theory for two rotating cylinders in contact, *ASME, Journal of Basic Engineering*, Vol. 84, N1, 166–174; Part II—Fluid lubrication theory applied to roller bearing, *Journal of Basic Engineering, ASME*, **84**, 175–180.

*Vichard, J.P.* (1971), Transient effects in lubrication of hertzian contacts, *Journal. of Mech. Engineering, Science*, **13**, 173–189.

*Wedeven, L.D., Evans, D. and Cameron, A.* (1971), Optical analysis of ball bearing starvation, *ASME Journal of Lubrication Technology*, **93**, 349–363

Chapter 13

# Elastohydrodynamic Lubrication

## 13.1 | Introduction

Machine components such as rolling-element bearings, gears, cams have nonconformal contacts. Under heavy load condition, the pressure generated within the contact zone is much higher, generally in the order of giga Pascals (GPa). The surfaces which carry this load are likely to deform. Due to hydrodynamic action, the lubricant is drawn into the contact zone. When the lubricant is exposed to such a high pressure, the viscosity of the lubricant is increased exponentially. It is due to this increase in viscosity of lubricant and elastic deformation of contact surfaces that the components are able to carry the load and perform satisfactorily. This type of lubrication is called elastohydrodynamic lubrication (EHL).

Probably, the first work on the theory of EHL started with Martin (1916). The formula of load capacity derived by Martin really does not consider the effect of film thickness. It is interesting to notice that it took about half a century before Ertel (1939) and Gurbin (1949) combined both effects of elastic deformation and hydrodynamic action into what is known as elastohydrodynamic lubrication (EHL). Since then, however, considerable progress has been made in theory as well as in experiment.

Petrusevich (1951) was the first to present numerical solution that satisfied both the Reynolds equation and the equation describing the elastic deformation. Many researchers like Weber and Saalfeld (1954), and then by Archard *et al.* (1961) using detailed analysis solved EHL line contact problems. Dowson and Higginson (1959) presented numerical solutions for a wide range of the parameters involved and combined these solutions to the first film thickness formula. Archard and Cowking (1965) studied EHL point contacts. Cameron (1952) and Archard and Kirk (1963) carried out experiments on EHL contacts.

Up to the mid-1970s elliptical contacts in ball bearings were treated as circular contacts or equivalent line contacts to avoid computational time and cost. But afterwards availability of faster computers made the way for researchers to work on actual elliptical contacts. A number of techniques have been proposed to solve the set of equations describing EHL contacts. For instance, iteration is performed by means of Gauss–Seidel iteration or by Newton–Raphson algorithms, to solve the Reynolds and the elasticity equation simultaneously.

During this period, Hertnett (1979) presented a novel method for the estimation of elastic deformation profiles, computationally effective to implement. Hamrock and Dowson (1976 a,b; 1977a,b) presented a comprehensive range of numerical solution for fully flooded point contacts in which the influence of ellipticity ratio was considered. The empirical expressions presented in their work are widely used in the design and analysis of machine element representing elliptical, lubricated conjunctions. Similar work was reported by Chittenden *et al.* (1985). An inverse method, which includes the method of Dowson and Higginson (1959), has been proposed. This method was applied to circular contacts by Evans and Snidle (1981). With the introduction of multilevel methods in lubrication by Lubrecht (1987) and with the later developments by Venner (1991), the efficiency and the stability of the numerical methods improved dramatically. Based on numerical calculations, function fit formula for the prediction of film thickness were proposed by Moes, Nijienbanning, and Venner (1992; 1994), using a minimum number of similarity parameters and validation for a wide range of operating conditions.

In the late 1950s, Crook (1957) made a disc machine and experiments were performed. The film thickness was measured by capacitance and oil flow methods. Martin's theory was verified by these experiments for isoviscous fluid. Crook found that the influence of surface temperature of disc was one of the important considerations in determining the film thickness. Crook also investigated the rolling friction under EHL and concluded that rolling friction was independent of load and proportional to film thickness. Sibley and Orcut (1961) gave a detailed experiment investigation based on x-ray transmission technique. The film thickness measured by them was in agreement with Crook's experimental and theoretical results. Cameron and coworkers (1966) were the pioneers of optical interferometry in lubrication.

EHL analysis is generally carried out with line and point contacts. The machine elements like cylindrical roller bearings, spur gears are the cases of line contact, whereas ball bearings are the example of point contact. In the following section, EHL line contact and point contact analysis will be presented.

## 13.2 | Line Contact Analysis

In case of the line contact problem, the contacting elements are assumed to be infinitely long in one direction. In fact, the radius of curvature of the paraboloid approximating the surface in this direction is infinitely large. In the unloaded dry contact situation, the surfaces touch along a straight line. If a load is applied a strip-shaped contact region is formed because of the elastic deformation. Figure 13.1(a) shows the approximation of the line contact situation. $R_1$ and $R_2$ are the radii of solid 1 and 2, respectively. Figure 13.1(b) shows the reduced geometry of the contact. The reduced radius of curvature $R$ is given by

$$\frac{1}{R}=\frac{1}{R_1}+\frac{1}{R_2} \tag{13.1}$$

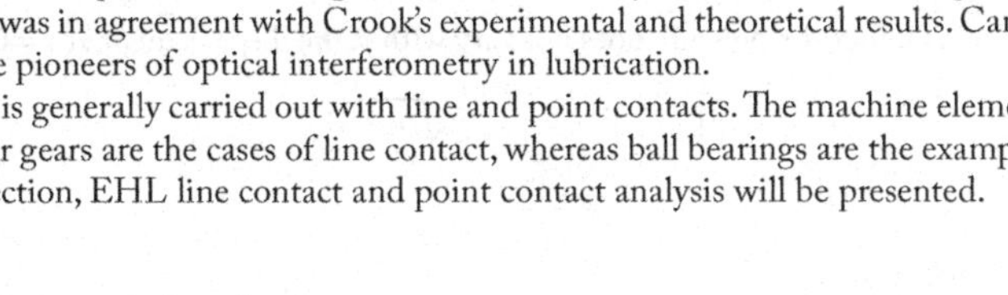


**Figure 13.1** | The EHL Line Contact and Reduced Geometry

### 13.2.1 | Elastic Deformation

Consider deflection $V$ at a point $B$ due to application of a point load $W$ at $A$ as shown in Fig. 13.2, where the distance between $B$ and $A$ is $l$. This deflection is

$$V = \frac{1-\nu^2}{\pi E}\frac{W}{l} \tag{13.2}$$

where $\nu$ = Poisson's ratio and $E$ = Young's modulus.

When the point load is distributed over a small area, then $W$ can be replaced by $pdrds$. Equation (13.2) can, thus be written as

$$V = \frac{1-\nu^2}{\pi E}\iint \frac{pdrds}{l} \tag{13.3}$$

If two long cylinders are in contact, the pressure distribution occurs over a rectangular area of sides $x = \pm b$ and $y = \pm a$. The deformation at a point $B$ due to a load $pdrds$ can be written as

$$V = \frac{1-\nu^2}{\pi E}\int_{-a}^{a}\int_{-b}^{b} \frac{pdsdr}{l} \tag{13.4}$$

For long (infinite) cylinder the pressure does not vary with $y$. the displacement at $y = 0$ can be considered,

$$l = \left((x-r)^2 + s^2\right)^{1/2} = \sqrt{\rho^2 + s^2}$$

where $\rho^2 = (x-r)^2$

Let us take the integral

$$\int_{-a}^{a} \frac{pdr}{\sqrt{\rho^2+s^2}} = 2\int_{0}^{a} \frac{pdr}{\sqrt{\rho^2+s^2}}$$

$$= 2p\left|\ln\left(s+\sqrt{\rho^2+s^2}\right)\right|_0^a$$

$$= 2p\ln\left(\frac{a+\sqrt{\rho^2+a^2}}{\rho}\right)$$

If $a$ is very large as compared to $\rho$, $\rho^2$ can be neglected with respect to $a^2$, giving

$$\int_{-a}^{a} \frac{pdr}{\sqrt{\rho^2+s^2}} = 2p\ln(2a) - 2p\ln(\rho)$$

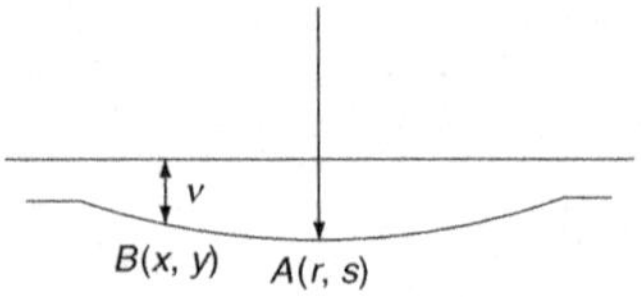


**Figure 13.2** | Deflection due to Point Load

Thus, Equation (13.4) can be written as

$$V = \frac{1-\nu^2}{\pi E} 2\int_{-b}^{b} \left(p\ln(2a) - 2p\ln(\rho)\right)ds$$

$$= \frac{1-\nu^2}{\pi E}\left(\frac{2W}{L}\ln(2a) - 2\int_{-b}^{b} p\ln(\rho)ds\right) \tag{13.5}$$

where $L$ is the length of the contact.

As $a \rightarrow \infty$, the first term in the bracket tends to infinity and constant deformation. This constant is ignored and the constant term vanishes, as the difference of the deflection is needed. Hence the deflection is,

$$V = -\frac{2(1-\nu^2)}{\pi E}\int_{-b}^{b} p\ln|x-s|ds \tag{13.6}$$

If two surfaces of $\nu_1, E_1$ and $\nu_2, E_2$ are considered, the deformation $V$ can be written as

$$V = -\frac{4}{\pi E'}\int_{-s}^{s} p\ln|x-s|ds \tag{13.7}$$

where

$$\frac{1}{E'} = \frac{1}{2}\left(\frac{1-\nu_1^2}{E_1} + \frac{1-\nu_2^2}{E_2}\right) \tag{13.8}$$

From Equation (13.7) it is evident that one cannot find the integration by a straightforward numerical integration, because at $x = s$ the integral goes to infinity. Various methods have been used by researchers to overcome this difficulty. Some of the methods are discussed below.

Dowson and Higginson (1959) calculated the deformation $V$ by expressing the pressure as a function which permits integration, so that the singularity is avoided. It is assumed that the pressure distribution can be represented by a polynomial,

$$p = p_0 \sum_{n=0,1.} A_n s^n \tag{13.9}$$

Let $\int_{s_1}^{s_2} s^n \ln(x-s)ds$ be $f_n$, then

$$f_n = \frac{1}{n+1}\left\{\begin{matrix}\left(s_2^{n+1} - x^{n+1}\right)\ln|x-s_2| - \left(s_1^{n+1} - x^{n+1}\right)\ln|x-s_1| - \frac{1}{n+1}\left(s_2^{n+1} - s_1^{n+1}\right) \\ -\frac{1}{n}x\left(s_2^n - s_1^n\right)............-\frac{1}{2}x^{n-1}\left(s_2^2 - s_1^2\right) - x^n\left(s_2 - s_1\right)\end{matrix}\right\} \tag{13.10}$$

The expression for deformation is

$$V = -\frac{4}{\pi E'}\int_{s_1}^{s_2} p\ln(x-s)\,ds$$

$$= \sum_{n=0,1..} A_n f_n \tag{13.11}$$

If the first three terms of the series are used, it gives fairly accurate results.

Cameron (1966) calculated the elastic deformation in the following way. If the pressure distribution is considered to be of the same form as that of dry contact, the pressure inside the contact zone will be

$$p = p_{\max}\left(1-\left(\frac{x}{b}\right)^2\right)^{1/2} \tag{13.12}$$

where $$p_{\max} = \frac{2W}{\pi Lb}$$

and $b$ is the half the length of Hertzain contact, which is given by the expression,

$$b = \left(\frac{8WR}{\pi E'L}\right)^{1/2} \tag{13.13}$$

Under this condition, the singularity at $x = s$ has been avoided by Cameron (1966), and the deformation within the contact zone was found as

$$V = -\frac{2}{E'}p_{\max}b\left(\frac{x^2}{b^2}+\ln\left(\frac{b}{2}\right)-\frac{1}{2}\right) \tag{13.14}$$

Majumdar *et al.* (1981), has evaluated the deformation $V$ in the following way. The pressure curve is divided into small mesh size $(\Delta = 0.0004 \text{ with } R = 1)$ and treating the pressure as constant in this interval, the deformation equation is integrated numerically between $s_1$ and $s_2$ excluding the point $x = s$. The value of $V$ at $x = s$ is taken as the average deflection of the two neighboring points. The accuracy of this method has been verified by comparing deflection at the central point for Hertzian pressure distribution. The agreement is within 1%.

Venner (1991) calculated the deformation $V$ by approximating the pressure profile by a piecewise constant function with value $p_j = p(x_j)$ in the region $x_j - h/2 \le x' \le x_j + h/2$ on a uniform grid with mesh size $h$. To avoid the singularity, the integration is performed as

$$V(x_i) = -\frac{4p_j}{\pi E'}\int_{x_j-h/2}^{x_j+h/2}\ln(x_i - s)\,ds$$

$$= \frac{4}{\pi E'}\sum_{j=1}^{n} p_j K_{ij} \tag{13.15}$$

where $K_{ij}$ is defined by,

$$K_{ij} = (x_i - x_j - h/2)\left(\ln\left|x_i - x_j - h/2\right| - 1\right) - (x_i - x_j + h/2)\left(\ln\left|x_i - x_j + h/2\right| - 1\right) \tag{13.16}$$

This method is convenient in implementing multilevel multi-integration technique.

### 13.2.2 | Hydrodynamic Equation

The governing Reynolds equation for the steady state isothermal EHL line contact for smooth surfaces with incompressible Newtonian lubricant, can be written as,

$$\frac{\partial}{\partial x}\left(h^3\frac{\partial p}{\partial x}\right) = 12\eta\frac{(u_1+u_2)}{2}\frac{\partial h}{\partial x} \tag{13.17}$$

where $u = u_1 + u_2$, $u_1$ and $u_2$ are velocities of two surfaces.

The pressure inside the contact zone is likely to be of considerable magnitude. Thus, the viscosity in this case cannot be treated as constant. The variation of viscosity with pressure is given by Barus (1893) relation,

$$\eta = \eta_0 \exp(\alpha p) \tag{13.18}$$

where $\eta$ = viscosity at pressure $p$

$\eta_0$ = viscosity at atmospheric pressure

and $\alpha$ = pressure viscosity coefficient

Substituting Equation (13.18) in Equation (13.17), we get

$$\frac{\partial}{\partial x}\left(h^3 \frac{\partial q}{\partial x}\right) = 6\eta_0 u \frac{\partial h}{\partial x} \tag{13.19}$$

where $q$ is called as modified pressure which is given as

$$q = \frac{1 - \exp(-\alpha p)}{\alpha} \tag{13.20}$$

### 13.2.3 | Film Thickness Equation

The equation for film thickness reads to be,

$$h(x) = h_0 + S(x) + V(x) - V(0)$$

where $h_0$ = central film thickness

$S(x)$ = separation due to the geometry of solids which is approximated to be parabolic

$V(x)$ = elastic deformation

$V(0)$ = elastic deformation at the center of contact

Therefore, for the cylindrical contacts

$$h(x) = h_0 + \frac{x^2}{2R} + V(x) - V(0) \tag{13.21}$$

where $R$ is the reduced radius of curvature.

### 13.2.4 | Force Balance Equation

The entire contact load exerted on the contacting elements is carried by the lubricant film. Hence, the integral of pressure in the film over the contact domain must be equal to the applied load. This condition is generally referred as the force balance equation and in the line contact problem it reads,

$$\int_{s_1}^{s_2} p\,dx = W \tag{13.22}$$

where $W$ = external load per unit length.

The solution for the pressure $p$ and film thickness $h$ must simultaneously satisfy Equations (13.19), (13.21), and (13.22). Furthermore, the solution is subjected to cavitation condition. A full analytical close form solution is not available up to now. But many of the researchers have solved the problem numerically. Some of the solution methods are presented in the following section.

### 13.2.5 | Grubin Type Solution

The EHL line contact problem has been solved using a method originally given by Grubin (1949). He made the assumption that the deformed shape within the contact zone is constant. He also made another simplifying assumption that the shape outside the Hertzain zone is the same, irrespective of the lubricant.

From Equation (13.19), the integrated form of modified Reynolds equation will be

$$\frac{dq}{dx} = 6\eta_0 u \frac{h - h_0}{h^3} \tag{13.23}$$

$h = h_0 + h_s$, Equation (13.23) can be written as

$$\frac{dq}{dx} = 6\eta_0 u \frac{h_s}{\left(h_0 + h_s\right)^3} \tag{13.24}$$

The film thickness outside the contact zone is given by

$$h = h_0 + h_s \tag{13.25}$$

$$h_s = V + \frac{x^2}{2R}$$

$$= V + \frac{b^2}{2R}\frac{x^2}{b^2} \tag{13.26}$$

Using Equation (13.13) and (13.14), we get

$$h_s = \frac{2}{E'} p_{max} b \left( \frac{x}{b}\sqrt{\frac{x^2}{b^2} - 1} - \ln\left( \frac{x}{b} + \sqrt{\frac{x^2}{b^2} - 1} \right) \right) \tag{13.27}$$

Putting $p_{max} = \frac{2W}{\pi L b}$, Equation (13.27) can be written as

$$\frac{h_s \pi E'}{W/L} = 4\left( \frac{x}{b}\sqrt{\frac{x^2}{b^2} - 1} - \ln\left( \frac{x}{b} + \sqrt{\frac{x^2}{b^2} - 1} \right) \right) \tag{13.28}$$

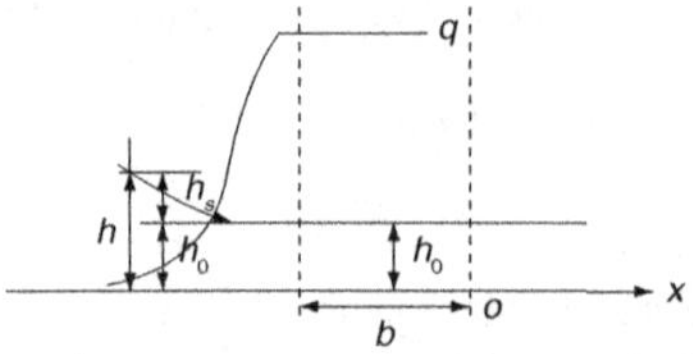


**Figure 13.3** | Modified Pressure and Film Thickness Curves

If we define $\bar{h} = \dfrac{h\pi E'}{W/L}$, $\bar{h}_0 = \dfrac{h_0 \pi E'}{W/L}$ and $\bar{h}_s = \dfrac{h_s \pi E'}{W/L}$

$$\bar{h} = \bar{h}_0 + \bar{h}_s = \frac{\pi E'}{W/L}\left(h_0 + h_s\right) \tag{13.29}$$

Equation (13.24) now can be written as

$$\frac{dq}{dx} = 6\eta_0 u \frac{h_s}{\left(h_0 + h_s\right)^3}$$

or

$$\frac{dq}{dx} = \frac{6\eta_0 u}{\left(\dfrac{W}{L\pi E'}\right)} \frac{\bar{h}_s}{\bar{h}^3} \tag{13.30}$$

putting $\bar{q} = \dfrac{\left(W/L\pi E'\right)^3}{6\eta_0 ub} q$, Equation (13.30) is nondimensionalized as

$$\frac{d\bar{q}}{d\bar{x}} = \frac{\bar{h}_s}{\bar{h}^3} \tag{13.31}$$

To find $\bar{q}$ one has to integrate Equation (13.31) numerically between limits $x = -\infty$ to $x = -b$ or $\bar{x} = -\infty$ to $\bar{x} = -1$. The integral will be

$$\bar{q} = \int_{-\infty}^{-1} \bar{h}_s \frac{d\bar{x}}{\bar{h}^3} \tag{13.32}$$

When $\bar{q}$ is solved numerically, it can be transformed to $p$ by using Equation (13.20).

From the above equation it will be shown that at $x/b = -1$, the pressure $\bar{q}$ can be approximated by

$$\bar{q} = 0.0986\bar{h}_0^{-11/8} \tag{13.33}$$

again at

$$\frac{x}{b} = -1, q = \frac{1}{\alpha}$$

or

$$q = \frac{6\eta_0 ub}{\left(W/L\pi E'\right)^2}\bar{q} = \frac{1}{\alpha}$$

or

$$h_0^{11/8} = 1.183\eta_0 u\alpha \frac{R^{1/2}}{\left(\dfrac{W}{L\pi E'}\right)^{1/8}} \tag{13.34}$$

If both surfaces are of same material, Equation (13.34) can be expressed in the following form

$$\frac{h_0}{R} = 1.183 \frac{\left(\eta_0 u\alpha / R\right)^{8/11}}{\left(\left(1-\nu^2\right)\dfrac{W}{L\pi E'R}\right)^{1/11}} \tag{13.35}$$

The above expression gives the film thickness equation for a line contact problem. Because of simplifying assumption, this $h_0$ is constant throughout the contact zone. This method of finding elastohydrodynamic film thickness is simple, but this does not give a correct value at the exit of the contact zone.

### 13.2.6 | Accurate Solution

Many researchers like Weber and Saalfeld (1954), Dowson and Higginson (1959), and Archard (1961) using detailed analysis solved EHL line contact problems. Dowson and Higginson (1959) give a single expression for minimum film thickness as

$$\frac{h_{min}}{R} = \frac{(\alpha E')^{0.6} (\eta_0 u / E'R)^{0.7}}{\left(\frac{W}{LE'R}\right)^{0.13}} \tag{13.36}$$

The Dowson–Higginson formula for minimum film thickness is based on numerical solution of Equation (13.23) using a different approach for finding deformation as mentioned in Section 13.2.1. However, their analysis has made two assumptions, which may be revealed from the pressure distribution shown in Fig. 13.4. They used central pressure as the maximum Hertzian pressure and during the process of calculation; this pressure was kept at a constant value. The other assumption that the pressure at $x = b$ was put equal to zero. This is really not so at the exit end. Majumdar *et al.* (1981) has made a similar analysis by removing those two assumptions. The two plots given in Fig. 13.4 show the difference of the pressure distribution and location of peak pressures.

A sudden peak pressure near the outlet is a characteristic of EHL for nonconformal contact. In the contact zone, the pressure is nearly Hertzain and the film thickness is almost parallel. The gap diverges rapidly at $x = b$. For matching flow continuity, the pressure should end nearly at this point. Prior to this, there is a large negative pressure gradient causing a marked increase of oil flow. The continuity of flow can be maintained if there is a local restriction of flow. Due to this reason, there is a sharp decrease in film thickness at the outlet. The EHL film thickness thus is dependent on material parameters $\nu$ and $E$, the speed of roller, the composite radius of curvature, and the magnitude of applied load.

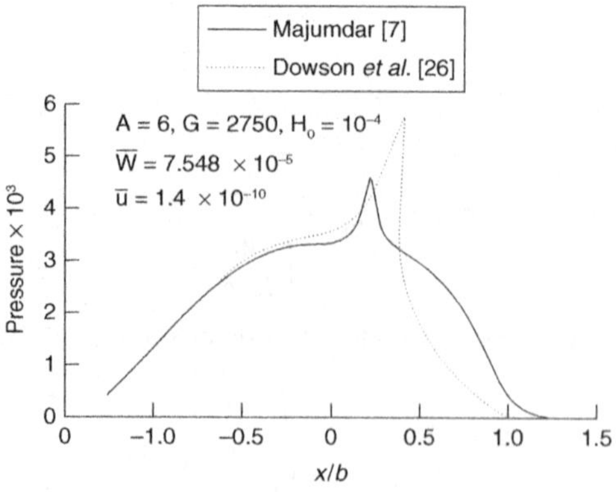


**Figure 13.4** | Comparision of Pressure Distribution

EHL line contact problem is solved by using the relaxation process in accordance to Venner (1991). From Equations (13.17) and (13.18) the integrated form of Reynolds equation in nondimensional form will be

$$\frac{d}{d\bar{x}}\left(\xi\frac{d\bar{p}}{d\bar{x}}\right)=\frac{d\bar{h}}{d\bar{x}} \tag{13.37}$$

where $\bar{x}=x/R, \bar{p}=p/E', \bar{h}=h/R, \xi=\frac{\bar{h}^3}{12\bar{U}\exp(G\bar{p})}$,

$$G=\alpha E', \bar{U}=\eta_0 u/(E'R), u=\frac{u_1+u_2}{2}$$

From Equation (13.21), film thickness in nondimensional form is given as

$$\bar{h}=\bar{h}_0+\frac{\bar{x}^2}{2}+\bar{V}-\bar{V}(0) \tag{13.38}$$

where $\bar{V}=V/R$

From Equation (13.22), the force balance equation in nondimensional form is written as

$$\int \bar{p}d\bar{x}=\bar{W} \tag{13.39}$$

where $\bar{W}=W/(E'RL)$

The boundary conditions are $\bar{p}(\bar{x}_{in})=0, \bar{p}(\bar{x}_{out})=0$

$$\bar{p}(\bar{x})\geq 0 \quad (\bar{x}_{in}<\bar{x}<\bar{x}_{out})$$

The elastic deformation $V$ is calculated as per Equation (13.15). The calculation domain for numerical analysis is taken as $\bar{x}_{in}=-4.5\bar{b}$, $\bar{x}_{out}=1.5\bar{b}$, where $\bar{b}=b/R$. The domain is divided into 305 nodes. Hertzain pressure distribution is taken as initial guess values of the pressure. Using the finite difference method, a steady-state pressure distribution is obtained by simultaneous solution of the pressure governing Equation (13.37), the film thickness Equation (13.38), and the force balance Equation (13.39), satisfying the boundary conditions. To satisfy the force balance equation, the constant $\bar{h}_0$ is adjusted in the following way,

$$\bar{h}_0=\bar{h}_0\left(1+C\left(\int \bar{p}d\bar{x}-\bar{W}\right)/\int \bar{p}d\bar{x}\right) \tag{13.39a}$$

where $C$ is the suitably chosen constant.

An under-relaxation factor varying from 0.5 to 1 is used for Gauss–Seidel relaxation changes and similarly an under-relaxation factor varying from 0.3 to 0.6 is used for the Jacobi dipole changes. The results of central film thickness and minimum film thickness are in close agreement with that of the central film thickness and central minimum film thickness values provided by Dowson and Toyoda (1977). From the results obtained by the authors using the analysis of Venner (1991), the effect of nondimensional parameters, load ($\bar{W}$), speed ($\bar{U}$), and material parameter ($G$), on pressure and film thickness are shown in Figs 13.5–13.7.

Figure 13.5 shows that an increase in load will produce a decrease in central and minimum film thickness. Pressure spike is seen at the exit end of the contact for moderate load, whereas for higher load, the pressure

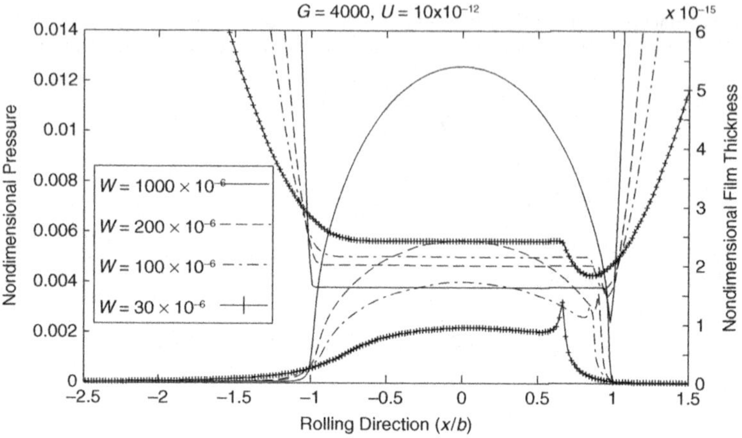


**Figure 13.5** | Effect of Load Parameter on Pressure and Film Thickness (Line Contact)

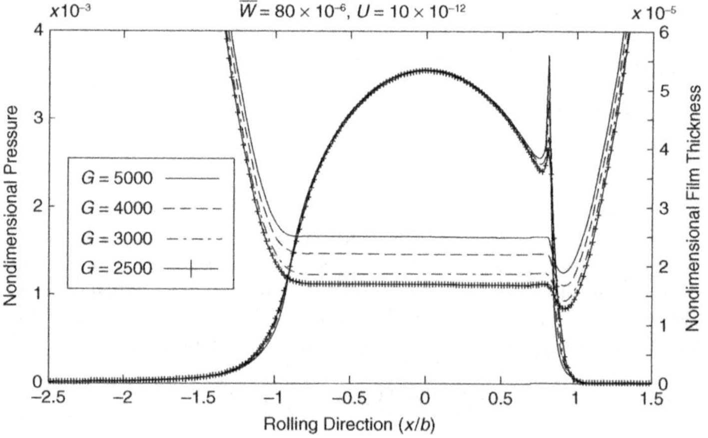


**Figure 13.6** | Effect of Material Parameter (G) on Pressure and Film Thickness (Line Contact)

profile approaches to Hertzian distribution is observed. Figure 13.6 indicates the effect of material parameter $G$ on pressure profile and film thickness. With the increase in material parameter $G$, film thickness increases. However, there is no much change in pressure profile. Figure 13.7 shows the effect of speed parameter on pressure and film thickness. With increase in the speed parameter, the pressure profile departs from Hertzain distribution. Central and minimum film thickness increases with increase in speed parameter.

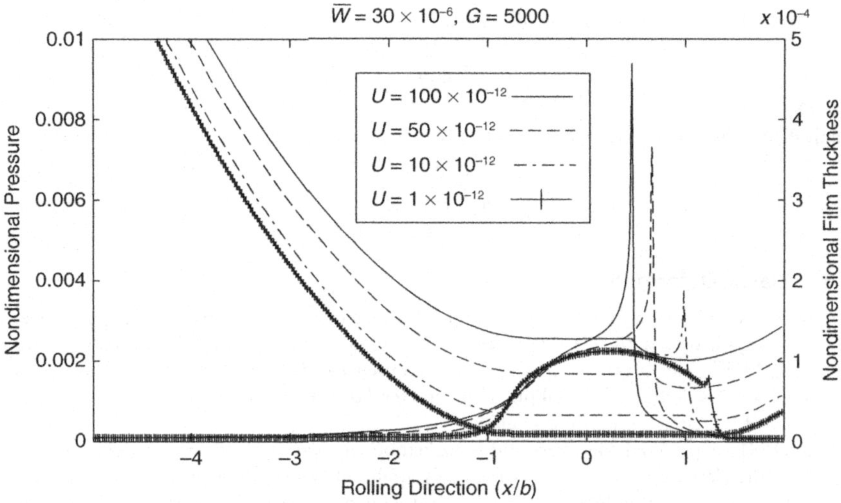


**Figure 13.7** | Effect of Speed Parameter ($U$) on Pressure and Film Thickness (Line Contact)

## 13.3 | Point Contact Analyses

The generalized undeformed geometry of contacting solids can be represented by two ellipsoids. The two solids with different radii of curvature in principal planes $x$ and $y$ make contact at a single point, owing to the concept of 'point contact' as shown in Fig. 13.8. The point contact spreads to form a circular or elliptical contact due to deformation of mating surfaces under normal load depending on radii of curvature. The radii of curvature are denoted by $r$'s.

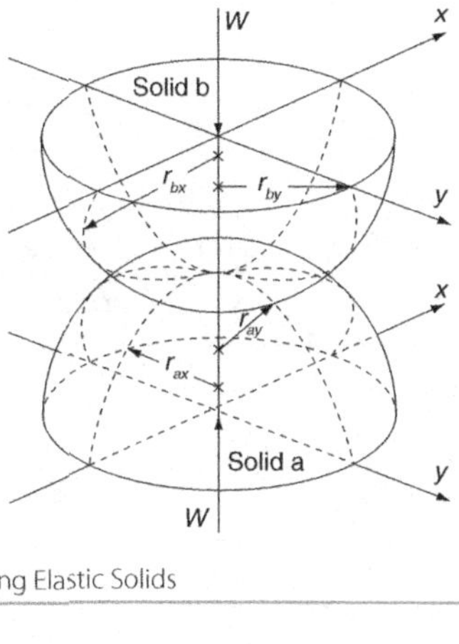


**Figure 13.8** | Geometry of Contacting Elastic Solids

Curvatures in $x$ and $y$ directions are defined as

$$\frac{1}{R_x}=\frac{1}{r_{ax}}+\frac{1}{r_{bx}} \quad \text{and} \quad \frac{1}{R_y}=\frac{1}{r_{ay}}+\frac{1}{r_{by}} \tag{13.40}$$

and equivalent radius of curvature of the contact is

$$\frac{1}{R}=\frac{1}{R_x}+\frac{1}{R_y} \tag{13.41}$$

### 13.3.1 | Elastic Deformation

When two elastic solids are brought together under a load, a contact area forms, the shape and size of which depend on the applied load, the elastic properties of materials, and the curvatures of the surfaces. This contact area (Fig. 13.9) typically attains the shape of an ellipse with $a$ being the semimajor and $b$ the semiminor lengths and $k = a/b$ is defined as the ellipticity parameter. For the special case where $r_{ax} = r_{ay}$ and $r_{bx} = r_{by}$, the resulting contact is a circle rather than an ellipse.

Hertz (1881) considered the stress and deformations in two perfectly smooth, ellipsoidal, contacting elastic solids. The classical theory of Hertz is equally applicable for machine elements such as ball and roller bearings, gears, and cams. The theory is based on the following assumptions:

- The materials are homogeneous and yield stress is not exceeded.
- No tangential forces are induced between the solids.
- Contact is limited to a small portion of the surface such that the dimensions of the contact region are small compared with the radii of the ellipsoids.
- The solids are at rest and in equilibrium.

According to Hertz, the pressure within the ellipsoidal contact is (Fig. 13.9)

$$p = p_{\max}\left[1-\left(\frac{x}{b}\right)^2-\left(\frac{y}{a}\right)^2\right]^{1/2} \tag{13.42}$$

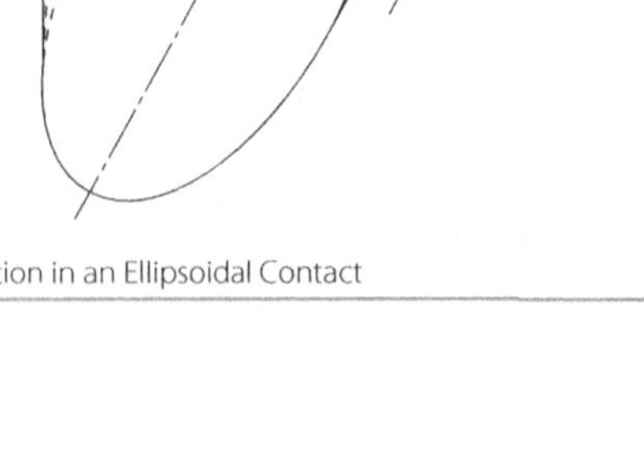


**Figure 13.9** | Pressure Distribution in an Ellipsoidal Contact

where $p_{max}$ is referred as the maximum pressure inside the contact envelope and given by

$$p_{max} = \frac{3F}{2\pi ab} \tag{13.43}$$

Figure 13.9 shows the distribution of pressure or compressive stress on the common interface, it is clearly a maximum at the center of the contact and decreases to zero at the periphery. When the ellipticity parameter $k$, the normal applied load $F$, Poisson's ratios $\nu$, and the moduli of elasticity $E$ of the contacting solids are known, the parameters $a$, $b$ , and $\delta$, the deformation at the center of the contact can be written from the analysis of Hertz as

$$a = \left(\frac{6k^2\zeta FR}{\pi E'}\right)^{1/3} \tag{13.44}$$

$$b = \left(\frac{6\zeta FR}{\pi kE'}\right)^{1/3} \tag{13.45}$$

$$\delta = \Im\left[\left(\frac{9}{2\zeta R}\right)\left(\frac{F}{\pi kE'}\right)^2\right]^{1/3} \tag{13.46}$$

where $\Im$ and $\zeta$ are elliptic integrals of first and second kinds, respectively, and given by

$$\Im = \int_0^{\pi/2}\left[1-\left(1-\frac{1}{k^2}\right)\sin^2\varphi\right]^{-1/2} d\varphi \tag{13.47}$$

$$\zeta = \int_0^{\pi/2}\left[1-\left(1-\frac{1}{k^2}\right)\sin^2\varphi\right]^{1/2} d\varphi \tag{13.48}$$

and

$$E' = \frac{2}{\left(1-\nu_a^2\right)/E_a + \left(1-\nu_b^2\right)/E_b}$$

Brewe and Hamrock (1977) used a linear regression analysis by the method of least squares to obtain simplified equations for $k$, $\Im$, and $\zeta$ as

$$k = 1.0339\left(\frac{R_y}{R_x}\right)^{0.6360} \tag{13.49}$$

$$\Im = 1.5277 + 0.6023\ \ln\left(R_y/R_x\right) \tag{13.50}$$

$$\zeta = 1.0003 + \frac{0.5968}{R_y/R_x} \tag{13.51}$$

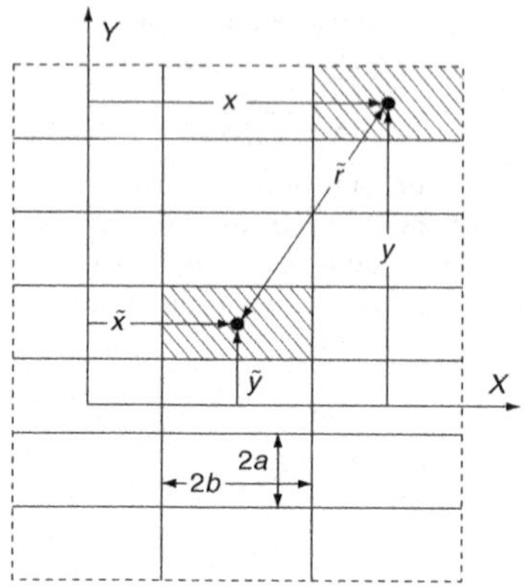


**Figure 13.10** | Discretization of the Contact Region

Although these simplified relations give a quick estimation of deformation at the center of the contact, the same over the contact envelope cannot be found. This problem can be solved making use of the formulation given by Timoshenko and Goodier (1951). Figure 13.10 shows a rectangular area of uniform pressure with the coordinate system to be used. The elastic deformation at a point (*x*, *y*) of a semi-infinite solid subjected to uniform pressure *p* at the point ($\tilde{x}, \tilde{y}$) can be written as

$$dV(x, y) = (1 - \nu^2)\frac{p(\tilde{x}, \tilde{y})\, d\tilde{x}\, d\tilde{y}}{\pi E \tilde{r}} \tag{13.52}$$

Then the combined elastic deformation of the two mating ellipsoidals at a point (*x*, *y*) due to the uniform pressure over the entire contact envelope is thus

$$V(x, y) = \frac{2}{\pi E'}\int_A \frac{p(\tilde{x}, \tilde{y})\, d\tilde{x}\, d\tilde{y}}{\left\{(x - \tilde{x})^2 + (y - \tilde{y})^2\right\}^{1/2}} \tag{13.53}$$

Equation (13.53) possesses singularity at $x = \tilde{x}$ and $z = \tilde{z}$. One has to avoid this singularity to obtain elastic deformation. While the numerical solution of this equation is possible following a flexibility method of solution given by Hertnett (1979). The integral can be evaluated over the contact envelope as shown in Fig. 13.8 after dividing to small segments of area $2\tilde{a} \times 2\tilde{b}$ and assuming constant pressure over each small segment. That makes the Equation (13.53) as,

$$V_i = \frac{2}{\pi E'}\sum_{j=1}^{n} p_j f_{ij} \quad \text{for } i, j = 1, 2, \ldots\ldots n = \text{total no. of elements.} \tag{13.54}$$

where $p_j$ is the distributed pressure over segment *j* and $f_{ij}$ is the influence coefficient representing the deflection of segment *i* because of uniform pressure over *j*. The influence coefficient is given by

$$f_{ij} = (\tilde{x}+\tilde{b})\ln\left(\frac{(\tilde{y}+\tilde{a})+\sqrt{(\tilde{y}+\tilde{a})^2+(\tilde{x}+\tilde{b})^2}}{(\tilde{y}-\tilde{a})+\sqrt{(\tilde{y}-\tilde{a})^2+(x+\tilde{b})^2}}\right)+(\tilde{y}+\tilde{a})\ln\left(\frac{(\tilde{x}+\tilde{b})+\sqrt{(\tilde{y}+\tilde{a})^2+(\tilde{x}+\tilde{b})^2}}{(\tilde{x}-\tilde{b})+\sqrt{(\tilde{y}+\tilde{a})^2+(\tilde{x}-\tilde{b})^2}}\right)$$

$$+(\tilde{x}-\tilde{b})\ln\left(\frac{(\tilde{y}-\tilde{a})+\sqrt{(\tilde{y}-\tilde{a})^2+(\tilde{x}-\tilde{b})^2}}{(\tilde{y}+\tilde{a})+\sqrt{(\tilde{y}+\tilde{a})^2+(\tilde{x}-\tilde{b})^2}}\right)+(\tilde{y}-\tilde{a})\ln\left(\frac{(\tilde{x}-\tilde{b})+\sqrt{(\tilde{y}-\tilde{a})^2+(\tilde{x}-\tilde{b})^2}}{(\tilde{x}+\tilde{b})+\sqrt{(\tilde{y}-\tilde{a})^2+(\tilde{x}+\tilde{b})^2}}\right)$$

Use of this flexibility method facilitates to avoid the aforesaid singularity in the solution of the integral Equation (13.53). From Equation (13.54) deformation over the contact envelope can be found using the influence coefficient. Influence coefficient $f_{ij}$ is a geometrical property and can be generated in advance once the domain for the solution is set.

### 13.3.2 | Hydrodynamic Equation

The governing Reynolds equation for the steady-state isothermal EHL elliptical contact for smooth surfaces with incompressible Newtonian lubricant can be written as,

$$\frac{\partial}{\partial x}\left(\frac{h^3}{12\eta u}\frac{\partial p}{\partial x}\right)+\frac{\partial}{\partial y}\left(\frac{h^3}{12\eta u}\frac{\partial p}{\partial y}\right)=\frac{\partial h}{\partial x} \tag{13.55}$$

In nondimensional form, Equation (13.55) is written as,

$$\frac{\partial}{\partial \bar{x}}\left(\bar{h}^3\frac{\partial \bar{p}}{\partial \bar{x}}\right)+\frac{\partial}{\partial \bar{y}}\left(\bar{h}^3\frac{\partial \bar{p}}{\partial \bar{y}}\right)=\frac{\partial \bar{h}}{\partial \bar{x}} \tag{13.56}$$

where $\bar{x}=\frac{x}{R_x}, \bar{y}=\frac{y}{R_x}, \bar{h}=\frac{h}{R_x}, \bar{p}=\frac{p}{E'}$

$$\bar{\eta}=\frac{\eta}{\eta_0}, U=\frac{\eta_0 u}{E'R_x}, u=\frac{u_1+u_2}{2}$$

Considering the effect of pressure on viscosity by using Equation (13.18), Reynolds equation is written as

$$\frac{\partial}{\partial \bar{x}}\left(\bar{h}^3\frac{\partial \bar{q}}{\partial \bar{x}}\right)+\frac{\partial}{\partial \bar{y}}\left(\bar{h}^3\frac{\partial \bar{q}}{\partial \bar{y}}\right)=\frac{\partial \bar{h}}{\partial \bar{x}} \tag{13.57}$$

where $\bar{q}$ is the modified reduced pressure as defined by Equation (13.20)

### 13.3.3 | Film Thickness Equation

The equation for film thickness reads to be,

$$h(x,y)=h_0+S(x,y)+V(x,y)-V_0$$

where $h_0$ = central film thickness

$S(x, y)$ = separation due to the geometry of solids

$V(x, y)$ = elastic deformation

$V_0$ = elastic deformation at the center of contact

Therefore,

$$h(x,y) = h_0 + \frac{x^2}{2R_x} + \frac{y^2}{2R_y} + V(x,y) - V_0 \tag{13.58}$$

In nondimensional form, the film thickness equation is written as,

$$\bar{h} = \bar{h}_0 + \frac{\bar{x}^2}{2} + \frac{\bar{y}^2}{2(R_y/R_x)} + \bar{V} - \bar{V}_0 \tag{13.59}$$

where $\bar{h} = h/R_x$, $\bar{V} = V/R_x$

## 13.3.4 | Force Balance Equation

The force balance equation is given as

$$\iint p\,dx\,dy = W \tag{13.60}$$

where $W$ is the external applied load.

In the nondimensional form, Equation (13.60) is given as

$$\iint \bar{p}\,d\bar{x}\,d\bar{y} = \bar{W} \tag{13.61}$$

where,
$$\bar{W} = W/(E'R_x^2)$$

The boundary conditions used are at the edges of the rectangular envelop of computation, the pressure is zero at the cavitation boundary, $\bar{p} = \frac{\partial \bar{p}}{\partial \bar{x}} = \frac{\partial \bar{p}}{\partial \bar{y}} = 0$.

## 13.3.5 | Numerical Solution

Using the finite difference method, a steady-state pressure distribution is obtained by simultaneous solution of the pressure governing Equation (13.57), film thickness Equation (13.59), and force balance Equation (13.61), by satisfying the boundary condition. To satisfy the force balance equation, the constant $\bar{h}_0$ is adjusted in the following way,

$$\bar{h}_0 = \bar{h}_0\left(1 + C\left(\iint \bar{p}\,d\bar{x}\,d\bar{y} - \bar{W}\right) / \iint \bar{p}\,d\bar{x}\,d\bar{y}\right) \tag{13.61a}$$

where $C$ is the suitably chosen constant. The calculation domain is considered as $x_{in} = -4\bar{b}$, $x_{out} = 1.65\bar{b}$, $y_{in} = -1.75\bar{a}$, and $y_{out} = 1.75\bar{a}$, where $\bar{a} = a/R_x$, $\bar{b} = b/R_x$. A mesh size of 69 points in rolling direction and 33 points in transverse direction is used in the numerical solution. Nonuniform mesh (finer in the contact zone) is used in the computation method. Mesh sizes of $\Delta x/b = 0.055$ and $\Delta y/a = 0.081$ are used in the contact area, while rest of the portion uses mesh sizes of $\Delta x/b = 0.165$ and $\Delta y/a = 0.245$. The adopted convergence criteria for pressure distribution $\bar{p}$ and desired input load $\bar{W}$ are $10^{-5}$ and $10^{-3}$, respectively.

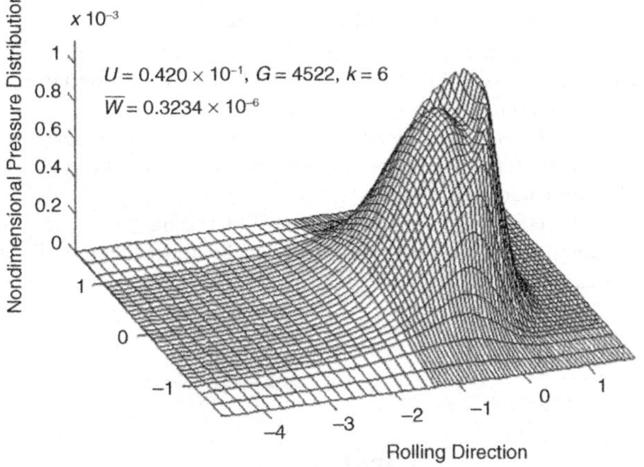


**Figure 13.11** | Steady State EHL Pressure Distributin ($U = 0.4208 \times 10^{-11}$, $G = 4522$, $k = 6$, $\overline{W} = 0.3234 \times 10^{-6}$)

Figure 13.11 shows the nondimensional pressure distribution as obtained under steady-state condition. The nondimensional minimum film thickness ($0.11878 \times 10^{-4}$) obtained by the authors is close to that of Hamrock and Dowson (1981) empirical relation ($0.12072 \times 10^{-4}$).

$$\overline{h}_{\text{central}} = 2.69U^{0.67}G^{0.53}W^{-0.067}\left(1 - 0.61e^{-0.73k}\right) \tag{13.62}$$

$$\overline{h}_{\text{minimum}} = 3.63U^{0.68}G^{0.49}\overline{W}^{-0.073}\left(1 - e^{-0.68k}\right) \tag{13.63}$$

Equations (13.62) and (13.63) are the empirical relations for central film thickness and minimum film thickness given by Hamrock and Dowson (1981).

EHL point contact problem is solved by the authors using the technique proposed by Venner (1991), as discussed in the previous section of line contact solution.

From Equations (13.55) and (13.18), Reynolds equation in nondimensional form will be

$$\frac{\partial}{\partial \overline{x}}\left(\xi \frac{\partial \overline{p}}{\partial \overline{x}}\right) + \frac{\partial}{\partial \overline{y}}\left(\xi \frac{\partial \overline{p}}{\partial \overline{y}}\right) = \frac{\partial}{\partial \overline{x}}(\overline{h}) \tag{13.64}$$

where

$$\xi = \frac{\overline{h}^3}{12U\exp(G\overline{p})}$$

Using the multigrid method, a steady-state pressure distribution is obtained by simultaneous solution of the pressure governing Equation (13.64) and the film thickness Equation (13.59) and force balance Equation (13.61) by satisfying the boundary conditions. Elastic deformation Equation (13.54) is solved by multilevel multi-integration method as proposed by Brandt and Lubrecht (1990). A Jacobi distributive line relaxation is used in the contact region. Whereas in the noncontact zone, a simple Gauss–Seidel line relaxation method is used. The calculating domain is considered as $x_{in} = -4.5\overline{b}$, $x_{out} = 1.5\overline{b}$, $y_{in} = -2\overline{a}$, and $y_{out} = 2\overline{a}$. The numbers of nodes in $x$-direction is 513 and in $y$-direction is 513. The adopted convergence criteria for pressure distribution $\bar{p}$ and desired input load $\bar{W}$ are $10^{-5}$ and $10^{-3}$, respectively.

Figure 13.12 shows the nondimensional pressure distribution as obtained under steady-state condition. The results of central film thickness and minimum film thickness are in close agreement with Hamrock and Dowson (1981) empirical relation.

From the results obtained by this analysis, the effect of nondimensional parameters, load ($\bar{W}$), speed ($U$), ellipticity parameter ($k$), and material parameter ($G$), on pressure and film thickness are shown in Figs. 13.13–13.16, in the rolling direction along the midplane.

Figure 13.13 shows the effect of load parameter on film thickness and pressure profile. Peak pressure increases with load because the lubricant has to carry more load with smaller film thickness. Central and minimum film thickness decreases with increase in load parameter. Pressure spike is visible at the exit of the contact

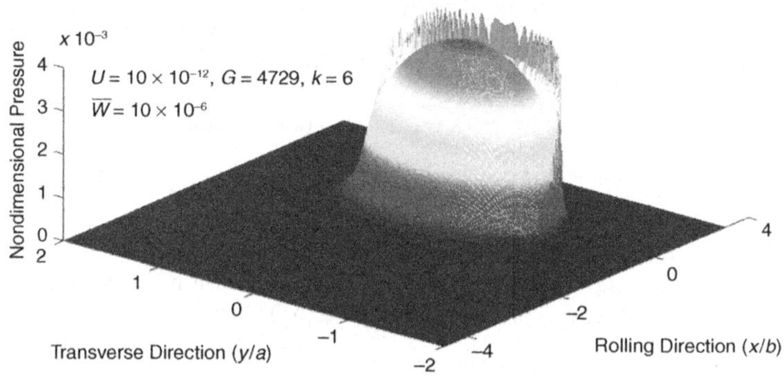


**Figure 13.12** | Steady State EHL Pressure Distribution

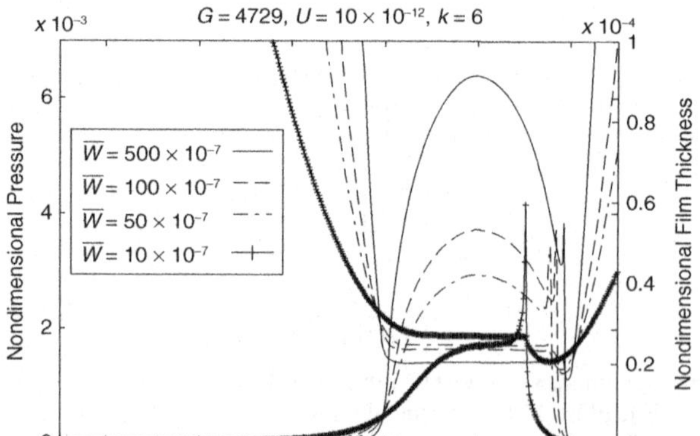


**Figure 13.13** | Effect of Load Parameter on Pressure and Film Thickness (Rolling Direction)

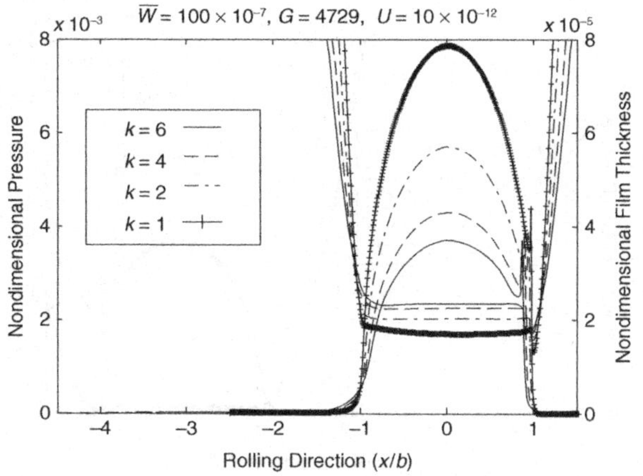


**Figure 13.14** | Effect of Elliptical Parameter (*k*) on Pressure and Film Thickness (Rolling Direction)

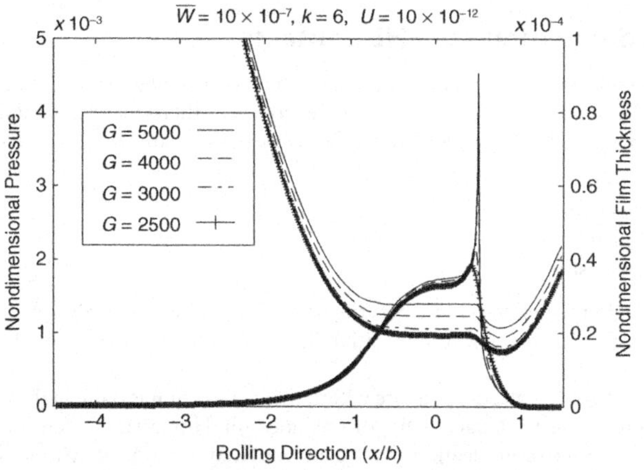


**Figure 13.15** | Effect of Material Parameter (*G*) on Pressure and Film Thickness (Rolling Direction)

for moderate loads. However, the pressure profile approaches the Hertzain distribution for higher values of load parameter. The variation of pressure and film thickness with ellipticity parameter is shown in Fig. 13.14. Peak pressure increases with decrease in ellipticity parameter, whereas the central and minimum film thickness increases. Figure 13.15 show the effect of material parameter on film thickness and pressure profile. Central and minimum film thickness increases with increase in material parameter. Figure 13.16 show the effect of speed parameter on pressure and film thickness. For higher values of speed parameter, the pressure profile departs from the Hertzain distribution. Central and minimum film thickness increases with increase in speed parameter.

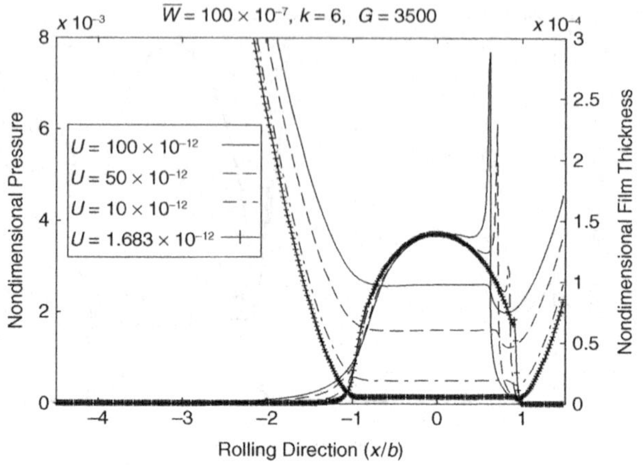


**Figure 13.16** | Effect of Speed Parameter ($U$) on Pressure and Film Thickness (Rolling Direction)

## 13.4 | Different Regimes in EHL Contacts

The lubrication of concentrated contact is normally influenced by two major physical effects such as elastic deformation and the increase in fluid viscosity with pressure. There may be several regimes of lubrication depending on the magnitude of the above two effects. The regimes are

- isoviscous-rigid,
- piezoviscous-rigid,
- isoviscous-elastic, and
- piezoviscous-elastic.

In isoviscous-rigid regime, the elastic deformation of the solids can safely be neglected because it is very insignificant in magnitude. This type of lubrication is seen to occur in circular arc thrust bearing pads and in industrial coating processes.

In piezoviscous-rigid regime, the pressure within the conjunction is quite high and it may be necessary to consider the pressure-viscosity characteristics of the lubricant. However, the deformation within the contact zone is neglected. This type of lubrication may be encountered in moderately loaded cylindrical taper roller, and between the piston rings and cylinder liners.

In isoviscous-elastic regime, the elastic deformation of the surfaces is adequate to warrant inclusion of elastic equation along with hydrodynamic equation. This form of lubrication can be seen in the materials of low elastic modulus or seals and human joints.

In heavily loaded concentrated contacts, the elastic deformation of solids as well as the variation of viscosity with pressure must be considered. The lubrication regime which considers both these two aspects is called piezoviscous regime. This type of lubrication is typically encountered in ball and roller bearings, gears, and cams.

The various approximate equation for film thickness of the above four regimes have been found by Moes (1965–1966), Theyse (1966), Archard (1968), Greenwood (1969), Johnson (1970), and Hooke (1977). This can be listed as follows:

***For Isoviscous-rigid Regime***

$$\left(\hat{h}_{\min}\right)_{IR} = 128\alpha_a \lambda_b^2 \left(0.131\tan^{-1}\left(\frac{\alpha_a}{2}\right) + 1.683\right)^2 \tag{13.65}$$

where $$\alpha_a = \frac{R_y}{R_x} \cong \left(\frac{k}{1.03}\right)^{1/0.64}$$

and $$\lambda_b = \left(1 + \frac{2}{3\alpha_a}\right)^{-1}$$

***For Piezoviscous-rigid Regime***

$$\left(\hat{h}_{\min}\right)_{PVR} = 1.66 g_V^{2/3}\left(1 - e^{-0.68k}\right) \tag{13.66}$$

***For Isoviscous-elastic Regime***

$$\left(\hat{h}_{\min}\right)_{IE} = 8.70 g_E^{0.67}\left(1 - 0.85 e^{-0.31k}\right) \tag{13.67}$$

***For Piezoviscous-elastic Regime***

$$\left(\hat{h}_{\min}\right)_{PVE} = 3.42 g_V^{0.49} g_E^{0.17}\left(1 - e^{-0.68k}\right) \tag{13.68}$$

The dimensionless film thickness parameter $\hat{h}$, viscosity parameter $g_V$ and elasticity parameter $g_E$, respectively, given by Archard (1968) is defined as

$$\hat{h} = \bar{h}\left(\frac{\bar{W}}{U}\right)^2, \quad g_V = \frac{G\bar{W}^3}{U^2} \quad \text{and} \quad g_E = \frac{\bar{W}^{8/3}}{U^2}$$

For the first two cases (i.e., for isoviscous-rigid regime and piezoviscous-rigid regime) the central film thickness is equal to the minimum film thickness. However, for the last two cases, the central film thickness is given by

$$\left(\hat{h}_c\right)_{IE} = 11.15 g_E^{0.67}\left(1 - 0.72 e^{-0.28k}\right) \tag{13.69}$$

and $$\left(\hat{h}_c\right)_{PVE} = 3.61 g_V^{0.53} g_E^{0.13}\left(1 - 0.61 e^{-0.73k}\right) \tag{13.70}$$

The procedure for mapping different lubrication regimes is given by Hamrock and Dowson (1981), and the maps of lubrication regimes for various ellipticity parameters can also be found in this reference.

## 13.5 | Mixed Lubrication

The concept of mixed lubrication is relatively new in the field of elastohydrodynamic lubrication. Under this condition, film thickness in the contact region is of the order of few microns. This does not ensure pure elastohydrodynamic lubrication. Apart from the lubricant film, a part of the load is shared by asperities under direct contact between two contacting surfaces, when the surfaces are rough. Hence, surface roughness should be modeled accurately to describe the asperity contact area and contact load along with hydrodynamic load.

A typical stochastic analysis for rough surfaces was developed by Zhu and Cheng (1988). They employed Patir and Cheng's average flow model (1978) for hydrodynamic lubrication and Greenwood and Tripp's load compliance relation (1970–1971) for asperity contacts. Hydrodynamic and contact pressure were separately obtained and then simply superimposed to balance the applied load. Later on, mixed lubrication model by Jiang *et al.* (1999) represented a significant advancement in the area of mixed lubrication for the first time solving hydrodynamic and asperity contact pressure simultaneously. In their analysis, three-dimensional rough surface profile is used and the Reynolds equation is solved with the multigrid scheme. Effect of rough surface topography and orientation on the characteristic of EHD and mixed lubrication were investigated by Zhu and Hu (2001). Few developments considering surface topography in the field of mixed lubrication have been reported by Qiu and Cheng (1998) and Wang *et al.* (2004).

In another approach to the development of mixed lubrication model, Zhu (2002) has published a series of papers presenting a novel method to handle mixed lubrication theory. The theory deals with an extended parameter range. According to the theory, when load is very high, film thickness reduces to support the load, theoretically it will approach zero at contact. Then hydrodynamic pressure in the contact is absent and pressure differential in the Reynolds equation can be neglected, this facilitates to solve only remaining film thickness differential terms. Contradiction to the theory was reported by Holmes *et al.* (2003), who showed that some of the results could not be validated with the experimental values.

In the following section, Patir and Cheng's average flow model (1978) is used to consider the surface roughness and asperity contact using Greenwood and Tripp's load compliance relation (1970–1971).

### 13.5.1 | Theory of Lubricated Rough Surfaces

Although the derivation of modified Reynolds equation considering surface roughness effect is available elsewhere, Patir and Cheng (1978, 1979), it is summarized here for completeness.

### 13.5.2 | Film Thickness Equation

The local film thickness $h_T$ between any two rough surfaces as shown in Fig. 13.17 will be of the form

$$h_T = h + \delta_a + \delta_b \tag{13.71}$$

where $h$ is the nominal film thickness (compliance) and is equal to the distance between the mean levels of the two surfaces. $\delta_a$ and $\delta_b$ are the random roughness amplitudes of the two surfaces measured from their mean levels. It is assumed that $\delta_a$ and $\delta_b$ have a Gaussian distribution of heights with zero mean and standard deviations $\sigma_a$ and $\sigma_b$, respectively. The combined roughness $\delta = \delta_a + \delta_b$ has a variance of $\sigma^2 = \sigma_a^2 + \sigma_b^2$. The average gap $h_{Tav}$ is defined as:

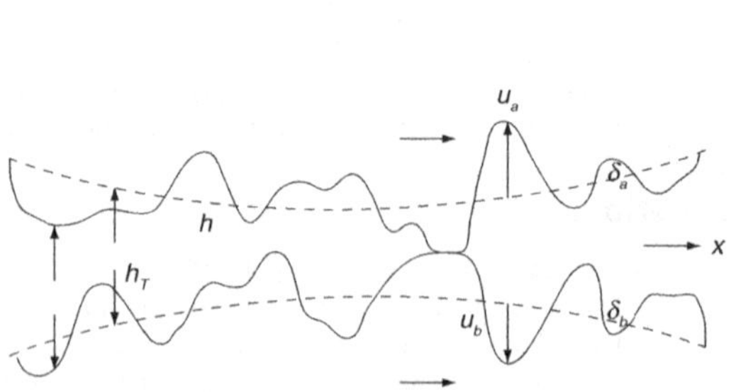


**Figure 13.17** | Film Thickness Function

$$h_{Tav} = E(h_T)$$

$$h_{Tav} = \int_{-\infty}^{\infty} (h+\delta) f(\delta) d\delta \tag{13.72}$$

where $E$ is the expectancy operator and $f(\delta)$ the probability density function of $\delta$. In the case of noncontact $h_{Tav} = h$ since $E(\delta_a) = E(\delta_b) = 0$. Now $h_{Tav}$ can be written as:

$$h_{Tav} = \int_{-h}^{\infty} (h+\delta) f(\delta) d\delta \tag{13.73}$$

since $h_T = 0$ at the contact points.

### 13.5.3 | Surface with Directional Patterns

Most engineering surfaces have directional patterns resulting from different manufacturing processes or because of running-in. These directional patterns are mostly in the longitudinal or transverse directions. The directional properties of roughness are described by a surface pattern parameter $\gamma$, first introduced by Peklenik (1967–1968). If $\lambda_{0.5}$ is the length at which the autocorrelation function of a profile reduces to 50% of its initial value, the surface pattern parameter $\gamma$ is defined as the ratio of $\lambda_{0.5}$ lengths of $x$ and $z$ profiles, *i.e.*,

$$\gamma = \frac{\lambda_{0.5x}}{\lambda_{0.5z}} \tag{13.74}$$

$\gamma$ can be visualized as the length-to-width ratio of a representative asperity. Purely transverse, isotropic, and longitudinal roughness patterns correspond to $\gamma = 0, 1$ and $\infty$, respectively, as shown in Fig. 13.18.

### 13.5.4 | Hydrodynamic Equation

Hydrodynamic pressure and the mean film thickness are obtained from the numerical solution of hydrodynamic equation. The governing hydrodynamic equation for rough surfaces for steady-state condition by using Patir and Cheng model (1978) can be written as

$$\frac{\partial}{\partial x}\left(\phi_x \frac{h^3}{12\eta}\frac{\partial p}{\partial x}\right) + \frac{\partial}{\partial y}\left(\phi_y \frac{h^3}{12\eta}\frac{\partial p}{\partial y}\right) = \frac{U_a + U_b}{2}\frac{\partial h_T}{\partial x} + \frac{U_a - U_b}{2}\sigma\frac{\partial \phi_s}{\partial x} \tag{13.75}$$

with film thickness

$$h = h_c - V_c + \frac{x^2}{2R_x} + \frac{y^2}{2R_y} + V$$

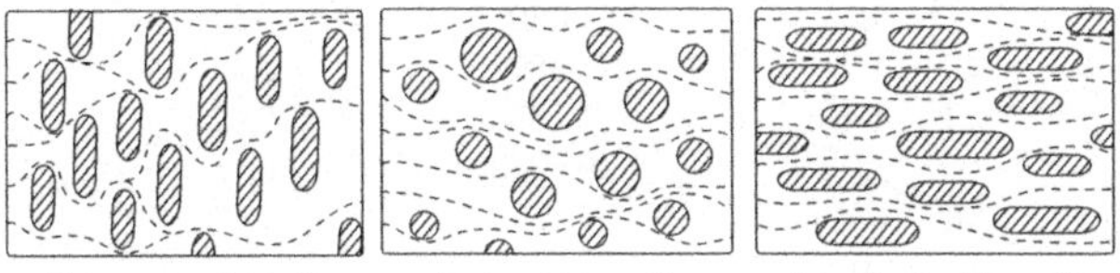


**Figure 13.18** | Surfaces with Directional Patterns

where $\phi_x$, $\phi_y$ = pressure flow factors
$\phi_s$ = shear flow factor
$\sigma$ = combined standard deviation of rough surfaces
and $h_T$ is the average gap height, given by

$$h_T = \int_{-h}^{\infty} (h+\delta) f(\delta) d\delta$$

$f(\varepsilon) = \frac{1}{\sigma\sqrt{2\pi}} e^{-\frac{\delta^2}{2\sigma^2}}$ is the probability density function of combined roughness $\delta$.

After performing integration and differentiating $h_T$ with respect to $x$ or $t$, we get

$$\frac{\partial h_T}{\partial(x)} = \frac{1}{2}\left[1+\operatorname{erf}\left(\frac{h}{\sqrt{2}\sigma}\right)\right]\frac{\partial h}{\partial(x)}, \quad \text{where} \quad \operatorname{erf}(x) = \frac{2}{\sqrt{\pi}}\int_0^x e^{-t^2} dt \tag{13.76}$$

The expressions of flow factors given by Patir and Cheng (1978) are

$$\phi_x = 1 - c_1 e^{-g(h/\sigma)} \quad \text{for} \quad \gamma \le 1$$

$$\phi_x = 1 + c_1 \left(\frac{h}{\sigma}\right)^{-g} \quad \text{for } \gamma > 1 \tag{13.77}$$

and

$$\phi_x\left(\frac{h}{\sigma}, \gamma\right) = \phi_y\left(\frac{h}{\sigma}, \frac{1}{\gamma}\right)$$

where $c_1$ and $g$ are constants and $\gamma$ is defined as the ratio of length at which autocorrelation functions of the $x$ and $y$ profiles reduce to 50% of the initial value. This can be thought of as the length-to-width ratio of a representative asperity. As per definition, transverse, isotropic, and longitudinal roughness patterns correspond to $\gamma < 1, \gamma = 1$, and $\gamma > 1$, respectively. The constants $c_1$ and $g$ are given in Table 13.1 for different values of $\gamma$.

Depending on the type of lubricant, a suitable viscosity variation function can be assumed. For piezoviscous lubricant, the following exponential pressure viscosity relationship is used.

$$\eta = \eta_0 e^{\alpha p}. \tag{13.78}$$

**Table 13.1** | Coefficients of Pressure Flow Factors

| $\gamma$ | $cw_1$ | g |
|---|---|---|
| 1/9 | 1.480 | 0.42 |
| 1/6 | 1.380 | 0.42 |
| 1/3 | 1.180 | 0.42 |
| 1 | 0.900 | 0.56 |
| 3 | 0.225 | 1.50 |
| 6 | 0.520 | 1.50 |
| 9 | 0.870 | 1.50 |

Then pressure can be represented with viscosity variation as

$$q = \frac{1 - e^{-\alpha p}}{\alpha}. \tag{13.79}$$

For the case of pure rolling condition, shear contribution term, second in the R.H.S. of Equation (13.75) vanishes. Making use of Equations (13.75–13.79) and considering pure rolling, the hydrodynamic governing equation can be written in nondimensional form as

$$\frac{\partial}{\partial \bar{x}}\left(\phi_x \bar{h}^3 \frac{\partial \bar{q}}{\partial \bar{x}}\right) + \frac{\partial}{\partial \bar{y}}\left(\phi_y \bar{h}^3 \frac{\partial \bar{q}}{\partial \bar{y}}\right) = 6\bar{U}\left[1 + \operatorname{erf}\left(\frac{\Lambda}{\sqrt{2}} \frac{\bar{h}}{\bar{h}_c}\right)\right]\frac{\partial \bar{h}}{\partial \bar{x}} \tag{13.80}$$

with film thickness

$$\bar{h} = \bar{h}_c - \bar{V}_c + \frac{\bar{x}^2}{2} + \frac{\bar{y}^2}{2\left(R_y/R_x\right)} + \bar{V} \tag{13.81}$$

where $\Lambda = h_c/\sigma$ quantifies the severity of roughness. It can be seen from the expressions for $\phi_x$ and $h_T$ that when $\Lambda$ approaches a large value, $\phi_x$ approaches 1 and $\partial h_T/\partial x$ approaches $\partial h/\partial x$. Then Equation (13.80) becomes classical two-dimensional Reynolds equation used for smooth surfaces. Generally, $\Lambda$ varies from '1' (rough surface) to '6' (smooth surface). Now the solution of Equation (13.80) with appropriate boundary conditions will give the pressure distribution and film thickness in the contact zone.

### 13.5.5 | Asperity Contact Pressure

Although the force–compliance relation has been investigated by many researchers, a unified theory applicable to all roughness configurations is not available. The asperity contact pressure is usually small fraction of the total pressure in the region $\Lambda > 0.5$. This facilitates one to use an approximate relationship. The asperity pressure distribution can be calculated using Greenwood and Tripp model (1970–1971) as given by:

$$p_a = K'E'F_{5/2}(\lambda) \tag{13.82}$$

where $$K' = \frac{8}{15}\sqrt{2}\pi\left(N\rho_a\sigma\right)\sqrt{\sigma/\rho_a}$$

N = number of asperities per unit area

$\rho_a$ = mean radius of curvature of asperities

$$F_{5/2}(\lambda) = \frac{1}{\sqrt{2\pi}}\int_{\lambda}^{\infty}(s-\lambda)^{5/2} e^{\left(-s^2/2\right)} ds$$

The value of $K'$ given by Majumdar and Hamrock (1981) varies from 0.003 to 0.0003 for the range of $\sigma/\rho_a$ between 0.01 and 0.0001. The nondimensional contact pressure can be written as

$$\bar{p}_a = K'\frac{1}{\sqrt{2\pi}}\int_A\left\{\int_{\Lambda}^{\infty}\left(\bar{s} - \Lambda\frac{\bar{h}}{\bar{h}_c}\right)^{5/2} e^{\left(-\bar{s}^2/2\right)} d\bar{s}\right\} d\bar{x}\, d\bar{y} \tag{13.83}$$

The boundary conditions used are at the edges of the rectangular envelop of computation, the pressure is zero at the cavitation boundary, $\bar{p} = \frac{\partial \bar{p}}{\partial \bar{x}} = \frac{\partial \bar{p}}{\partial \bar{y}} = 0$ .

### 13.5.6 | Numerical Solution and Results

The steady-state pressure distribution is obtained by simultaneous solution of elastic deformation Equation (13.54), lubricant film pressure governing Equation (13.82), film thickness Equation (13.81), and asperity contact pressure Equation (13.84) satisfying the boundary conditions and using finite-difference method with successive relaxation scheme (Equation (13.85)). The numbers of mesh points used in the rolling and transverse directions are 69 and 33, respectively. Nonuniform mesh size (finer in the contact zone) is used

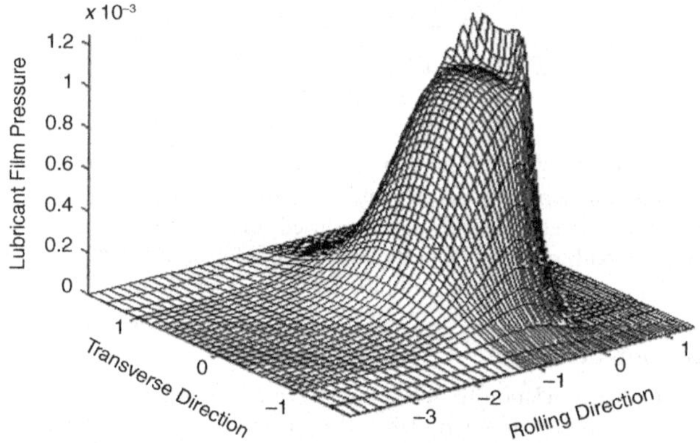


(a) Nondimensional lubricant pressure distribution

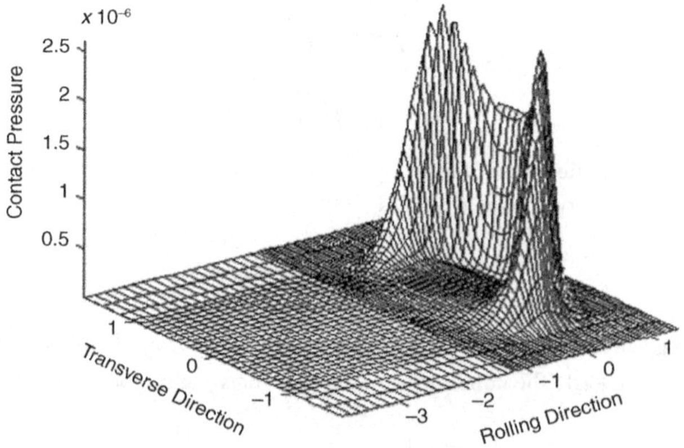


(b) Nondimensional contact pressure distribution

**Figure 13.19** | Steady-State EHD Mixed Lubricated Pressure Distributions

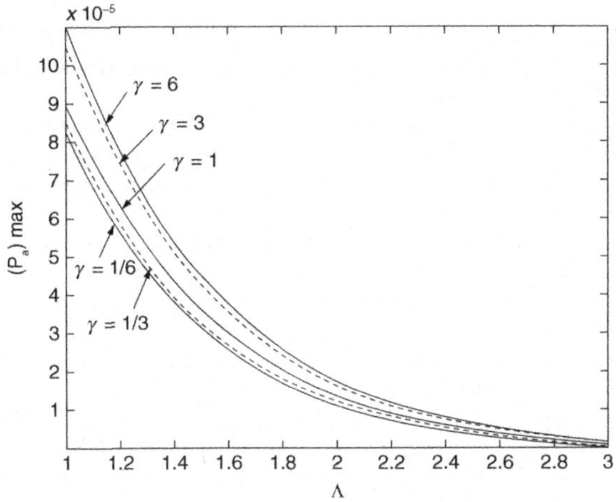


**Figure 13.20** | Maximum Contact Pressure Variation with Surface Roughness Parameters

in the computation method. The convergence criteria for pressure distribution $\bar{q}$ and desired input load $\bar{W}_i$ adopted are $10^{-5}$ and $10^{-3}$, respectively. A direct solution method is employed providing relaxation for the change of central film thickness based on the convergence of desired input load.

$$\begin{aligned} \bar{q}_{new} &= \bar{q}_{old} + orfq \cdot \left(\bar{q}_{new} - \bar{q}_{old}\right) && \text{with, } orfq = 1.2 \text{ to } 1.5 \\ \bar{h}_{cnew} &= \bar{h}_{cold}\left[1 + orfh \cdot \left(\frac{\bar{W}_{inew} - \bar{W}_i}{\bar{W}_i}\right)\right] && \text{with, } orfh = 0.01 \text{ to } 1 \end{aligned} \tag{13.84}$$

$\bar{W}_i$ the total load carrying capacity of EHL contact is

$$\bar{W}_i = \iint_A \bar{p}\, d\bar{x}d\bar{y} + \iint_A \bar{p}_a\, d\bar{x}d\bar{y} \qquad \text{where, } \bar{p} = -\frac{\ln(1 - G\bar{q})}{G}$$

The nondimensional lubricant film pressure and contact pressure distributions as obtained from present solution are shown in Fig. 13.19 for the set of values, $G = 4522$, $\bar{U} = 0.1683 \times 10^{-11}$, $\bar{W} = 0.0567 \times 10^{-6}$, $k = 1.75$, $\Lambda = 4$, $\gamma = 6$, and $K' = 0.003$. It can be seen from Fig. 13.19(b) that the contact pressure is maximum at the outlet region where a pressure spike exists in Fig. 13.19(a) because of more asperities are in direct contact in the minimum film thickness zone. Effect of surface roughness parameter $\Lambda$ on maximum contact pressure increases is shown in Fig. 13.20 for various values of roughness pattern parameter $\gamma$. This indicates that more and more asperities are in direct contact when the surface roughness increases.

## REFERENCES

*Archard, G.D., Gair, F.C. and Hirst, W.* (1961) The elastohydrodynamic lubrication of rollers., *Proce. Royal Soc.*, London, A **262**, 57—72.

*Archard, J.F.* (1968) Nondimensional parameters in isothermal theories of elastohydrodynamic lubrication, *J. Mech. Eng. Sci.*, **10**, 165–167.

*Archard, J.F. and Cowking, E.W.* (1965) Elastohydrodynamic lubrication of point contacts., *Proc. Inst. Mech. Engrs.*, **180** (3B), 47–56.

*Archard, J.F. and Kirk, M.T.* (1963) Lubrication at point contacts., *Proce. Royal Soc.*, London, A **261**, 532–550.

*Brandt, A. and Lubrecht, A.A.* (1990) Multilevel matrix multiplication and fast solution of integral equations. *J Comput Phys.*, **90,** 348–70.

*Brewe, D.E. and Hamrock, B.J.* (1977) Simplified solution for elliptical-contact deformation between two elastic solids, *J. Lubr. Technol.*, **99** (4), 485–487.

*Cameron, A.* (1952) Hydrodynamic theory in gear lubrication., *J. Inst. Petrol*, **38**, 614–618.

*Cameron, A.* (1966) *The Principle of Lubrication*, Longmans, London, 196–198

*Cameron, A. and Gohar, R.* (1966) Theoretical and experimental studies of the oil film in lubricated point contacts, *Proceedings of the Royal Society*, Series A, **291**, 520–536.

*Chittenden, R.J., Dowson, D., Dunn, J.F. and Taylor, C.M.* (1985) A theoretical analysis of the isothermal elastohydrodynamic lubrication of concentrated contacts II. general case, with lubricant entrainment along either principal axis of the Herzian contact ellipse or at some intermediate angle., *Proc. R. Soc.*, London, A. **397**, 271–294.

*Crook, A.W.* (1957) Simulated gear-tooth contact:some experiments upon their lubrication and sub-surface deformations, *Proc. Instn. Mech. Engrs.* London, **171**, 187–199.

*Dowson, D. and Higginson, G.R.* (1959) A numerical solution to the elastohydrodynamic problem., *Journal of Mechanical Engineering Science*, **1**, 7–15.

*Dowson, D. and Toyoda, S.* (1977) A central film thickness formula for elastohydrodynamic line contacts. *Proceedings of 5th Leeds-Lyon Symp.*, Mechanical Engineering Publication, Leeds, England, 60–67.

Engineering Science Data Unit (1978) Contact phenomena, 1. Stresses, deflection and contact dimensions for normally-loaded unlubricated elastic components, *ESDU* 78035, **5**, 35–40.

*Ertel, A.M.* (1939) Hydrodynamic lubrication based on new principles., *Akad. Nauj SSSR Prikadnaya Mathematica i Mekhanika*, **3** (2), 41–52.

*Evans, H.P. and Snidle, R.W.* (1981) Inverse solution of reynolds equation of lubrication under point-contact elastohydrodynamic conditions., *ASME Journal of Tribology*, **103**, 539–546.

*Greenwood, J.A.* (1969) Presentation of elastohydrodynamic film thickness results, *J. Mech. Engg. Sci.*, **11**, 128–132.

*Greenwood, J.A. and Tripp, J.H.* (1970–71) The contact of two nominally flat rough surfaces. *Proc. of the IMechE*, **185**, Part 1, No. 48, 625–633.

*Grubin, A.N.* (1949) *Fundamentals of the hydrodynamic theory of lubrication of heavily loaded cylindrical surfaces. Investigation of the Contact of Machine Components.*, K1. F. Ketova, ed., Transl. of Russian Book no. 30, *Central Scientific Research Institute for Technology and Mechanical Engineering*, Moscow, Chapt. 2.

*Hamrock, B.J. and Dowson, D.* (1976a) Isothermal elastohydrodynamic lubrication of point contacts, Part I: Theoretical formulation., *Journal of Lubrication Technology*, **98**, 223–229.

*Hamrock, B.J. and Dowson, D.* (1976b) Isothermal elastohydrodynamic lubrication of point contacts, Part II: Ellipticity parameter results., *Journal of Lubrication Technology*, **98**, 375–383.

*Hamrock, B.J. and Dowson, D.* (1977a) Isothermal elastohydrodynamic lubrication of point contacts, Part III: Fully flooded results., *Journal of Lubrication Technology*, **99**, 264–276.

*Hamrock, B.J. and Dowson, D.* (1977b) Isothermal Elastohydrodynamic Lubrication of Point Contacts, Part IV: Starvation Results., *Journal of Lubrication Technology*, **99**, 264–276.

*Hamrock, B.J. and Dowson, D.* (1981) *Ball Bearing Lubrication*, John Wiley & Sons.

*Hertnett, M.J.* (1979) The analysis of contact stresses in rolling element bearings., *ASME, Journal of lubrication Technology*, **101**, 105–109.

*Hertnett, M.J.* (1979) The Analysis of Contact Stresses in Rolling Element Bearings., *ASME, Journal of lubrication Technology*, **101**, 105–109.

*Hertz, H.* (1881) On the Contact of Elastic Solids, *J. Reine Angew. Math.*, **92**, 292–302.

*Holmes, M.J.A., Evans, H.P. and Snidle, R.W.* (2003) Discussion of elastohydrodynamic lubrication in extended parameter ranges. *STLE Tribology Transactions*, **46** (2), 282–288.

*Hooke, C.J.* (1977) The elastohydrodynamic lubrication of heavily loaded contacts, *J. Mech. Engg. Sci.*, **19**, 149–156.

*Jiang, X., Hua, D.Y., Cheng, H.S., Ai, X. and Lee, S.C.* (1999) A mixed elastohydrodynamic lubrication model with asperity contact. *ASME Journal of Tribology*, **121**.

*Johnson, K.L.* (1970) Regimes of elastohydrodynamic lubrication, *J. Mech. Engg. Sci.*, **12**, 9–16.

*Lubrecht, A.A.* (1987) *The Numerical Solution of the Elastohydrodynamically Lubricated Line and Point Contact Problem using Multigrid Techniques.*, PhD. Thesis, University of Twente, Enschede, ISBN 90-9001583-3.

*Majumdar, B.C. and Hamrock, B.J.* (1981), Effect of surface roughness on elastohydrodynamic line contact, *NASA* TM 81753.

*Martin, H.M.* (1916) Lubrication of gear teeth. *Engineering.*, **102**, 119–121.

*Moes, H.* (1965–66) Communication, elastohydrodynamic lubrication, Proc. Instn. Mech. Engrs. London, 3B, **180**.

*Moes, H.* (1992) Optimum similarity analysis, an application to elastohydrodynamic lubrication, *Wear*, **159**, Part 1, 57–66.

*Nijenbanning, G., Venner, C.H. and Moes, H.* (1994) Film thickness in elastohydrodynamic lubricated elliptical contacts, *Wear*, **176**, 217–229.

*Patir, N. and Cheng H.S.* (1978) An average flow model for determining effects of three-dimensional roughness on partial hydrodynamic lubrication. *Transactions of ASME*, **100**, 12–17.

*Petrusevich, A.I.* (1951) Fundamental conclusions from the contact-hydrodynamic theory of lubrication., *Izv. Akad. Nauk. SSSR. (OTN)*, **2**, 209.

*Qiu, L. and Cheng, H.S.* (1998) Temperature rise simulation of three-dimensional rough surfaces in mixed lubrication contact. *ASME Journal of Tribology*, **120**, 310–318.

*Sarangi, M., Majumdar, B.C. and Sekhar, A.S.* (2004) Stiffness and damping characteristics of lubricated ball bearings considering the surface roughness effect. Part 1: theoretical formulation. *Proc. Instn Mech. Engrs, Part J, Journal of Engineering Tribology*, , **218**, 529–538

*Sarangi, M., Majumdar, B.C. and Sekhar, A.S.* (2005) On the dynamics of Elastohydrodynamic mixed lubricated ball bearing. Part I: Formulation of stiffness and damping coefficients *Proc. Instn Mech. Engrs, Part I, Journal of Engineering Tribology*, **219**, 411–421

*Sibley, L.B and Orcutt, F.K.* (1961) Elasto-hydrodynamic lubrication of rolling contact surfaces, *ASLE Trans.*, **4**, 234–240.

*Theyse, F.H.* (1966) Some aspects of the influence of hydrodynamic film formation on the contact between rolling/sliding surfaces, *Wear*, **9**, 41–59.

*Timoshenko, S. and Goodier, J.N.* (1951) *Theory of Elasticity*, 2nd ed., McGraw-Hill, New York.

*Venner, C.H.* (1991) Multilevel solution of the EHL line and point contact problems: PhD thesis, University of Twente, Netherlands.

*Wang, W-Z., Liu, Y-C., Wang, H. and Hu, Y-Z.* (2004) A computer thermal model of mixed lubrication in point contacts. *ASME Journal of Tribology*, **126**, 162–169.

*Weber, C. and Saalfeld, K.* (1954) Schmierfilm bei Walzen mit Verforming., *Zeits, ang. Math. Mech.*, **34** (1-2), 54–64.

*Zhu, D.* (2002) Elastohydrodynamic lubrication in extended parameter range-part i: speed effect. *STLE Tribology Transactions*, **45**, 540–548.

*Zhu, D.* (2002) Elastohydrodynamic lubrication in extended parameter range-part ii: load effect. *STLE Tribology Transactions*, **45**, 549–555.

*Zhu, D.* (2004) Elastohydrodynamic lubrication in extended parameter range-part iv: effect of material properties. *STLE Tribology Transactions*, **47**, 7–16.

*Zhu, D. and Cheng, H.S.* (1988) Effect of surface roughness on the point contact EHL. *Journal of Tribology, ASME*, **110**, 32–37.

*Zhu, D. and Hu, Y-Z.* (2001) Effects of rough surface topography and orientation on the characteristics of EHD and mixed lubrication in both circular and elliptical contacts. *STLE Tribology Transactions*, **44** (3).

Chapter 14

# Vibration Analysis with Lubricated Ball Bearings

## 14.1 | Introduction

In many industries the demand for high power and high speed together with uninterrupted and reliable operation is increasingly important. The accurate prediction and control of dynamic behavior (unbalance response, critical speeds, and instability) is another vital requirement. The bearings clearly constitute a vital component in any turbomachine. Today's high precision ball bearing draws more interest toward the noise and vibration-free operation in industrial as well as household appliances. Over a decade-long research and development on the linear and nonlinear vibration analysis of dry contact ball bearings put some effort into the understanding of basic phenomenon.

Gupta (1975, 79a–d) has published a series of papers, describing the dynamics of rolling element bearings. He presented the analysis with transient ball motion (1975). Here a set of generalized equations of motion of ball operating under elastohydrodynamic traction conditions is formulated and solved numerically, in which the traction in the EHL contact was predicted by semiempirical model of Gu (1973) describing the relationship of traction with slip. In an another approach (1979a–d), the generalized equations of motion describing the rolling element, cage, and race motion were formulated and analyzed in detail for both roller and ball bearings. Gupta *et al.* (1977) presented the numerical solution of the differential equations describing the motion of the ball in an angular contact thrust loaded ball bearing. Two characteristic natural frequencies were identified and their existence was verified with experimental data. Work of Gupta in the field giving detailed vibration analysis of ball bearing is highly recognized by intellectuals, but such described phenomenon never validated experimentally. This is perhaps due to complexity of experimental facility that needs to describe the vibrational characteristics, hardly possible to isolate of each individual components of a ball bearing. Datta and Farhang (1997a,b) presented a nonlinear model for structural vibration in rolling element bearings with Hertz contact stiffness. Recently, a study of ball size variation on the dynamics of ball bearing has been reported by Harsha (2004). In his work, the contact stiffness was modeled by nonlinear springs considering Hertzian elastic contact deformation theory. Similar studies on vibration of ball bearing are also available, Lim and Singh (1990a,b) and Tiwari *et al.* (2000a,b). Aktürk *et al.* (1997) suggested from the experimental investigation that the preload and the number of balls in a ball bearing are two of the important governing parameters affecting the dynamic behavior and have to be considered at the design stage. They found that the vibration characteristics of the shaft and its bearings change when the bearings operates in different regions of their nonlinear load-deflection characteristics.

Dareing and Johnson (1975) experimentally evaluated the damping characteristics of two steel discs under contact. In their analysis, damping generated by elastohydrodynamic lubrication was separated from other form of contact damping, indicating EHL damping contributes significantly to the total contact damping and is dependent on lubricant viscosity. El-Sayed (1980) derived an equation for predicting the stiffness of deep-groove ball bearings and expressed it in terms of available bearing dimension for practical use. The experimental investigation of Walford and Stone (1983) concluded that joints in a rolling element bearing was a source of damping and should not be neglected in theoretical analysis. Kraus *et al.* (1987) and Mitsuya *et al.* (1998) investigated the effects of speed and preload on radial and axial bearing damping in deep-groove ball bearings, and arrived at the same conclusion that damping decreases with speed and preload. Zeillinger *et al.* (1994) presented an experimental work on the calculation of damping coefficients of a ball bearing. A theoretical work of Dietl (1997) followed subsequently. In this theoretical study, an elliptical EHL contact was converted to an equivalent line contact considering major-axis of the ellipse and the mixed-lubrication stiffness and damping coefficients are evaluated. Relatively high axial load was applied which allows one to assume equal load distribution of each ball in the race. Measurement and theoretical estimation of damping coefficients and related rotor dynamics behavior are available, Yhland and Johansson (1970), Elsermans *et al.* (1976), Braun and Datner (1979), Igarashi *et al.* (1982), Wijnant (1998), Hendrikx *et al.* (1998), and Wensing (1998).

Recently, Sarangi *et al.* (2004a,b; 2005a,b) presented the linear as well as nonlinear prediction of vibrational characteristics of ball bearing depends on the stiffness and damping capability of individual ball having contact with raceways transmitting vibration to the supporting members. Based on his work, a relatively simpler linear vibration analysis of rotor supported on lubricated bearings using finite element method (FEM) and nonlinear structural vibration analysis has been presented in this chapter.

## 14.2 | Rotor Supported on Lubricated Ball Bearings

Finite element procedures are at present very widely used in engineering analysis. Each finite element possesses a mathematical formula which is associated with a simple geometric description, irrespective of the overall geometry of the structure. Thus, the method leads to the construction of a discrete system of matrix equations to represent the mass, damping, and stiffness effects of a continuous structure. The overall system equation of the structure is represented by global stiffness, damping, and mass matrices assembled from the individual constituents. Therefore, the formulation of stiffness, damping, and mass matrices of individual constituents is a prerequisite.

The empirical relationship of single lubricated contact stiffness and damping coefficients were given by Sarangi *et al.* (2004b). The overall equivalent bearing stiffness and damping directional matrices are obtained from static load distribution using these empirical relations for individual ball contact stiffness and damping coefficients. Thereafter, the bearing stiffness and damping matrices are used directly in the FEM analysis of rotor-bearing system, and dynamic response is predicted.

### 14.2.1 | Overall Equivalent Stiffness and Damping Matrices of Ball Bearings

Overall stiffness and damping of a ball bearing are the result of inner and outer race contact stiffness and damping of individual load sharing balls, which vary with contact geometry and load carrying capacity.

A radially loaded ball bearing with radial clearance $p_d$ is shown in Fig. 14.1(a), the inner ring makes contact under static and no load condition. It is noticeable that the clearance at the load line is zero and increases with the angle $\varphi$, which can be written as

$$c = \frac{p_d}{2}(1 - \cos\varphi)$$

Now application of load causes elastic deformation of balls over the arc $2\varphi_l$, and with dynamic condition there will be lubricant film to support this external load. Then total interference along the load line ($\varphi = 0$) is given by, (Fig. 14.1 b)

$$\delta^0 = V_c^0 - h_c^0$$

where $V_c^0$ is the combined inner and outer race elastic deformation of the ball along load line ($\varphi = 0$). $h_c^0$ is the combined inner and outer race lubricant film thickness in the load line contact ($\varphi = 0$).

Interference at any angular position from the load line can be represented in terms of total radial distance of inner ring or shaft from the concentric position $\delta$.

$$\delta_\varphi = \left(\delta\cos\varphi - \frac{p_d}{2}\right) \qquad \text{where, } \delta = \delta^0 + \frac{p_d}{2} \tag{14.1}$$

Using Hertz elastic deformation and Hamrock and Dowson (1981) film thickness empirical relations, one can write

$$\delta_\varphi = K_H W_\varphi^{2/3} - K_{EHL} W_\varphi^{-0.067} \tag{14.2}$$

where $K_H$ is the proportionality of load deformation constant from Hertz contact theory.

$$K_{EHL} = 2.69\bar{U}_i^{0.67} G_i^{0.53}\left(1 - 0.61e^{-0.73k_i}\right) + 2.69\bar{U}_o^{0.67} G_o^{0.53}\left(1 - 0.61e^{-0.73k_o}\right)$$

$K_{EHL}$ is the proportionality constant for load and central film thickness (Hamrock and Dowson 1977a), suffixes '$i$' and '$o$' correspond to the inner and outer race contacts, respectively.

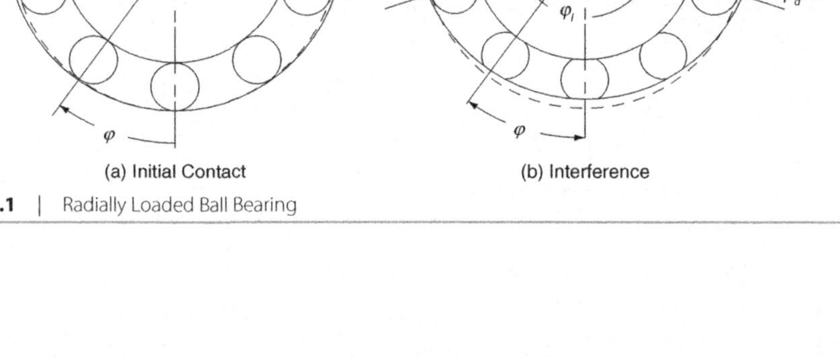


(a) Initial Contact (b) Interference

**Figure 14.1** | Radially Loaded Ball Bearing

$K_H$ can be found using the empirical relations as given by Brewe and Hamrock (1977) for elliptical integrals $\zeta$ and $\Im$ of Hertz contact,

$$\zeta = 1.0003 + \frac{0.5968}{R_y / R_x} \qquad \text{and} \qquad \Im = 1.5277 + 0.6023 \ln\left(R_y / R_x\right)$$

The deformation at the center of contact is expressed as

$$\delta = \left(\frac{W}{K}\right)^{2/3} \qquad \text{where} \qquad K = \pi k E' \left(\frac{R\zeta}{4.5\Im^3}\right)^{1/2}$$

and combining the deformation due to inner and outer race contacts

$$\delta = K_H W^{2/3} \qquad \text{with} \qquad K_H = \left(\frac{1}{K_i}\right)^{2/3} + \left(\frac{1}{K_o}\right)^{2/3}$$

Now the total load carrying capacity of the bearing is

$$W_x = \sum_0^{\varphi_l} W_\varphi \cos\varphi \qquad \text{where,} \qquad \varphi_l = \cos^{-1}\left(\frac{P_d}{2\delta}\right) \tag{14.3}$$

Equation (14.3) is for pure radial load. The radial, axial loads, and moment due to misalignment for a bearing having contact angle $\beta$ are given by

$$W_x = \sum^{\varphi_l} W_\varphi \cos\varphi \cos\beta$$
$$W_a = \sum_0^{\vartheta_l} W_\varphi \sin\beta$$
$$M = -r_e \sum_0^{\varphi_l} W_\varphi \cos\varphi \sin\beta$$

Once the radial load is known, the other two components of load $W_a$, $M$ can be found. Equations (14.1) are (14.3) are solved using nonlinear least-square method and iteratively with successive over-relaxation with the change of $\delta$ as follows.

$$\delta_{new} = \delta_{old}\left[1 + orfd \cdot \left(\frac{W_{xnew} - W_x}{W_x}\right)\right] \qquad \text{with } orfd = 0.1 \text{ to } 1$$

Having obtained the load sharing of each ball contact on race, the empirical relations (Equation 14.4) from Sarangi *et al.* (2004b) are used to obtain the stiffness and damping coefficients of inner race and outer race contact of each ball.

$$\bar{K} = k_1 G^{k_2} \bar{W}^{k_3} \bar{U}^{k_4} \left(k_5 - \left(ke^k\right)^{k_6}\right) \Lambda^{k_7} e^{-\Lambda^{k_8}} \gamma^{k_9}$$
$$\bar{C} = \frac{c_1}{R} G^{c_2} \bar{W}^{c_3} \bar{U}^{c_4} \left(c_5 - k^{c_6} e^{k^{c_7}}\right) \Lambda^{c_8} e^{-\Lambda^{c_9}} \gamma^{c_{10}} \tag{14.4}$$

where the constants are given in Table 14.1.

**Table 14.1** | Constants Used for the Empirical Relations (Equation 14.4)

| $k_1$ | $k_2$ | $k_3$ | $k_4$ | $k_5$ | $k_6$ | $k_7$ | $k_8$ | $k_9$ | |
|---|---|---|---|---|---|---|---|---|---|
| −5.284 | 1.237 | −0.671 | −0.263 | −0.475 | −0.344 | −0.267 | −0.397 | −0.021 | |
| $c_1$ | $c_2$ | $c_3$ | $c_4$ | $c_5$ | $c_6$ | $c_7$ | $c_8$ | $c_9$ | $c_{10}$ |
| −304.105 | 1.155 | −0.406 | −0.773 | 1.978 | 0.204 | −0.316 | −0.276 | −0.385 | −0.013 |

If two linear spring-damper combinations are connected in series as in Fig. 14.2, the resulting equivalent frequency dependent stiffness and damping can be determined with the help of complex equation (Dietl 1997),

$$K^b + j\omega C^b = \left( \frac{1}{K_i + j\omega C_i} + \frac{1}{K_o + j\omega C_o} \right)^{-1} \tag{14.5}$$

where $\omega$ is the angular frequency, which is generally of the inner race, outer race, cage, or ball pass frequencies or combinations of two or more of these (Goodwin 1989). It is appropriate to take the combination of inner and outer race ball pass frequencies, as they contribute more in dynamics of the system. Replacing a ball with equivalent spring-damper system (Fig. 14.2b), one may be able to find the overall stiffness and damping matrices from compatibility and coordinate transformation matrix.

$$\begin{aligned} K_b &= \sum_{i=1}^{z} K_i^b N_i^T N_i \\ C_b &= \sum_{i=1}^{z} C_i^b N_i^T N_i \end{aligned} \tag{14.6}$$

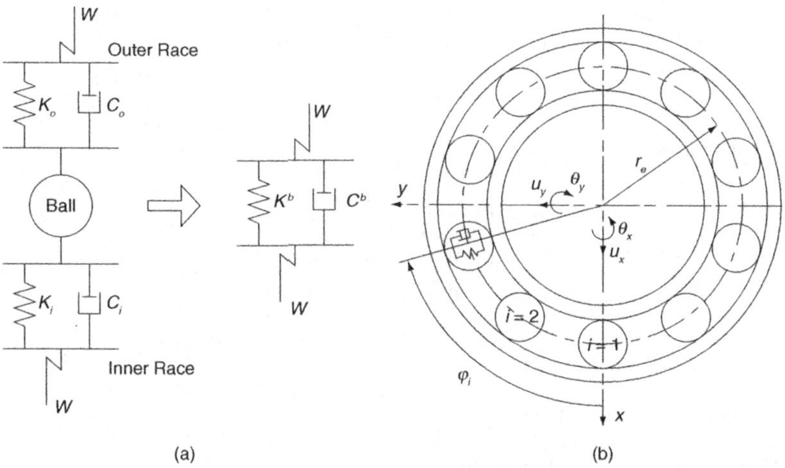


**Figure 14.2** | Equivalent Stiffness and Damping Model of Ball Bearing

where $z$ is the number of balls under contact (within arc $2\varphi_l$) and $N$ is the transformation matrix with global displacement vector $u = \{u_x, u_y, \theta_x, \theta_y\}^T$. The displacement along the axial direction is not considered for simplicity.

$$N_i = \begin{Bmatrix} \cos\varphi_i \cos\beta \\ \sin\varphi_i \cos\beta \\ r_e \sin\varphi_i \sin\beta \\ -r_e \cos\varphi_i \sin\beta \end{Bmatrix}^T$$

Similarly, stiffness matrix of dry contact ball bearing is obtained, where $K_{EHL} = 0$ in the Equation (14.2); Equations (14.4) and (14.6) are replaced by

$$K^b = \frac{3}{2}\left[\left(\frac{1}{K_i}\right)^{2/3} + \left(\frac{1}{K_o}\right)^{2/3}\right]^{-1} W^{1/3}$$

Stiffness and damping matrices obtained for lubricated ball bearing (Table 14.2) with a rotational speed of 3600 rpm of journal and the speed independent stiffness of dry contact ball bearing are given as example.

### Case-I

Bearing radial load = 420 N
Bearing axial load = 0 N
Lubricated ball bearing

$$K_b = \begin{bmatrix} 5.3786 & 2.6419 & 0 & 0 \\ 2.6419 & 3.5310 & 0 & 0 \\ 0 & 0 & 0 & 0 \\ 0 & 0 & 0 & 0 \end{bmatrix} 10^8 \quad C_b = \begin{bmatrix} 4.4599 & 2.2210 & 0 & 0 \\ 2.2210 & 3.0065 & 0 & 0 \\ 0 & 0 & 0 & 0 \\ 0 & 0 & 0 & 0 \end{bmatrix} 10^3$$

Dry contact ball bearing

$$K_b = \begin{bmatrix} 1.1176 & 0.5786 & 0 & 0 \\ 0.5786 & 0.7964 & 0 & 0 \\ 0 & 0 & 0 & 0 \\ 0 & 0 & 0 & 0 \end{bmatrix} 10^8$$

### Case-II

Bearing radial load = 420 N
Bearing axial load = 50 N
Moment due to misalignment = −1.7896 Nm
Ball contact angle $\beta$ = 5°

Lubricated ball bearing

$$K_b = \begin{bmatrix} 5.337\cdot10^8 & 2.622\cdot10^8 & 1.231\cdot10^6 & -2.505\cdot10^6 \\ 2.622\cdot10^8 & 3.505\cdot10^8 & 1.645\cdot10^6 & -1.231\cdot10^6 \\ 1.231\cdot10^6 & 1.645\cdot10^6 & 7.724\cdot10^3 & -5.778\cdot10^3 \\ -2.505\cdot10^6 & -1.231\cdot10^6 & -5.778\cdot10^3 & 1.176\cdot10^4 \end{bmatrix}$$

$$C_b = \begin{bmatrix} 4.405\cdot10^3 & 2.194\cdot10^3 & 10.299 & -20.677 \\ 2.194\cdot10^3 & 2.970\cdot10^3 & 13.943 & -10.299 \\ 10.299 & 13.943 & 0.065 & -0.048 \\ -20.677 & -10.299 & -0.048 & 0.097 \end{bmatrix}$$

Dry contact ball bearing

$$K_b = \begin{bmatrix} 1.1091\times10^8 & 5.743\times10^7 & 2.695\times10^5 & -5.206\times10^5 \\ 5.743\times10^7 & 7.903\times10^7 & 3.710\times10^5 & -2.695\times10^5 \\ 2.695\times10^5 & 3.710\times10^5 & 1.741\times10^3 & -1.265\times10^3 \\ -5.206\times10^5 & -2.695\times10^5 & -1.265\times10^3 & 2.444\times10^3 \end{bmatrix}$$

### 14.2.2 | Application to Rotor-bearing Systems

The stiffness and damping matrices as formulated in Equation (14.6) can be used in rotor-bearing systems. Characteristic equations are summarized for the case of a rotor-bearing system to understand the influence of lubricated ball bearings on the dynamics of rotor systems (Nelson and McVaugh 1976). Rotor-bearing system model as shown in Fig. 14.3 has been simulated with bearings model considering EHL with surface roughness effect.

The system characteristic equation will be assembled considering individual components using finite element method (Nelson 1980; Ozguven and Ozkan 1984).

#### Finite Rotor Element

A two-dimensional beam element shown in Fig. 14.4 having four degrees of freedom per node is used for modeling the shaft.

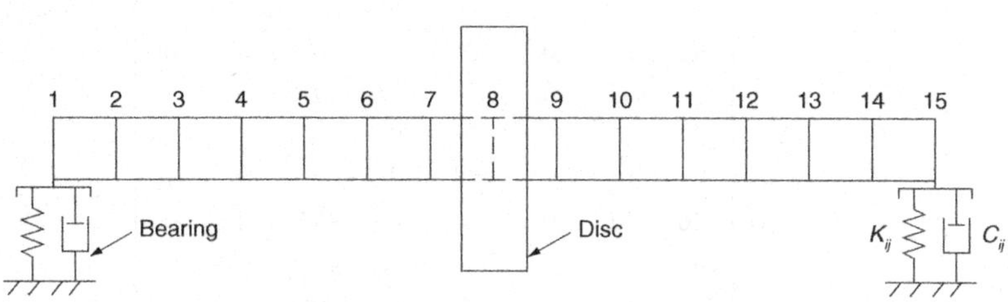


**Figure 14.3** | Rotor-bearing System Models

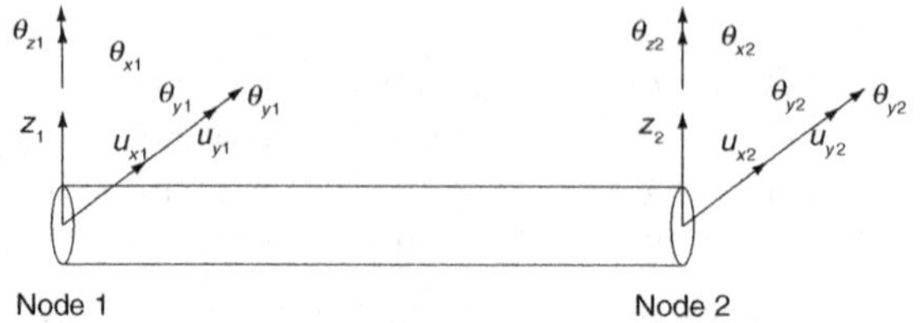


**Figure 14.4** | Rotor Element

When the rotor is very thin, the effect of rotary inertia and shear deformation can be neglected. In other words, Euler–Bernoulli beam theory is accurate enough as compared to the Timoshenko beam theory which considers the rotations due to shear deformation and is therefore more appropriate for thick beams. A Timoshenko beam element with four degrees of freedom per node is considered for present analysis (Ozguven and Ozkan 1984), whose equation of motion is given as

$$\left(M_R + M_T\right)_e \ddot{u}_e + \left(B - \Omega G\right)_e \dot{u}_e + \left(K_B - K_A\right)_e u_e = F_e \tag{14.7}$$

where $M_R$ is the rotational mass matrix
$M_T$ is the translational mass matrix
$K_B$ is the bending stiffness matrix of Timoshenko beam
$K_A$ is axial stiffness matrix due to axial load
$B$ is the damping and $G$ is the gyroscopic matrices, respectively
$u = \{u_x, \theta_y, u_y, \theta_x\}^T$ is the generalized degrees of freedom
$F$ is the external excitation force
'$e$' stands for the element degrees of freedom

$$M_T = \frac{\rho AL}{420}\begin{bmatrix} 156 & 22L & 0 & 0 & 54 & -13L & 0 & 0 \\ 22L & 4L^2 & 0 & 0 & 13L & -3L^2 & 0 & 0 \\ 0 & 0 & 156 & -22L & 0 & 0 & 54 & 13L \\ 0 & 0 & -22L & 4L^2 & 0 & 0 & -13L & -3L^2 \\ 54 & 13L & 0 & 0 & 156 & -22L & 0 & 0 \\ -13L & -3L^2 & 0 & 0 & -22L & 4L^2 & 0 & 0 \\ 0 & 0 & 54 & -13L & 0 & 0 & 156 & 22L \\ 0 & 0 & 13L & -3L^2 & 0 & 0 & 22L & 4L^2 \end{bmatrix}$$

$$M_R = \frac{\rho I}{30L}\begin{bmatrix} 36 & 3L & 0 & 0 & -36 & 3L & 0 & 0 \\ 3L & 4L^2 & 0 & 0 & -3L & -L^2 & 0 & 0 \\ 0 & 0 & 36 & -3L & 0 & 0 & -36 & -3L \\ 0 & 0 & -3L & 4L^2 & 0 & 0 & 3L & -L^2 \\ -36 & -3L & 0 & 0 & 36 & -3L & 0 & 0 \\ 3L & -L^2 & 0 & 0 & -3L & 4L^2 & 0 & 0 \\ 0 & 0 & -36 & 3L & 0 & 0 & 36 & 3L \\ 0 & 0 & -3L & -L^2 & 0 & 0 & 3L & 4L^2 \end{bmatrix}$$

$$K_B = \frac{EI}{(1+a)L^3}\begin{bmatrix} 12 & 6L & 0 & 0 & -12 & 6L & 0 & 0 \\ 6L & 4L^2+a & 0 & 0 & -6L & 2L^2-a & 0 & 0 \\ 0 & 0 & 12 & -6L & 0 & 0 & -12 & -6L \\ 0 & 0 & -6L & 4L^2+a & 0 & 0 & 6L & 2L^2-a \\ -12 & -6L & 0 & 0 & 12 & -6L & 0 & 0 \\ 6L & 2L^2-a & 0 & 0 & -6L & 4L^2+a & 0 & 0 \\ 0 & 0 & -12 & 6L & 0 & 0 & 12 & 6L \\ 0 & 0 & -6L & 2L^2-a & 0 & 0 & 6L & 4L^2+a \end{bmatrix}$$

$$K_A = \frac{F}{30L}\begin{bmatrix} 36 & 3L & 0 & 0 & -36 & 3L & 0 & 0 \\ 3L & 4L^2 & 0 & 0 & -3L & -L^2 & 0 & 0 \\ 0 & 0 & 36 & -3L & 0 & 0 & -36 & -3L \\ 0 & 0 & -3L & 4L^2 & 0 & 0 & 3L & -L^2 \\ -36 & -3L & 0 & 0 & 36 & -3L & 0 & 0 \\ 3L & -L^2 & 0 & 0 & -3L & 4L^2 & 0 & 0 \\ 0 & 0 & -36 & 3L & 0 & 0 & 36 & 3L \\ 0 & 0 & -3L & -L^2 & 0 & 0 & 3L & 4L^2 \end{bmatrix}$$

$$K_T = \frac{T}{L}\begin{bmatrix} 0 & 0 & 0 & -1 & 0 & 0 & 0 & 1 \\ 0 & 0 & -1 & -0.5L & 0 & 0 & 1 & 0.5L \\ 0 & -1 & 0 & 0 & 0 & 1 & 0 & 0 \\ -1 & 0.5L & 0 & 0 & 1 & -0.5L & 0 & 0 \\ 0 & 0 & 0 & 1 & 0 & 0 & 0 & -1 \\ 0 & 0 & 1 & -0.5L & 0 & 0 & -1 & 0.5L \\ 0 & 1 & 0 & 0 & 0 & -1 & 0 & 0 \\ 1 & 0.5L & 0 & 0 & -1 & -0.5L & 0 & 0 \end{bmatrix}$$

$$G = \frac{\rho I \Omega}{15L}\begin{bmatrix} 0 & 0 & 36 & -3L & 0 & 0 & -36 & -3L \\ 0 & 0 & 3L & -4L^2 & 0 & 0 & -3L & L^2 \\ -36 & -3L & 0 & 0 & 36 & -3L & 0 & 0 \\ 3L & 4L^2 & 0 & 0 & -3L & -L^2 & 0 & 0 \\ 0 & 0 & -36 & 3L & 0 & 0 & 36 & 3L \\ 0 & 0 & 3L & L^2 & 0 & 0 & -3L & -4L^2 \\ 36 & 3L & 0 & 0 & -36 & 3L & 0 & 0 \\ 3L & -L^2 & 0 & 0 & -3L & 4L^2 & 0 & 0 \end{bmatrix}$$

All system matrices are represented conventional notations, they can be found in Bettig (1996), Nelson and McVaugh (1976). Values of $a$, $F$, and $T$ are considered to be zero for simplicity.

### Bearing Element

Most rotors are supported on oil-film bearings or in rolling contact bearings. The bearings influence the rotor vibrations to a certain degree by their dynamic properties. This influence originates essentially from the ratio of the rotor stiffness to the bearing stiffness. The bearings are modeled by stiffness and damping elements horizontal as well as in vertical plane as shown in Fig. 14.3, where $i, j = 1, 2..., 4$ for four degrees of freedom. Neglecting the mass of bearing, the equation of motion for bearing can be written as,

$$C_b \dot{u}_b + K_b u_b = F_b \tag{14.8}$$

where suffix '*b*' corresponds to the bearing degrees of freedom, for example, in Fig. 14.3 suffix '*b*' is referred to node numbers 1 and 15.

### Rigid Disc

The disc is assumed to have an effect like a concentrated mass and so it can be characterized solely by kinetic energies. Mass and inertia properties can be concentrated on the corresponding node on the shaft (Nelson and McVaugh 1976). Hence, disc element with a single node and four degrees of freedom (as in Fig. 14.3) has been used for modeling the disc.

$$M_d \ddot{u}_d - \Omega G_d \dot{u}_d = F_d \tag{14.9}$$

where $M_d$ and $G_d$ are the mass and gyroscopic matrices of the disc. Suffix '*d*' corresponds to the disc nodal degrees of freedom, for example, in Fig. 14.3 suffix '*d*' is referred to the node number 8.

$$M_d = \begin{bmatrix} M_d & 0 & 0 & 0 \\ 0 & I_d & 0 & 0 \\ 0 & 0 & M_d & 0 \\ 0 & 0 & 0 & I_d \end{bmatrix}, \quad G_d = \begin{bmatrix} 0 & 0 & 0 & 0 \\ 0 & 0 & 0 & I_p \\ 0 & 0 & 0 & 0 \\ 0 & -I_p & 0 & 0 \end{bmatrix}$$

### System Equation of Motion

After assembling Equations (14.6) to (14.8), the system equation of motion becomes

$$M_s \ddot{u}_s + D_s \dot{u}_s + K_s u_s = F_s \tag{14.10}$$

where '*s*' stands for system, $M_s$ is the global mass matrix of the system that includes the mass matrices of shaft and disc. $D_s$ matrix includes damping matrices of shaft and bearings as well as gyroscopic matrices of shaft and disc. The global stiffness matrix $K_s$ is the contribution of stiffness matrices of shaft and bearings.

#### *Assembly of System Matrices*

The component equations are assembled for the rotor system. The assembly process is explained for the stiffness matrix in the following way. The procedure is repeated for the other matrices similarly.

For example, consider two elements connected as shown in Fig. 14.5. The first element no. 1 has the degrees of freedom (D.O.F) as $(q_1 - q_4)$ at node 1 and $(q_5 - q_8)$ at node 2. Using the conditions of compatibility, the second element will have the D.O.F. of $(q_5 - q_8)$ at 2 same as that of the node 2. At the node 3, the second node of the element 2, the D.O.F are $(q_9 - q_{12})$. When both the elements are combined as in Fig. 14.5, the total D.O.F of the assembly becomes $3 \times 4 = 12$ (which is number of nodes X 4 D.O.F at each node).

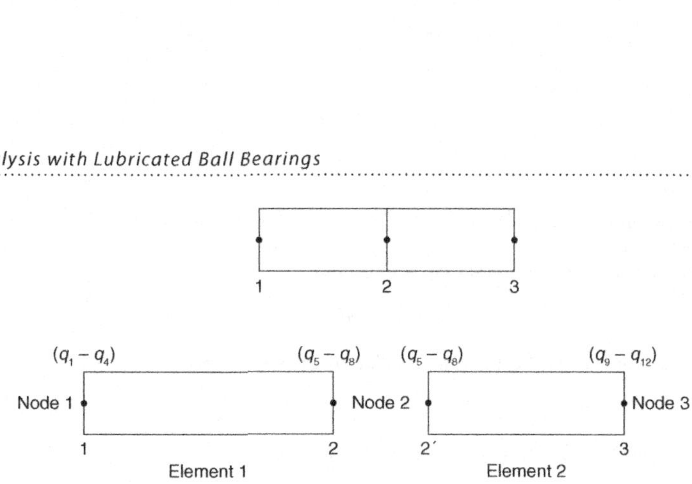


**Figure 14.5** | Joining of Finite Beam Elements

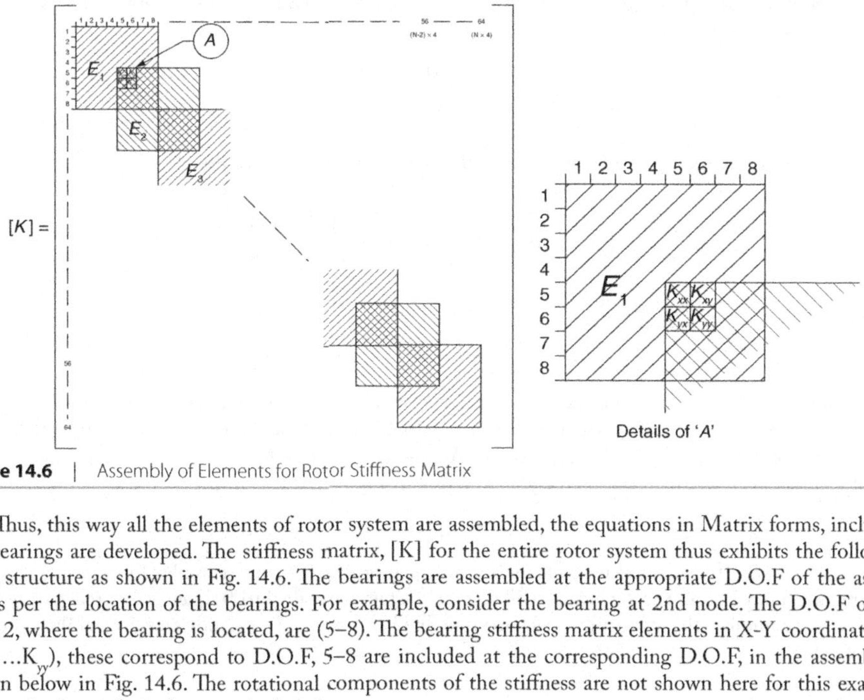


**Figure 14.6** | Assembly of Elements for Rotor Stiffness Matrix

Thus, this way all the elements of rotor system are assembled, the equations in Matrix forms, including the bearings are developed. The stiffness matrix, [K] for the entire rotor system thus exhibits the following band structure as shown in Fig. 14.6. The bearings are assembled at the appropriate D.O.F of the assembly as per the location of the bearings. For example, consider the bearing at 2nd node. The D.O.F of this node 2, where the bearing is located, are (5–8). The bearing stiffness matrix elements in X-Y coordinates are ($K_{xx}$,....$K_{yy}$), these correspond to D.O.F, 5–8 are included at the corresponding D.O.F, in the assembly as shown below in Fig. 14.6. The rotational components of the stiffness are not shown here for this example, whereas in the analysis those are included.

Equation (14.10) can be solved using state-space method in the absence of external excitation for the eigen analysis. The forced vibration characteristics can be studied with inclusion of external excitation force.

### *Eigenvalue Problem*

The eigenvalue problem becomes complex due to the inclusion of damping in the equation of motion. Such problems are not amenable to direct numerical applications as the computational costs involved would be higher than for the simple undamped case (Ewins 2000). Therefore, the equations will normally be transformed into the *state-space* form.

The equation of motion for free vibration is given by

$$M_s\ddot{u} + D_s\dot{u} + K_s u = 0 \tag{14.11}$$

Assuming a harmonic response, the rotor displacement can be represented by,

$$u = \hat{u}e^{i\omega t} \tag{14.12}$$

Substituting Equation (14.12) with Equation (14.11) leads to

$$\left[-\omega^2 M_s + i\omega D_s + K_s\right] = 0 \tag{14.13}$$

which constitutes a complex eigenvalue problem. To transform the equation (14.13), into *state-space* form, a new state vector is defined as

$$r = \begin{Bmatrix} u \\ \dot{u} \end{Bmatrix} \tag{14.14}$$

Finally from Equations (14.11) and (14.14), the state space form is given as,

$$\begin{bmatrix} D_s & M_s \\ M_s & 0 \end{bmatrix}\begin{Bmatrix} \dot{u} \\ \ddot{u} \end{Bmatrix} + \begin{bmatrix} K_s & 0 \\ 0 & -M_s \end{bmatrix}\begin{Bmatrix} u \\ \dot{u} \end{Bmatrix} = \begin{Bmatrix} 0 \\ 0 \end{Bmatrix}$$

or in general

$$B_1\dot{r} + B_2 r = 0 \tag{14.15}$$

The formulation is often called the state-space analysis, by contrast with usual vector-space analysis. $B_1$ and $B_2$ are $2N \times 2N$ real-symmetric matrices. Equation (14.15) is in a standard eigenvalue form which can be easily solved for $2N$ eigenvalues and eigenvectors. Assuming a solution form, $r = r_0 e^{\lambda t}$ for the homogeneous case of Equation (14.10), the associated eigenvalue problem is

$$\left|A - \lambda I\right| = 0 \tag{14.16}$$

where $A = -B_1^{-1}B_2$. The eigenvalues evaluated from Equation (14.16) are in complex form, $\lambda_i = \zeta_i(\Omega) + j\,\omega_i(\Omega)$. The imaginary part of the eigenvalue will give the system natural whirl frequency and real part will give about system damping. The logarithmic decrement defined as $\delta_i = -2\pi\,\zeta_i/\omega_i$, indicates the stability threshold when $\delta_i < 0$.

### *Unbalance Response*

Considering only the unbalance of disc, force for Equation (14.10) can be of the form

$$F_s = f_c\cos\Omega t + f_s\sin\Omega t \tag{14.17}$$

A steady-state solution of the same form,

$$u_s = a\cos\Omega t + b\sin\Omega t \tag{14.18}$$

is assumed and substituted in Equation (14.10), which yields the solutions

$$\begin{Bmatrix} a \\ b \end{Bmatrix} = \begin{bmatrix} (K_s - \Omega^2 M_s) & \Omega D_s \\ -\Omega D_s & (K_s - \Omega^2 M_s) \end{bmatrix}^{-1} \begin{Bmatrix} f_c \\ f_s \end{Bmatrix} \tag{14.19}$$

The back substitution of Equation (14.19) into Equation (14.18) provides system unbalance response.

### 14.2.3 | Numerical Solution and Results

Stiffness and damping of bearings can alter the dynamics of rotor-bearing system. To understand the effect of lubricated ball bearings, dynamic study of rotor system supported on these bearings was carried out and compared with that of other two different bearing types, viz., dry contact ball bearing and plain journal bearing supports. Three bearings were tested numerically on a rotor system as shown in Fig. 14.3 for the values given in Table 14.2. The journal bearing data have chosen in such a way that the rotor has nearly critical speeds more or less as that of rotor with ball bearing supports. All the three bearing types support same load for a given rotor-bearing system.

Dynamic equation of the rotor-bearing system is solved using FEM, where stiffness and damping matrices generated for bearings are assembled into the global stiffness and damping matrices of rotor. Stiffness and

**Table 14.2** | Parameters Selected for Numerical Simulation

| Parameter | Value |
|---|---|
| ***Rotor and disc data*** | |
| Length of shaft | 1.04 m |
| Shaft radius | 0.035 m |
| Disc radius | 0.176 m |
| Disc width | 0.070 m |
| Young's modulus | $2.1 \cdot 10^{11}$ N/m$^2$ |
| Density | 7800 kg/m$^3$ |
| Unbalance | $10^{-4}$ kg-m |
| Unbalance phase | $60^0$ |
| ***Ball bearing data: Deep groove ball bearing - 6214*** | |
| Ball diameter | 0.0195 m |
| No. of balls | 10 |
| Oil viscosity | 0.06 Pa s |
| Pressure viscosity coefficient | $2 \cdot 10^{-8}$ m$^2$/N |
| ***Plain journal bearing data*** | |
| Radial clearance ratio | 0.001 |
| Length to diameter ratio | 1 |
| Oil viscosity | 0.01 Pa s |

damping matrices of plain journal bearing are evaluated numerically from theory (Majumdar *et al.*, 1988), given in Chapter 8. In case of the dry contact ball bearing, the dry contact damping is neglected, as this is very small compared to lubricated contact.

The mode shapes of the rotor supported on a dry contact ball bearing is shown in Fig. 14.7. Four modes have been plotted with corresponding natural frequencies in both horizontal and vertical directions, where the fundamental natural frequency of the system is 57.8 Hz.

Figures 14.8 and 14.9 show the frequency response and unbalance response for all the three kinds of bearing. The response characteristics of rotor with lubricated ball bearing are in between those of the rotor with dry contact and fluid film bearing. It can be interpreted from the result that lubricated ball bearing retains both the properties of fluid film bearing and dry contact ball bearings. In other way, lubricated ball bearing gives a damping capability comparable to fluid film bearing.

From the response plots, it is noticed that the nature of the lubricated ball bearing is close to the plain journal bearing. Therefore, in case of lubricated ball bearing, an attempt has been made to search for the existence of oil whirl or whip instability phenomenon as present in case of a plain journal bearing. Oil whirl and whip are the self-induced instabilities caused by small unbalance masses. Apart from the spinning, the journal center exhibits an eccentric motion about the centroidal axis, this is known as whirling of the journal. Generally, whirling occurs at a frequency close to the half of the spinning speed of the journal. Therefore, it is sometimes called as half frequency whirl. As the spinning speed of journal increases, the whirling frequency also increases maintaining a ratio close to 0.5. However, when the journal spinning speed approaches the value twice of the system's 1st natural frequency, the frequency of whirl coincides with the 1st natural frequency of the system. This is known as oil whip. Frequency of whip does not change further with the spinning speed of journal. The system fails at this stage due to large vibration caused by oil whip mechanism.

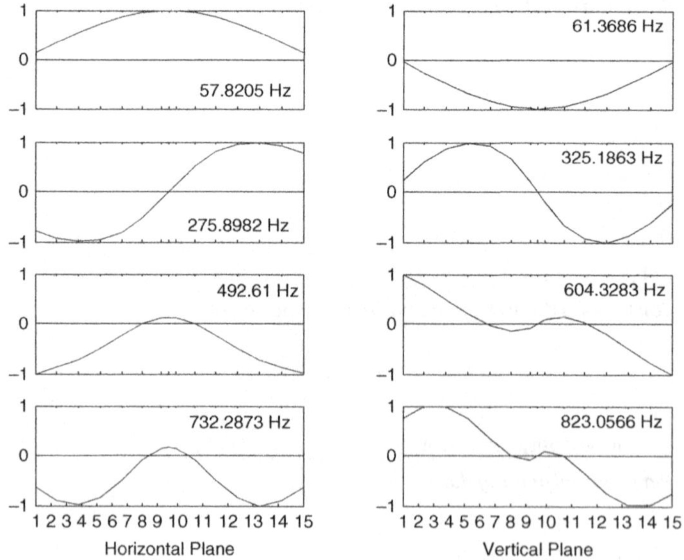


**Figure 14.7** | Mode Shapes of the Rotor

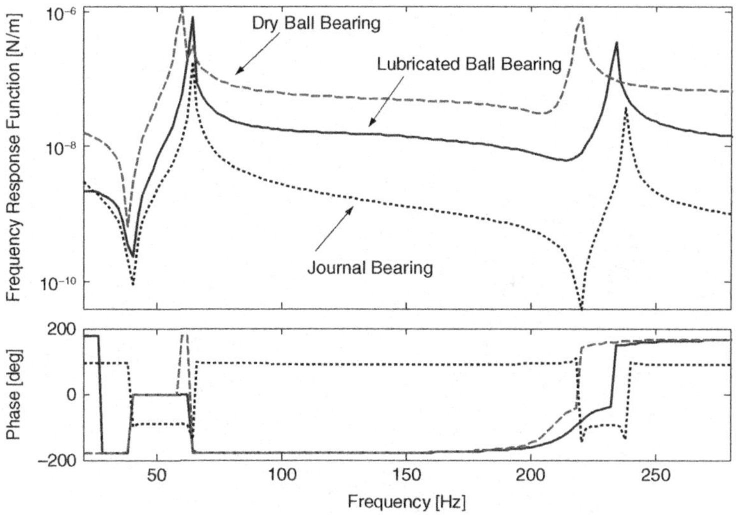


**Figure 14.8** | Frequency Response Function at Left End Bearing

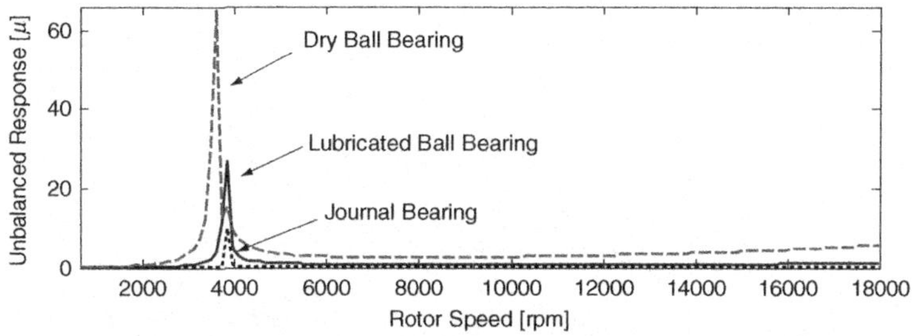


**Figure 14.9** | Unbalanced Response at Left End Bearing

Figure 14.10 shows the waterfall plot for the plain journal bearing and lubricated ball bearing, respectively. These waterfall plots have been plotted with the frequencies of vibration on $x$-axis, speed of the rotor on $y$-axis. Oil whirl and whip mechanisms can be identified easily from a waterfall plot and clearly visible in case of a plain journal bearing (Fig. 14.10a), whereas these mechanisms are not present in case of a lubricated ball bearing as shown in Fig. 14.10(b).

The stability threshold speed of the journal bearing is found to be 9984 rpm (166.4 Hz) and shown in the Fig. 14.10(a). The stability threshold speed is the speed at which whip occurs and the system vibrate violently. This speed is referred to as the value of the rotor speed above which the system becomes unstable. The stability threshold speed has been evaluated from the eigenvalue analysis, as the rotor speed at which the real value of the natural frequency changes sign from negative to positive.

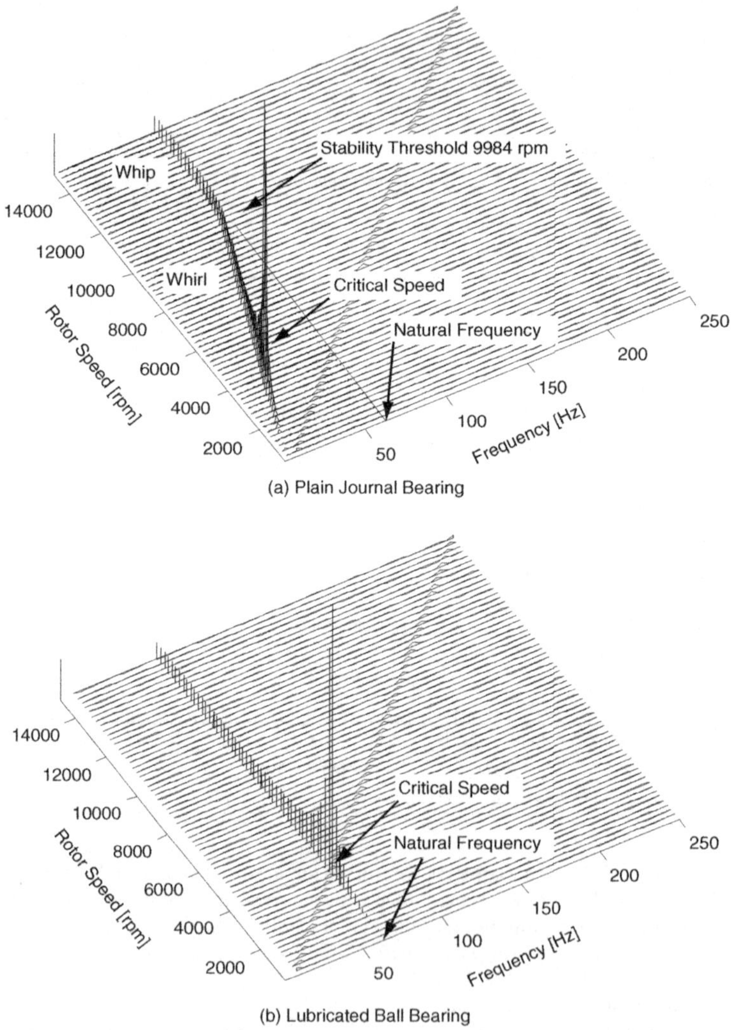


**Figure 14.10** | Waterfall Plot

## 14.3 | Nonlinear Structural Vibration Analysis in Lubricated Ball Bearings

The two non-dimensional empirical relations for the stiffness and damping coefficients were obtained Sarangi *et al.* (2005a) from the curve fitted numerical results. Following the work of Sarangi *et al.* (2005b), these empirical relations are used to develop a nonlinear contact model. The motion of balls as well as cage, inner race, and outer race for a lubricated ball bearing has been investigated using this nonlinear contact model. In the derivation of equation of motion, it is assumed that only rolling contact exists between the races and rolling elements. A system of nonlinear differential equations describing the motion of the bearing system is derived applying Lagrange's equations and solved using the fourth order Runge–Kutta integration technique.

### 14.3.1 | Formulation of Equation of Motion

A model for studying structural vibration in ball bearing has been developed following a procedure given by Datta and Farhang (1997a, b). First the expression for energies of the individual components of the bearing is formulated. Using these energies, the equations of motion are derived with the help of Lagrange's equation. The mathematical model developed is based on the following assumptions.

- The contact is of pure rolling type and any form of sliding is absent in ball to race contact.
- All the elements of the bearing have motion in the plane of bearing only. That means there is no axial motion.
- The motion of the balls is restricted by small elastic deformations and thin lubricant film present in the contact zone avoiding any plastic deformation.
- All balls in the race separated by an equal angular gap with the help of cage and maintains throughout the operation. The interaction of cage to that of balls is isolated and neglected.
- The bearing is assumed to operate under isothermal condition. Thereby variation of lubricant properties with temperature is ignored.
- All the components of the bearing are rigid enough to undergo bending.

#### Energy Expressions

Figure 14.11 shows a schematic diagram of a ball bearing containing balls, inner race, outer race, and cage. Point 'A' and 'B' are the centers of inner and outer races, respectively, under loaded condition. The total energy of this system is considered to be the sum of kinetic energy, potential energy, strain energy of the springs representing contact, and dissipation energy due to contact damping. Figure 14.12 shows the contact model of the ball on races represented by nonlinear springs and dampers.

#### Kinetic Energy

Kinetic energy of the system is the sum of individual kinetic energies of each element and can be formulated separately. The total kinetic energy of balls is

$$T_e = \sum_{i=1}^{N} T_i = \sum_{i=1}^{N} \frac{1}{2} m_i \left(\dot{\bar{\rho}}_i + \dot{\bar{R}}_a\right) \cdot \left(\dot{\bar{\rho}}_i + \dot{\bar{R}}_a\right) + \frac{1}{2} I_i \dot{\phi}_i^2 \tag{14.20}$$

where $N$ is the number of balls in the bearing and $\phi_i$ is the angular displacement of the ball about its center, then the displacements in vectorial form are

$$\bar{\rho}_i = \left(\rho_i \cos\theta_i\right)\hat{i} + \left(\rho_i \sin\theta_i\right)\hat{j} \tag{14.21}$$

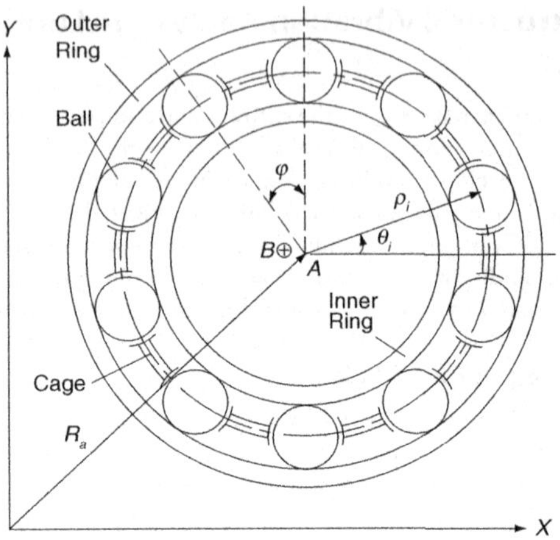


**Figure 14.11** | A Schematic Diagram of a Ball Bearing

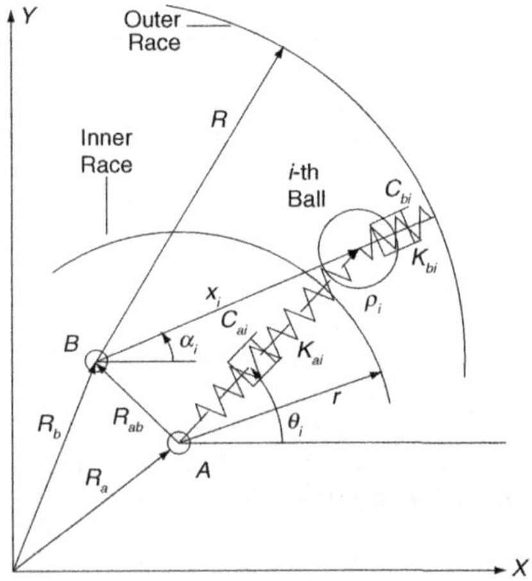


**Figure 14.12** | Spring-damper Contact Model of the Ball Bearing

and

$$\overline{R}_a = x_a\hat{i} + y_a\hat{j} \tag{14.22}$$

Considering pure rolling at ball inner race contact as shown in Fig. 14.13, the linear velocity relation of the contact point is given by

$$\rho_r\left(\dot{\phi}_i - \dot{\theta}_i\right) = -r\left(\dot{\phi}_a - \dot{\theta}_i\right)$$

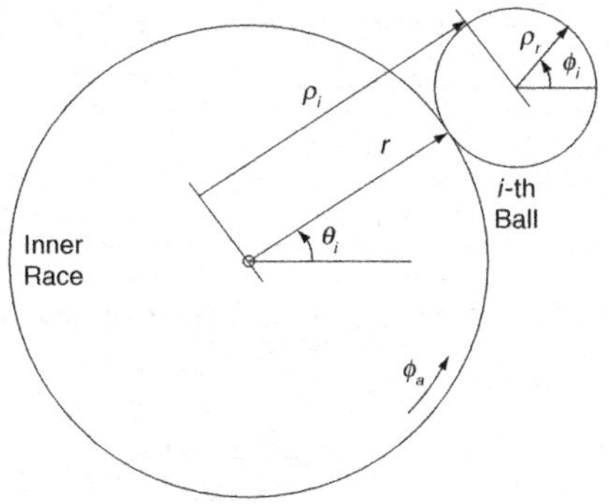


**Figure 14.13** | Inner Race to Ball in Pure Rolling Contact

then

$$\dot{\phi}_i = \dot{\theta}_i - \frac{r}{\rho_r}\left(\dot{\phi}_a - \dot{\theta}_i\right) \tag{14.23}$$

Similarly, considering pure rolling at ball outer race contact

$$\dot{\phi}_b = \dot{\theta}_i\left(1 + \frac{r}{R}\right) - \frac{r}{R}\dot{\phi}_a \tag{14.24}$$

Now kinetic energy given by Equation (14.20) of the balls can be simplified using Equations (14.21–14.23) as

$$T_e = \sum_{i=1}^{N}\left\{\frac{1}{2}m_i\begin{bmatrix}\dot{\rho}_i^2 + \rho_i^2\dot{\theta}_i^2 + \dot{x}_a^2 + 2\dot{x}_a\dot{\rho}_i\cos\theta_i \\ -2\dot{x}_a\rho_i\dot{\theta}_i\sin\theta_i + 2\dot{y}_a\dot{\rho}_i\sin\theta_i + 2\dot{y}_a\rho_i\dot{\theta}_i\cos\theta_i + \dot{y}_a^2\end{bmatrix} + \frac{1}{2}I_i\left[\dot{\theta}_i - \frac{r}{\rho_r}\left(\dot{\phi}_a - \dot{\theta}_i\right)\right]^2\right\} \tag{14.25}$$

Kinetic energy of the inner race is

$$T_a = \frac{1}{2}m_a\left(\dot{\bar{R}}_a \cdot \dot{\bar{R}}_a\right) + \frac{1}{2}I_a\dot{\phi}_a^2 \tag{14.26}$$

Substituting Equation (14.22) with Equation (14.26), the kinetic energy expression for the inner race will become

$$T_a = \frac{1}{2}m_a\left(\dot{x}_a^2 + \dot{y}_a^2\right) + \frac{1}{2}I_a\dot{\phi}_a^2 \tag{14.27}$$

The kinetic energy associated with the outer race is

$$T_b = \frac{1}{2}m_b\left(\dot{\bar{R}}_b \cdot \dot{\bar{R}}_b\right) + \frac{1}{2}I_b\dot{\phi}_b^2 \tag{14.28}$$

The displacement of outer race can be considered relative to the inner race

$$\bar{R}_b = \bar{R}_a + \bar{R}_{ab} \tag{14.29}$$

where,

$$\bar{R}_{ab} = x_b \hat{i} + y_b \hat{j}$$

Using Equations (14.24, 14.28–14.29), the expression for the kinetic energy of the outer race can be written as

$$T_b = \frac{1}{2} m_b \left[ \left( x_a + x_b \right)^2 + \left( y_a + y_b \right)^2 \right] + \frac{1}{2} I_b \left[ \dot{\theta}_i \left( 1 + \frac{r}{R} \right) - \frac{r}{R} \dot{\phi}_a \right]^2 \Bigg\} \tag{14.30}$$

It is assumed that the center of the cage remains coincident with that of the inner race that facilitates one to write the kinetic energy expression of cage as

$$T_c = \frac{1}{2} m_c \left( \dot{x}_a^2 + \dot{y}_a^2 \right) + \frac{1}{2} I_c \dot{\theta}^2 \tag{14.31}$$

Total kinetic energy of the bearing is the sum of individual components (Equations 14.25, 14.27, 14.30 and 14.31)

$$T = T_e + T_a + T_b + T_c \tag{14.32}$$

### Potential Energy

Potential energy of all the individual elements in the bearing can be formulated considering the datum as the horizontal plane through the global origin. The total potential energy of balls is the sum of potential energy of each ball that maintains different elevation from the datum.

$$V_e = \sum_{i=1}^{N} m_i g \left( \rho_i \sin \theta_i + y_a \right) \tag{14.33}$$

Potential energy for inner race

$$V_a = m_a g y_a \tag{14.34}$$

for outer race

$$V_b = m_b g \left( y_a + y_b \right)$$

and for cage considering the center to be coincident with inner race

$$V_c = m_c g y_a$$

Then the total potential energy of the system is

$$V = V_e + V_a + V_b + V_c \tag{14.35}$$

### Strain Energy

Strain energy of the contact is thought to be conserved due to the nonlinear springs. Generally, the nonlinearity in the load-deflection relation is expressed as

$$F = k\eta^{n_1} \tag{14.36}$$

where $n_1$ is the power index representing the nonlinearity, $\eta$ is the total approach of two contacting ellipsoids and $k$ is the stiffness constant, then the strain energy associated with the load-deflection is

$$U = \int_0^x k\eta^{n_1}\,dx = \frac{k}{n_1+1}\eta^{n_1+1} \tag{14.37}$$

Using Equation (14.37), the total strain energy of all contacts can be written as

$$U = \sum_{i=1}^{N} \frac{k_{ai}}{n_1+1}\left(\rho_i - L_a\right)^{n_1+1} + \sum_{i=1}^{N} \frac{k_{bi}}{n_1+1}\left(R - x_i - L_b\right)^{n_1+1} \tag{14.38}$$

where suffices '*a*' and '*b*' correspond to the inner and outer race ball contacts and $L$ is the unstressed length of the springs. Hence, first part of Equation (14.38) is referred as the strain energy of the inner-race-ball contact, and last part as that of outer-race-ball contact. This strain energy of the spring is conservative and can be summed up with the total potential energy.

### Dissipation Energy

Dissipation energy of the contact is nonconservative caused by nonlinear relation of load and velocity and generalized as

$$F = c\dot{\eta}^{n_2} \tag{14.39}$$

where $n_2$ is the power index representing the nonlinearity of load velocity relation and $c$ is the damping constant, then the associated dissipation energy is

$$D = \int_0^x c\,\dot{\eta}^{n_2}\,dx = \frac{c}{n_2+1}\dot{\eta}^{n_2+1} \tag{14.40}$$

Using Equation (14.40), the total dissipation of the contact can be written as

$$D = \sum_{i=1}^{N} \frac{c_{ai}}{n_2+1}\left(\dot{\rho}_i\right)^{n_2+1} + \sum_{i=1}^{N} \frac{c_{bi}}{n_2+1}\left(-\dot{x}_i\right)^{n_2+1} \tag{14.41}$$

## 14.3.2 | Equations of Motion

Knowing all aforesaid energies a set of equations in generalized coordinates can be written with the help of Lagrange's equations

$$\frac{d}{dt}\left(\frac{\partial T}{\partial \dot{q}_k}\right) - \frac{\partial T}{\partial q_k} + \frac{\partial (V+U)}{\partial q_k} + \frac{\partial D}{\partial \dot{q}_k} = Q_k \tag{14.42}$$

where $q_k$ and $Q_k$ are the $k$-th generalized coordinates and corresponding generalized forces, respectively. The total number of generalized coordinates is $N$+2, $N$ numbers of balls ($\rho_{i=1,2,\ldots,N}$) and the relative outer race displacement to that of inner race $x_b$ and $y_b$. It is assumed that the balls are separated with a constant angular gap and maintains throughout the operation, hence

$$\theta_i = \theta_1 + (N-1)\varphi \tag{14.43}$$

where $\varphi = 2\pi/N$, the angular velocities of all the balls are then reduced to be same

$$\dot{\theta}_1 = \dot{\theta}_2 = \ldots = \dot{\theta}_N = \dot{\theta}$$

Using Equation (14.42) for generalized coordinates of balls, $\rho_{i=1,2,\ldots,N}$, the equations of motion are

$$\ddot{\rho}_i + \ddot{x}_a \cos\theta_i + \ddot{y}_a \sin\theta_i - \rho_i\dot{\theta}^2 + g\sin\theta_i + \frac{k_{ai}}{m_i}(\rho_i - L_a)^{n_1} - \frac{k_{bi}}{m_i}(R - x_i - L_b)^{n_1}\frac{\partial x_i}{\partial \rho_i} \tag{14.44}$$
$$+\frac{c_{ai}}{m_i}(\dot{\rho}_i)^{n_2} - \frac{c_{bi}}{m_i}(-\dot{x}_i)^{n_2}\frac{\partial \dot{x}_i}{\partial \dot{\rho}_i} = 0$$

Equation of motion for generalized coordinate $x_b$ is

$$\ddot{x}_b + \ddot{x}_a - \sum_{i=1}^{N}\frac{k_{bi}}{m_b}(R - x_i - L_b)^{n_1}\frac{\partial x_i}{\partial x_b} - \sum_{i=1}^{N}\frac{c_{bi}}{m_b}(-\dot{x}_i)^{n_2}\frac{\partial \dot{x}_i}{\partial \dot{x}_b} = \frac{F_x}{m_b} \tag{14.45}$$

and for the generalized coordinate $y_b$

$$\ddot{y}_b + \ddot{y}_a + g - \sum_{i=1}^{N}\frac{k_{bi}}{m_b}(R - x_i - L_b)^{n_1}\frac{\partial x_i}{\partial y_b} - \sum_{i=1}^{N}\frac{c_{bi}}{m_b}(-\dot{x}_i)^{n_2}\frac{\partial \dot{x}_i}{\partial \dot{y}_b} = \frac{F_y}{m_b} \tag{14.46}$$

where $F_x$ and $F_y$ are the generalized forces in $X$ and $Y$ directions acting on the outer race corresponding to $x_b$ and $y_b$ generalized coordinates, respectively. The deformation of the outer-race-ball contact is defined with the help of $x_i$, which can be found from the relations

$$\begin{aligned} x_i\cos\alpha_i + x_b &= \rho_i\cos\theta_i \\ x_i\sin\alpha_i + y_b &= \rho_i\sin\theta_i \end{aligned} \tag{14.47}$$

From these relations, we can get $x_i$ as

$$x_i = \left[(\rho_i\cos\theta_i - x_b)^2 + (\rho_i\sin\theta_i - y_b)^2\right]^{1/2} \tag{14.48}$$

Now we can obtain the partial derivatives of $x_i$ appear in the equations of motion

$$\frac{\partial \dot{x}_i}{\partial \dot{\rho}_i} = \frac{\partial x_i}{\partial \rho_i} = \frac{\rho_i - x_b\cos\theta_i - y_b\sin\theta_i}{x_i} \tag{14.49}$$

$$\frac{\partial \dot{x}_i}{\partial \dot{x}_b} = \frac{\partial x_i}{\partial x_b} = \frac{x_b - \rho_i\cos\theta_i}{x_i} \tag{14.50}$$

$$\frac{\partial \dot{x}_i}{\partial \dot{y}_b} = \frac{\partial x_i}{\partial y_b} = \frac{y_b - \rho_i \sin\theta_i}{x_i} \tag{14.51}$$

This is a system of $N$+2 second order nonlinear ordinary differential equations (14.44–14.46) and can be solved using Runge–Kutta fourth order numerical integration for known stiffness and damping constants and corresponding indices.

### 14.3.3 | Nonlinear Model of the Lubricated Contact

The equations of motion formulated in the previous section can be solved if the power indices $n_1$ and $n_2$ of the contact nonlinear stiffness and damping properties are known. These indices can be found using the empirical relations for stiffness and damping given by Sarangi *et al.* (2005a). For the case of elastohydrodynamically mixed lubricated contacts, the lubricant stiffness, lubricant damping, and contact stiffness coefficients are expressed in Equations (14.52–14.54).

#### Lubricant Film Stiffness

$$\bar{K}_l = a_1 G^{a_2} U^{a_3} W^{a_4} \left(a_5 - \left(ke^k\right)^{a_6}\right) K'^{a_7} \Lambda^{a_8} e^{-\Lambda^{a_9}} \gamma^{a_{10}} \tag{14.52}$$

where

$$a_1 = -0.4053,\ a_2 = -0.2521,\ a_3 = -0.6995, a_4 = 1.1678, a_5 = -0.5891,$$
$$a_6 = -0.3102, a_7 = -0.0358,\ a_8 = -0.3368,\ a_9 = -0.928, a_{10} = -0.0399$$

#### Lubricant Film Damping

$$\bar{C}_l = \frac{c_1}{R} G^{c_2} U^{c_3} W^{c_4} \left(c_5 - \left(k^{c_6} e^{(k^{c_7})}\right)\right) K'^{c_8} \Lambda^{c_9} e^{-\Lambda^{c_{10}}} \gamma^{c_{11}} \tag{14.53}$$

where

$$c_1 = -0.7300, c_2 = -0.7511, c_3 = -0.5871, c_4 = 1.1963,$$
$$c_5 = -0.5188\ , c_6 = 0.1275, c_7 = -0.4548, c_8 = \ 0.0013,$$
$$c_9 = -0.3521, c_{10} = -0.8472,\ c_{11} = -0.0366$$

#### Contact Stiffness

$$\bar{K}_a = b_1 G^{b_2} U^{b_3} W^{b_4} \left(b_5 - k^{b_6} e^{(k^{b_7})}\right) K'^{b_8} \Lambda^{b_9} e^{-\Lambda^{b_{10}}} \gamma^{b_{11}} \tag{14.54}$$

where

$$b_1 = -0.1828,\ b_2 = 0.1065,\ b_3 = -0.4931,\ b_4 = 0.7239,$$
$$b_8 = 1.0418,\ b_9 = 0.5396,\ b_{10} = 1.6221,\ b_{11} = 0.0861$$

for $k <= 4$     for $k > 4$

$$b_5 = \quad -0.1263 \quad 0.0588$$
$$b_6 = \quad 0.2377 \quad 0.4420$$
$$b_7 = \quad 0.2462 \quad 0.1081$$

Figure 14.14 shows the equivalent nonlinear stiffness and damping model representing a rough surface lubricated contact. It is assumed that damping is caused by the squeezing effect and predominant at the inlet zone rather than the center of the contact region where stiffnesses are higher. Therefore, damping from the lubricated contact is considered to be parallel to the stiffnesses. Here $h$ and $\delta$ are the lubricant film thickness and elastic deformation at the contact region, respectively. All the stiffness and damping terms in the figure are shown corresponding to the equivalent force system. $K_l$ and $C_l$ are the lubricant film stiffness and damping, respectively, whereas $K_a$ denotes the asperity contact stiffness and $K_h$ denotes the Hertz stiffness constant. Using these quantities an equivalent force system with nonlinear load-deflection relation representing stiffness constant $k$ and damping constant $c$ can be formulated as:

The total approach of two ellipsoids ($\eta$) under the lubricated contact is governed by both elastic deformation ($\delta$) and lubricant film thickness ($h$) as

$$\eta = \delta - h \tag{14.55}$$

From Fig. 14.14, same load is carried by elastic deformation stiffness as well as the lubricant stiffness and asperities contact stiffness, hence

$$F = K_h \delta^{3/2} + C_l \dot{h} \tag{14.56}$$

and

$$F = \left(K_l + K_a\right) h + C_l \dot{h} \tag{14.57}$$

Using complex relation $\dot{h} = i\omega h$ in Equations (14.55 and 14.56) and substituting $\delta$ and $h$ in Equation (14.55) one obtains

$$\eta = \left(\frac{F}{K_h}\right)^{2/3} \left[1 - \frac{i\omega C_l}{\left(K_l + K_a\right) + i\omega C_l}\right]^{2/3} - \frac{F}{\left(K_l + K_a\right) + i\omega C_l} \tag{14.58}$$

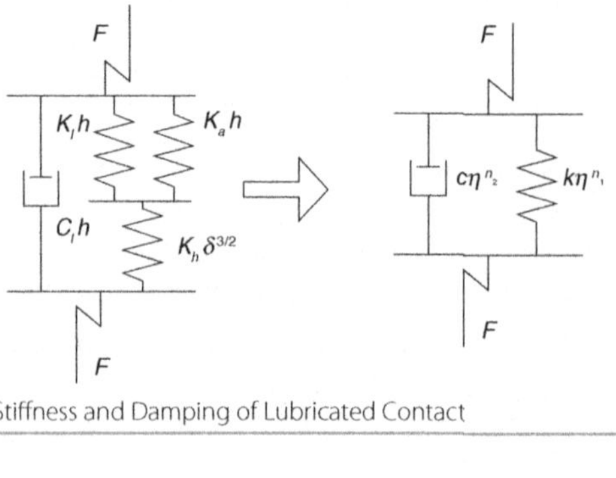


**Figure 14.14** | Equivalent Stiffness and Damping of Lubricated Contact

Equation (14.58) is the required nonlinear load-deflection relationship, where $\omega$ is the angular frequency, which is generally of the inner race, outer race, cage or ball pass frequencies or combinations of two or more of these [19]. For known bearing geometry and operating conditions this load-deflection relation can be reduced to more suitable form as follows.

The geometrical and operating properties of the bearing selected for the present simulation are given in Table 14.3. Using these values, the stiffness and damping present in Equation (14.58) can be found for both inner-race-ball and outer-race-ball contacts. Now we have one Equation (14.58) with two unknowns $F$ and $\eta$. For a operating load range of $F = 0$ to $20\ N$ of each contact, corresponding displacement $\eta$ can be found, which will give a nonlinear relational curve of $F$ Vs $\eta$. Then a suitable curve can be fitted using the generalized nonlinear Equation (14.59).

$$F = k\,\eta^{n_1} + ic_1\,\eta^{n_2} \tag{14.59}$$

In this expression, $k$ and $c_1$ are equivalent stiffness and damping constants with power indices $n_1$ and $n_2$, respectively. This has been derived using nonlinear least-square curve fitting technique and the expression is found to be

### *For Inner-race-ball Contact*

$$F_a = 9.447 \cdot 10^8\,\eta^{1.26} + i\,2.511 \cdot 10^6\,\eta^{1.26} \tag{14.60}$$

### *For Outer-race-ball Contact*

$$F_b = 9.268 \cdot 10^8\,\eta^{1.26} + i\,5.726 \cdot 10^6\,\eta^{1.26} \tag{14.61}$$

The curve fitted results are shown in Figs 14.15 and 14.16, where the real and imaginary parts of $\eta$ signify the amount of equivalent stiffness and damping present in the system, respectively. A close agreement has been achieved. These two Equations (14.60 and 14.61) can be further approximated using relation $\dot{\eta} = i\omega\eta$ as

### *For Inner-race-ball Contact*

$$F_a = 9.447 \cdot 10^8\,\eta^{1.26} + 1247.3\,\dot{\eta}^{1.26} \tag{14.62}$$

### *For Outer-race-ball Contact*

$$F_b = 9.268 \cdot 10^8\,\eta^{1.26} + 2844.7\,\dot{\eta}^{1.26} \tag{14.63}$$

The constants and indices from these two Equations (14.61 and 14.62) can be substituted in Equations (14.44–14.46) of motion such as $k_{ai} = 9.447 \cdot 10^8$, $k_{bi} = 9.268 \cdot 10^8$, $c_{ai} = 1247.3$, $c_{bi} = 2844.7$, and $n_1 = n_2 = 1.26$.

## 14.3.4 | Numerical Solution and Results

The properties of the ball bearing selected for the analysis are similar to those of Datta and Farhang (1997a,b) with additional roughness and lubricant properties as given in Table 14.3 and 14.4. When the model of the lubricated contact is known, the equations of motion (Eqns. 14.44–14.46) can be solved utilizing Runge–Kutta numerical integration method. The equations are modeled and solved in Simulink with MATLAB functions. The mass of each ball is assumed to be same as $m$ and the initial values are set as given in Table 14.4. Keeping

**Table 14.3** | Geometric and Operational Parameters Used for the Ball Bearing

| Parameters | Value |
|---|---|
| Outside diameter of the bearing | 68 mm |
| Radius of inner race at point of contact with the ball ($r$) | 23 mm |
| Radius of inner race at point of contact with the ball ($R$) | 31 mm |
| Radius of each ball ($\rho$) | 4 mm |
| Radius of inner race groove ($r_i$) | 4.31 mm |
| Radius of outer race groove ($r_o$) | 4.35 mm |
| Contact angle ($\beta$) | $0^0$ |
| Angular velocity of the inner ring ($\dot{\phi}_a$) | 1000 rpm |
| Angular velocity of the outer ring ($\dot{\phi}_b$) | 130 rpm |
| Surface roughness parameter ($\Lambda$) | 2 |
| Roughness pattern parameter ($\gamma$) | 1 |
| Pressure viscosity coefficient ($\alpha$) | 2E-8 $m^2$/N |
| Ambient viscosity ($\mu_0$) | 0.068 Pas |
| Equivalent Young's modulus ($E'$) | 2.7E+11 N/$m^2$ |
| Constant for asperity contact stiffness ($K'$) | 0.0008 |
| Number of balls | 8 |

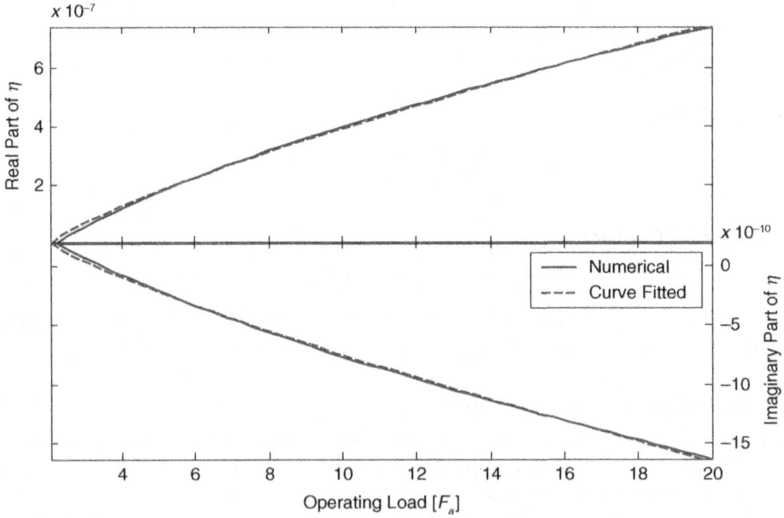


**Figure 14.15** | Curve Fitted Load-Deflection Relation for Inner-Race-Ball Contact

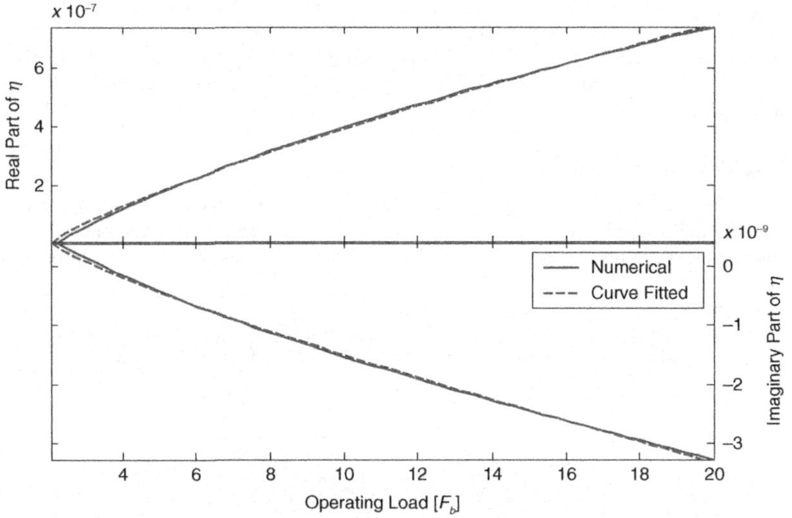


**Figure 14.16** | Curve Fitted Load-Deflection Relation for Outer-Race-Ball Contact

**Table 14.4** | Additional Physical and Operational Model Parameters Used for the Ball Bearing

| Parameters | Value |
|---|---|
| Mass of the bearing | 0.22 Kg |
| Mass of each ball ($m$) | 0.009 Kg |
| Mass of the inner ring ($m_a$) | 0.060 Kg |
| Mass of the outer ring ($m_b$) | 0.065 Kg |
| Mass of the cage ($m_c$) | 0.023 Kg |
| Moment of inertia of each ball ($I$) | $7.2 \times 10^{-8}$ Kgm$^2$ |
| Moment of inertia of inner ring ($I_a$) | $2.7 \times 10^{-5}$ Kgm$^2$ |
| Moment of inertia of outer ring ($I_b$) | $6.9 \times 10^{-5}$ Kgm$^2$ |
| Moment of inertia of cage ($I_c$) | $1.7 \times 10^{-8}$ Kgm$^2$ |
| Unstressed length of inner-race-ball contact ($L_a$) | 27 mm |
| Unstressed length of outer-race-ball contact ($L_b$) | 4 mm |
| Initial radial position of the $i$-th ball ($\rho_i$) | 27 mm |
| Initial angular position of 1st ball ($\theta_1$) | $0^0$ |
| Other initial conditions ($\dot{\rho}_i, \ddot{\rho}_i, x_b, \dot{x}_b, \ddot{x}_b$) are assumed | 0 |

in mind a problem of stiff system, variable time step size is adopted with a maximum permissible time step of $10^{-6}$ and the initial time step could be as small as $10^{-20}$. The system equations are integrated over time until the steady-state solution is reached. Two cases have been analyzed covering free and forced vibration with both undamped and damped conditions. Due to space limitation, only vibration of alternative balls is given for all the cases analyzed.

### Case-1: Free Vibration Analysis

Although vibrating structure without external load is uncommon in real-life operations, it is more suitable in predicting the natural behavior. Figure 14.17 shows the undamped free vibrating response of four alternative balls and outer race with respect to the inner race. The resulting signals are extracted for a prolonged period of

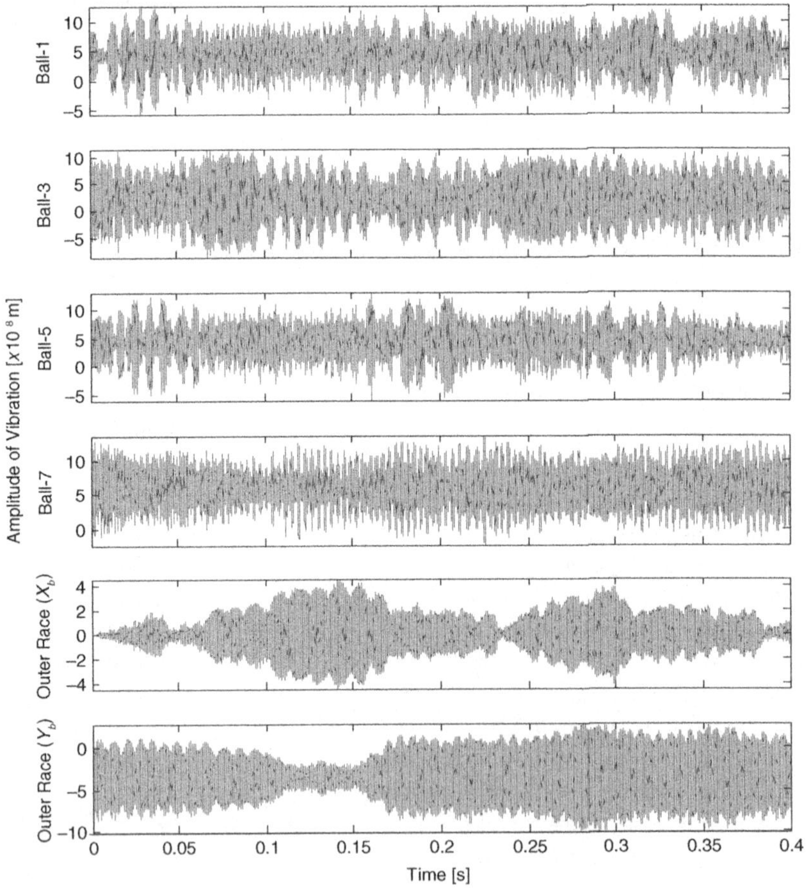


**Figure 14.17** | Free Undamped Vibration for Prolonged Period of Time

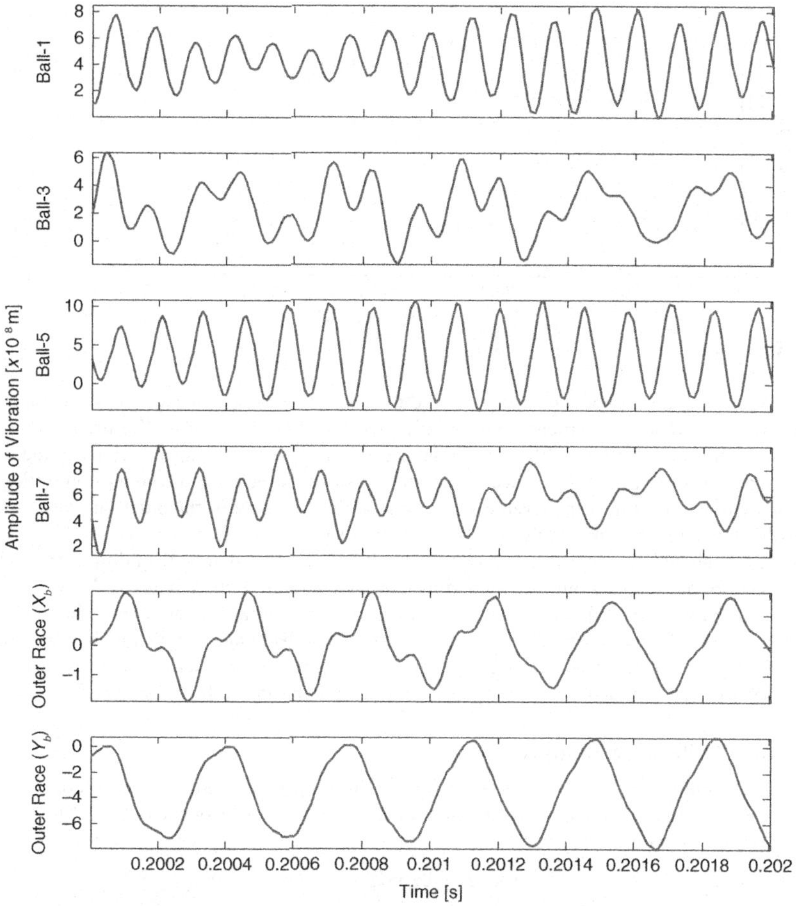


**Figure 14.18** | Free Undamped Vibration for 2 Milliseconds

time 0.4s concerning the stability of the system. Without the consideration of damping, the system is said to be conservative, hence all the individual elements in the bearing share the total energy which is constant. Hence, the vibrating amplitudes of some elements increase corresponding to the reduction of others which can be noticed clearly between the vibrating amplitudes of outer race. Figure 14.18 shows vibrating response for 2 milliseconds, which has been extracted from Fig.14.17. This gives a clear picture of vibrating amplitudes and frequencies involved. All the responses reveal the existence of two characteristic frequencies as reported by Gupta (1977).

The associated two characteristic frequencies of individual elements of the bearing are identified from the FFT spectrum of the vibrating signals. Two frequencies close to 2804 Hz and 8468 Hz as identified from the FFT of outer race horizontal signal ($x_b$) are shown in Fig. 14.19. All the characteristic frequencies

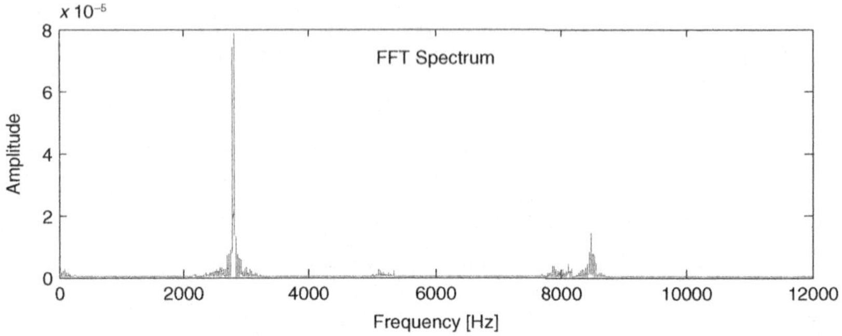


**Figure 14.19** | FFT Spectrum of Outer Race Response ($x_b$)

of individual components are more or less same, and the second characteristic frequency is close to the thrice of first one. On the whole, the elements of the bearing have oscillatory response. The amplitudes of vibration are not uniform, however, the rms of the amplitudes are uniform over a prolonged period of time. The motion of outer race with respect to the inner race is more or less of the same order as the motion of the balls. Due to the presence of gravity, the average value of the vertical displacement of the outer race always falls on the fourth quadrant, whereas the mean value of horizontal displacement is close to zero.

Figure 14.20 shows the vibrating response of bearing under free damped condition. It can be noticed that the oscillatory motion dies out rapidly within 2 milliseconds time. The energy is damped out from the system with the damping provided by the lubricant present in the contact region leaving the bearing with the static defection resulting from gravitational force. Under single ball angular position, ball has two contacts with one to inner and outer races which gives more damping to the balls rather than the races. This could be the explanation that the oscillatory vibrating amplitude of balls dies faster as compared to the outer race.

### Case-2: Forced Vibration Analysis

This is the most common application of bearing. A radial rotating load is allowed to act on the bearing mostly caused by unbalance mass. The radial load is considered on the outer race of the bearing while rotating synchronously as given in Equation (14.64).

$$F_x = A\cos\phi_b$$
$$F_y = A\sin\phi_b \tag{14.64}$$

where $A$ is the amplitude of the radial force which is equal to the product of eccentric mass and square of the angular velocity of the outer race. For the present analysis, value of $A$ is considered to be 0.0027 N and the forcing frequency is same as that of outer race (2.1667 Hz) given in Table 14.3. The vibrating responses as obtained under the forced undamped condition are shown in Fig. 14.21. It is noticeable that apart from the high frequencies one low frequency exists in all the vibration responses. This frequency is calculated and found nearly equal to 2.1 Hz corresponds to the forcing frequency synchronous with outer race.

Figure 14.22 shows the forced damped vibration response of the bearing. The energy associated with the higher frequencies is damped out rapidly within 2 milliseconds, whereas the low frequency amplitude response to the external force remains. The existing vibration shows clearly a vibrating frequency nearly 2.1 Hz and the amplitude of vibration is less than the mean amplitude as observed under undamped condition in

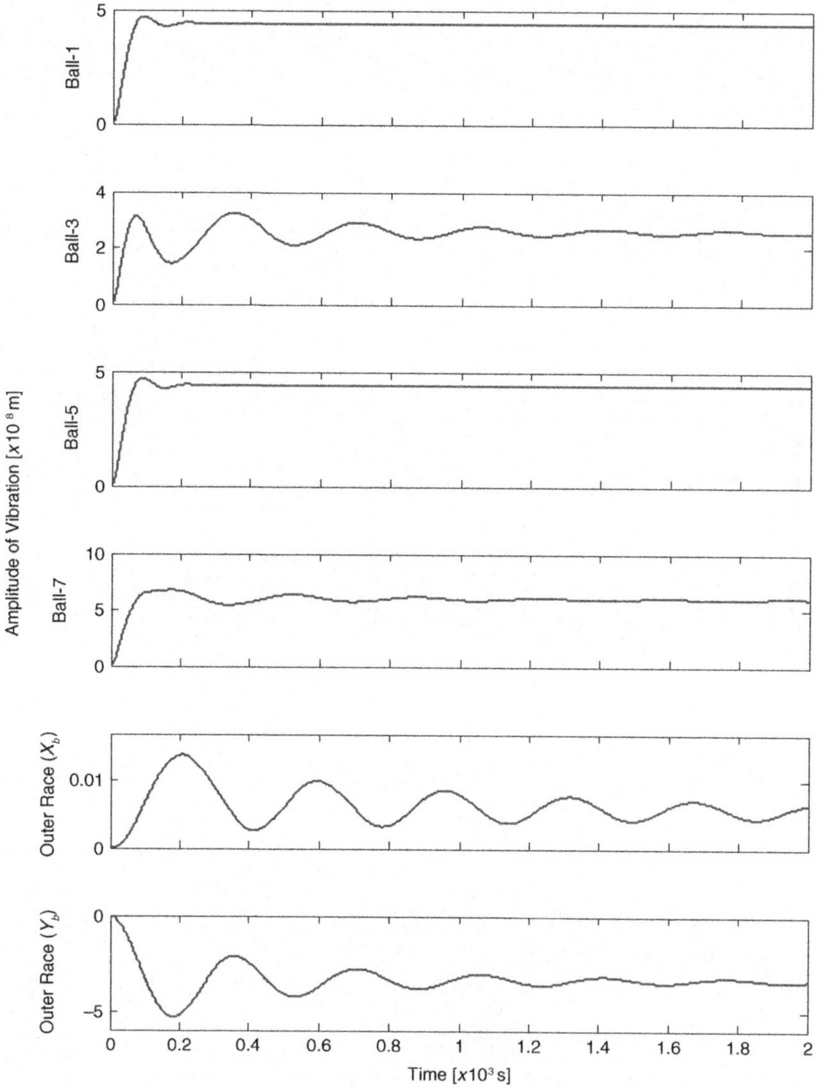


**Figure 14.20** | Free Damped Vibration

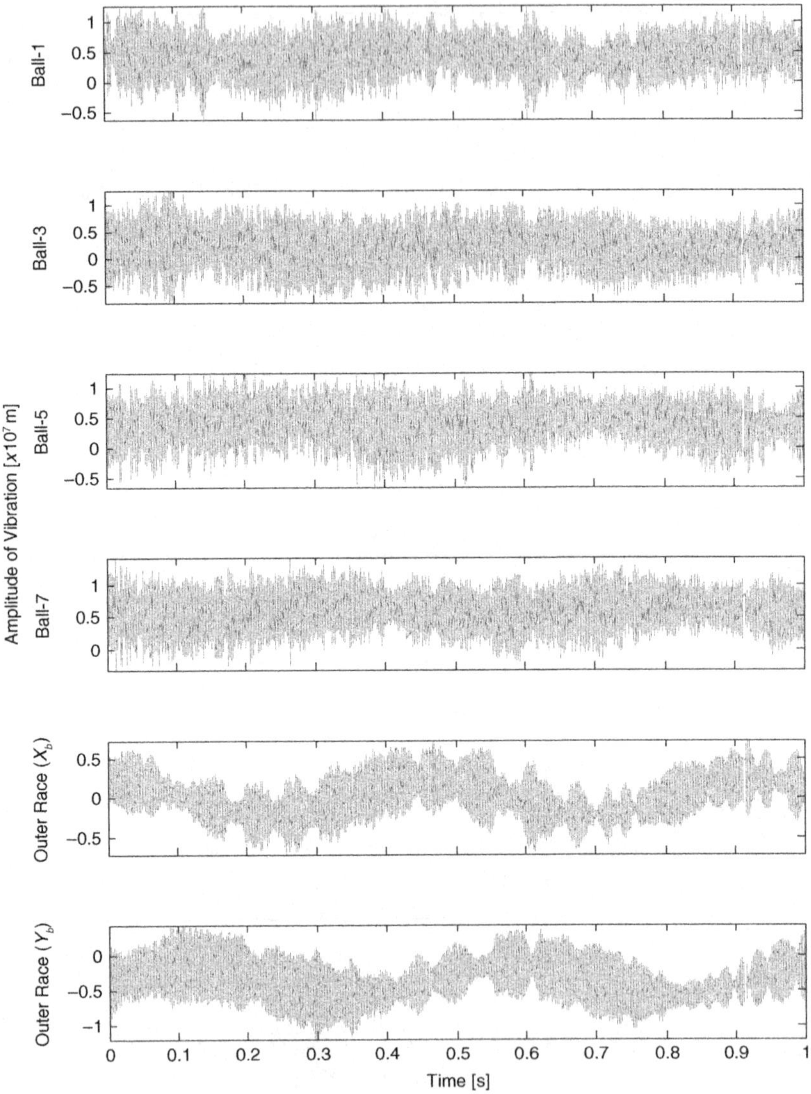


**Figure 14.21** | Forced Undamped Vibration

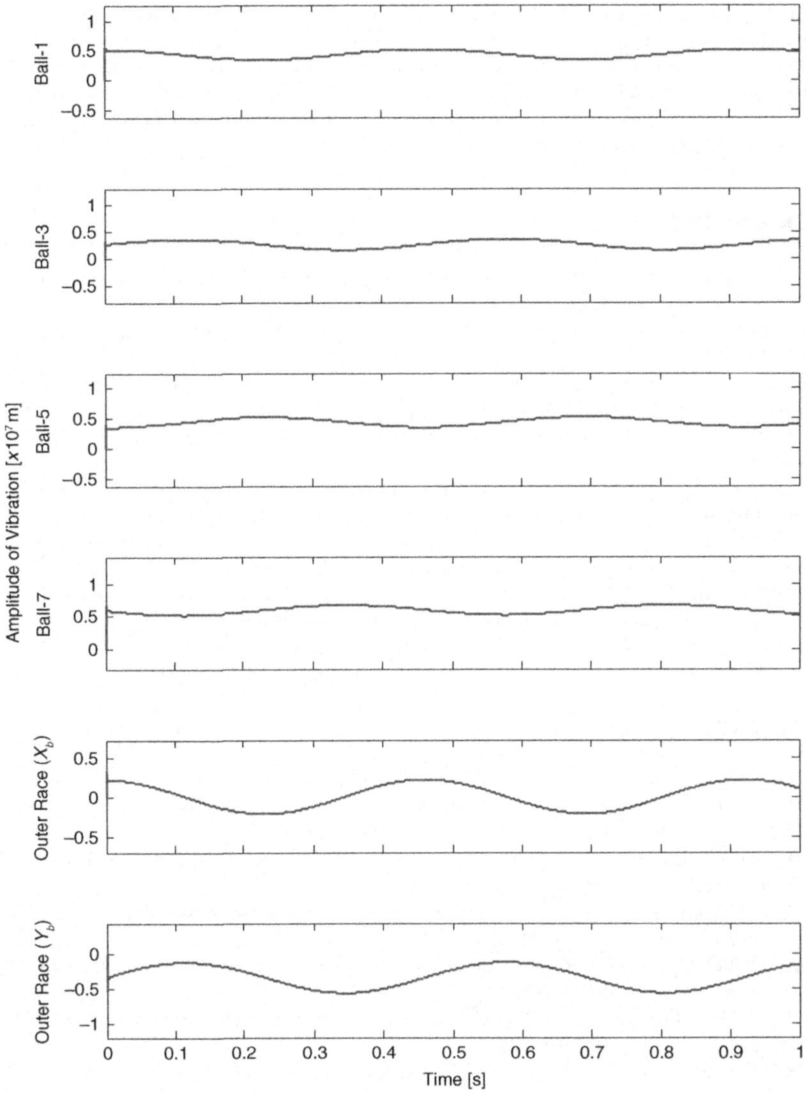


**Figure 14.22** | Forced Damped Vibration

Fig. 14.21. The higher frequency vibrations are generally induced due to the contact stiffness while the low frequency vibration is caused by the external load. Thereby the contact damping is effective in reducing the higher frequency vibrations completely and also it reduces the low frequency vibration up to some extent.

Damping present in a lubricated bearing may reduce sometimes consequence to the interruption in lubrication then the higher frequency amplitudes will appear in the signature similar to Fig. 14.21. Monitoring of such signature will enable to predict the condition of the bearing.

## REFERENCES


*Aktürk, N., Uneeb, M. and Gohar, R.* (1997), The effects of number of balls and preload on vibrations associated with ball bearings, *ASME Journal of Tribology*, **119**, 747–753.

*Braun, S. and Datner, B. (1979),* Analysis of roller/ball bearing vibrations. *ASME Journal of Mechanical Design*, 1979, **101**, 118–125.

*Brewe, D.E. and Hamrock, B.J.* (1977), Simplified solution for elliptical-contact deformation between two elastic solids, *J. Lubr. Technol.*, **99** (4), 485–487.

*Dareing, D.W. and Johnson, K.L.* (1975), Fluid film damping of rolling contact vibrations, *Journal Mechanical Engineering Science, IMechE.*

*Datta, J. and Farhang, K.* (1997a), A nonlinear model for structural vibrations in rolling element bearings: Part I—Derivation of governing equations, *ASME Journal of Tribology*, **119**, 126–131.

*Datta, J. and Farhang, K.* (1997b), A nonlinear model for structural vibrations in rolling element bearings: Part II—Simulation and results, *ASME Journal of Tribology*, **119**, 323–331.

*Dietl, P.* (1997), Damping and stiffness characteristics of rolling element bearings, theory and experiments., *PhD Thesis, Mechanical Engineering*, Technical University of Vienna.

*El-Sayed, H.R.* (1980), Stiffness of deep-groove ball bearings, *Wear*, **63**, 89–94.

*Elsermans, M., Hongerloot, M. and Snoeye, R.* (1976), Damping in taper roll bearings, *Proceedings of 16th MTDR Conference*, London, UK, 223–229.

*Ewins, D.J.* (2000), *Modal Testing: Theory, Practice and Application*, Second Edition, Research Studies Press Ltd., England.

*Goodwin, M.J.* (1989), *Dynamics of Rotor-bearing Systems*, Unwin Hyman, London.

*Gu, A.* (1973), An important method for calculating the spin torque in fully lubricated ball race contact., *ASME Journal of Lubrication Technology*, Series F, **95**, 106–108.

*Gupta, P.K.* (1979a), Dynamics of rolling-element bearing-part i: cylindrical roller bearing analysis, *ASME Journal of Lubrication Technology*, **101**, 293–304.

*Gupta, P.K.* (1979b), Dynamics of rolling-element bearing-part ii: cylindrical roller bearing results, *ASME Journal of Lubrication Technology*, **101**, 304–311.

*Gupta, P.K.* (1979c), Dynamics of Rolling-Element Bearing-Part III: Ball Bearing Analysis, *ASME Journal of Lubrication Technology*, **101**, 312-318.

*Gupta, P.K.* (1979d), Dynamics of rolling-element bearing-part iv: ball bearing results, *ASME Journal of Lubrication Technology*, **101**, 319–326.

*Gupta, P.K., Winn, L.W. and Wilcock, D. F.* (1977), Vibration characteristics of ball bearings, *ASME Journal of Lubrication Technology*, **99**, 284–289.

*Hamrock, B.J. and Dowson, D.* (1976a), Isothermal elastohydrodynamic lubrication of point contacts, part i—theoretical formulation., *Journal of Lubrication Technology*, **98**, 223–229.

*Hamrock, B.J. and Dowson, D.* (1976b), Isothermal elastohydrodynamic lubrication of point contacts, part ii—ellipticity parameter results., *Journal of Lubrication Technology*, **98**, 375–383.

*Hamrock, B.J. and Dowson, D.* (1977a), Isothermal elastohydrodynamic lubrication of point contacts, Part III—Fully flooded results., *Journal of Lubrication Technology*, **99**, 264–276.

*Hamrock, B.J. and Dowson, D.* (1981), *Ball Bearing Lubrication*, John Wiley & Sons.

*Harsha, S.P.* (2004), The effect of ball size variation on nonlinear vibrations associated with ball bearings, *J. Multi-body Dynamics, Proc. Inst. Mech. Engr. Part-K*, **218**, 191–210.

*Hendrikx, R.T.W.M., Nijen, G.C.V. and Dietl, P.* (1998), Vibration in household appliances with rolling element bearings., *Proc. ISMA23 Noise and Vibration Engineering*, **3**, 1537–1544.

*Igarashi, T. and Hamada, H.* Studies on the vibration and sound of defective roller bearings (First Report: Vibration of ball bearings with one defect). *Bulletin of the JSME*, 1982, **25**, No. 204, 994–1001.

*Igarashi, T. and Kato, J.* Studies on the vibration and sound of defective roller bearings (Third Report: Vibration of ball bearings with multiple defects). *Bulletin of the JSME*, 1985, **28**, No. 237, 492–499.

*Kraus, J., Blech, J.J. and Braun, S.G.* (1987), Determination of rolling bearing stiffness and damping by modal analysis., *Journal of Vibration, Acoustics, Stress and Reliability in Design*, **109** (7).

*Lim, T.C. and Singh, R.* (1990a), Vibration transmission through rolling element bearings, Part I: Bearing stiffness formulation, *Journal of Sound and Vibration*, **139** (2), 179–199.

*Majumdar, B.C., Brewe, D.E. and Khonsari, M.M.* (1988), Stability of a rigid rotor supported on flexible oil journal bearings. *ASME, Journal of Tribology*, **110**, 181–187.

*Mitsuya, Y., Sawai, H., Shimizu, M. and Aono, Y.* (1998), Damping in vibration transfer through deep-groove ball bearings, *ASME Journal of Tribology*, **120**, 413–420.

*Nelson, H.D.* (1980), A finite rotating shaft element using Timishenko beam theory. *Transactions of ASME, Journal of Mechanical Design*, **102**, 793–803.

*Nelson, H.D. and McVaugh, J.M.* (1976), The dynamics of rotor bearing systems using finite elements. *ASME, Journal of Engineering for Industry*, 593–600.

*Ozguven, H.N. and Ozkan, Z.L.* (1984), Whirl speeds and unbalance response of multi bearing rotor using finite elements. *Transactions of ASME, Journal of Vibration, Acoustics, Stress and Reliability in Design*, **106**, 72–79.

*Prabhakar, S., Mohanty, A.R. and Sekhar, A.S.* Application of discrete wavelet transform for detection of ball bearing race faults. *Tribology International*, 2002, **35**, 793–800.

*Sarangi, M., Majumdar, B.C. and Sekhar, A.S.* (2005a), On the dynamics EHD mixed lubricated ball bearings. Part I—Formulation of stiffness and damping coefficients, *Journal of Engineering Tribology, IMechE Part-J*, 2005.

*Sarangi, M., Majumdar, B.C. and Sekhar, A.S.* (2004a), Stiffness and damping characteristics of lubricated ball bearings, part-i: theoretical formulation, *Journal of Engineering Tribology, IMechE Part-J*, **218**, 529–538.

*Sarangi, M., Majumdar, B.C. and Sekhar, A.S.* (2004b), Stiffness and damping characteristics of lubricated ball bearings, part-ii: numerical results and application, *Journal of Engineering Tribology, IMechE Part-J*, **218**, 539–547.

*Sarangi, M., Majumdar, B.C. and Sekhar, A.S.* (2005), Nonlinear structural vibration in lubricated ball bearings, *Journal of Sound and Vibration*, May 2005.

*Sarangi, M., Majumdar, B.C. and Sekhar, A.S.* (2005b), On the dynamics EHD mixed lubricated ball bearings. Part II—Nonlinear structural vibration, *Journal of Engineering Tribology, IMechE Part-J*, 2005.

*Sunnersjö, C.S.* **(1978)**, Varying compliance vibrations of rolling bearings. *Journal of Sound and Vibration*, 1978, **58** (3), 363–373.

*Tiwari, M., Gupta, K. and Prakash, O.* (2000a), Effect of Radial Internal Clearance of a Ball Bearing on the Dynamics of a Balanced Horizontal Rotor, *Journal of Sound and Vibration*, **238** (5), 723–756.

*Tiwari, M., Gupta, K. and Prakash, O.* (2000b), Dynamic response of an unbalanced rotor supported on ball bearings, *Journal of Sound and Vibration*, **238** (5), 757–779.

*Walford, T.L.H. and Stone, B.J.* (1983), The source of damping in rolling element bearings under oscillating conditions., *Proc. of the IMechE Part C*, **197**, 225–232.

*Wensing, J.A.* (1998), On the dynamics of ball bearings., *PhD Thesis*, University of Twente, Netherlands, ISBN 90-36512298.

*Wijnant, Y.H.* (1998), Contact dynamics in the field of elastohydrodynamic lubrication., *PhD Thesis*, University of Twente, Netherlands, ISBN 90-36512239.

*Yhland, E. and Johansson, L.* **(1970)**, Analysis of bearing vibration. *Aircraft Engineering*, December, 1970, 18–20.

*Yhland, E.*(1992), A linear theory of vibrations caused by ball bearings with form errors operating at moderate speed. *ASME Journal of Tribology*, 1992, **114**, 348–359.

*Zeillinger, R., Springer, H. and Köttritsch, H.* (1994), Experimental determination of damping in rolling bearing joints., 39th *ASME International Gas Turbine and Aeroengine Congress*/Users Symposium and Exposition, The Hague.

Chapter 15

# Thermal Effect in Rolling–Sliding Contacts

At high rolling speeds, viscous shear heating can be very significant which may result in substantial reduction in the minimum film thickness in the contact. It can also reduce rolling friction and friction due to sliding when rolling is associated with sliding. A few percentage of sliding is always associated with rolling and in gears sliding can be quite high. It is therefore, necessary to look into thermal effect at high rolling speeds both in pure rolling and rolling with sliding.

Thermal problem requires coupled solution of thermal Reynolds equation and energy equation satisfying proper boundary conditions at the inlet and exit of lubricant film.

## 15.1 | Thermal Analysis of Rigid Rolling–Sliding Contacts

The theory now presented is based on the procedure developed for the solution of thermohydrodynamic lubrication problems under laminar flow by Elrod and Brewe (1986) and Elrod (1991). The momentum equation for noninertial laminar lubricating films and the corresponding energy equation are, respectively, as follows:

$$\nabla p = \frac{\partial}{\partial z}\left(\eta \frac{\partial u}{\partial z}\right), \tag{15.1}$$

$$\rho c_p u \frac{\partial T}{\partial x} = \frac{\partial}{\partial z}\left(k \frac{\partial T}{\partial z}\right) + \phi, \tag{15.2}$$

where $\phi = \eta(\partial u / \partial z)^2$ is the viscous dissipation function.

Along with Equations (15.1) and (15.2) following mass continuity equation for an incompressible fluid must be satisfied:

$$\nabla \cdot u = 0 \tag{15.3}$$

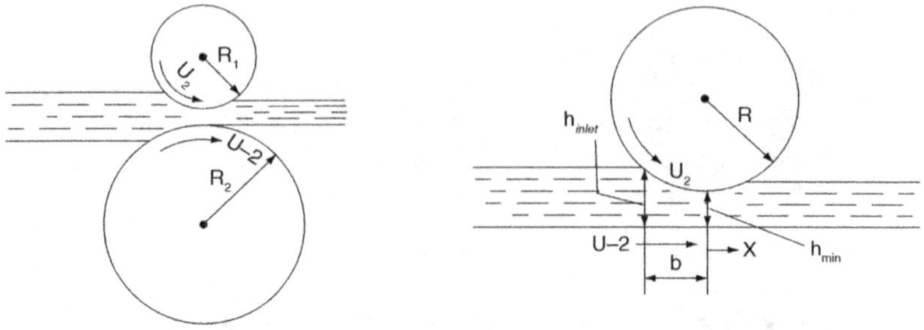


**Figure 15.1** | Geometry of Line Contact

Numerical solution to the flow field is sought by sampling the velocities, pressures, and temperatures over chosen grid points, and the appropriate physical laws mentioned above are satisfied through an algorithm to which these values are interlinked. For a given geometry of line contact shown in Fig. 15.1, which represents an equivalent roller bearing, the temperature variation across the film is represented by a Legendre polynomial of order $N$, $P_N(\zeta)$ and the sampling points then are $N$ Lobatto points. It can be shown that $N$ such internally selected points permit exact numerical integration of a polynomial of order $2N+1$ over range $-1<\zeta<1$. Thus,

$$\int_{-1}^{1} T(\zeta)\,d\zeta = \sum w_k T_k$$

and for $N = 2$ the Lobatto location $\zeta_k$ and weight factor $w_k$ are as follows and includes end-point values as given in Table 15.1. Therefore, if end-point temperatures are known, then it requires only two interior Lobatto point temperatures to be determined. The fluidity $\xi = \frac{1}{\eta}$ is also collocated to its Lobatto point values by the series $\zeta_k = \zeta(T_k)$. A Galerkin style analysis used here involves the expansion of the temperature in a truncated series of Legendre polynomials. Satisfaction is required of as many moments of the energy equation as there are unknowns in this series. The ensuing partial differential equations for the Legendre components are then solved. In the present solution, only two unknowns are used, and for these it is feasible to carry out explicit integration as follows:

**Table 15.1** | Lobatto Locations and Weight Factors for $N = 2$

| Location | Weight factor |
|---|---|
| $-1$ | 1/6 |
| $-1/\sqrt{5}$ | 5/6 |
| $1/\sqrt{5}$ | 5/6 |
| 1 | 1/6 |

$$\int u\frac{\partial T}{\partial x}dz = k\left\{\left(\frac{\partial T}{\partial z}\right)_2 - \left(\frac{\partial T}{\partial z}\right)_{-2}\right\} + \frac{1}{\rho c_p}\int \phi dz,$$

or

$$\int_{-1}^{1}\frac{h}{2}u\frac{\partial T}{\partial x}d\zeta = k\left\{\left(\frac{\partial T}{\partial z}\right)_2 - \left(\frac{\partial T}{\partial z}\right)_{-2}\right\} + \frac{1}{\rho c_p}\int_{-1}^{1}\frac{h}{2}\phi d\zeta \tag{15.4}$$

and for the second moment

$$\int u\frac{\partial T}{\partial x}zdz = k\frac{h}{2}\left\{\left(\frac{\partial T}{\partial z}\right)_2 + \left(\frac{\partial T}{\partial z}\right)_{-2}\right\} + \frac{1}{\rho c_p}\int \phi z dz,$$

or

$$\int_{-1}^{1}\frac{h^2}{2}u\frac{\partial T}{\partial x}d\zeta = k\frac{h}{2}\left\{\left(\frac{\partial T}{\partial z}\right)_2 + \left(\frac{\partial T}{\partial z}\right)_{-2}\right\} + \frac{1}{\rho c_p}\int_{-1}^{1}\frac{h^2}{4}\phi\zeta d\zeta \tag{15.5}$$

The temperature distribution which passes through the Lobatto points expressed in Legendre polynomials as

$$T(\zeta) = \sum_{k=0}^{3}\tilde{T}_k P_k(\zeta), \tag{15.6}$$

Then the Legendre coefficients are easily evaluated by integration

$$\int_{-1}^{1}T(\zeta)P_k(\zeta)d\zeta = \frac{2}{2k+1}\tilde{T}_{k,}$$

or

$$\tilde{T}_k = \frac{2k+1}{2}\sum_{k=0}^{3}w_i T_i P_k(\zeta), \tag{15.7}$$

The above linear set of equations can be solved for the $T_i$. For $N = 2$

$$\tilde{T}_0 = \frac{1}{12}\left\{T_{-2} + 5T_{-1} + 5T_1 + T_2\right\}, \tag{15.8a}$$

$$\tilde{T}_1 = \left\{T_2 - T_{-2} + \sqrt{5}\left(T_1 + T_{-1}\right)\right\}/4, \tag{15.8b}$$

$$\tilde{T}_0 = \frac{5}{12}\left\{T_{-2} + T_{-2} - \left(T_1 + T_{-1}\right)\right\}, \tag{15.8c}$$

$$\tilde{T}_3 = \left\{T_2 - T_{-2} + \sqrt{5}\left(T_1 + T_{-1}\right)\right\}/4, \tag{15.8d}$$

The wall or surface temperatures $T_2$ and $T_{-2}$ are considered known for purposes of film calculation. Thus,

$$\tilde{T}_2 = \left(T_2 + T_{-2}\right)/2 - \tilde{T}_{0,} \tag{15.8e}$$

$$\tilde{T}_3 = (T_2 - T_{-2})/2 - \tilde{T}_1, \tag{15.8f}$$

Similar expressions are obtained for fluidity with the Lobatto point temperatures. Thus,

$$\xi_0 = \frac{1}{12}\{\xi - 2 + 5\xi_{-1} + \xi_1 + \xi_2\} \tag{15.9}$$

and fluidity distribution is

$$\xi = \sum_{k=0}^{3} \tilde{\xi}_k P_k(\zeta), \tag{15.10}$$

***Velocity Distribution and Mass Flux***

A double integration of Equation (15.1) with $\zeta = 1/\xi$ gives the tangential velocity vector

$$u = u_{-2} + \bar{A}\int_{-1}^{\zeta} \xi d\zeta + \bar{B}\int_{-1}^{\zeta} \xi\zeta d\zeta \tag{15.11}$$

where

$$\bar{A} = \frac{u_2 - u_{-2} - \bar{B}\int_{-1}^{1} \xi\zeta d\zeta}{\int_{-1} \xi d\zeta} \tag{15.12}$$

and

$$\bar{B} = \left(\frac{h}{2}\right)^2 \nabla p.$$

The linear mass flux is obtained as

$$\frac{\dot{m}}{\rho} = \int_{-1}^{1} \frac{h}{2} u d\zeta, \tag{15.13}$$

$$\frac{\dot{m}}{\rho} = (u_2 + u_{-2})\frac{h}{2} - \frac{2}{3}\tilde{\xi}_1 \bar{A}\frac{h}{2} - \frac{2}{3}\bar{B}\left(\xi_0 + \frac{2}{5}\xi_2\right)\frac{h}{2}. \tag{15.14a}$$

For symmetric cross-film temperature distribution arithmetic averaging of fluidities at Lobatto points can be done and therefore, mass flux is given by

$$\frac{\dot{m}}{\rho} = (u_2 + u_{-2})\frac{h}{2} - \frac{h^3}{12}\nabla p\left\{\frac{3}{2}\int_{-1}^{1} \xi\zeta^2 d\zeta\right\}. \tag{15.14b}$$

The cross-film temperature distribution would be symmetric in pure rolling conditions when the surface velocities $u_2$ and $u_{-2}$ are equal and so also the surface temperatures $T_2$ and $T_{-2}$.

Mass continuity given by Equation (15.3) when applied to mass flux leads to the generalized Reynolds equation, as follows:

$$\nabla \cdot \frac{\dot{m}}{\rho} = \frac{d}{dx}\{(u_2 + u_{-2})\}\frac{h}{2} - \frac{d}{dx}\left\{\frac{2}{3}\xi_1 \bar{A}\frac{h}{2}\right\} - \frac{d}{dx}\left\{\frac{2}{3}\bar{B}\left(\tilde{\xi}_0 + \frac{2}{5}\xi_2\right)\frac{h}{2}\right\} = 0 \tag{15.15}$$

The temperature equation with the aid of the Legendre series for temperature and fluidity integrals in the zeroth and first moment of energy Equations (15.4) and (15.5) can be evaluated as follows:

Equation (15.4) becomes

$$\frac{12k}{h}\tilde{T}_0+\frac{h}{4}\alpha_7\frac{d}{dx}(\tilde{T}_0)+0.\tilde{T}_1+\frac{h}{4}\alpha_8\frac{d}{dx}(\tilde{T}_1)=\frac{12k}{h}\left\{\frac{T_2+T_{-2}}{2}\right\}+\frac{1}{\rho c_p}\int_{-h/2}^{h/2}\phi dz. \tag{15.16}$$

The temperature $\tilde{T}_0$ is $\zeta$— space mean temperature where the integral of the dissipation function is

$$\int_{-h/2}^{h/2}\phi dz=\frac{2}{h}\left\{2\xi_0\bar{A}^2+\frac{2}{3}\tilde{\xi}_1(2\bar{A}\bar{B})+\frac{2}{3}\left(\tilde{\xi}_0+\frac{2}{5}\tilde{\xi}_2\right)\bar{B}^2\right\} \tag{15.17}$$

and Equation (15.5) becomes

$$0.\tilde{T}_0+\frac{h^2}{8}\beta_1\frac{d}{dx}(\tilde{T}_0)+10k\tilde{T}_1+\frac{h^2}{8}\beta_4\frac{d}{dx}(\tilde{T}_1)=10k\left\{\frac{T_2-T_{-2}}{2}\right\}+\frac{1}{\rho c_p}\int_{-1}^{1}\left(\frac{h^2}{4}\phi\zeta d\zeta\right) \tag{15.18}$$

where the moment of the dissipation function is

$$\int_{-1}^{1}\phi\zeta d\zeta=\frac{2}{3}\tilde{\xi}_1\bar{A}^2+\frac{2}{3}\left(\tilde{\xi}_0+\frac{2}{5}\tilde{\xi}_2\right)(2\bar{A}\cdot\bar{B})+\frac{2}{5}\left(\tilde{\xi}_1+\frac{2}{7}\tilde{\xi}_3\right)\bar{B}^2. \tag{15.19}$$

Constants of Equations (15.16) and (15.18) are given in the Appendix. Two simultaneous partial differential Equations (15.16) and (1.18) with two variables $\tilde{T}_0$ and $\tilde{T}_1$ are obtained by eliminating $\tilde{T}_2$ and $\tilde{T}_3$ via Equations (15.8e) and (15.8f) coupled with the generalized Reynolds Equation (15.15) provide the solution to the thermohydrodynamic lubrication problem for laminar films.

Boundary conditions for the generalized Reynolds equation are

$$p=0\ at\ x=x_i\ \text{ and }\ p=\frac{dp}{dx}=0\text{ at }x=x_e \tag{15.20}$$

Isothermal pressure distribution required to initiate the solution procedure is written following reference [13] as:

$$p(\gamma)=\frac{12u_r\eta_0\sqrt{2Rh_{min}}}{h_{min}^2}\left[\gamma+\frac{\pi}{2}+\frac{\sin 2\gamma}{2}-1.226\left\{\frac{3}{4}\left(\gamma+\frac{\pi}{2}\right)+\frac{\sin 2\gamma}{2}+\frac{\sin 4\gamma}{6}\right\}\right] \tag{15.21}$$

where

$$\tan\gamma=\frac{x}{\sqrt{2Rh_{min}}},$$

$u_r$, $\eta_0$ are average rolling speed and viscosity of the lubricant at inlet temperature.

### 15.1.1 | Computational Procedure

The solution to the thermohydrodynamic line-contact problem begins with the known pressure distribution within the contact as obtained from Equation (15.21) for isothermal films which assumes that the temperature in the entire fluid film is equal to the inlet oil temperature at 311°K.

Fluid film is discretized in the flow direction and the term $(d\tilde{T}_0/dx)$ and $(d\tilde{T}_1/dx)$ in Equations (15.16) and (15.18) are expressed in finite-difference form using backward differencing. Separating the terms, simultaneous algebraic equations in terms of two unknowns $\tilde{T}_0$ and $\tilde{T}$ are obtained which are solved to determine them. This is done starting from the inlet to the outlet of the film in a forward marching manner. Iterations are done to obtain a converged solution. For known surface temperatures $T_2$ and $T_{-2}$ equal to the inlet oil temperature 311° K, the temperature distribution within the film region $T_1$ and $T_{-1}$ are thus obtained. Reverse flow situations, whenever they occur, are handled by resorting to upwind differencing for the terms in Equations (15.16) and (15.18).

Once the Lobatto point temperatures are known over the entire film domain, fluidity functions given in the Appendix are evaluated afresh. Then the generalized Reynolds Equation (15.15) is solved iteratively for pressure distribution following the finite-difference method with appropriate boundary conditions satisfied at the inlet and exit of the film.

For the pressure distribution thus known, we return to determine the temperature distribution within the entire film region by solving Equations (15.16) and (15.18) for $\tilde{T}_0$ and $\tilde{T}_1$

As discussed earlier, this process is repeated until both pressure and temperature converge simultaneously.

Pressure and temperature are treated as converged when the following convergence criteria are satisfied:

***For Pressure***

$$\frac{\left|\left(\Sigma p_i\right)_{N-1} - \left(\Sigma p_i\right)_N\right|}{\left|\left(\Sigma p_i\right)_N\right|} \le 0.0001.$$

***For Temperature***

$$\frac{\left|\left(\Sigma T_i\right)_{N-1} - \left(\Sigma T_i\right)_N\right|}{\left|\left(\Sigma T_i\right)_N\right|} \le 0.0001.$$

where $N$ is the number of iterations

However, convergence is quick and is obtained in only four or five iterations.

### 15.1.2 | Load Capacity and Rolling Traction

Load capacity per unit length of the roller is

$$\frac{W}{L} = \int_{x_i}^{x_e} p\,dx \tag{15.22}$$

and the friction force per unit length of roller is

$$\frac{F}{L} = \int_{x_i}^{x_e} \eta\left(\frac{\partial u}{\partial z}\right)_{0,h} dx. \tag{15.23}$$

When expressed in dimensionless form these are

$$\overline{W} = \frac{Wh_{min}}{24LRu_r\eta_0} \tag{15.24}$$

$$\overline{F} = \frac{Fh_{min}}{24LRu_r\eta_0} \tag{15.25}$$

Coefficient of friction, $f = \dfrac{\overline{F}}{\overline{W}}$.

This procedure was adopted by Ghosh and Gupta (1998) to investigate thermal effect on film thickness and traction coefficient in a rigid rolling sliding contact operating at high rolling speeds. Rolling speeds from 5 to 40 m/s and dimensionless film thickness between $10^{-4}$ and $10^{-3}$ were used as data to investigate thermal effect. It was observed that there is a marked influence of viscous shear heating on load carrying capacity, film thickness, and rolling traction at high rolling speeds. Lubricant properties and geometric parameters of the contact are given in Table 15.2. Results of calculations are presented for temperature and pressure distributions, film thickness, and traction coefficients in the graphical form.

### 15.1.3 | Temperature and Pressure Distribution

Variation of mid-film temperature within the contact region is plotted for various minimum film thicknesses in Fig. 15.2 for a rolling speed of 25 m/s. Mid-film temperature variation for different rolling speeds are shown in Figs 15.3 and 15.4 for two different values of dimensionless minimum film thickness of $2.5 \times 10^{-4}$ and $10^{-3}$, respectively. It is observed in Fig. 15.3 that for a given minimum film thickness of $2.5 \times 10^{-4}$, mid-film temperature increases considerably as the rolling speed increases due to increase in viscous shear heating.

Peak temperature of approx. 400° *K* is obtained which indicates an increase of 89° *K* from the inlet oil temperature of 311° *K*. As the film thickness increases, however, the temperature rise is relatively low. The temperature distribution also changes and the peak temperature is observed near the exit. This is due to the heat being convected away from the inlet and contact regions toward the exit.

When the film thickness is low, a temperature peak is observed in the inlet region, conduction across the film playing a dominant role. The energy equation considers the mechanism of the generation as due to viscous shear heating, while the heat removal is due to conduction across the film and convection by the flowing

**Table 15.2** | Lubricant Properties and Geometric Parameters of the Contact

| | |
|---|---|
| Inlet temperature of lubricant, *K* | 311.11 |
| Inlet viscosity of lubricant, *Pa s* | 0.13885 |
| Temperature-viscosity coefficient of lubricant $K^{-1}$ | 0.045 |
| Thermal diffusivity of lubricant, $m^2/s$ | $7.306 \times 10^{-8}$ |
| $c_p$ of the lubricant, $J/m^3K$ | $1.7577 \times 10^{-6}$ |
| Radius of equivalent roller on plane, *cm* | 1.11125 |
| Dimensionless inlet film thickness | 0.035 |
| Thermal conductivity of lubricant, $W/m^{-k}$ | 0.1284 |

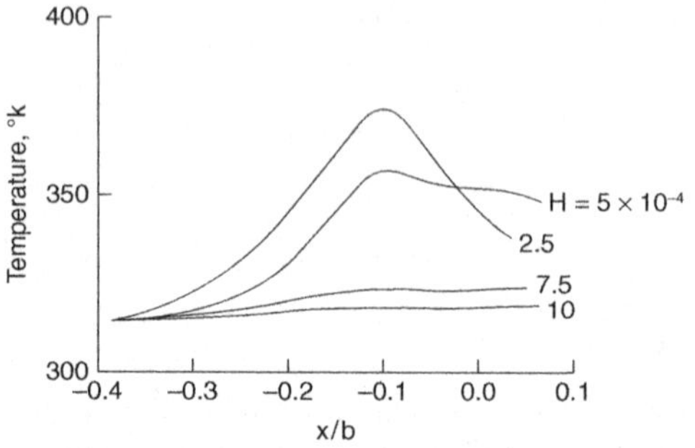


**Figure 15.2** | Variation of Mid-film Temperature for Rolling Speed of 25 m/s

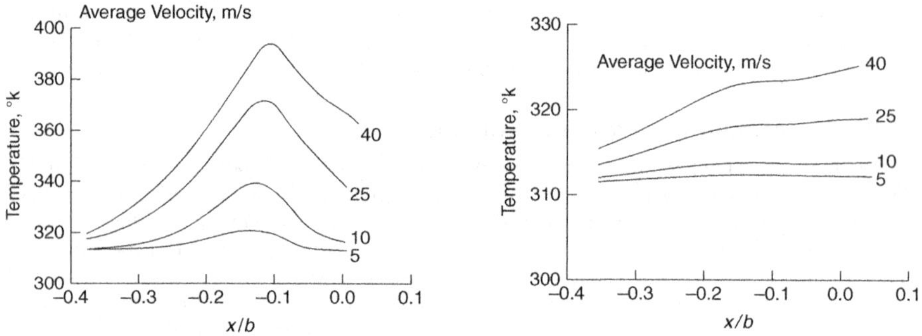


(a) For Dimensionless Minimum Film Thickness = $2.5 \times 10^{-4}$

(b) For Dimensionless Minimum Film Thickness = $10 \times 10^{-4}$

**Figure 15.3** | Variation of Mid-film Temperature for Different Rolling Speeds

fluid along the rolling direction. However, in high-speed hydrodynamic conjunctions the major mechanism of heat removal is convection by the fluid and these results in insignificant rise of the solid surface temperatures. Thus, roller surface temperatures are assumed constant in this analysis and equal to the inlet oil temperature, i.e., 311° *K*. Compression heating and decompression cooling were neglected too. This assumption is generally valid for hydrodynamic lubrication and is therefore invariably adopted. Figure 15.4 shows corresponding thermal pressure distribution for various film thicknesses at a rolling speed of 25 m/s. Pressure within the contact rises considerably as the film thickness decreases. It is also seen that for a given film thickness, increase in rolling speed results in higher pressures.

Pressure distributions so obtained are typical of hydrodynamic lubrication. Peak pressure obtained for the minimum film thickness of $2.5 \times 10^{-4}$ is $7.5 \times 10^{7}$ Pa when the rolling speed is 40 m/s. This is the highest pressure that has been obtained in the conjunction. At this pressure, perhaps some piezoviscous effect might be observed. Some calculations done to estimate piezoviscous effects revealed that it was not significant. At these pressures, elastic deformation would also be negligible. Therefore, neglecting piezoviscous and elastic effects in the analysis appears to be well justified. The highest value of shear rate is obtained at a sliding speed of 20 m/s (i.e., for slide-to-roll ratio of 0.5 and rolling speed of 40 m/s) and the minimum film thick-

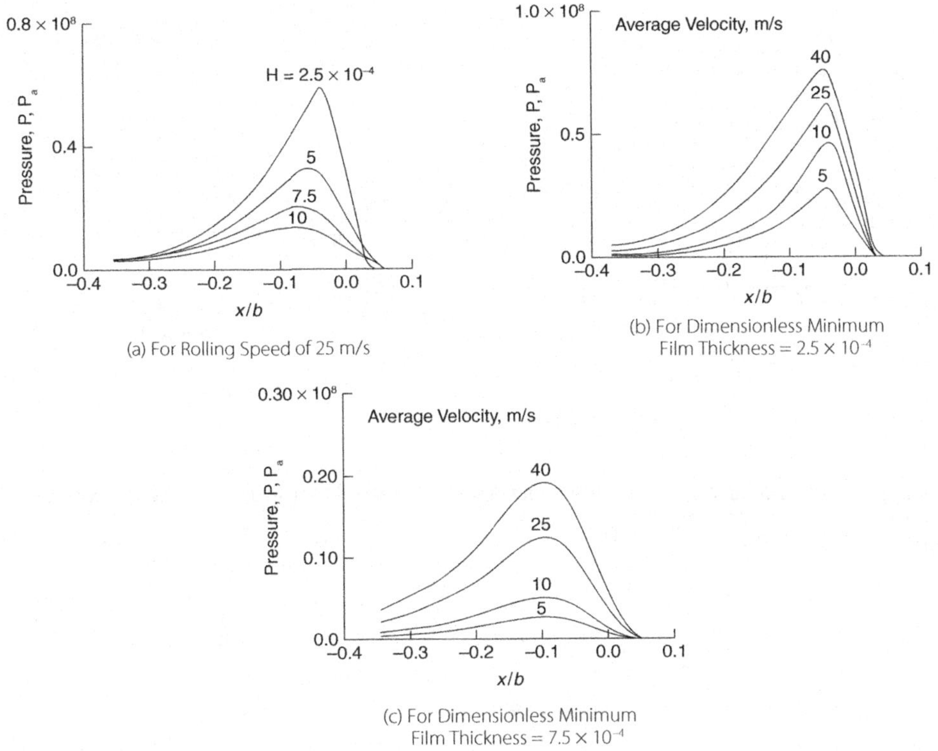


**Figure 15.4** | Variation of Peak Pressure

ness (dimensionless) of $2.5 \times 10^{-4}$ is of the order of $10^5/s$. At such values of shear rates and pressures of the order 50–75 MPa, the non-Newtonian effects would also be negligible. Therefore, neglecting non-Newtonian effects in the analysis is also justified. Higher rolling speeds of the order of 30–40 m/s are usually obtained in large bearings of 50–100 mm diameters operating at speeds in the range of 5000–10 000 rpm.

### 15.1.4 | Thermal Load Capacity and Rolling Traction

It is important to estimate the reduction in the film thickness and rolling friction due to viscous shear heating for a given operating condition of load and rolling speed. For a given load, the film thickness reduction due to viscous shear heating increases with rolling speed. Therefore, it can be said that isothermal calculation at 311° K would result in gross overestimation of film thickness at high rolling speeds and thermal effect must be taken into account to get an accurate estimate of film thickness and rolling traction.

Variations of dimensionless load capacity and rolling friction versus rolling speed are shown in Fig. 15.5 for various film thickness values. Thermal effect reduces the load carrying capacity considerably. For film thickness of $2.5 \times 10^{-4}$, the reduction in load capacity is much higher at high rolling speeds.

For a constant film thickness, the coefficient of friction is observed to increase with rolling speed as can be seen in Fig. 15.5 for film thickness of $2.5 \times 10^{-4}$ and $5.0 \times 10^{-4}$. This is due to the larger reduction in load

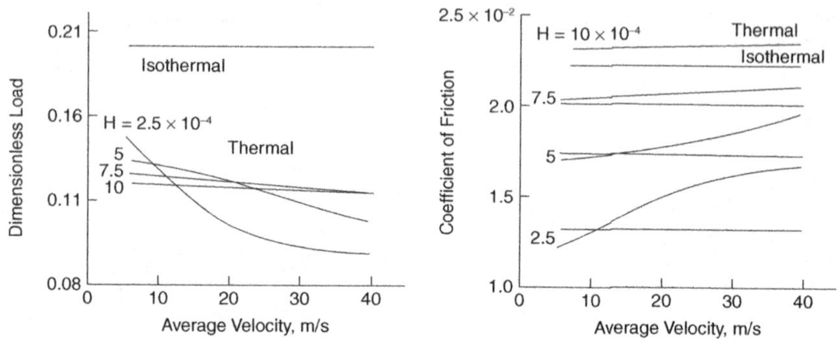


**Figure 15.5** | Variations of Dimensionless Load Capacity and Rolling Friction

**Table 15.3** | Isothermal and Thermal Minimum Film Thickness

| Rolling speed (m/s) | $W = 15000$ N/m | | | $W = 17000$ N/m | | | $W = 19000$ N/m | | | $W = 21000$ N/m | | |
|---|---|---|---|---|---|---|---|---|---|---|---|---|
| | $H_{iso} \times 10^4$ | $H_{th} \times 10^4$ | $\frac{H_{iso}}{H_{th}}$ | $H_{iso} \times 10^4$ | $H_{th} \times 10^4$ | $\frac{H_{iso}}{H_{th}}$ | $H_{iso} \times 10^4$ | $H_{th} \times 10^4$ | $\frac{H_{iso}}{H_{th}}$ | $H_{iso} \times 10^4$ | $H_{th} \times 10^4$ | $\frac{H_{iso}}{H_{th}}$ |
| 5 | 2.2656 | 1.777 | 0.7846 | 1.999 | 1.555 | 0.7781 | 1.7886 | 1.3704 | 0.7662 | 1.6183 | 1.222 | 0.7552 |
| 10 | 4.5312 | 2.8518 | 0.6294 | 3.9981 | 2.4814 | 0.6206 | 3.5772 | 2.1851 | 0.6108 | 3.2386 | 1.9259 | 0.5950 |
| 15 | 6.7968 | 4.1111 | 0.6048 | 5.9971 | 3.5555 | 0.5929 | 5.3659 | 3.051 | 0.5686 | 4.8548 | 2.7037 | 0.5569 |
| 20 | 9.0624 | 5.3704 | 0.5926 | 7.9962 | 4.6666 | 0.5836 | 7.1545 | 4.037 | 0.5643 | 6.4731 | 3.5555 | 0.5493 |
| 25 | 11.3279 | 6.6296 | 0.5852 | 9.9952 | 5.7777 | 0.5780 | 8.9431 | 5.000 | 0.5591 | 8.0914 | 4.4074 | 0.5447 |
| 30 | 13.5935 | 7.8148 | 0.5749 | 11.9943 | 6.8148 | 0.5682 | 10.7317 | 5.963 | 0.5556 | 9.7097 | 5.2591 | 0.5416 |
| 35 | 15.8591 | 9.0000 | 0.5675 | 13.9933 | 7.8889 | 0.5638 | 12.5203 | 6.9259 | 0.5531 | 11.3289 | 6.1111 | 0.5394 |
| 40 | 18.1247 | 10.1852 | 0.5615 | 15.9924 | 8.9260 | 0.5581 | 14.3090 | 7.8888 | 0.5513 | 12.9462 | 6.963 | 0.5378 |

Note: $H_{iso}$ = Dimensionless isothermal film thickness.
$H_{th}$ = Dimensionless thermal film thickness.

capacity and altered pressure distribution with higher pressures and a steeper pressure gradient. Thermal effect on load capacity and rolling friction are shown in Tables 15.3 and 15.4, respectively. Figures 15.6 and 15.7 show the variation of load capacity and rolling friction coefficient, respectively, against film thickness for various rolling speeds. It can be seen that at lower values of film thickness rolling friction increases with rolling speed. However, rolling friction is observed to approach a constant value with increase in film thickness irrespective of rolling speed. Thermal effect on load capacity is observed to be highly dependent on film thickness for low rolling speeds.

Effects of slide to roll ratio on the load capacity and friction coefficient are shown in Figs 15.8 and 15.9, respectively. Slip result in further reduction of the load capacity and an increase in friction coefficient.

**Table 15.4** | Isothermal and Thermal Friction Coefficient

| Rolling speed (m/s) | *W* = 15000 *N/m* | | *W* = 17000 *N/m* | | *W* = 19000 *N/m* | | *W* = 21000 *N/m* | |
|---|---|---|---|---|---|---|---|---|
| | $f_{iso} \times 10^2$ | $f_{th} \times 10^2$ | $f_{iso} \times 10^2$ | $f_{th} \times 10^2$ | $f_{iso} \times 10^2$ | $f_{th} \times 10^2$ | $f_{iso} \times 10^2$ | $f_{th} \times 10^2$ |
| 5 | 1.2433 | 1.0494 | 1.1743 | 0.9898 | 1.1168 | 0.9322 | 1.0728 | 0.8833 |
| 10 | 1.6488 | 1.3648 | 1.5731 | 1.2879 | 1.2030 | 1.2059 | 1.4433 | 1.1437 |
| 15 | 1.9200 | 1.6231 | 1.8322 | 1.5420 | 1.7555 | 1.4778 | 1.6931 | 1.4298 |
| 20 | 2.1055 | 1.8035 | 2.0254 | 1.7268 | 1.9488 | 1.6601 | 1.8843 | 1.6047 |
| 25 | 2.2627 | 1.9798 | 2.1681 | 1.8881 | 2.0977 | 1.8176 | 2.0342 | 1.7429 |
| 30 | 2.4064 | 2.1154 | 2.3043 | 2.0221 | 2.2243 | 1.9446 | 2.1555 | 1.8823 |
| 35 | 2.5255 | 2.2518 | 2.4285 | 2.1382 | 2.3421 | 2.0550 | 2.2658 | 1.9901 |
| 40 | 2.6242 | 2.3695 | 2.5364 | 2.2443 | 2.4464 | 2.1455 | 2.3701 | 2.0770 |

Note: $f_{iso}$ = Dimensionless isothermal friction coefficient.
$f_{th}$ = Dimensionless thermal friction coefficient.

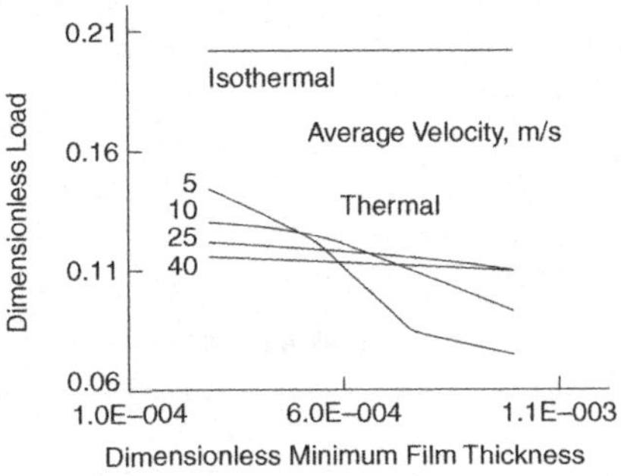


**Figure 15.6** | Variation of Load Capacity With Dimensionless Minimum Film Thickness

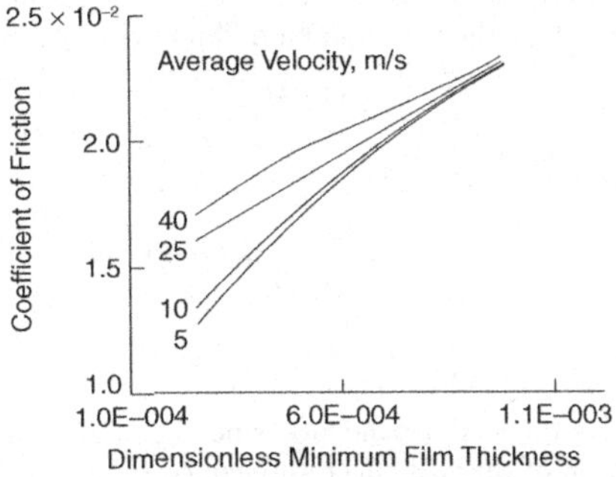


**Figure 15.7** | Variation of Coefficient of Friction With Dimensionless Minimum Film Thickness

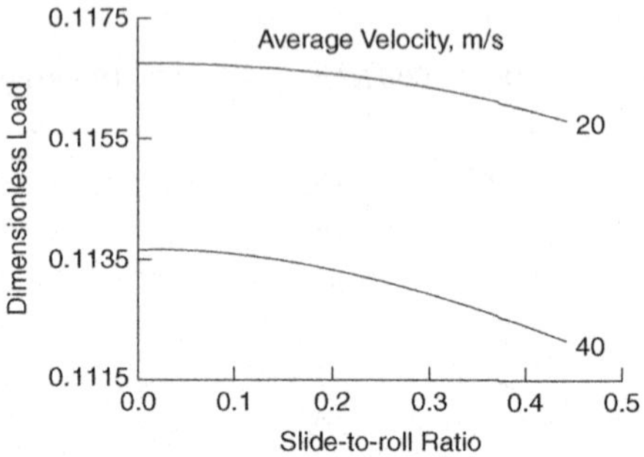


**Figure 15.8** | Variation of Slide-to-roll Ration on Load Capacity for Dimensionless Film Thickness = $10 \times 10^{-4}$

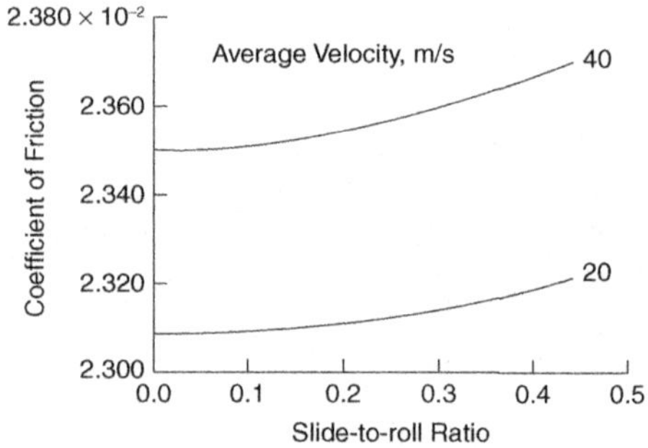


**Figure 15.9** | Variation of Slide-to-roll Ration on Coefficient of Friction for Dimensionless Film Thickness = $10 \times 10^{-4}$

However, thermal effect due to the slip is relatively less significant in hydrodynamic rolling conjunction. A formula was also proposed for the thermal correction factor ($C$) to calculate the minimum film thickness under thermal condition in terms of isothermal minimum film thickness as follows:

$$H_{min,\,Th} = C \times H_{min,\,I} \tag{15.26}$$

where the thermal correction factor $C$ is given as:

$$C = \frac{4}{4 + \log_e Q} \text{ for } Q > 1$$

Thermal loading parameter, $Q = \gamma \eta_0 u_r^2 / k.$

As can be seen in Table 15.3, the load dependence is not significant and, therefore, has been neglected in the above formula, Comparison of thermal effect as obtained from Equation (15.26) and the computed results can be obtained from the values of $C$ given in Table 15.5.

**Table 15.5** | Thermal Correction Factor for Minimum Film Thickness

| **Rolling speed (m/s)** | **5** | **10** | **15** | **20** | **25** | **30** | **35** | **40** |
|---|---|---|---|---|---|---|---|---|
| Thermal loading parameter (*Q*) | 1.2166 | 4.8660 | 10.9500 | 19.4660 | 30.4150 | 43.8000 | 59.6 | 77.86 |
| Thermal correlation factor (*C*) | 0.9533 | 0.7170 | 0.6266 | 0.5740 | 0.5395 | 0.5142 | 0.4946 | 0.4787 |

The above analysis and parametric investigation of thermal effect in fluid film lubricated roller bearings operating at high speeds reveal that thermal effect due to viscous shear heating is very significant and therefore must be accounted for to get a correct estimate of film thickness and rolling traction for a known operating load. Thermal effect on film thickness due to the slip is relatively less significant in hydrodynamic rolling-sliding rigid conjunctions.

## 15.2 | Thermal Analysis of Elastohydrodynamic Lubrication of Line Contacts

It is important to estimate the reduction in the fully flooded minimum film thickness in EHD lubricated contacts due to viscous shear heating at high rolling speeds. Thermal effect on the film thickness were investigated by Dowson and Whitaker (1965), Cheng and Sternlicht (1965), Cheng (1965, 1967), Murch and Wilson (1975), Ghosh and Hamrock (1983, 1985). Dowson and Higginson (1959), and Hamrock and Jacobson (1983) presented a solution for isothermal problem of line contacts wherein computations were carried out from inlet to the outlet of the conjunction as one complete solution. While the analysis of Murch and Wilson (1975) and Cheng (1967) were essentially restricted to the inlet zone only, the others presented a complete solution of fully flooded line contact problem.

An inlet zone analysis using Lobatto Quadrature method was done by Pandey and Ghosh (1998) to evaluate thermal effect on fully flooded minimum film thickness in line contacts. To evaluate the thermal effect, it is necessary to seek simultaneous solutions of the fluid flow and elasticity equations along with the energy equation in the contact. The following sections will deal with thermal analysis of fully flooded line contact problem.

### 15.2.1 | Thermal Analysis of Fully Flooded Elastohydrodynamic Lubrication of Line Contacts (Ghosh and Hamrock 1985)

The generalized thermal Reynolds in written for the line contact problem is written as:

$$\frac{\partial}{\partial x}\left(m_2\frac{\partial p}{\partial x}\right)=u_2\frac{\partial}{\partial x}(m_3)+\frac{\partial}{\partial x}\left[\frac{m_1}{f_0}(u_1-u_2)\right] \tag{15.27}$$

$$m_2=\frac{f_1}{f_0}m_1-\int_0^h\rho\left(\int_0^z\frac{z}{\eta}dz\right)dz$$

$$m_1=\int_0^h\rho\left(\int_0^z\left(\frac{dz}{\eta}\right)dz\right),m_3=\int_0^h\rho dz$$

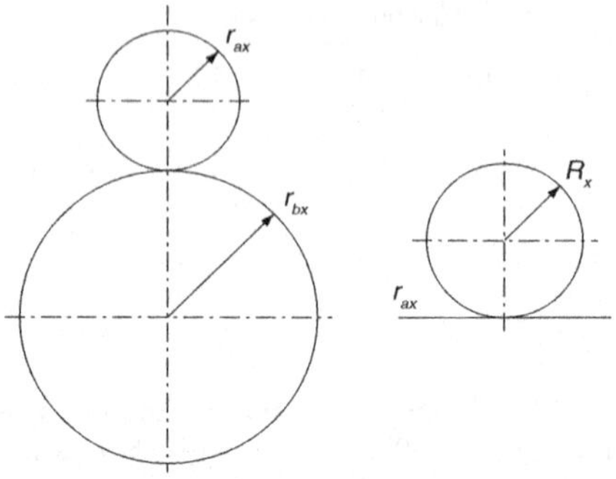


**Figure 15.10** | Undeformed Geometry of the Rollers

$$f_1 = \int_0^h \frac{z}{\eta} dz,\ f_0 = \int_o^h \frac{dz}{\eta}$$

Boundary conditions for Equation (15.27) are

$$p = 0;\ x = -x_i \text{ at film inlet};\quad p = \frac{dp}{dx} = 0 \text{ at } x = x_e \text{ at film exit}$$

Figure 15.10 show the undeformed geometry of the rollers. It has been assumed that convex surfaces exhibit positive curvature and concave surfaces negative curvature. Therefore, if the center of curvature lies within the solid, the radius of curvature is positive or otherwise it is negative.

Film shape in the contact is given as:

$$h(x) = h_0 + s(x) + \delta(x) \tag{15.28}$$

where $s(x) = x^2 / 2R_x$, separation due to geometry of undeformed solids

$$\delta(x) = -\frac{2}{\pi \bar{E}} \int_{-x_i}^{x_e} p \ln(x - x_1)^2 dx_1 \tag{15.29}$$

where

$$\bar{E} = \frac{2}{\left( \frac{1-\nu_1^2}{E_1} + \frac{1-\nu_1^2}{E_2} \right)}$$

$p$, is the pressure at $x_1$, $p(x_1)$ varies in the film domain from $-x_i$ to $x_e$, i.e., from inlet to exit. $\delta(x)$ is elastic deformation at a point $x$ due to pressure distribution in the entire fluid film.

The temperature distribution within the lubricant film is determined from the solution of the energy equation which is written for the lime contact problem as:

$$\frac{\partial}{\partial z}\left( k \frac{\partial T}{\partial z} \right) + \rho c_p u \frac{\partial T}{\partial x} - T \varepsilon u \frac{\partial p}{\partial x} - \eta \left( \frac{\partial u}{\partial z} \right)^2 \tag{15.30}$$

$T$ is lubricant film temperature

$$u = u_2 + \frac{\partial p}{\partial x}\int_0^z \frac{z}{\eta}dz - \frac{\partial p}{\partial x}\frac{f_1}{f_0}\int_0^z \frac{dz}{\eta} + \frac{(u_1 - u_2)}{f_0}\int_0^z \frac{dz}{\eta} \tag{15.31}$$

$$\frac{\partial u}{\partial z} = \frac{1}{\eta}\left[\frac{(u_1 - u_2)}{f_0} + \left(z - \frac{f_1}{f_0}\right)\frac{\partial p}{\partial x}\right] \tag{15.32}$$

The boundary conditions for the above equations are given below as:

$$p = 0;\ x = -\infty \text{ i.e. at the inlet } x_i$$

$$p = \frac{dp}{dx} = 0; \text{ at the film exit i.e. } x = x_e \tag{15.33}$$

and

$$T = T_0 \text{ at } x = -\infty$$

$$T(x,0) = T_2 \quad \text{and} \quad T(x,h) = T_1$$

where $T_1$ and $T_2$ are surface temperatures of solids which are evaluated by solving the transient heat conduction equation of Carlsaw and Jaeger (1969).

Surface temperatures are given as:

$$T_1 = \frac{1}{(\pi k_s \rho_s u_1 c_{ps})^{1/2}}\int_{-\infty}^{x} k\left(\frac{\partial T}{\partial z}\right)_{z=h}\frac{d\xi}{(x-\xi)^{1/2}} \quad \text{For solid 1} \tag{15.34}$$

and

$$T_2 = \frac{1}{(\pi k_s \rho_s u_2 c_{ps})^{1/2}}\int_{-\infty}^{x} k\left(\frac{\partial T}{\partial z}\right)_{z=0}\frac{d\xi}{(x-\xi)^{1/2}} \quad \text{For solid 2} \tag{15.35}$$

$\zeta$, is a dummy variable used for integration. $k_s$, $\rho_s$ and $c_{ps}$, are thermal conductivity, density, and specific heat at constant pressure for the solid surfaces (rollers), $k$, $\rho$ and $c_p$ are thermal conductivity, density and specific heat of the lubricant. These equations are valid when $ul\left(\frac{\rho c_p}{k}\right)_s > 10$ which is usually satisfied and $l$ is a characteristic length. Viscosity ($\eta$) and density ($\rho$) of the lubricant are functions of both pressure and temperature according to following description:

$$\eta = \eta_0 e^{\alpha p};\ \eta = \eta_0 e^{\{\alpha p + \beta(T_0 - T)\}}$$

$$\rho = \rho_0\left(1 + \frac{1.16p}{1 + 1.7p}\right)\left[1.0 - \varepsilon(T - T_0)\right]. \tag{15.36}$$

ε, is thermal expansivity of the lubricant and is defined as $\varepsilon = \frac{1}{V}\left(\frac{\partial V}{\partial T}\right)$, $V$ is the volume, $\alpha$ and $\beta$ are pressure and temperature–viscosity coefficients of the lubricant.

The heat generated due to viscous shear heating in the lubricant is removed due to convection by the flow of the lubricant through the contact and conduction to the solid surfaces in contact with the lubricant.

Ghosh and Homrock adopted a numerical procedure using finite difference method to simultaneously solve the equations in an iterative manner satisfying the required boundary conditions. Flowchart for the computational procedure is shown in Fig. 15.11. Fluid properties and properties of solid materials are given in Table 15.6. Results of computation are given in terms minimum film thickness, mid film temperature rise, and traction coefficients in the Tables 15.7 to 15.9.

Typical pressure, film thickness, and mid-film temperature variations in the contact are shown in Figs 15.12 to 15.14. Traction coefficients in pure rolling and rolling/sliding condition are also shown in Fig. 15.15. Temperature rise in the contact in pure rolling is usually not very high, whereas with sliding temperature distribution in the conjunction is altered totally. Both mid-film and solid surface temperatures rise considerably due to sliding. Sliding results in different surface temperatures, the slower surface attiring higher temperature than the faster moving surface.

Ghosh and Hamrock developed a relationship to determine the ratio of thermal minimum film thickness to isothermal minimum film thickness determined using Equation (15.39) as:

$$\frac{H_{min,\,Th}}{H_{min,\,I}} = \frac{10}{10+Q^{0.4}} \tag{15.37}$$

where $Q$ is the thermal loading parameter defined as:

$$Q = \beta\eta_0 u_s^2 / k \tag{15.38}$$

$\eta_0$-Viscosity of the lubricant at inlet temperature $T_0$.

From the data generated through computer solutions for wide range of speed and load parameters for hard materials, Hamrock and Jacobson generated relationship to determine isothermal minimum film thickness in line contacts as given below:

$$H_{min,\,I} = 3.07\ U^{0.71}\ G^{0.57}\ W^{-0.11} \tag{15.39}$$

Hamrock and Dowson (1976, 1977) developed equations to determine isothermal minimum and central film thickness for point contacts following rigorous numerical analysis as given below:

$$H_{\min,\,I} = 3.63\ U^{0.68}\ G^{0.49}\ W^{-0.073}(1-e^{-0.067k}) \tag{15.40}$$

$$H_{c,I} = 2.69\ U^{0.67}\ G^{0.53}\ W^{-0.067}(1-e^{-0.73k}) \tag{15.41}$$

where $k = a/b$ is ellipticity parameter. Theoretical results of Hamrock and Dowson have been verified experimentally by Koye and Winer (1980) using optical interferometer to measure film thickness. Measurements have also shown that starvation or inadequate supply of lubricant results in significant reduction of minimum film thickness (Wedeven 1971). Measurements of film thickness using capacitance method by Coy and Zaretsky (1981) have revealed that at high speeds, the film thickness in ball bearings reduce significantly due to viscous shear heating effect. Kinematics starvation due to high speed can reduce film thickness and also lead to failure of lubrication in ball bearings as revealed by the theoretical analysis of the inlet zone of Bonneau and Frene (1983).

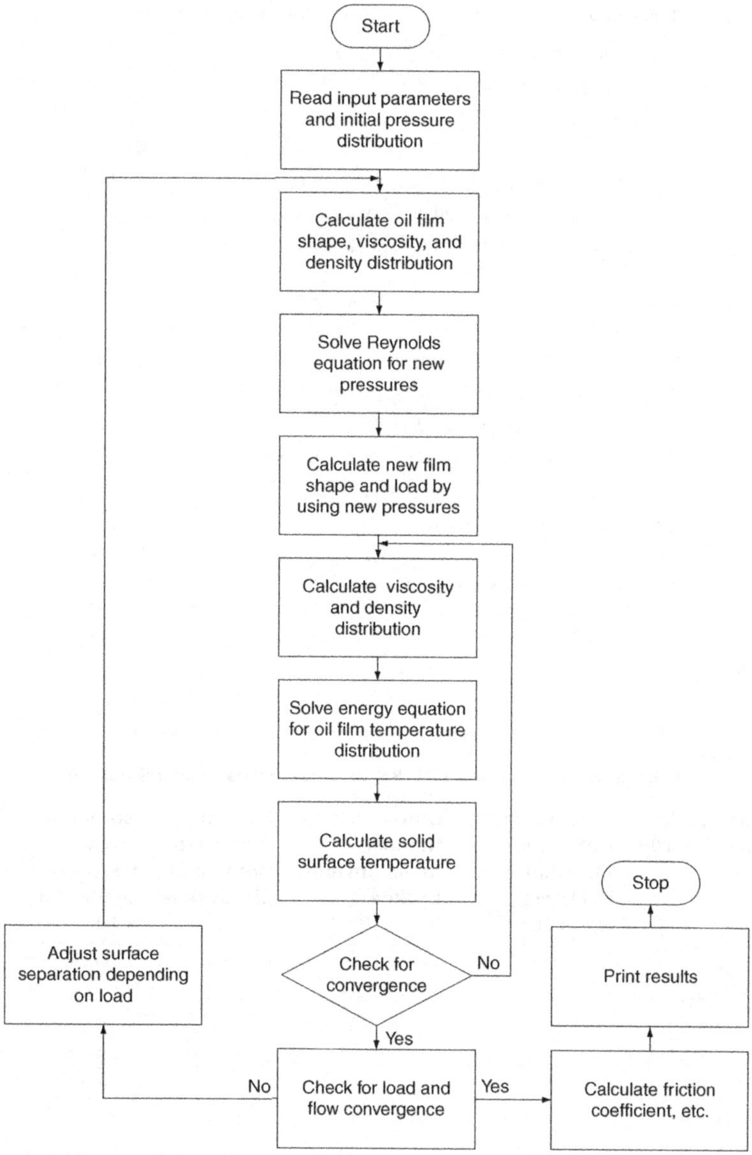


**Figure 15.11** | Flowchart for Computation

**Table 15.6** | Input Data, Material and Lubricant Properties

| | |
|---|---|
| Radius of equivalent roller on plane, *cm* | 1.11125 |
| Inlet temperature of lubricant, *K* | 313 |
| Inlet viscosity of lubricant, *pa-s* | 0.0411 |
| Inlet density of lubricant, $kg/m^3$ | 866 |
| Pressure-viscosity coefficient of lubricant, $\alpha$, $GPa^{-1}$ | 22.76 |
| Temperature-viscosity coefficient of lubricant, $K^{-1}$ | 0.04666 |
| Pressure-density coefficient of lubricant, $GPa^{-1}$ | |
| A | 0.582744 |
| C | 1.68348 |
| Thermal expansivity of lubricant, $K^{-1}$ | $6.5 \times 10^{-4}$ |
| Thermal conductivity of lubricant, $W/m-K$ | 0.12 |
| Specific heat of lubricant, $kJ/kg-K$ | 2.0 |
| Thermal conductivity of steel rollers, $W/m-K$ | 52 |
| Specific heat of steel rollers, $J/kg-K$ | 460 |
| Density of steel rollers, $kg/m^3$ | 7850 |
| Elastic modulus of steel rollers, *GPa* | 200 |
| Poissons ratio of steel rollers | 0.3 |

**Table 15.7** | Isothermal and Thermal Minimum Film Thickness, Rolling Traction Coefficient

**[Dimensionless load, $W = 2.0478 \times 10^{-5}$, material parameter $G = 5000$]**

| Case | Dimensionless speed, $U \times 10^{11}$ | Dimensionless isothermal minimum film thickness, $H_{min} \times 10^6$ | Dimensionless thermal minimum film thickness, $H_{min} \times 10^6$ | Ratio of thermal to isothermal film thickness | Isothermal rolling traction coefficient $\times 10^3$ | Thermal rolling traction coefficient $\times 10^3$ |
|---|---|---|---|---|---|---|
| 1 | 1.000 | 20.015 | 20.015 | 1.000 | 0.537 | 0.537 |
| 2 | 2.000 | 335.384 | 34.581 | 0.9773 | 0.683 | 0.683 |
| 3 | 3.000 | 44.978 | 43.884 | 0.9773 | 0.836 | 0.836 |
| 4 | 5.000 | 65.144 | 61.995 | 0.9516 | 0.968 | 0.968 |
| 5 | 7.575 | 87.943 | 83.123 | 0.9452 | 1.074 | 1.051 |
| 6 | 10.000 | 113.66 | 105.240 | 0.9259 | 1.7097 | 1.563 |
| 7 | 13.466 | 138.06 | 125.74 | 0.9108 | 1.9097 | 1.678 |
| 8 | 20.000 | 177.88 | 156.71 | 0.8810 | 2.1610 | 1.733 |

**Table 15.8** | Midfilm and Solid Surface Temperatures, Coefficient of Sliding Friction for Rolling-Sliding Contact

**[Dimensionless rolling speed, $U = 1 \times 10^{-11}$, material parameter $G = 5000$ dimensionless load, $W = 2.0478 \times 10^{-5}$]**

| Slide-to-roll ratio | Sliding coefficient of friction | Dimensionless mid-film temperature rise at center of contact $\times 10^3$ | Dimensionless solid surface temperature rise at center of contact $\times 10^3$ | |
|---|---|---|---|---|
| | | | Faster surface | Slower surface |
| 0 | 0 | 0 | 0 | 0 |
| 0.04 | 0.0145 | 0.528 | 0.124 | 0.124 |
| 0.08 | 0.0319 | 3.552 | 1.615 | 1.638 |
| 0.16 | 0.0384 | 13.799 | 6.416 | 6.736 |
| 0.2 | 0.046 | 20.217 | 9.359 | 9.975 |
| 0.3 | 0.0547 | 38.000 | 17.46 | 19.32 |
| 0.5 | 0.0677 | 73.142 | 33.04 | 39.49 |

**Table 15.9** | Comparison of Computed Thermal and Isothermal Film Thickness with Results of Kaludjercic and Murch and Wilson

**[Dimensionless load, $W = 2.0478 \times 10^{-5}$, material parameter $G = 5000$]**

| Dimensionless speed parameter, $U \times 10^{11}$ | Dimensionless thermal loading parameter, $Q \times 10^2$ | Dimensionless peclet group parameter, $Pe\ W^{3/2} \times 10^2$ | Ratio of thermal to isothermal film thickness | | |
|---|---|---|---|---|---|
| | | | Present formula | Kaludjercic *et al.'s*(1980) formula | Murch and Wilson's (1975) formula |
| 1.000 | 0.564 | 0.883 | 0.9875 | 0.994 | 0.9890 |
| 2.000 | 2.256 | 1.766 | 0.9785 | 0.9876 | 0.9764 |
| 3.000 | 5.076 | 2.649 | 0.9705 | 0.9812 | 0.9616 |
| 5.000 | 14.1 | 4.415 | 0.9563 | 0.9682 | 0.9299 |
| 7.575 | 32.36 | 6.688 | 0.9401 | 0.9517 | 0.8880 |
| 10.000 | 56.4 | 8.298 | 0.9263 | 0.9361 | 0.8489 |
| 13.466 | 102.27 | 11.89 | 0.9083 | 0.9150 | 0.7953 |
| 20.000 | 225.6 | 17.66 | 0.8784 | 0.8767 | 0.7040 |

### 15.2.2 | Inlet Zone Analysis for Thermal Film Thickness in Elastohydrodynamic Lubrication of Line Contact

Ghosh and Pandey (1998) used Lobatto Quadrature method to analyze the inlet zone of the line contact problem similar to what Murch and Wilson (1975) and Cheng (1967) had done to determine thermal effect on film thickness. A detailed analysis was carried out, which also incorporated effect of slide to roll ratio on

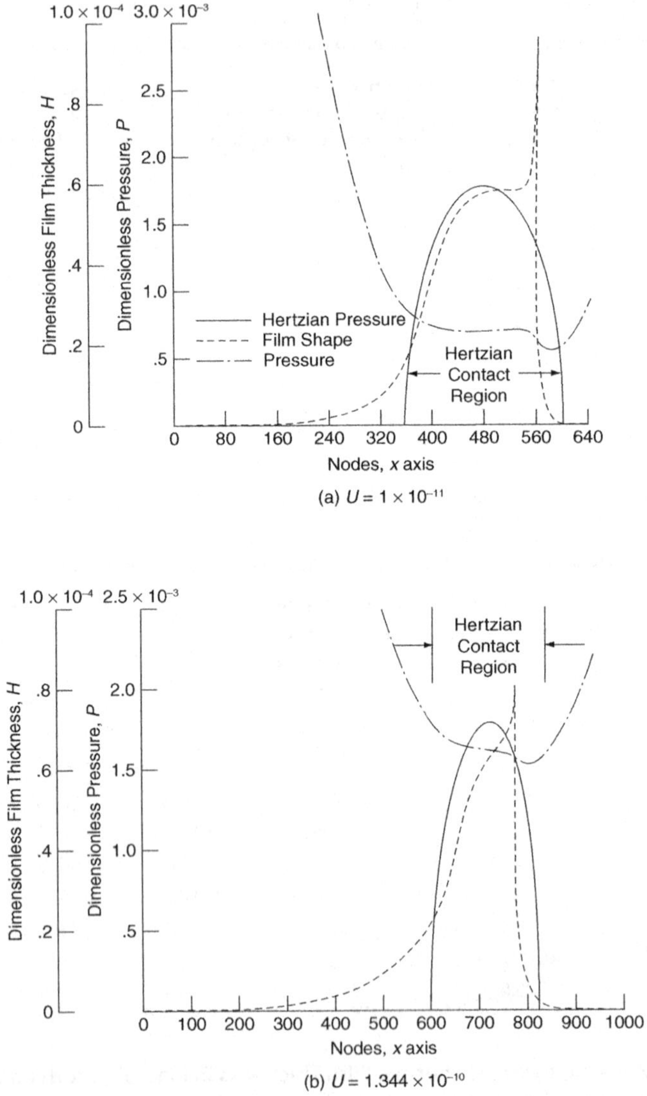


**Figure 15.12** | Variation of Dimensionless Film Thickness and Pressure for Different Speed

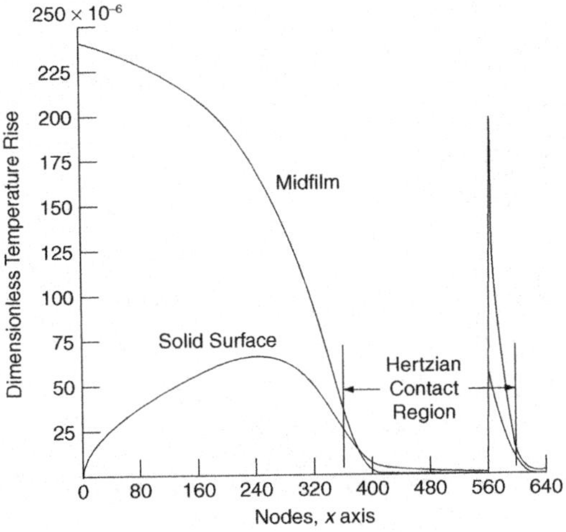


(a) $U = 1 \times 10^{-11}$

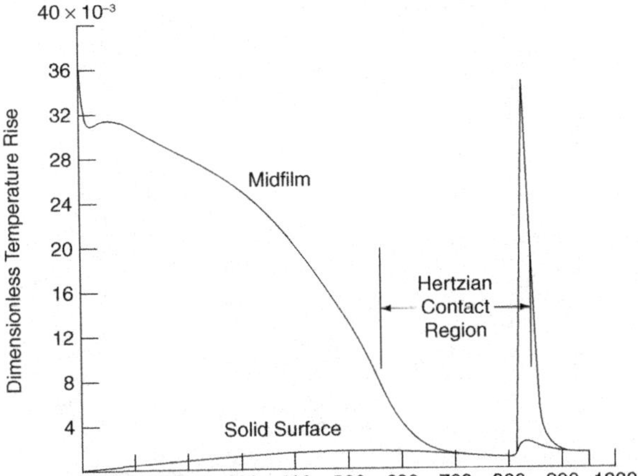


(b) $U = 1.344 \times 10^{-10}$

**Figure 15.13** | Variation of Dimensionless Temperature for Different Speed

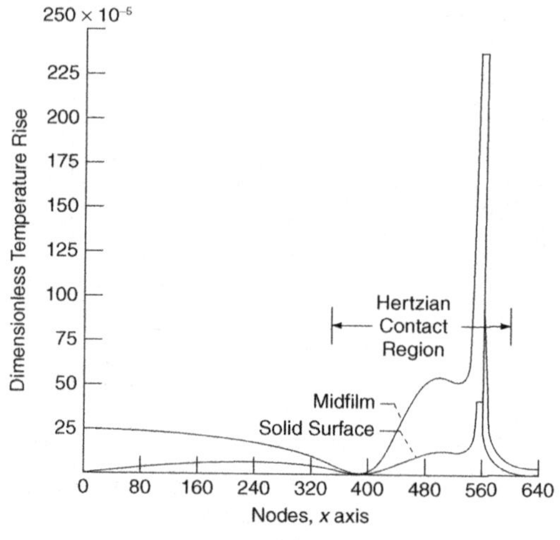


(a) $S = 0.04$

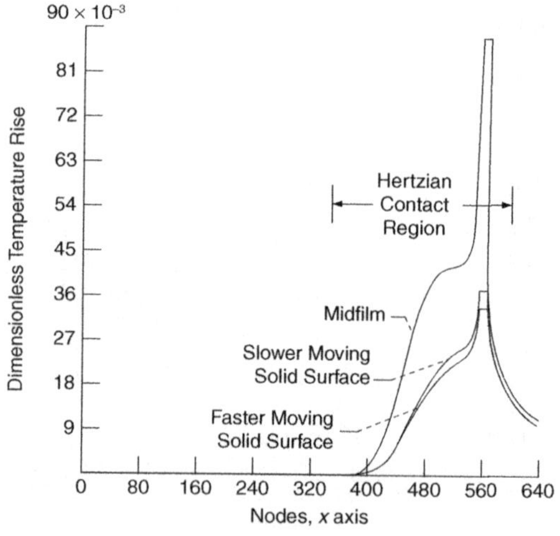


(b) $S = 0.3$

**Figure 15.14** | Variation of Dimensionless Temperature Rise for Different Slide-to-roll Ratio

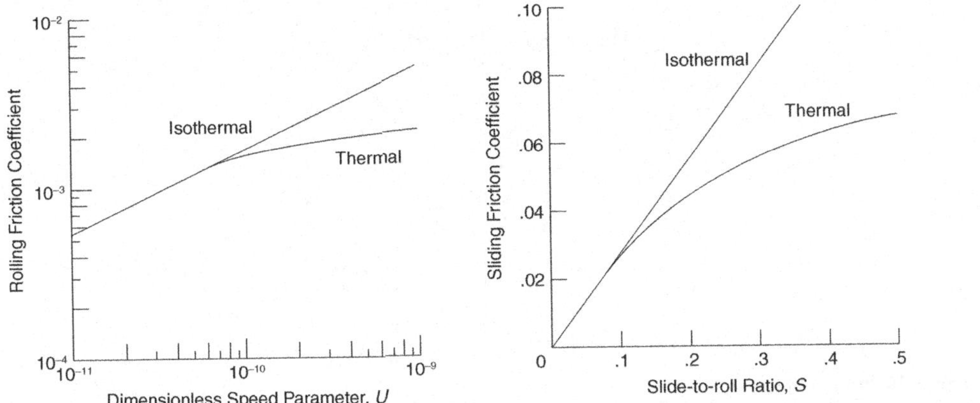


(a) Variation of Pure Rolling Friction Coefficient with Dimensionless Speed Parameter for Material Parameter $G = 5000$ and Dimensionless Load Parameter

(b) Variation of Sliding Friction Coefficient with Slide-to-roll Ratio for Material Parameter $G = 5000$, Dimensionless Load Parameter $W = 2.0478 \times 10^{-5}$ and Dimensionless Speed Parameter $U = 1 \times 10^{-11}$

**Figure 15.15** | Variation of Rolling and Sliding Friction Coefficient With Speed Parameter and Slide-to-roll Ratio

minimum film thickness. Figure 15.16 shows the contact configuration. Results of the analysis are shown in Figs 15.17 and 15.18 which clearly depict thermal effect on film thickness at various rolling speeds and also influence of sliding, respectively.

Ghosh and Pandey developed the following formula for estimation of the thermal film thickness more accurately in comparison to what Murch and Wilson's formula does.

$$C = \frac{H_{min,Th}}{H_{min,I}} = \frac{1.0}{1.0 + 1.6^{0.152} Q^{0.359} \left(1 + 3.96 S^{3..96}\right)} \tag{15.41}$$

Murch and Wilson (1975) also developed similar relationship using their inlet zone analysis as given below:

$$C = \frac{H_{min,Th}}{H_{min,I}} = \frac{3.94}{3.94 + Q^{0.62}} \tag{15.42}$$

However, this relationship significantly overestimates the thermal effect on the minimum film thickness. The method adopted by Ghosh and Pandey is more accurate and computationally very efficient to analyze thermohydrodynamic lubrication problems.

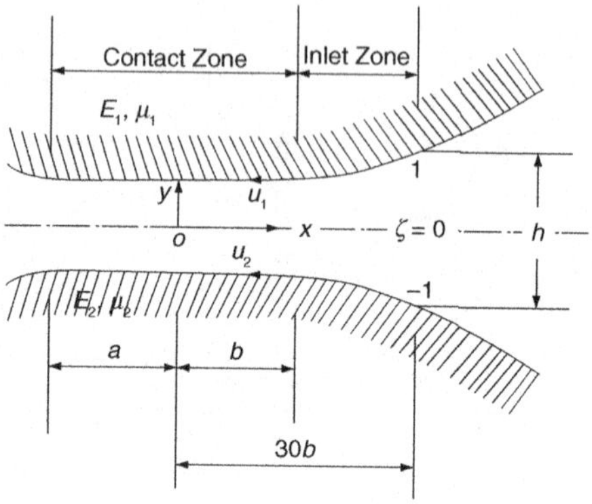


**Figure 15.16** | Contact Configuration

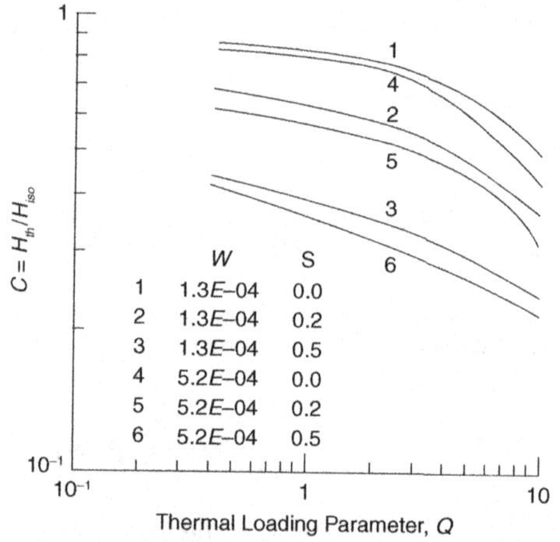


**Figure 15.17** | Effect of Thermal Loading Parameter on the Ratio of Thermal Film Thickness to Isothermal Film Thickness

## 15.3 | Thermal Traction and Temperature Rise in the Eastohydrodynamic Line Contacts

Thermal effect on film thickness in EHL contacts have been discussed in the previous section. It has been seen that pure rolling traction coefficient is generally very low, whereas with sliding associated along with rolling, traction coefficient increases very significantly. Sliding traction is mainly governed by the contact region where the pressure distribution is similar to Hertzian pressure distribution in the case of heavily loaded contacts. Sliding causes severe heating in the contact zone and temperature of the fluid film increases

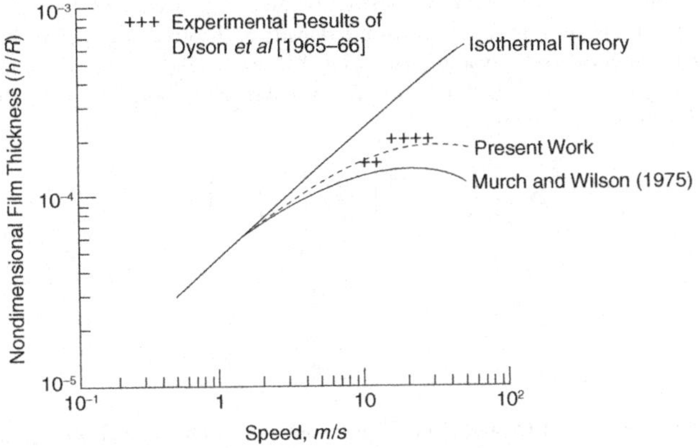


**Figure 15.18** | Effect of Temperature on Nondimensional Film Thickness for Various Rolling Speed

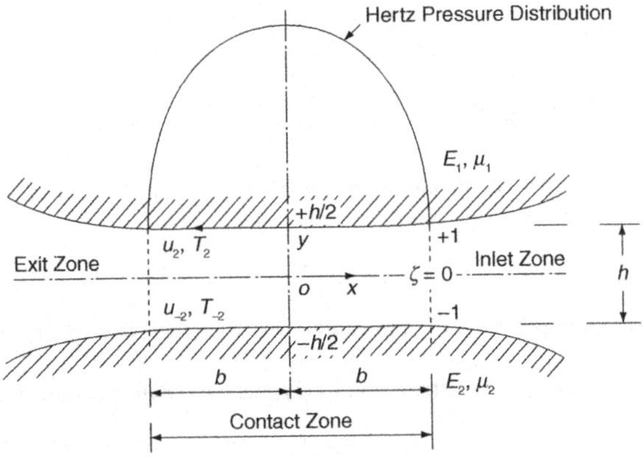


**Figure 15.19** | Theoretical Model of Contact Zone of EHL Line Contact

very significantly which reduces traction coefficient. Therefore, for contacts operating under heavy loads, an accurate analysis is necessary to determine traction coefficient due to sliding. Characteristics of traction coefficient in EHL contacts were investigated experimentally by Crook (1963), Johnson and Cameron (1967–68), and Smith (1965). On the other hand, theoretical predictions were done by Cheng and Sternlicht (1965), Gupta *et al.* (1981), and Kannel and Walowit (1971). Pandey and Ghosh (1998) presented accurate theoretical analysis of traction and temperature rise due to sliding in heavily loaded EHL line contacts.

Theoretical model of the typical contact zone of elastohydrodynamic line contact is shown in Fig. 15.19.

Based on the assumptions of Kannel and Walowit, the contact zone problem was analyzed using an efficient numerical analysis procedure developed by Elrod and Brewe (1986).

The line contact problem for temperature distribution takes into account convective heat transfer in the lubricant film along with variations in the viscosity and density of the lubricant with pressure and temperature using a more realistic model for viscosity and density, i.e., Roeland's relationship. The relevant equations to be solved are, viz., the energy equation, which is written neglecting convection across the film, conduction along the film, and heat of compression term as:

$$\rho c_p u\left(\frac{\partial T}{\partial x}\right) = \frac{\partial}{\partial z}\left[k\frac{\partial T}{\partial z}\right] + \eta\left(\frac{\partial u}{\partial z}\right)^2 \tag{15.43}$$

where

$$\frac{\partial u}{\partial z} = \frac{1}{\eta}\left[\frac{u_2 - u_{-2}}{f_0} + \left(z - \frac{f_1}{f_0}\right)\frac{\partial p}{\partial x}\right]$$

Roeland's viscosity model and Dowson and Higginson's mass density relationships are given as:

$$\eta = \eta_0 \; exp\left[\left(\ln\eta_0 + 9.67\right)\left\{-1 + \left(1 + 5.1x10^{-9} p\right)^z\right\} - \beta\left(T - T_0\right)\right] \tag{15.44}$$

z-exponent in Roeland's viscosity model

$$\rho = \rho_0\left[1.0 + \frac{0.6x10^{-9} p}{1 + 1.7x10^{-9} p}\right]\left[1.0 - \varepsilon\left(T - T_0\right)\right] \tag{15.45}$$

$\beta$ and $\varepsilon$ are temperature-viscosity coefficient and thermal expensively of the lubricant, respectively.

Boundary conditions for the energy equation are as given below:

$$T\left(x, h/2\right) = T_1, T\left(x, -h/2\right) = T_2$$

where $T_1$ and $T_2$ are surface temperatures of upper and lower discs, respectively, which are given by Equations (15.34) and (15.35).

A simplified semielliptical pressure distribution has been assumed in the contact zone as expressed below:

$$p = p_H\left[1.0 - \left(1.0\left|\frac{x}{b}\right|\right)^2\right]^{1/2} \tag{15.46}$$

For heavily loaded line contacts, this assumption is more realistic. In such a situation, film thickness in the contact may be treated as constant and its value can be evaluated using the Dowson and Higginson formula for minimum film thickness as:

$$H_{min,l} = \frac{h_{min}}{R} = 1.6G^{0.6}W^{-0.13}U^{0.7} \tag{15.47}$$

A flowchart for computational procedure is given in Fig. 15.20.

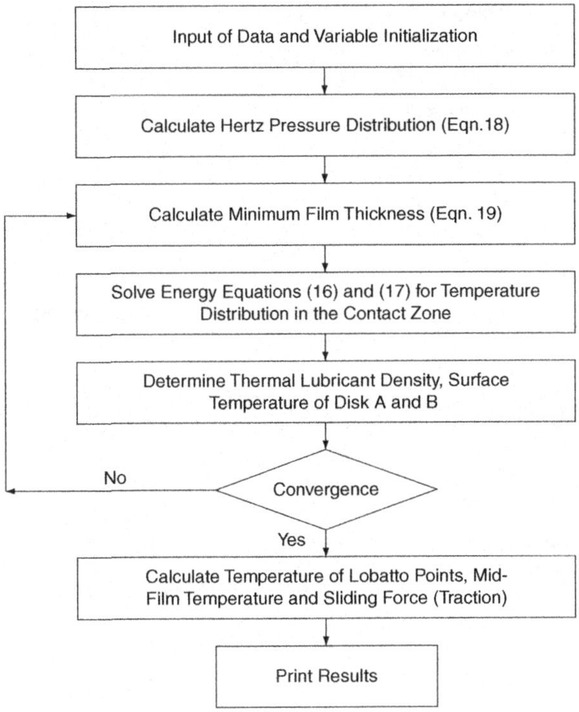


**Figure 15.20** | Flowchart for Thermo EHL Line Contact

The computation process is repeated until the convergence criteria for temperature as given below is satisfied.

$$\frac{\left|\left(\sum T\right)_{N-1} - \left(\sum T\right)_{N}\right|}{\left|\left(\sum T\right)_{N}\right|} \leq 0.001$$

where $N$ is the number of iterations. For further details, Pandey and Ghosh (1998) may be referred.

Disc and lubricant properties used in the calculations are given Tables 15.10 and 15.11, respectively. Results of traction coefficient are shown in Fig.15.21.

The variation of traction coefficient is quite similar to those reported by various investigators. The traction coefficient increases linearly for low slip or low slide to roll ratios, for high slip due to thermal effect it decreases very significantly. Figures 15.22 and 15.23 show comparison with the results of other investigators. Figure 15.24 shows comparison of mid-film and solid surface temperatures in the contact as obtained from

**Table 15.10** | Disc Properties

| | |
|---|---|
| Equivalent radius (*m*) | 0.0175 |
| Elastic modulus (*Pa*) | $2.0 \times 10^{11}$ |
| Density ($kgm^{-3}$) | 7850 |
| Thermal conductivity ($Wm^{-1} K^{-1}$) | 52 |
| Specific heat ($Jkg^{-1} K^{-1}$) | 460 |
| Poisson's ratio | 0.30 |

**Table 15.11** | Lubricant Properties

| | P-150 | Santotrac-50 | MIL-L-7808 |
|---|---|---|---|
| Inlet temperature (*K*) | 323.00 | 323.00 | 321.9 |
| Inlet viscosity (*Pa s*) | 0.01539 | 0.02513 | 0.0009 |
| Pressure viscosity coefficient ($Pa^{-1}$) | $22.89 \times 10^{-9}$ | $26.43 \times 10^{-9}$ | $1.088 \times 10^{-9}$ |
| Temperature viscosity coefficient($K^{-1}$) | 0.03 | 0.038 | 0.034 |
| Exponent in Roelands viscosity formula | 0.63 | 0.90 | 0.43 |
| Inlet density ($kgm^{-3}$) | 864 | 889 | 846 |
| Coefficient of thermal expansivity ($K^{-1}$) | $6.5 \times 10^{-4}$ | $6.5 \times 10^{-4}$ | $6.5 \times 10^{-4}$ |
| Thermal conductivity ($Wm^{-1} K^{-1}$) | 0.12 | 0.12 | 0.096 |
| Specific heat ($Jkg^{-1} K^{-1}$) | 2000 | 2332 | 1675 |

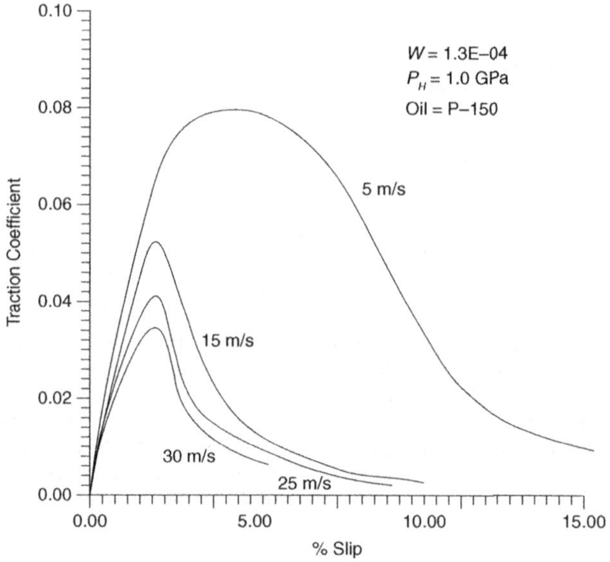


**Figure 15.21** | Variation of Traction Coefficient With % Slip

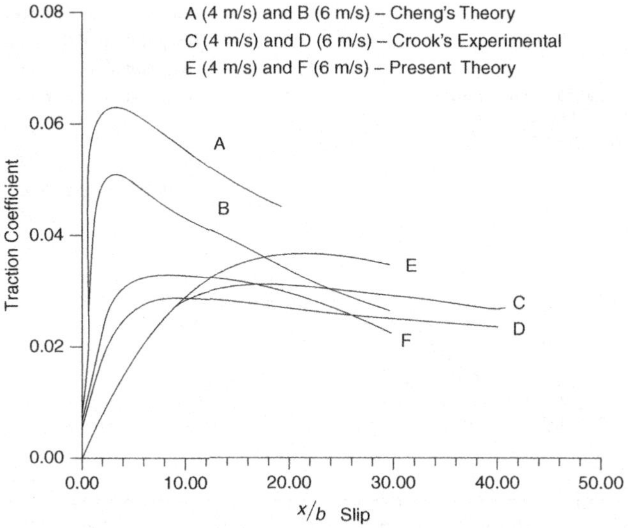


**Figure 15.22** | Variation of Traction Coefficient With % Slip (Comparison of the Obtained Results With the Available Experimental and Theoretical Results)

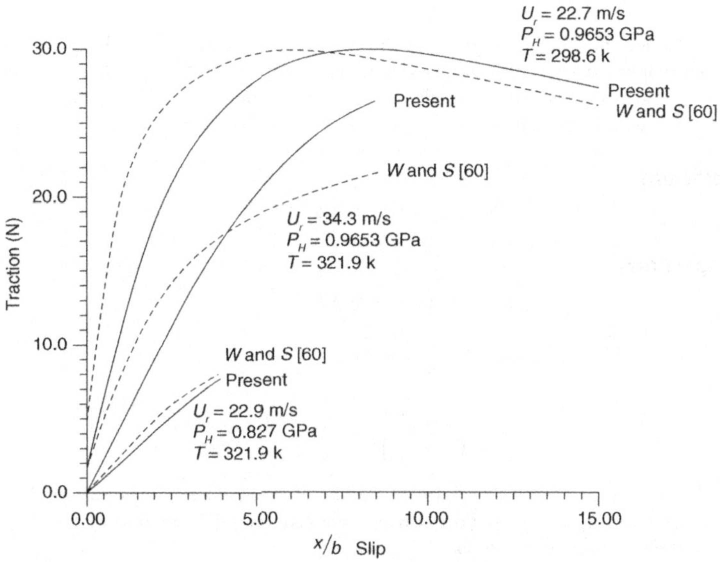


**Figure 15.23** | Variation of Traction Force With % Slip (Comparison of the Obtained Results With the Available Results in the Literature)

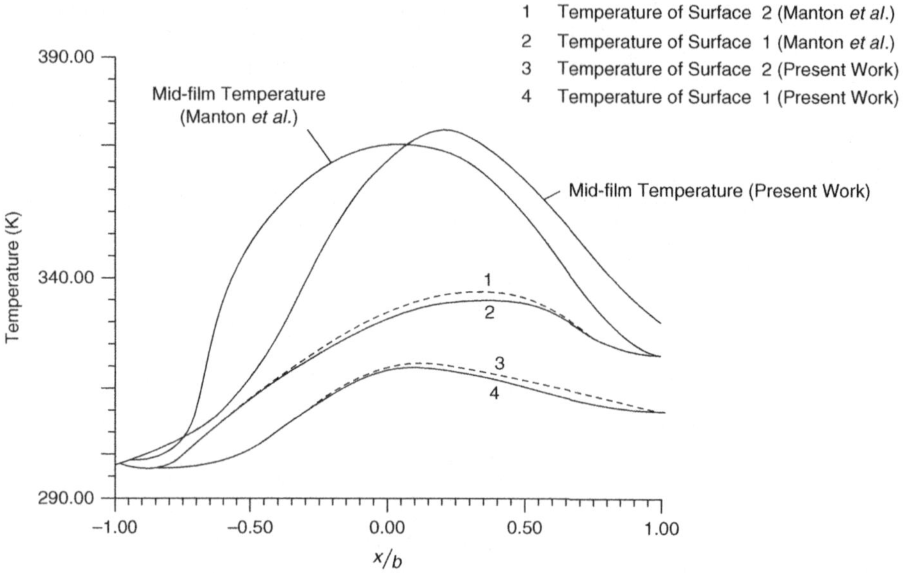


**Figure 15.24** | Mid-film and Solid Surface Temperatures in the Contact

analysis done by Pandey and Ghosh (1998) with those of Manton *et al.* (1968). The observed discrepancies are due to the fact that the analysis of Manton *et al.* is very approximate for the solution of energy equation. Pandey and Ghosh also developed empirical relationships to determine thermal traction coefficient and mid-film temperature rise using large number of data generated through computation as given below:

### *Traction Coefficient*

$$\mu_s = 2.31 W^{0.8} Q^{-0.66} S^{-0.72} \tag{15.48}$$

### *Mid-film Temperature*

$$\tilde{T}_{mf} = 11.08 W^{0.249} S^{0.0265} e^{0.201Q} \tag{15.49}$$

where
$$\tilde{T}_{mf} = T_{mf} / T_0$$

$$S = \frac{2(u_2 - u_{-2})}{(u_2 + u_{-2})}, \text{ slide to roll ratio}$$

$W$, is dimensionless load as defined earlier in the text and $Q$ is thermal loading parameter also defined in the text. These relationships are generally valid for heavily loaded contacts. Further information regarding this can be obtained from the papers cited in the text.

## REFERENCES

*Bonneau, D. and Frene, J.* (1983), Film formation and flow characteristics at the inlet of a starved contact-theoretical study, *ASME Journal of Lubrication Technology*, **105**, 178–186.

*Cheng, H.S.* (1965), A refined solution of thermal-elastohydrodynamic lubrication of rolling and sliding cylinders, *Trans. ASLE*, **8**, 397–410.

*Cheng, H.S.* (1967), Calculation of elastohydrodynamic film thickness in high speed rolling and sliding contacts, *Mechanical Tech. Inc.*, New York Report No. AD65292.

*Cheng, H.S. and Sternlicht, B.* (1965), A numerical solution for pressure, temperature and film thickness between two infinitely long, lubricated rolling and sliding cylinders under heavy loads, *ASME Journal of Basic Engineering*, **87**, 695–707.

*Coy, J.J. and Zaretsky, E.V.* (1981), Some limitations in applying classical EHD film thickness formulas to a high speed bearing, *ASME Journal of Lubrication Technology*, **103**, 2, 295–304.

*Crook, A.W.* (1963), The lubrication of rollers: measurement of friction and effective viscosity, *Philosophical Transaction Royal Society*, London, Ser A, **225**, 281–312.

*Dowson, D. and Higginson, H.R.*, (1959), A numerical solution for the elastohydrodynamic problem, *Journal of Mechanical Engineering Science*, **1**, 6–15.

*Dowson, D. and Whittaker, A.V.* (1965), A numerical procedure for the solution of the elastohydrodynamic problem of rolling and sliding contacts lubricated by a Newtonian fluid, *Proc. Inst. of Mech. Engineers*, London, U.K., **180**, 57–71.

*Elrod, H.G.* (1991) Efficient numerical method for computation of the thermohydrodynamics of laminar lubricating films, *Trans. ASME, J. of Tribology*, **113**, 506–511.

*Elrod, H.G. and Brewe, D.E.* (1986), Thermohydrodynamic analysis for laminar lubricating films, *NASA Technical Memorandum* No. 88845.

*Ghosh, M.K. and Gupta, K.* (1998), Thermal effect in hydrodynamic lubrication of line contacts-piezoviscous effect neglected, *Int. J. of Mechanical Sciences*, **40**(6), 603–616.

*Ghosh, M.K. and Hamrock, B.J.* (1983), Thermal elastohydrodynamic lubrication of line contacts, *NASA Technical* Memorandum No. 83424.

*Ghosh, M.K. and Hamrock, B.J.* (1985), Thermal elastohydrodynamic lubrication of line contacts, *Trans. ASLE*, **28**, 2, 159–171.

*Ghosh, M.K. and Pandey, R.K.* (1998), Thermal elastohydrodynamic lubrication of heavily loaded line contacts-an efficient inlet zone analysis, *ASME Journal of Tribology*, **120**, 119–125.

*Hamrock, B.J. and Dowson, D.* (1976), Isothermal elastohydrodynamic lubrication of point contacts, Part I, theoretical formulation, *ASME Journal of Lubrication Technology*, **98**, 223–229.

*Hamrock, B.J. and Dowson, D.* (1977), Isothermal elastohydrodynamic lubrication of point contacts, Part III, fully flooded results, *ASME Journal of Lubrication Technology*, **99**, 264–276.

*Hamrock, B.J. and Jacobson, B.O.* (1983), Elastohydrodynamic lubrication of rectangular contacts, *NASA Technical Paper* 2111.

*Johnson, K.L. and Cameron, R.* (1967–68), Shear behaviour of elastohydrodynamic oil films at high rolling contact pressures, *Proc. Inst. Mech. Engrs.* London, Part I, **182**, 307–319.

*Kannel, J.W. and Walowit, J.A.* (1971), Simplified analysis for traction between rolling/sliding elastohydrodynamic contacts, *ASME, Journal of Lubrication Technology*, **93**, 39–46.

*Koye, K.A. and Winer, W.O.* (1980), An evaluation of the Hamrock–Dowson minimum film thickness equation for fully flooded EHD point contacts, *ASME Journal of Lubrication Technology*, **102**, 284–294.

*Manton, S.M., O'Donoghue, J.P. and Cameron, A.* (1967), Temperatures at lubricated rolling/sliding contacts, *Proc. I. Mech. Engrs.*, London, **182**(417), 813–824.

*Murch, L.E. and Wilson, W.R.D.* (1975), A thermal-elastohydrodynamic inlet zone analysis, *ASME, JOLT*, **97**, 2, 212–216.

*Pandey, R.K. and Ghosh, M.K.* (1998), A thermal analysis of traction in elastohydrodynamic rolling/sliding line contacts, *Wear*, **216**, 106–114.

*Pandey, R.K. and Ghosh, M.K.* (1998), Temperature rise due to sliding in rolling/sliding elastohydrodynamic lubrication line contacts: an efficient numerical analysis for contact zone temperatures, *Tribology International*, **31**(12), 745–752.

*Wedeven, L.E., Evans, D. and Cameron, A.* (1971), Optical analysis of ball bearing starvation, *ASME Journal of Lubrication Technology*, **93**, 349–363.

## APPENDIX 15.1

$$\tilde{\xi}_1 = \tilde{\xi}(T_1);\ \tilde{\xi}_0 = (\xi_2 + 5\tilde{\xi}_1 + 5\tilde{\xi}_{-1} + \tilde{\xi}_{-2})\big/12$$

$$\tilde{\xi}_2 = \tilde{\xi}(T_2);\ \tilde{\xi}_1 = [\xi_2 - \xi_{-2} + \sqrt{5}(\tilde{\xi}_1 - \xi_{-1})]\big/4$$

$$\bar{\xi}_{-1} = \tilde{\xi}(T_{-1});\ \tilde{\xi}_2 = 5(\tilde{\xi}_2 + \tilde{\xi}_{-2} - \xi_1 - \bar{\xi}_{-1})\big/12$$

$$\tilde{\xi}_{-2} = \xi(T_{-2});\tilde{\xi}_3 = \left[\tilde{\xi}_2 - \tilde{\xi}_{-2} - \sqrt{5}(\tilde{\xi}_1 - \xi_{-1})\right/4$$

$$\alpha_1 = \tilde{\xi}_0 - \xi_1 / 2 + \tilde{\xi}_3 / 8$$

$$\alpha_2 = -\tilde{\xi}_0 / 2 + \tilde{\xi}_1 / 3 - \xi_2 / 8$$

$$\alpha_3 = 2\tilde{\xi}_1 / 15 - \frac{9}{70}\tilde{\xi}_3 + \frac{4}{3}\alpha_1$$

$$\alpha_4 = \frac{2}{15}\tilde{\xi}_0 - \frac{1}{42}\tilde{\xi}_2 + \frac{4}{3}\alpha_2$$

$$\alpha_5 = \frac{4}{15}\tilde{\xi}_0 - \frac{8}{105}\tilde{\xi}_2$$

$$\alpha_6 = \frac{4}{105}\tilde{\xi}_1 - \frac{8}{315}\tilde{\xi}_3$$

$$\alpha_7 = 4\bar{u}_{-2} + 3\alpha_3\bar{A} + 3\alpha_4\bar{B}$$

$$\alpha_8 = 5\alpha_5\bar{A} + 5\alpha_6\bar{B}$$

$$\beta_1 = \frac{3}{5}\alpha_8$$

$$\beta_2 = \frac{2}{35}\tilde{\xi}_1 - \frac{29}{630}\tilde{\xi}_3 + \frac{4}{15}\alpha_1$$

$$\beta_3 = \frac{2}{35}\tilde{\xi}_0 - \frac{1}{210}\tilde{\xi}_3 + \frac{4}{15}\alpha_2$$

$$\beta_4 = \frac{4}{3}\bar{u} - 2 + 5\beta_2\bar{A} + 5\beta_3\bar{B}$$

# Index

# Authors' Profiles

**Prof. M.K. Ghosh** received his Bachelor of Science (Mechanical Engineering) degree in 1966 from Banaras Hindu University, Varanasi, India, M.Tech (1968) and Ph.D. (1979) in Mechanical Engineering respectively from Indian Institute of Technology (IIT), Kharagpur, India. Professor Ghosh joined as a faculty member in the Department of Mechanical Engineering at IIT BHU, Varanasi in 1969 and subsequently worked as Professor of Mechanical Engineering since 1993. He has worked as a NRC Research Associate during 1982–1984 and as a Senior NRC Research Associate during 1989–90 at the NASA Lewis (Presently NASA Glenn), Research Center, Cleveland, OH (USA). He was a Professor of Mechanical Engineering in the Department of Mechanical Engineering at IIT Kharagpur during 1987–88. Professor Ghosh is a member of the American Society of Mechanical Engineers since 1984, a Life Member of the Tribology Society of India (TSI) and served as a member of the Executive Committee of the TSI during 1995–97. He is a member of the honorary editorial board of the journal "Advances in Vibration Engineering", the scientific journal of the "Vibration Institute of India". He is a recipient of the University Grants Commission of India "Career Award" in Engineering in 1984. Professor Ghosh's areas of research interests lie in Tribology, Vibration and Dynamics & Control of Mechanical Systems. He has published about 60 papers in peer reviewed international journals. Professor Ghosh has supervised about 33 graduate theses and 7 PhD theses.

**Professor B. C. Majumdar** received his Ph.D. in 1970 from Indian Institute of Technology, Kharagpur in Mechanical Engineering (Specialisation: Gas-lubricated Bearings). He joined as a Professor of Mechanical Engineering at IIT, Kharagpur in 1978, held Alexander von Humboldt-Foundation Fellowship, at various universities in Germany for more than two and a half years, held Senior Research Associateship of National Research Council, USA at NASA Lewis Research Centre, Cleveland, Ohio, USA for two years. He was visiting professor at Cranfield University, UK, Lulea Technical University, Sweden, Queensland University of Technology, Australia and Tokyo Denki University, Japan for short periods. Guided 18 Ph.D. candidates on various aspects of lubrication including hydrodynamic, hydrostatic, porous oil and gas bearings, surface roughness effect, elastodydrodynamically lubricated line and point contacts, published over 100 papers in reputed Journals abroad, written a text book on Tribology, has been a fellow of Indian National Academy of Engineering since 1997, Associate Editor of Advances in Tribology, USA and Member of Editorial Board, Journal of Engineering Tribology, Part J, Proc. IMECHE, UK.

**Dr. Mihir Sarangi**, B.E., M.Tech., Ph.D., is an assistant professor of Mechanical Engineering at Indian Institute of Technology, Kharagpur. Prior to the present assignment he served as a Sr. Lecturer in Department of Mechanical Engineering at Indian Institute of Technology, Bombay, held a post-doctoral fellowship at University of Kentucky, USA. He received the High Value Ph.D. Fellowship by IIT Kharagpur and DAAD Fellowship by Germany during his doctoral and master programmes, respectively. He has contributed 7 papers in reputed Journals abroad. His area of research interests are Tribology, Machine Design and Machine Dynamics.